The Industrial Revolutions

VOLUME 5

The Industrial Revolution in Europe

II

THE INDUSTRIAL REVOLUTIONS

General Editors R. A. Church and E. A. Wrigley

Volume 1
Pre-industrial Britain
Edited by J. A. Chartres

Volumes 2 and 3
The Industrial Revolution in Britain
Edited by Julian Hoppit and E. A. Wrigley

Volumes 4 and 5
The Industrial Revolution in Europe
Edited by P. K. O'Brien

Volume 6
Industrialization in North America
Edited by Peter Temin

Volume 7
The Industrialization of Japan
Edited by W. J. Macpherson

Volume 8
The Textile Industries
Edited by D. T. Jenkins

Volume 9
The Metal Fabrication and Engineering Industries
Edited by Sidney Pollard

Volume 10
The Coal and Iron Industries
Edited by R. A. Church

Volume 11
Commercial and Financial Services
Edited by R. C. Michie

The Industrial Revolution in Europe II

Edited with an introduction by

P. K. O'Brien

The Economic History Society

BLACKWELL
Oxford UK & Cambridge USA

First published 1994

Blackwell Publishers
108 Cowley Road, Oxford OX4 1JF
UK

238 Main Street
Cambridge, Massachusetts 02142
USA

British Library Cataloguing in Publication Data

A CIP catalogue record for this book is available from the British Library.

Library of Congress Cataloging-in-Publication Data

The Industrial Revoluton in Europe/edited, with an introduction, by P. K. O'Brien.
p. cm. — (The Industrial revolutions : v. 4-5)
"The Economic History Society."
Includes bibliographical references.
ISBN 0-631-18073-7 (v. 1). —— ISBN 0-631-18145-8 (v. 2)
ISBN 0-631 17486 9 (11-volume set).
1. Europe—Industries—History. 2. Technological innovations—Europe—History. 3. Europe—Economic conditions. I. O'Brien Patrick Karl. II. Economic History Society. III. Series.
HD2321.I533 1994 vol. 4-5
[HC240]
338.09 s—dc20
[338.094]

93-40550
CIP

Introduction typeset in 11 on 12 point Plantin by Typesetting Solutions, Slough, Berks., UK.
Printed in Great Britain by T.J. Press Ltd, Padstow, Cornwall.

Table of Contents

VOLUME 4

VOLUME 5

General Editors' Introduction

R. A. CHURCH and E. A. WRIGLEY

The aim of this series of volumes is to bring together journal articles and chapters from collections of essays that deal with some aspect of the industrial revolution and may be regarded as 'classics'. Some of the articles were published recently, but writings from each decade since the Second World War are well represented and there is a scattering of articles that appeared in the interwar period. Very few libraries contain all the journal series, collections of essays, and research monographs from which the material reproduced in these volumes was drawn.

As well as indicating the latest trends in research, these volumes provide an insight into the development of thinking about the issues that have informed work on the industrial revolution over the last half-century. Each piece is reproduced with its original pagination, but with continuous pagination added for ease of reference. In every case but one the original was reproduced photographically. In the one remaining case (volume 4, article 7), the text was reset. Each volume contains an introduction which explains the choice of material reproduced, describes the main debates to which they relate, and provides a guide to further reading that should make it possible to pursue particular questions effectively.

The series consists both of topical volumes dealing with an individual industry or area of economic activity (for example, textiles or metal fabrication and engineering industries), and of volumes that cover the occurrence of the industrial revolution in a particular country or continent (Britain or North America).

By 'industrial revolution', a loose and imprecise term, we mean principally what is often referred to as the first industrial revolution, the period of the cotton mill and the steam engine, rather than that of electrical engineering, fine chemicals, and the internal combustion engine (though the first volume deals with the period immediately before the conventional dating of the industrial revolution in Britain). The industrial revolution was an event of such profound and continuing importance for all societies, both those initially affected and those which were remote from the main centres of early growth, that achieving a better understanding of its nature should be a prominent element in the reflections of all those concerned about society today as well as in the past. We hope that this series may prove of use to those who are persuaded of the truth of this assertion.

The Industrial Revolution in Europe

*Capitalism and Bureaucracy in German Industrialization before 1914**

By JÜRGEN KOCKA

I

Capitalism and bureaucracy are key concepts in Max Weber's analysis of western civilization. Nevertheless, in discussing the relation between them, and their contribution to western "rationalization", his conclusions were ambivalent. According to him, bureaucratic administration is

> characterized by formal employment, salary, pension, promotion, specialized training and functional division of labor, well-defined areas of jurisdiction, documentary procedures, and hierarchical sub- and super-ordination". [It is] capable of attaining the highest degree of efficiency and is in this sense formally the most rational known means of exercising authority over human beings. It is superior to any other form in precision, in stability, in the stringency of its discipline, and in its reliability. It thus makes possible a particularly high degree of calculability of results for the heads of the organization and for those acting in relation to it. It is finally superior both in intensive efficiency and in the scope of its operations, and is formally capable of application to all kinds of administrative tasks . . . . [There is a] continual spread of bureaucratic administration. This is true of church and state, of armies, political parties, economic enterprises, interest groups, endowments, clubs, and many others.[1]

On this basis Weber argued that modern capitalism could not do without bureaucratic administration, both inside the large firms and in the society at large, and stressed the anti-traditional rationality of both capitalism and bureaucracy.

On the other hand, Weber was well aware of the basic differences between capitalism and bureaucracy, and of the possible conflicts between them. In capitalist economies, it is primarily the invisible hand of the market which allocates resources and provides for co-ordination; in bureaucratic organizations resource allocation and co-ordination are performed by the visible hand of hierarchies. While officials are appointed on the basis of formal qualifications and examinations, according to general rules, workers and employees in the market economy are hired according to less formalized and more market-oriented criteria. While officials hold tenure and enjoy seniority rights in matters of promotion and income, the position and rewards of both employers and employees in the market economy are less secure, less predictable, more dependent on the market, and on measurable achievements. While capitalist success implies risks, innovation, individual initiatives, and quick adoption to

* A revised version of the Tawney Memorial Lecture, 1980.

[1] Max Weber, *Economy and Society. An Outline of Interpretive Sociology* (New York, 1968), III, p. 1393; I, p. 223.

change, bureaucratic administrations are successful if they handle individual constellations as cases of general and well established rules, react as expected, and fulfill the criterion of appropriateness. While Weber was fascinated by the superior efficiency of bureaucratic organization, he also feared that the universal spread of bureaucratic structures could stifle capitalist dynamics and entrepreneurial innovation—besides political leadership and individual self-realization, in general.[2]

German industrialization before 1914 would seem to be a suitable context within which the relationship between capitalism and bureaucracy can be discussed in a more concrete way. German industrialization was undoubtedly capitalist in character, as were all other instances of industrialization before 1917. Most strategic decisions about the allocation of factors of production were made by private entrepreneurs, using profit and capital accumulation as major yardsticks. They managed privately owned firms which were connected with each other principally through market mechanisms. Wage work on a contractual basis clearly dominated, and the tensions and conflicts between capital and labour structured many aspects of German society.[3]

On the other hand, and in sharp contrast to the situation in Britain and the United States, the rise of public bureaucracies preceded capitalist industrialization in Germany. Leaving regional differences aside, industrialization in Germany (measured by accelerated growth, the rise of the factory system and wage labour, accelerated re-distribution of the labour force, urbanization, and the formation of a wage workers' class) started in the 1840s, more than half a century later than in Britain.[4] At this time, however, the large German states, particularly Prussia, Saxony, and Bavaria, had well established, powerful public bureaucracies, strikingly similar to the type later analysed by Max Weber. They had originated under absolutist rule, in the seventeenth and eighteenth centuries. No revolution had broken their continuity. In the late eighteenth and early nineteenth centuries the king's servants had curbed the monarchical powers of patronage by making appointment conditional on formal qualifications, and by acquiring seniority rights, tenure, and pension rights. They had thereby turned themselves into civil servants similar to the type described by Weber. Constitutional reforms at the beginning of the nineteenth century increased the power of the high civil service, by limiting the monarch's power without creating strong parliaments or other representative institutions. The higher civil servants gained power, status, and some exclusiveness of a post-aristocratic type, which was based not directly on birth, but on delegated authority and formalized education—access to which was, of course, not available to the lower classes, but not restricted to the offspring of the traditional élites either. Given the relative economic weakness of most of Germany, the immaturity of her bourgeoisie, and the underdevelopment of her representative institutions, it was the higher civil service which took the lead in the "reforms from above" which moved the larger

[2] M. Weber, 'Parlament und Regierung im neugeordneten Deutschland', in *Gesammelte politische Schriften* (Tübingen, 2nd ed. 1959), pp. 294-431, esp. 308-39. Cf. W. J. Mommsen, *The Age of Bureaucracy. Perspectives on the Political Sociology of Max Weber* (New York, 1974), pp. 47-71, 95-115.

[3] There is a good introduction in English by K. Borchardt in *The Fontana Economic History of Europe*, IV: *The Emergence of Industrial Societies* (1973), pt. 1, pp. 76-160.

[4] For a short chronology see R. H. Tilly, 'Capital Formation in Germany in the Nineteenth Century', in *Cambridge Economic History of Europe*, VII, 1 (Cambridge, 1978), pp. 386-7.

German states from the corporate-absolutist Ancien Régime to the nineteenth-century civil society, which even in twentieth-century Germany would still display the effects of its bureaucratically guided genesis. Of course, there were other proponents of change, notably the emerging bourgeoisie, particularly in the west. And it is also worth emphasizing that higher civil servants, contrary to what they often thought and said of themselves, were not merely representatives of the "common good": they had special interests of their own, and they were not independent from powerful social classes, in particular the land-owning class. It is also true that the process of reform which they promoted had important limits: in the last analysis, emancipation cannot be attained by decree; and in the long run the policy of the bureaucracy tended to slow down and hinder processes of change whose dynamics threatened to get out of bureaucratic control. Yet in spite of all this the fact remains that the higher civil servants, under pressure from a competitive international system, impressed by West European models, and concerned to enhance their state's and their own positions, became the sponsors of bureaucratically led reforms from above, whose relative success was mutually related to the failures of revolution from below in nineteenth-century Germany. At the same time, the trend was not uniform throughout the nineteenth century: the bureaucratic impact on economic and social change was strongest in the early decades, and in the last thirty years or so, it started to grow again.[5]

In 1852, in addition to some 134,000 members of the armed forces, there were about 63,000 civil servants in Prussia, or nearly 1 per cent of the labour force. Around 1910, there were 1·2 million public employees in the German Reich (including military persons), or about 4 per cent of the labour force. Rough comparison seems to show that about 1890 the proportionate importance of goverment employees in Germany was about twice as great as in Great Britain.[6] Of course, only a small minority of these civil servants had the political power and educational background stressed above. But most of them, down to the postal employees and policemen, shared, albeit in varying degrees, important aspects of the civil service status: a specific legal status under public law, special privileges in and loyalities to the state, high esteem, and a favourable image associated with power, a sense of duty, hierarchy, formalized procedures and security. They were, so to speak, part of the state and its authority, not merely private citizens.

[5] H. Rosenberg, *Bureaucracy, Aristocracy, and Autocracy. The Prussian Experience, 1660-1815* (Cambridge, Mass. 1958); R. Koselleck, *Preußen zwischen Reform und Revolution. Allgemeines Landrecht, Verwaltung und soziale Bewegung von 1791-1848* (Stuttgart, 1967; 2nd ed. 1975); J. R. Gillis, *The Prussian Bureaucracy in Crisis, 1840-1860. Origins of an Administrative Ethos* (Stanford, 1971); J.Kocka, 'Preußischer Staat und Modernisierung im Vormärz: marxistisch-leninistische Interpretationen und ihre Probleme', in H.-U. Wehler, ed. *Sozialgeschichte heute* (Göttingen, 1974), pp. 211-27. On Bavaria: E. Weis, *Montgelas, 1759-1799. Zwischen Revolution und Reform* (München, 1971). A classic treatment is O. Hintze, 'Der Beamtenstand', in *Soziologie und Geschichte* (Göttingen, 2nd ed. 1964), pp. 66-125. For the later part of the nineteenth century see H. Rosenberg, *Große Depression und Bismarckzeit. Wirtschaftsablauf, Gesellschaft und Politik in Mitteleuropa* (Berlin, 1967); H.-U. Wheler, 'Der Aufstieg des organisierten Kapitalismus und Interventionsstaates in Deutschland', in H. A. Winkler, ed. *Organisierter Kapitalismus. Voraussetzungen und Anfänge* (Göttingen, 1974).

[6] Figures for 1852 in G. F. Kolb, *Handbuch der vergleichenden Statistik der Völkerzustands- und Staatenkunde* (Leipzig, 1860), p. 160; for 1910: Hintze, 'Der Beamtenstand'. p. 68. For German-British comparisons see P. Flora, 'Quantitative Historical Sociology', *Current Sociology*, XXII (1975), p. 99, based on M. Abramovitz and V. Eliasberg, *The Growth of Public Employment in Great Britain* (New York, 1957); J. P. Cullity, 'The Growth of Government Employment in Germany, 1882-1950', *Zeitschrift für die gesamte Staatswissenschaft*, CXXIII (1967), pp. 201-17.

Althoughly poorly paid, civil service positions were much sought after, at all levels of the hierarchy and demand for these positions usually exceeded supply.[7]

Bureaucratization preceded industrialization, and bureaucratic structures, processes, and values therefore profoundly shaped the process and character of industrialization in Germany, in contrast to Great Britain and the United States. This can be demonstrated in terms of the role of public economic policy, the government's share of capital stock and of the total product, the role of businessmen and officials in local reform movements, the role of public authorities in the reaction to social protests, and in many other ways.[8] This article, however, is restricted to two related issues, two small aspects of German industrialization, which are representative of much broader processes: industrial organization and management on the one hand, and the emergence of a white-collar employee class on the other. Considering these two themes, what difference did the bureaucratic tradition make? How did it influence the economic and social change occurring with industrialization? Through which mechanisms and with what results?[9]

II

It is well known that entrepreneurs and managers in the early factory system, were faced with severe problems of coordination and control, selection and motivation, discipline and organization, which had been unknown to the masters of craft shops or to the merchants in the putting-out systems. Early entrepreneurs and managers, nevertheless, looked back to traditional models in order to solve those problems.[10] In Germany, it was logical to use the bureaucratic models which belonged to the stock of tradition.

[7] Besides the titles in note 5 above, see W. Naudé, 'Zur Geschichte des preußischen Subalternbeamtentums', *Forschungen zur brandenburgischen und preußischen Geschichte*, XVIII (1905), pp. 364-86; A. Lotz, *Geschichte des deutschen Beamtentums* (Berlin, 2nd ed. 1914). Also see J. Caplan, ' "The Imaginary Universality of Particular Interests": the "Tradition" of the Civil Service in German History', *Social History*, IV (1979), pp. 299-317.

[8] According to W. W. Rostow, *Politics and the Stages of Growth* (Cambridge, 1971), p. 15, government expenditures as percentages of GNP were 13·2 in Germany (1891), but only 8·9 in the United Kingdom (1890). This is supported by the comparative figure in W. L. Woytinsky, *Die Welt in Zahlen*, VI (Berlin, 1927), pp. 123-6: in 1913, annual government expenditures per capita amounted to $25·8 in Germany, $18·3 in Great Britain, and $12·7 in the USA (US Dollars). But such a British-German difference is not apparent in S. Kuznets, *Modern Economic Growth* (New Haven, Conn., 1966), pp. 236-7. The government's share of the total German capital stock was 18% in 1850/4, 24% in 1880/4, and 20% in 1910/13, according to W. G. Hoffmann et al. *Das Wachstum der deutschen Wirtschaft seit der Mitte des 19. Jahrhunderts* (Berlin, 1965), p. 44. On the bureaucracy, the military, and social protests see A. Lüdtke, 'The Role of State Violence in the Period of Transition to Industrial Capitalism: the example of Prussia from 1815 to 1848', *Social History*, IV (1979), pp. 175-221; R. H. Bowen, 'The Roles of Government and Private Enterprise in German Industrial Growth, 1870-1914', *Journal of Economic History*, X (1950), pp. 68-81; W. Fischer, *Wirtschaft und Gesellschaft im Zeitalter der Industrialisierung* (Göttingen, 1972), pp. 60-213.

[9] I have dealt with single aspects of this problem in different places: J. Kocka, *Unternehmensverwaltung und Angestelltenschaft am Beispiel Siemens, 1847-1914* (Stuttgart, 1969); Kocka, 'Vorindustrielle Faktoren in der deutschen Industrialisierung. Industriebürokratie und "neuer Mittelstand"' in M. Stürmer, ed. *Das Kaiserliche Deutschland. Politik und Gesellschaft, 1870-1918* (Düsseldorf, 1970), pp. 265-86; Kocka, 'Family and Bureaucracy in German Industrial Management, 1850-1914: Siemens in Comparative Perspective', *Business History Review*, XLV (1971), pp. 133-56; Kocka, 'Bildung, soziale Schichtung und soziale Mobilität im Deutschen Kaiserreich' in D. Stegmann, et al. eds. *Industrielle Gesellschaft und politisches System* (Bonn, 1978), pp. 297-314. The following remarks try to summarize some results and to put them into a larger context.

[10] See S. Pollard, *The Genesis of Modern Management. A Study of the Industrial Revolution in Great Britain* (1965), pp. 37 ff.

There were, in practice, many channels through which bureaucratic patterns spread to the developing factory system and its management. Various interdependencies between government agencies and early enterprises continued after the mercantilistic period. Government departments staffed by technically trained men, responsible for building and administering roads and canals, continued to exist and grow. Civil servants acted as entrepreneurs, and the government continued to run some enterprises, especially in mining, and later in the railroad sector. Civil servants played a leading role in the system of technical and industrial education begun in the 1820s, and also in early scientific and industrial associations. Engineering expertise was concentrated in the Prussian technical administration and in special military units. In 1850 the Prussian engineering corps consisted of 4,000 men and 220 officers. Competent civil servants and military men were hired by the private railroads and other firms which paid higher salaries than the government.[11] No doubt they brought bureaucratic patterns, styles, and values with them into the growing private enterprises. In addition, the early entrepreneurs and their employees often shared the preferences and stereotypes which dominated the social climate and the culture of their time. The broad recognition and, by and large, favourable public image of German civil servants were mentioned above.

The bureaucratic impact on industrial organization can be usefully examined in terms of the experience of one firm. The Siemens & Halske electrical manufacturing firm was founded in Berlin in 1847; it was not only the pioneer in the field of electrical engineering and installations, but has succeeded in staying ahead of the competitors until the present day. The number of the employees in Germany (i.e. excluding its foreign branches), was 600 in 1872, 4,000 in 1895, and 57,000 in 1912.[12] Bureaucratic traditions were clearly visible within the Siemens management. A substantial minority of the salaried employees were former Prussian civil servants. The founder, Werner Siemens, had received part of his training in a technical military school in Berlin and had spent 15 years in a military career before starting his own business. As early as 1855 shop rules were formulated and written down. (Such written and general rules of shop discipline were applied in other German factories at least as early as the 1830s.[13]) The

[11] For the active role of Prussian civil servants around 1800, see F. Redlich, 'The Leaders of the German Steam-Engine Industry During the First Hundred Years', *Journal of Economic History*, IV (1944), pp. 121 ff.; F. Zunkel, 'Beamtenschaft und Unternehmertum beim Aufbau der Ruhrindustrie, 1849-1880', *Tradition*, IX (1964), pp. 261-76. For technical and industrial schools and associations, see P. Lundgreen, *Techniker in Preußen während der frühen Industrialisierung. Ausbildung und Berufsfeld einer entstehenden sozialen Gruppe* (Berlin, 1975); K.-H. Manegold, *Universität, Technische Hochschule und Industrie* (Berlin, 1970); *Chronik der Kgl. Technischen Hochschule zu Berlin, 1799-1899* (Berlin, 1899); R. Rürup, 'Die Technische Universität Berlin, 1879-1979', id., ed., *Wissenschaft und Gesellschaft. Beiträge zur Geschichte der Technischen Universtät Berlin, 1879-1979* (Berlin, 1979), pp. 3-47; P. Borscheid, *Naturwissenschaft, Staat und Industrie in Baden* (1848-1914) (Stuttgart, 1976); F. Schnabel, *Die Anfänge des Technischen Hochschulwesens* (Karlsruhe, 1925); C. Matschoss, 'Geschichte der Königlich-Preussischen Technischen Deputation für Gewerbe', *Beiträge zur Geschichte der Technik und Industrie*, III (1911), pp. 239-53. For an example of a civil servant hired by a private railroad company see V. v. Unruh, *Erinnerungen aus dem Leben*, H. v. Poschinger, ed. (Stuttgart, 1895). U. v. Bonin, *Geschichte des Ingenieurkorps und der Pioniere in Preußen* (Berlin, 1877 and 1878).

[12] For a history of the Siemens enterprises see G. Siemens, *History of the House of Siemens*, 2 vols (Freiburg/München, 1957). The following paragraphs on the basis of Kocka, 'Family and Bureaucracy'.

[13] For an example from 1837 see W. Fischer, *Der Staat und die Anfänge der Industrialisierung in Baden, 1800-1850* (Berlin, 1961), p. 357. For an earlier example of written office rules, see W. Köllmann, *Friedrich Harkort* (Düsseldorf, 1964), I, p. 187.

Siemens company quickly developed a system of written and generalized instructions which provided fixed lines of communication within and between the offices. The sources show a well developed sense of hierarchy, and sometimes read like the files of a contemporary administrative agency. This high degree of bureaucratization cannot be explained merely as managerial response to the operational requirements of the enterprise; it also resulted from the acceptance of traditional organizational models developed outside industry.

The influence of bureaucratic patterns from outside the firm was also evident in the status and self-image of the early salaried employees at Siemens. Their remuneration by monthly salaries, which were based in part on seniority, their actual job security, vacation privileges, and the non-manual nature of their work differentiated them clearly from the wage earners. In these respects they were comparable to civil servants. Indeed, they were called *Privatbeamte*, and they regarded themselves, somewhat inaccurately, as a private kind of civil servants bearing delegated authority and deserving certain privileges.[14]

It should be noted, however, that the bureaucratic tendencies within the management of the early Siemens company were clearly limited by several interrelated factors, more so than in later and larger companies. The strength of family traditions within this organization set a limit to its bureaucratic character.[15] The power of the owner-entrepreneur and his closest aides was such that they could cut through the hierarchical lines and break through established patterns of communication. The relatively small size and rapidly changing nature of the young enterprise set limits to the repetitiveness of its operations, to the generalization of its processes, and to the institutionalization of its functions.

As far as bureaucratic patterns were adopted, however, they made a direct contribution to the success of the business. Bureaucratic controls stressing accuracy, punctuality, and regularity tended to check the more traditional, irregular, and slow performance of the still prevailing artisan-type first-generation factory workers, and thus helped to increase the efficiency of work. In this respect, capitalist industrial needs and bureaucratic rationality joined hands in suppressing and changing more traditional forms of behaviour.[16] Furthermore, the bureaucratic impact manifested itself in the civil-service ideology of the salaried employees, and thus served the success of the enterprise. Especially since sufficient instruments of direct control (sophisticated accounting techniques, easy communication over long distances) did not exist, the civil-servant ethos of employees was in the interest of management. This ethos implied, as a sympathetic observer has put it, "integrity, a sense of duty, unselfish

[14] For details see J. Kocka, 'Industrielle Angestelltenschaft in frühindustrieller Zeit', in O. Büsch, ed. *Untersuchungen zur Geschichte der frühen Industrialisierung vornehmlich im Wirtschaftsraum Berlin-Brandenburg* (Berlin, 1971), pp. 317-67.

[15] The first clerks and administrative employees at Siemens were relatives of the founder. The three branches of the firm, in Berlin, St. Petersburg, and London, were led by three brothers: Werner, Carl, and William Siemens. Co-ordination of this multi-national structure, in the first decades, was achieved through the private correspondence of the three brothers. Family cohesion was used to solve managerial problems, in a quite unbureaucratic way. On the relations between capitalist enterprises and entrepreneurial families in general see E. J. Hobsbawm, *The Age of Capital* (New York, 1975), pp. 230-2, 237-40; J. Kocka, 'Familie, Unternehmer und Kapitalismus. An Beispielen aus der frühen Industrialisierung', *Zeitschrift für Unternehmensgeschichte*, XXIV (1979), pp. 99-135 – English translation in *German Yearbook on Business History*, I (1981, forthcoming).

[16] Similar problems, though without reference to the bureaucratic impact, are discussed in E. P. Thompson, 'Time, Work-Discipline and Industrial Capitalism', *Past & Present*, 38 (1967), pp. 56-97.

diligence, public spirit, an unbending sense of justice, and unpretentious loyalty".[17] To the extent that this was more than mere rhetoric, such attitudes and self-images of the early employees fulfilled a function which, during the early industrialization in Britain, was partly performed by professional ethics.[18] They checked the widespread unreliabilities and fraudulent activities of poorly controlled employees whose loyalty was decisive for the early company. Moreover, their civil-service self-images made it impossible for the white collar employees to consider joining hands with protesting wage earners. They clearly identified themselves with management and regarded the blue collar force as different and inferior. In later years at least, management maintained some of the white collar privileges just because that helped to provide for stability and loyalty.

This is not the place to follow the history of the Siemens firm in any detail. There were crises of growth in the 1880s, when the disadvantages of bureaucratic organization became more visible: a certain slowness of response, organizational bottle-necks, and over-formalization. But under the leadership of the second generation, after 1890, new growth set in; the firm started to select professional managers, largely graduates from the technical universities; and thorough reforms of the management structure took place, which *both* built on the bureaucratic traditions of the firm *and* altered them.

Not only did the top management adopt a more systematic approach and subject itself to some impersonal, general rules. Also on the middle management levels, within the huge white collar departments and in the management of the factory, bureaucratic tendencies became more manifest than ever before. The sales departments and field offices, while behaving flexibly in the market, internally worked according to most detailed, centrally issued regulations. They were organized like public administrations, and most of the activities performed in them were highly specialized and routinized. By 1910, Siemens & Halske introduced a revised shop organization. The planning and control of the factory work now took place in new planning offices in advance, clearly separated from the operations on the shop floor. Production and operations were increasingly standardized. A painstaking system of written prescriptions and controls, using forms and cards of different colours to an unprecedented extent, was introduced to rationalize the production process. It seems that Siemens, like other large German enterprises, adopted elements of "Taylorism"—without using the name—even before these American principles of industrial organization were propagated in Germany in the last years before World War I.[19]

The number of salaried employees increased both in absolute and in relative terms. The ratio of non-manual to manual workers was 1:11 in 1865, 1:7 in 1890, and 1:3·5 in 1912. In many respects the status of the 12,500 Siemens salaried employees of 1912 had become more similar than before to the status of employees in public bureaucracies. Most of them (except those at the top) received salaries in which achievement criteria played a smaller and seniority a larger role than in previous years. They were treated according to general rules about recruitment, remuneration, promotion, fringe benefits, and controls.

[17] Hintze, 'Der Beamstenstand', p. 77.

[18] Pollard, *Genesis of Modern Management*, pp. 129 ff.

[19] H. Homburg, 'Anfänge des Taylorsystems in Deutschland vor dem Ersten Weltkrieg', *Geschichte und Gesellschaft*, IV (1978), pp. 170-94.

Generalized qualifications (degrees from technical, commercial, and general schools) had gained some emphasis in the process of recruitment. Most of them performed highly specialized, routinized functions within a rigid network of regulations and in hierarchically structured departments and offices.

Such changes indicated increasing bureaucratization, which resulted from different factors: from the mere expansion of the enterprise as such; from the technological refinement and the accuracy required by expensive and complicated machineries; from the increased application of science; and from the requirements of more sophisticated accounting and sales methods. Although influences from outside public bureaucracies continued to play some role, this development around the turn of the century was largely the product of changes occurring within the enterprise. While up to 1890 the bureaucratic features of the Siemens management had largely resulted from outside influences, they were now reinforced by an internally generated process of industrial bureaucratization.

There remained, however, many limits to the bureaucratization at Siemens. On the top management level informal and personal factors continued to play a role. The top decision makers were not constrained by the network of bureaucratic roles which they gradually imposed on the administrative departments and shops. Many important decision-making processes took place through informal channels of communication. Non-hierarchical patterns of co-operation between departments were consciously stressed. An internal price system was established and elements of the market were incorporated into the organization of the firm. Financial, success-related incentives determined the income of the highest ranking officials. Conscious policies of decentralization were applied. And there remained clear differences between the status of the employees and a fully developed civil service status—whatever the employees wished and demanded. Tenure and seniority rules were ultimately dependent on the firm's success. Some employees experienced lay-offs in depressions, even at Siemens. Graduation from certain schools did not guarantee an "appropriate" position in the firm, while careers remained possible on a self-made basis (though they seem to have become less frequent). Nepotism in promoting employees was possible, and could be very functional.

Of course, these details derive from a single case study. The Siemens firm may have been influenced by the bureaucratic tradition more than others, due to the biography of its founder, its close contacts with such large institutions as the post office and railroads, and needs of electrical engineering for particular accuracy and scientific preparation. But what we know about other firms points into a similar direction, though often to a lesser degree. This is especially true for the railways, which in the United States pioneered modern management methods, and which in Germany were under strong government influence from the beginning.[20] Starting in the 1870s there is a lively and growing literature on

[20] For the United States, see A. D. Chandler, Jr. 'The Railroads: Pioneers in Modern Corporate Management', *Business History Review*, XXXIX (1965), pp. 16-40. The history of German Railroad management in comparative perspective still needs to be written. Cf. C. W. F. Schmeidler, *Geschichte des Deutschen Eisenbahnwesens* (Leipzig, 1871); B. Stumpf, *Geschichte der deutschen Eisenbahnen* (Mainz/Heidelberg, 3rd. ed. 1960). For a case study, see *Das Bergisch-Märkische Eisenbahn-Unternehmen in seiner Entwicklung während der ersten 25 Jahre des Betriebes* (Elberfeld, 1875). A Cologne electrotechnical firm is studied in depth by G. Schulz, *Die Arbeiter und Angestellten bei Felten & Guilleaume* (Wiesbaden, 1979). Also see W. Treue, *Die Geschichte der Ilseder Hütte* (Peine, 1960).

business management and administration, and a reading of the manuals, journals and the like tends to support the picture I have drawn from the Siemens case study.[21] Nevertheless, our conclusions must remain somewhat hypothetical in character.

As a result of the specific conditions of German industrialization, bureaucratic patterns strongly influenced the development of industrial management. First, they were largely induced from the outside, although they directly contributed to the managerial success of the early firms. Later on, under new technological and commercial conditions, bureaucratization continued, more as a result of changes within the enterprise than as a result from external influences. While being modified and adjusted, existing bureaucratic structures and dispositions could be utilized by German large-scale enterprises in later periods of growth, even though they had to be adjusted to new strategies and operations.[22] Within these structures, the need for systematic management, for increasingly professional personnel with formal training from outside schools, for accuracy and rational organization could more easily be met.[23]

In countries without strong bureaucratic traditions, this building of adequate large-scale organizations in an advanced stage of industrial capitalism must have been a much more difficult process. Recently, the hundred largest German manufacturing and mining firms in 1887 and 1907 were compared with a similar British list. The differences of size, degree of diversification, and extent of functional integration are striking.[24] There are many reasons for this much discussed German-British difference. One of them might be the easier availability of organizational and managerial skills on the German side—due to the structures and traditions discussed in this article. But it should be stressed again that bureaucratic structures and techniques always were just one dimension of industrial management. In order to be functional they had to be checked by and amalgamated with other patterns, especially with personal, family-based managerial techniques and with market incentives and mechanisms. The top management positions in particular continued to work according to pre-bureaucratic patterns. In the last analysis, all the limits to bureaucratization in the industrial enterprise resulted from its market dependence and particular achievement orientation, and the fact that private property remained its major source of legitimate power which could successfully refuse to yield to the pressures of bureaucratization.

III

The distinction between manual (wage) workers and non-manual (salaried)

[21] J. Kocka, 'Industrielles Management. Konzeptionen und Modelle in Deutschland vor 1914', *Vierteljahrschrift für Sozial- und Wirtschaftsgeschichte*, LVI (1969), pp. 332-72.

[22] For the concepts see A. D. Chandler, Jr. *Strategy and Structure* (Cambridge, Mass. 1962), pp. 7-17.

[23] J. Kocka, 'Les entrepreneurs salariés dans l'industrie allemande à la fin du XIXe et au début du XXe siècle', M. Levy-Leboyer, ed. *Le patronat de la seconde industrialisation* (Paris, 1979), pp. 85-100.

[24] J. Kocka and H. Siegrist, 'Die hundert größten deutschen Industrieunternehmen im späten 19. und frühen 20. Jahrhundert. Expansion, Diversifikation und Integration im internationalen Vergleich', in N. Horn and J. Kocka, eds. *Recht und Entwicklung der Großunternehmen im 19. und frühen 20. Jahrhundert* (Göttingen, 1979), pp. 55-122, esp. 84-9; J. Kocka, 'The Rise of the Modern Industrial Enterprise in Germany' in A. D. Chandler, Jr. and H. Daems, eds. *Managerial Hierarchies. Comparative Perspectives on the Rise of the Modern Industrial Enterprise* (Cambridge, Mass. 1980), pp. 77-116, esp. 99-110. On England see the contributions in the same volume by L. Hannah.

employees can be observed in all industrializing societies. And a good deal of sociological literature has been devoted to the distinction between blue collar and white collar, *Arbeiter* and *Angesttellte*, *ouvriers* and *employés*.[25] I want to argue that this virtually universal distinction was particularly sharp and socially relevant in nineteenth- and early twentieth-century Germany, and that the particular sharpness of this "collar line" was partly due to the bureaucratic impact on German industrialization.

It has already been mentioned that, in the early stages of industrialization and *within factories*, management differentiated clearly between salaried persons and wage earners, with respect to type and amount of payment, job security, and early welfare benefits. In the German industrial enterprises so far investigated, salaried employees and wage workers differed more clearly than in comparable American enterprises of the same period. The distinction between wage workers and salaried employees may have been equally visible in nineteenth-century English firms, but neither in the United States nor in Britain did technicians, supervisory personnel, salesmen, clerks, or office employees conceive of themselves as '*Privatbeamte*' ("private civil servants").[26]

To explore the significance of the emerging distinction between blue- and white-collar work and workers, it is necessary to consider it in general terms, rather than as a manifestation of the experience of individual enterprises. And in what follows I shall take draftsmen, technicians, and engineers as examples. Bureaucratic traditions contributed to the formation of these groups, and the school system served as an intermediating link.

Starting in the 1820s, Prussia and other German states had established a series of public industrial and technical schools, *Gewerbeinstitute* and *Gewerbeschulen*. There were several motives behind this initiative, but most important was the government aim of promoting industry, the national wealth and the state's power. In the early years, access to these schools was easy, due to the virtual lack of entrance examinations and to the many available scholarships. In the first fifty years of the Berlin *Gewerbeinstitut*, for example, about 3,500 technicians graduated there, and most of them went into salaried positions within private firms, as foremen, draftsmen, supervisors, engineers, and managers.

But around the middle of this century some of these schools experienced a thorough process of academization. Practical training, until then a central part of the courses, was reduced; mathematics became more important, specialization increased. The teachers sought the relatively autonomous status of faculties. Entrance qualifications were raised: elementary school background had sufficed in the 1830s, but by 1870 *Abitur*—graduation from a *Gymnasium* or a similar secondary school—was a condition for application. Stipends were reduced, and fess were raised. Access for the sons of lower class families became more difficult. The characteristic name of the institution evolved inexorably from industrial

[25] F. Croner, *Soziologie der Angestellten* (Cologne, 1972); S. Braun, *Zur Soziologie der Angestellten* (Frankfurt, 1964); G. S. Bain and R. Price, 'Who is a White Collar Employee?', *British Journal of Industrial Relations*, x (1972), pp. 329-39; S. Pollard, 'The Rise of the Service Industries and White-Collar Employment', in B. Gustafsson, ed. *Post-Industrial Society* (London, 1979), pp. 17-43.

[26] J. Kocka, *White Collar Workers in America, 1890-1940. A Social-Political History in International Perspective* (1980), pp. 92 ff., 268 ff.

school to industrial institute to industrial academy, and finally, in Berlin in 1879, to technical college or technical university (*Technische Hochschule*).[27]

All these changes were part of a complex process in which industrial production techniques became more scientific. By raising standards and altering their courses, the schools responded to increasing demands from industry, and they supplied qualified employees who then introduced more systematic and partly scientific methods to the firms which hired them. Thus, from the 1850s on we find construction bureaux in the large German machine building firms, and laboratories in the metal producing and chemical industries. But on the other hand, detailed research suggests that the supply of academically trained graduates ran ahead of manifest demands from industry: schools did not just respond to existing demands but created them. Behind the academization of those schools we partly see at work the teachers', the administrators' and the graduates' wishes for social upgrading.

In the early stages of German industrialization the social prestige of businessmen, managers, and employees in the private economy seems to have been low. Admittedly, the low status of early businessmen, compared with more traditional positions and élites, was not peculiar to Germany. But perhaps the remedy chosen was. In Germany, education was a very important criterion of social standing; formalized, examination-proved qualifications served as an important platform for claims to positions, status, income, and power. There existed a close fit between the type of graduation a person had achieved, and the type of career to which he or she was entitled; in fact, the great importance of titles, so often observed, and ridiculed, by foreign visitors, symbolized this close fit which did not have a parallel in nineteenth-century Britain or America; in Germany, all this was part of the bureaucratic legacy. Consequently, a high degree of functional specialization and hierarchical differentiation characterized the German school system, which was in any case run by civil servants. Now, in the third quarter of the century, technical schools were integrated into this system, the teachers and graduates of those schools thus entered the system of titles and positions, which in Germany ranked besides landed property and noble birth as an important source of esteem and recognition. In contrast to becoming a landed rentier or a scholar, this route to social upgrading did not deprive industry of human resources, talents and energy—on the contrary.[28]

This bureaucratically guided German process of academization served as a basis on which the profession of engineers emerged, clearly separated from craftsmen, artisans, foremen, and the like. Until then there was no clear separation between mechanics, technicians, and engineers in terms of self-identification, job profiles, and semantic classification; the word "engineer" had usually referred simply to members of the specific military formation responsible

[27] P. Lundgreen, 'Industrialization and the Educational Formation of Man-Power in Germany', *Journal of Social History*, IX (1975/6), pp. 64-79; P. Lundgreen, *Techniker in Preußen während der frühen Industrialisierung. Ausbildung und Berufsfeld einer entstehenden sozialen Gruppe* (Berlin, 1975); Kocka, *Unternehmensverwaltung*, pp. 166 ff.

[28] Details and evidence in J. Kocka, 'Bildung, soziale Schichtung und soziale Mobilität', pp. 297-309. Most recently on the 'Berechtigungswesen' see R. v. Westphalen, *Akademisches Privileg und demokratischer Staat. Ein Beitrag zur Geschichte des Laufbahnwesens in Deutschland* (Stuttgart, 1979); K.-H. Manegold, 'Technology Academized: Education and Training of the Engineer in the 19th century', in W. Krohn et al. eds. *The Dynamics of Science and Technology* (Dordrecht, 1978), pp. 137-58.

for the construction of roads, bridges, and war machines. However, around the middle of the nineteenth century, the meaning of "engineer" changed. It became a collective name for non-military technical men with at least some theoretical training, engaged in non-manual work.[29]

The close connection between the academization of technical schools and the formation of the engineering profession is exemplified by the origins of the most influential engineering association, *Verein Deutscher Ingenieure (VDI)*. This originated in 1856 from an alumni association of the Berlin industrial academy. Its statutes stressed higher technical education and theoretical background as a major attribute of the *Ingenieur*.[30] There seems to have been no parallel development in nineteenth-century England. At least in the British machine tool industry, on-the-job apprenticeship remained at the centre of the engineer's training. And in that respect the separation between non-manual technical employees and skilled mechanics, was, in England, less clearly demarcated than in Germany. Again, semantic differences tend to corroborate this finding. While the German word "*Ingenieur*" like the French "*ingenièur*" clearly excluded manual workers, even the most skilled mechanics, the English word "engineer" could include certain types of skilled blue-collar workers as well.[31]

It would be easy to elaborate this line of argument by exploring the relationship between changes in the educational system and the changing pattern of group, strata, and class formation, in the late nineteenth and early twentieth centuries.[32] However, in terms of the present article, I should like, instead, to discuss one other mechanism through which bureaucratic influences contributed to the emergence of a distinct white collar class, and indirectly to the relatively clear delineation of the blue collar working class, in Germany.

By the 1880s, in spite of what has been said so far, the dividing line between blue collar and white collar workers, *outside* the single firms, remained fragmented and vague, peripherical to public opinion, and politically somewhat irrelevant. Different white collar groups—retail clerks, technicians, bank employees—did not share a sufficient range of experiences or attitudes to display a common consciousness, join common organizations, or engage in common actions. Outside individual enterprises, the semantic distinction between *Arbeiter* and *Angestellte* or *Arbeiter* and *Privatbeamte*, although not unknown in the 1880s, was not very common.

By the beginning of World War I, however, the picture had radically changed. Books, pamphlets, periodicals, and public debates dealt with the *Angestelltenfrage*; the term "*Angestellte*" had been fully established; associations of different white collar occupations had begun to join hands and had started

[29] H. Schimank, 'Das Wort "Ingenieur",' *Zeitschrift des VDI*, LXXXIII (1939), pp. 325-31. On early engineers in general, see U. Troitzsch, 'Die Rolle des Ingenieurs in der Frühindustrialisierung—ein Forschungsproblem', *Technikgeschichte*, XXXVII (1970), pp. 289-309.

[30] *Verein Deutscher Ingenieure, 1856-1926* (Berlin, 1926), p. 51; L. U. Scholl, 'Bürokratisierung und Professionalisierung. Zur Genesis technischen Beamtentums im Staatsdienst am Beispiel des Königreichs Hannover', *Technikgeschichte*, XLVI (1979), pp. 117-38; L. V. Scholl, *Ingenieure in der Frühindustrialisierung. Staatliche und private Techniker im Königreich Hannover und an der Ruhr (1815-1873)* (Göttingen, 1978).

[31] R. C. Floud, *The British Machine Tool Industry, 1850-1914* (Cambridge, 1976); P. L. Robertson, 'Technical Education in the British Ship-Building and Marine Engineering Industries, 1863-1914', *Economic History Review*, 2nd ser. XXVII (1974), pp. 222-35.

[32] Kocka, 'Bildung', pp. 303 ff.

common action; and something like a white collar movement had developed, encompassing white collar groups from different occupations and industries, setting them clearly apart from the blue collar workers and those self-employed. Of course, the different types of white collar workers continued to differ in many ways, and they continued to adhere to specific occupational or sectoral identities. Nevertheless, the concept "*Angestellte*" now began to refer to a real social group with some sort of common consciousness, shared interests, internal communication, common organization, and clear visibility.[33] What had brought this about?

Among other influences, changes in the occupational structure played a part, while the rising challenge of a blue collar working class movement also served as an additional stimulus for white collar employees to stress the one thing they had in common—the fact that they were not manual workers. But the major factor behind this early formation of a relatively distinctive white collar class was clearly connected with the bureaucratic tradition. It was very much in tune with the German tradition of reform from above, with the power and the self-consciousness of the civil service, and with the general values and expectations of many citizens that (at least since the 1870s and 1880s) the government increasingly intervened in the economy and in social relations in order to manage tensions and mollify conflicts which mainly resulted from the advancement of industrial capitalism. This increase in state intervention stimulated an articulation of interests along previously vague lines of social differentiation—a development which was designed to influence government decisions, and which in the process contributed to the clarification of those previously vague lines of differentiation and to the formation of social groups.

The public insurance system, enacted in the 1880s, was the most important issue with respect to which white collar workers developed something like a common identity apart from blue collar workers and employers alike. Bismarck's compulsory health, accident, and old age insurance schemes were designed to include both industrial blue collar workers and a large majority of white collar employees whose annual incomes were below a certain limit. Payments and benefits were graded according to income levels, length of membership, and other criteria, but the schemes did not differentiate between wage workers and salaried employees. However, the law permitted occupational groups to contract-out of some of the public schemes if they formed organizations of their own to set up voluntary insurance schemes which met certain criteria.

Not surprisingly, some groups of salaried employees felt sufficiently different from and superior to the mass of the workers to choose this alternative. And the first attempt to organize salaried employees from different occupations and sectors resulted from this: The "Deutscher Privatbeamten-Verein" was founded in 1881; it included bookkeepers, engineers, bank employees, office workers, and others, and it provided for an insurance scheme imitating as much as possible the benefits civil servants would enjoy after retirement.[34]

[33] For a recent over-view see D. Schulz, 'Die industriellen Angestellten. Zum Wandel einer sozialen Gruppe im Industrialisierungsprozeß', in H. Pohl, ed. *Sozialgeschichtliche Probleme in der Zeit der Hochindustrialisierung (1870-1914)* (Paderborn, 1979), pp. 217-66; G. Hartfiel, 'Germany', in A. Sturmthal, ed. *White-Collar Trade Unions* (Urbana, Ill. 1967), pp. 127-64; on the history of the concept 'Angestellte' see J. Kocka, 'Angesteller', in O. Brunner et al. eds. *Geschichtliche Grundbegriffe. Historisches Lexikon zur politisch-sozialen Sprache in Deutschland*, I (Stuttgart, 1972), pp. 110-28.

[34] Kocka, *Unternehmensverwaltung*, pp. 515 ff.

This initiative remained within narrow limits, but in the late 1890s a much larger movement with similar purposes and far reaching effects got off the ground. By then many white collar workers earning less than 2,000 marks a year had been included in the public old age pension scheme, together with the mass of industrial workers. The majority of employees' organizations, under the lead of three commercial employees' associations but also including organizations of technicians, foremen, bank, and general office employees, started to agitate in favour of taking all *Angestellte* out of the general insurance scheme. They demanded a separate public insurance system for *Angestellte*, more appropriate to their special needs, by which they allegedly differed from blue collar workers. They agitated for an insurance system modelled—as much as possible—after the pension schemes of civil servants. They were ready to pay a slightly higher premium, wanted higher benefits, and required specific conditions. In the course of more than ten years of increasing agitation in favour of such a law, a whole set of different ideologies and rhetoric was developed in order to justify the claim that white collar employees were different from and superior to blue collar workers. And in these years a common identity of *Angestellte* across occupational lines crystallized in spite of many continuing differences. As a result the separation between white collar workers and blue collar workers became more manifest. The concepts of "*Privatbeamte*" and "*Angestellte*" became fully accepted, sharply defined, and popular.

In 1911 the desired law was enacted, although financial exigencies meant that benefits remained far behind those of civil servants. At bottom, the demand for legal privilege succeeded because it secured support from different middle-class parties and the government, hoping thereby to ensure that the rapidly growing mass of salaried employees would remain outside the strong socialist camp and committed to the existing order.[35]

There were similar legal developments in Austria,[36] but not in England, France, or the United States. In Germany it was now of legal importance whether an employee qualified as white collar employee (*Angestellter*) or not. Labour and social legislation, enacted in the following years, frequently differentiated between *Arbeiter* and *Angestellte*, incorporating the definition of the 1911 law (slightly revised and enlarged in later years). The more often laws differentiated this way, the more relevant the line between blue and white collar workers became in terms of real life chances. At least until 1933, and to some extent even today, the distinction counts for more in Germany than in other countries, in terms of social and labour legislation, industrial relations, union structures, political behaviour, collective mentalities, public symbolism, political language, and in many other respects.[37]

[35] E. Lederer, *Die Privatangestellten in der modernen Wirtschaftsentwicklung* (Tübingen, 1912); Lederer, 'Angestelltenorganisationen und -sozialpolitik', *Archiv für Sozialwissenschaft und Sozialpolitik*, XXXV (1912), pp. 263-98; Kocka, *Unternehmensverwaltung*, pp. 516-19, 536-44.

[36] G. Ortruba, 'Zur Geschichte der "Angestellten" und ihrer wachsenden Bedeutung in Österreich bis 1918 (im Vergleich zu Deutschland)', *Österreich in Geschichte und Literatur*, XXI (1977), pp. 74-122.

[37] Kocka, *White Collar Workers in America*, pp. 251-84. Recently, some British historians have started to play down the role of pre-industrial structures and traditions (such as the bureaucratic one) in the history of German society. They also underestimate the resulting differences between German and British social, political, and cultural developments. See G. Eley, 'Capitalism and the Wilhelmine State', *Historical Journal*, XXI (1978), pp. 737-50; R. J. Evans, 'Introduction: Wilhelm II's Germany and the Historians', in R. J. Evans, ed. *Society and Politics in Wilhelmine Germany* (1978), pp. 11-39; D. Blackbourn and G. Eley,

This had far reaching social and political consequences. For example, there seems to have been a mutually reinforcing relationship between the sharpness of the blue collar/white collar line on the one hand, and the relative small importance of the division between skilled and unskilled blue collar workers on the other. Compared with Britain and the United States, it would seem that the division between an upper stratum of skilled manual workers and the mass of unskilled or little skilled blue collar workers was less pronounced in Germany, in the late nineteenth and early twentieth centuries, in terms of union structure, political behaviour, perhaps even life styles and cultures. It may not be just a historiographical peculiarity, but rather a function of social reality that there has been no debate about a German labour aristocracy. And I would argue that the German debate about the *Angestelltenfrage*, the socio-political division between white collar and blue collar workers (which was, of course a political debate—would the employees join the socialist camp, why did they not?) was a functional equivalent of the British labour aristocracy debate. It has been lively in Germany since about 1910 and nearly absent in Britain or the United States.[38]

This is not the place to discuss the problems of the labour aristocracy. It is, however, relevant to emphasize that Germany's early social welfare legislation, itself part of the German tradition of bureaucratic reform from above, created the circumstances in which different types of salaried employees stressed what they had in common as against the blue collar working class. The private civil servant ideology both motivated and legitimated their demands. This ideology of bureaucratic origin offered them a powerful if utopian model for collective self-identification which was neither dichotomic nor functional but hierarchical. It cut across functional and occupational differences between them, playing them down; at the same time it accentuated the gulf, the social distance, between them and the mass of the workers and lower class people in general; it defined the salaried employees as belonging to a middle stratum, clearly marked off from the lower class, and also differentiated from the employers and the élite. In addition, the private civil servant ideology stressed *specific* aims and claims. It was not accidental that salaried employees were mobilized by insurance issues. As private civil servants they did not only claim clearly delineated status, but also security, protection against market fluctuations, and against risks in general. Bureaucratic, rather than capitalist, traditions had imprinted their aspirations and mentalities. Later on, in the first third of the twentieth century, both their marked claims for status, and their quest for security beyond the market came into sharp conflict with the reality of advancing capitalism and its crises. The resulting resentments and protests were particularly strong, not without anti-

---

Mythen deutscher Geschichtsschreibung (Frankfurt, 1980). It seems that they have not taken into consideration the kind of evidence presented above, and the body of literature from which it derives. They claim to have destroyed various "myths of German historiography", but in reality have produced new ones.

[38] E. Lederer and J. Marschak, 'Der neue Mittelstand', *Grundriß der Sozialökonomik*, IX, 1 (Tübingen, 1926), pp. 120-41; E. Fehrmann and U. Metzner, *Angestellte in der Sozialwissenschaftlichen Diskussion. Ein Literaturbericht* (Köln, 1977). Also see the comparative remarks in E. J. Hobsbawm, 'The Labour Aristocracy in 19th-century Britain', reprinted in *Labouring Men. Studies in the History of Labour* (1968), pp. 296-97. Cf. the comparison between Germany, England and France by W. Mangold in J. Kocka ed. *Angestellte im europäischen Vergleich* (Göttingen, 1981, forthcoming). See also the articles on English white collar workers in the same volume by G. Anderson and G. Crossick.

capitalist overtones, and more pronounced than they would have been without the bureaucratic tradition.[39]

IV

This has been a very crude picture leaving aside many differentiations and qualifications. In particular, regional differences have been ignored, even though they have been stressed in the recent debate on industrialization.[40] But in the discussion the bureaucratic conditions of industrialization, the use of the national context and international comparisons may be more justified than in the analysis of some other dimensions of industrialization.

In the German case, and at the level of industrial organization and white collar workers, bureaucratic structures and traditions seemed to have facilitated economic rationality and effectiveness. And in this respect the experience would support Weber's fascination with the virtues of bureaucracy rather than his fears of bureaucratic ossification. On the social historical side however, the bureaucratic tradition contributed to a degree of inequality not fully explicable in terms of capitalist industrialization *per se*. And this conclusion would need to be examined by analysing other dimensions of German industrialization from similar viewpoints, and by a more sustained comparison with other countries.

University of Bielefeld

[39] C. Dreyfuß, *Beruf und Ideologie der Angestellten* (München/Leipzig, 1933), English translation: *Occupation and Ideology of the Salaried Employees* (New York, 1938); H. Speier, *Die Angestellten vor dem Nationalsozialismus. Ein Beitrag zum Verständnis der deutschen Sozialstruktur, 1918-1933* (Göttingen, 1977).

[40] S. Pollard, ed. *Region and Industrialisation. Studies on the Role of the Region in the Economic History of the Last Two Centuries* (Göttingen, 1980).

RICHARD TILLY

German Industrialization and Gerschenkronian Backwardness*

Germany played an important role in Gerschenkron's work. It supplied the subject of his first major book, *Bread and Democracy*[1] and subsequently served his celebrated typology of industrialization[2] as the principal case of "moderate backwardness"–that case in which banks supply crucial financial and entrepreneurial inputs. In addition, frequent references to Germany, to German industrialization, and especially to the German universal banks–which he at one point described as an innovation comparable in importance to the steam engine–are scattered throughout his writings.[3] These are bases for expecting to find important connections between the historiography of German economic history and Gerschenkron's work.

The connections, however, reflect a kind of historiographical multicollinearity. For Gerschenkron's reception in German economic historiography was part of the latter's interest in two major paradigms: 1) the "growth paradigm" of economics which came to dominate European economic history in the 1950s and 1960s, and according to which German industrialization and its causes–capital accumulation, technical innovation, labor force growth, etc.–represented a species of a larger genus encompassing many, if not all, countries of the world. For this purpose Gerschenkron proved relevant, though not more so than, say, W. W. Rostow or Simon Kuznets;[4] and 2) the "Sonderweg" paradigm, which stressed the need to interpret recent German history as a

* Paper prepared for the Conference on "Patterns of European Industrialization: Rethinking Gerschenkron's Hypotheses," Villa Serbelloni, Bellagio (Italy), October 3-6, 1988.

special case of political, socioeconomic and cultural development (or "modernization") occurring in many countries but which in Germany was marked by a sharp discrepancy between successful economic and incomplete political modernization. The norm here is the strong positive correlation between industrialization and democratization–Hobsbawm's "Dual Revolution" seems apt here–observed for such Western countries as Great Britain, the United States and France. In this "Sonderwegdebatte," as with Rostow or Gerschenkron, concern with Marxist interpretations of general history in terms of economic classes is paramount. Perhaps H.-U. Wehler's writings offer the best recent example of German work dealing with this triangle of issues: the general relationship between economic and political change in modern history, the challenge of improving upon Marxist versions of that connection, and the fitting of German history into the discussion of both.[5]

In the following, I wish to discuss Gerschenkron mainly in association with the first paradigm, but some passing remarks are directed to the second as well. We ask: to what extent does the history of Germany's industrialization reflect, confirm or refute Gerschenkron's view of European modern economic growth? Our answer adopts the primitive approach of running down a check list of the major elements of his "syndrome of backwardness."

1. *The timing of growth and the "big spurt" discontinuity*

Over the last 10 years or so the view that German's industrial breakthrough–in Gerschenkron's terminology, its "big spurt"–came in the 1840s has increasingly dominated the country's economic historiography.[6] This view supports Gerschenkron since it involves not only the notion of a clear acceleration of industrial production concentrated in producers' goods, but also the successful borrowing of technology from abroad, e.g. in ironmaking and railroad organization, and the development of investment banking on an unprecedently large scale. The empirical basis of this view derives from Spree's study of German growth cycles and from a number of studies of heavy industry. The former stresses the cycle making role of railroad investment and heavy industrial growth 1840-80, the latter the backward and forward linkages generated by railroad expansion beginning in the 1840s and rising dramatically in the 1850s.[7] As Fremdling has shown, thanks to a combination of tariffs and technological progress, continued growth of the German network in the 1850s induced rapid growth of iron output leading in turn to increased domestic coal production. We may see this as a case study in successful German borrowing of technology and import substitution aided by tariff policy.[8] Nevertheless, it is a case of successful adoption tied to the leading sector, domestic railway expansion. Further studies of the relations between railway demands for equipment and iron materials and the iron and metal-working firms show that railroad orders were a significant source of income for the largest and technologically most progressive firms and undoubtedly a major cause of their investment programs.[9] And work on early railroad finance shows that universal banking originated here–modernization of German banking thus being a kind of backward linkage derived from railroad building.[10]

But perhaps the clearest expression of this view is in Hans-Ulrich Wehler's modern classic volume 2 of *Deutsche Gesellschaftsgeschichte*. Here, the expansion of heavy industry in the 1840s marks the beginning of Germany's "industrial revolution" and of the end of the period of her tutelage based on imitation of British industrial technology.[11]

In my opinion this has been a useful way of looking at German industrialization. However, it has deficiencies, some of which are inherent in the Gerschenkronian idea of industrialization via a "big spurt" discontinuity. First, and least important, if one adheres to Gerschenkron's definition of a "big spurt" as an extended phase of expansion of industrial output covering at least two major business cycles which do not reflect international cyclical down turns, then Germany experienced no "big spurt" in the 19th century.[12] Second and more important, the "big spurt" view is based mainly on empirical study of German heavy industry and railroads, coupled to the leading sector theory of industrialization; it is *not* based on firm quantitative evidence covering other sectors and the *pre*-1840 period. Although the latter period is largely *terra incognita* from the econometric point of view, some quantitative data are available which raise doubts about the uniqueness of the 1840s: (1) Schremmer's recent reworking and extension of Hoffmann's series on industrial investment back to 1815 shows a slowdown in investment in the 1840s relative to the 1820s and 1830s; (2) the same holds for my own estimates of total net capital formation in Prussia over the 1816-49 period; (3) agricultural output estimates for the entire 1800-60 period identify the 1840s as a marked slowdown or even collapse preceded and followed by respectable growth trends; and (4) a careful reestimation of Spree's industry and social product series using 1850-54 weights (instead of 1913 ones) reversed the ordering of the 1840s and 1850s: the new estimate has the 1840s growing only slightly more than half as much as the 1850s (see Table 1).

Third, and most important, even if the quantitative underpinning of the "big spurt" of the 1840s were unambiguous a case could be made for seeing the latter as the result of gradual, cumulative changes in previous decades. An interesting indication of this possibility can be found in recent analyses of the German railroads, already referred to. The weakest link in the argument connecting railroads with the "big spurt" is the explanation of railway investment itself. Time series analyses show the latter to depend largely upon previous profits and profit expectations. Studies of the early companies, moreover, suggest that they were generally responses to pre-existing transportation demands; for this reason they yielded above-average returns to their owners–virtually as soon as they began operation.[13] According to these findings, the restrictive policies of the German states toward railways in the 1830s held up development, for such profits represented demands which could have been realized earlier. If that is so, we need to explain the emergence of those demands. That is, we need to explain pre-railway, pre-1840 development.

This need enhances the importance of a number of recently revised chapters in the history of Germany's early industrialization. One such chapter concerns the spread of export-oriented rural industry in various parts of Germany since the 16th century. This "protoindustrialization" involved the relative growth of proletaroid strata dependent on non-agricultural wage incomes,

Table 1

Estimated Annual Rates of Growth (per cent)

Net Investment Levels			Agricultural Output		Net Domestic Product		
	Schremmer	Tilly		Helling		Spree	Tilly
1815-30	4,68	3,72	1820-30	1,32	1840-50	3,43	1,74
1831-40	3,17	2,20	1830-40	2,58	1850-60	2,03	2,70
1841-50	1,46	1,72[a]	1840-50	0,12	1860-70	2,02	
			1850-60	1,72	1870-80	2,21	

[a] 1840-49.

Sources: G. Helling, "Berechnung eines Index der Agrarproduktion in Deutschland im 19. Jahrhundert," *Jahrbuch für Wirtschaftsgeschichte* (1965), part 4, pp. 125-43; E. Schremmer, "Die Badische Gewerbesteuer und die Kapitalbildung in gewerblichen Anlagen und Vorräten in Baden und in Deutschland, 1815 bis 1913," *Vierteljahrschrift für Sozial- und Wirtschaftsgeschichte*, 74 (1987), pp. 18-61; Spree, *Die Wachstumszyklen*; R. Tilly, "Some Problems in the Measurement of Economic Growth in Germany in the Nineteenth Century," paper at Workshop on Quantitative Economic History, University of Groningen, 1985.

the development of supraregional markets for agricultural and protoindustrial products such as cloth, and the accumulation of capital in the hands of merchants creating new needs as well as providing the basis for further expansion of rural industry. Thanks to the work of such scholars as Kriedte, Medick, Schlumbohm, Mager, Schulze, or Kisch, we are able to identify dynamic centers of craft-type industrial development clearly emerging as early as the 18th century in such Rhenish centers as Krefeld, Elberfeld and Barmen or in Saxon regions such as the county of Zwickau. Such regions not only produced the labor surpluses and merchant capitalists characteristic of "protoindustrialization," but also the cadres of skilled craftsmen and innovative industrial entrepreneurs which introduced factory-type industrial growth in the 19th century. Although research on this problem has not yet reached the point where quantitative evidence on the contribution of craft-type development (textiles and iron wares) to German industrialization can be easily summarized, it is clear enough that (1) individual, local examples of rapid industrial growth abound long before 1840 and that (2) their cumulative effect could conceivably have been weighty enough to have induced the investment of the 1840s.[14]

A second chapter focuses directly on agriculture–seen by Gerschenkron to play but a minor role in cases of backwardness. Most of the new contributions unfortunately continue to build on the older estimates of von Finckenstein, Helling and Hoffmann, and can only partially overcome the basic uncertainty surrounding them (particularly with respect to estimates of the total area under cultivation and of fallow land). On the one hand a visible expansion of agricultural production (and of land, seed, and labor productivity) can be traced, with interruptions, back to the 18th century. Indeed, for the first half of the 19th century one can identify productivity gains which could have contributed significantly to economic growth before the Take-off ("big spurt").[15] One important source has argued–on the basis of largely non-quantitative evidence–that agrarian households were far and away the most important source of increased demand for domestically produced non-agricultural goods and services in Germany, 1800-50.[16] This view is supported by old and new estimates of the demand of agricultural producers for iron products (implements, machinery)–a demand which turns out to be approximately equal in aggregate to that of the railroads in the 1840s and 1850s.[17] And further support comes from a recent reassessment of users of early pre-railway nineteenth-century transportation facilities: agricultural products clearly dominated.[18] These are grounds for encouraging more work on German agriculture in the early-period of industrialization.

The third relevant chapter takes up the growth and structure of German foreign trade and links it to the development of the country's internal trade. It first notes the structure of foreign trade:[19] imports of intermediate and colonial products and exports of primary and manufactured goods. It then discusses a triangular trading pattern. The Prussian eastern provinces enjoy a primary goods export surplus vis-à-vis Britain dominated by intermediate goods (such as iron and yarn) and an export surplus vis-à-vis the eastern provinces dominated by final manufactured goods. High transportation costs limit east-west trade to relatively high value commodities. Dumke attempts terms of trade and trade volume comparisons to show that his "system" is determined by British demand for primary products: an expansion of primary goods' exports raises eastern incomes and stimulates eastern German demand for western German manufactures, while this expansion of sales stimulates, in turn, enlarged imports of intermediate goods from Britain. Manufactured goods exports–particularly textiles and iron goods produced to a large extent under protoindustrial conditions–expand little before 1850, mainly due to British competition. The gradually rising level of internal trade draws attention to the need for internal transportation improvements–which are then carried out, beginning with the road investments of the 1820s and culminating with the railroad projects of the 1830s. The significance of this "model" is–apart from its interpretation of regional differences–that it offers an explanation of how a virtually stagnating export in manufactures is consistent with a gradually expanding internal trade leading to, indeed inducing, investment in transportation improvements. It thus offers German industrialization as a case in which export-led growth of one kind gradually turns into rapid industrial growth carried forward by investment and import substitution.[20]

These strictures raise the more general question of the timing and character of German industrialization. As suggested, Gerschenkron's "model" seems to best fit the "take-off" period from the 1840s to the 1870s. As already noted, that implicitly downgrades the importance of prior changes. It has another defect, however. It obscures the subsequent changes in German economic life which were also part of the industrialization process. Historians have argued that German industrialization entered its "second phase" sometime around the 1870s. This "second phase"–some have called it "high industrialization"–describes the period to 1914 and encompasses a number of important changes: the political organization of agriculture in response to its decline as the principal sector of the German economy; the related decline in economic liberalism as the dominant ideology of the country's economic policymakers; the related increase in state intervention designed to soften some of the effects of market competition on the less well-to-do population, e.g. social insurance; the development of scientific knowledge as a factor of production and its encouragement by government institutions; and the absolute and relative growth of very large industrial enterprises.

Some historians have found that the "second phase" of industrialization came quite abruptly, initiated by the crisis of the 1870s which in attenuated form lasted until the mid-1890s and has been called the "great depression" (of 1874-94). Most economic historians, however, have resisted this notion of a "great depression" 20 years in length, some of them on the grounds that the "great depression" was largely a price phenomenon–a "great deflation"–while the existing output data show no marked decline in volume well correlated with the relevant period.[21] In this view, sustained real output growth as observed in Germany, is semantically inconsistent with the term "great depression." But although I sympathize with this position and fully accept dismissal of the term "great depression," something with analogous connotations may be needed to replace it, e.g., the "prolonged slowdown." There are three reasons for this. First, Rainer Fremdling's recent work casts doubt on the reliability of Hoffmann's index of industrial production for the period, mainly its probable sensitivity to weighting assumptions.[22] Second, the Hoffmann or Hoffmann-based data *do* show a slowdown in real output even if one adjusts for position within the business cycle to obtain fully comparable periods (see Table 2). Note the sharp contrast ob-

Table 2

Comparable Growth Cycles and Periods in Germany, 1850-1913 (real output, annual rate of growth)

Industry and Crafts				Net Domestic Product	
Trough-to-Trough		Contrast Maximizing		Peak to Peak	
1848-1866	.0294	1850-1874	.0441	1857-1864	.0261
1866-1879	.0380			1864-1874	.0250
1879-1891	.0338	1874-1894	.0262	1874-1884	.0127
1891-1901	.0364			1884-1890	.0283
1901-1908	.0406	1894-1913	.0416	1890-1900	.0339
1908-1913	.0497			1900-1907	.0268
				1907-1913	.0321

Sources: W. G. Hoffmann et al., *Das Wachstum der deutschen Wirtschaft seit der Mitte des 19. Jahrhunderts* (Berlin, 1965); S. Solomou, *Phases of Economic Growth, 1850-1973. Kondratieff Waves and Kuznets Swings* (Cambridge, 1987).

tained by adhering to the older ("Contrast Maximizing") periodization. I am not convinced that this is illegitimate practice.

Third, some other indices of industrial activity also reflect a slowdown from the late 1870s to the mid-1890s, namely Hoffmann's estimates of net investment in crafts and industry and, moreover, an independent estimate of industrial fixed investment based on the experience of large enterprises confirms the periodization suggested by Hoffmann's investment estimates.[23]

The upshot of these observations is that one can make a case for a slowdown phase–possibly a transitional phase?–followed by a phase of re-acceleration around the mid-1890s. This latter period, however, has also been seen as sufficiently unique to deserve its own name, the "rise of organized capitalism." This involved big business and oligopoly, cartels, organized labor and strikes, increased government regulation and intervention which was seen in part as "stabilizer" of a system otherwise prone to anarchy and breakdown. This notion has not found favor with most economic historians, but it does raise the question of swings in industrialization which would not seem to fit in with the idea of relative backwardness as the "master process" driving economic development.

2. *The role of the banks*

Gerschenkron's well-known emphasis upon Germany's universal banks as developmental instruments built on an historiographical tradition going back to the beginning of this century and including such names as Hilferding and Schumpeter–illustrious names, I might add, strategically placed, as it were, to make an impression on Gerschenkron. The continuing flow of publications on the subject is thus obviously more than a response to his work.[24]

In general, the notion that these banks–at first private firms and from the 1850s increasingly joint-stock companies–provided much entrepreneurship and risk capital and thus positively contributed to industrial growth and especially to the growth of large-scale industrial enterprises in Germany has stood the test of time quite well. The contrast to the role of banks in Great Britain–Gerschenkron's basic standard–has been confirmed. The consensus can be described in the following four propositions.

First, despite some dissent, the historical literature generally attributes to the banks a positive and significant contribution to Germany's economic development in the 19th century. This contribution consisted in the financing of risky investments, particularly in heavy industry, and included entrepreneurial feats such as the formation of new enterprises, the implementation of mergers, and the organization of cartels.[25] It is unsurprising, but nevertheless interesting, that recent work–including some application of modern portfolio theory–suggests that the German institutional arrangements for capital market finance of risky industrial investments were significantly more effective in the 1870-1913 period than those of Great Britain at this same time.[26]

Second, there is more dissent but on balance a clear preference in the literature for the view that these "great" banks and their executives exercized considerable power in German economic life, power indicated by the large number of banks they took over and the resultant increase in their share of total bank deposits, by the large number of directorships they occupied in German business corporations, by their well-known ability to control strategic decisions through the institution of proxy voting in shareholders' meetings, by the close links between leading bankers and the political élite of the German Kaiserreich, and by a number of well-documented concrete cases of enterprise decision-making in which conflicts resolved themselves in favor of intervening banks.[27] This is related to the first point about financing risky investments, for such power reflected banker access to information which could lower their ex ante risk assessment.

Third, the literature agrees that by concentrating on the financing of small numbers of relatively large-scale projects and established enterprises, the "great banks" neglected large segments of the country's financial business: agricultural credit, housing, small business (especially new business enterprises), small savings, denote fields left to the municipal savings banks, the credit cooperatives, local small private bankers–so far as these fields attracted financial intermediaries at all. After all, then, as later, most industrial investment was self financed.[28] It is only in the 1890s that the large credit banks, spurred by the expansion of the savings banks and the credit cooperatives, began to utilize their branch systems to attract small savings and finance smaller business.[29] This is significant for two reasons: (1) the relatively élite character of their business operations, in fact, explains much of their willingness to engage in fairly risky investment finance and also the low cash reserves they maintained and (2) the overall effectiveness of Germany's financial institutions in promoting industrialization was by no means solely a function of the activities of universal banks. Fourth, much of the vitality of the universal banks rested on the complementary relationship developed between their own activities and those of the government institutions regulating the payments system, above all the central bank of issue. By the early 1860s note issue was virtually a government monopoly. The result was a division of labor which left most of pure payments and short-term trade credit business of the economy to the government bank of issue and most of the industrial credit and also the security issue business to the private bankers and their protégés, the corporate "mixed banks." In addition, the latter found themselves increasingly able to turn to the former for payments services, cash, and for short-term discount credits when the need arose. By establishing a thick network of branch offices covering most of Germany, the Prussian Bank contributed to a significant reduction in transaction costs of trade and finance, illustrated, for example, through the setting of one rate of interest and fixed exchange rates for bill payments anywhere in Prussia as early as 1848. Thus, to some extent, even before the 1870s, the "mixed banks" could build their growing business upon the security that the bank of issue could and would supply it with liquidity, if necessary.

However, from 1876 on the powerful support universal banks gave German industry and trade rested in part on the payments network and liquidity guarantee provided by the Reichsbank. By the 1880s this government institution (with its more than 200 branches) dominated the country's payment business. The credit banks used it extensively, eventually even when they were transferring funds between offices of their own systems. But most important in the present context was the virtually unlimited access to the Reichsbank's discounting facilities which the credit banks came to enjoy. The German banks could get by with less liquidity and "lend to the hilt" if demands warranted because the Reichsbank provided extremely liberal rediscounting facilities. In fact, they were so liberal–though not cheap–that bills of exchange held by the banks or their acceptances could be seen as substitutes for central bank notes. This was in contrast to the Bank of England, which frequently resorted to credit rationing.

The behavior just described tended to stabilize the business cycle in Germany, for at least since the 1880s, through its discount policy, the Reichsbank tended to attract business in upswings and repel it in downswings, which meant a substitution of central bank money or gold coin for short bills, or what was, in effect, potential central bank money. (The reason for this was that the Reichsbank discount rate was generally higher than the market rate, but it rose and fell proportionately less than the latter). With this stabilizing guarantee behind them, the German banks did not have to live with the fear of illiquidity which combining commercial and investment banking activity might otherwise–e.g., under the British set of arrangements–have dictated.

Of course, the "take-off" phase of German industrialization from the 1840s to the 1870s had in fact evidenced considerable instability. The crises of 1847-48, 1857-58 and 1873-76 brought down large numbers of firms and especially banks. The crisis of the 1870s was particularly severe as noted above. It forcibly called attention to the need for more comprehensive controls of financial activity. No doubt the Reichsbank's subsequent policies reflected this.

Nevertheless, the Reichsbank remained a gold standard creature to 1914 and did not underwrite more "on-trend" credit expansion than its gold reserve position allowed. Its contribution to the "German system," to repeat, lay in the flexibility/elasticity of its rediscount policy, especially in times of "liquidity pressure." The universal banks, in consequence, could hold more risky portfolios than would have otherwise been the case.

However, explaining this connection involves a good deal more than Gerschenkron's summary references to Germany's industrial backwardness and capital scarcity relative to Great Britain. It requires going back to the early nineteenth century and examining two related questions: a) the precarious fiscal position of most of the German states (and especially Prussia) and its connection with internal political and class conflict on the one hand, and with monetary and banking policy on the other; and b) the relationship between those fiscal needs and the development of the German customs union, on the one hand, and the need to coordinate currency policies among the several states, on the other. Briefly, the story is this. Its first part concerns the metallic currency. In the first half of the nineteenth century metal monies dominated Germany's currency stock. Owing to the historical heritage of multiple governments a wide variety of coins circulated, and fluctuations in their value are alleged to have hampered intra-German trade and capital flows. For rulers in some of the German states, the coinage was seen as a sovereign monopoly right and valuable independent source of revenue worth defending. However, given the initial difficulties, the problem was overcome with astounding speed. The catalyst was the Prussian government's customs union policy of the 1820s leading to the Zollverein. Here, the Prussian governement built on the success of its own customs union of 1818–which had already proven to be a useful generator of revenues. Its offer of a substantial share in anticipated customs revenues to the rulers of the smaller German states proved irresistible to a decisive number of them, for it promised tax revenues free from parliamentary restraints or concessions. But the practical problem of revenue collection and redistribution required agreement on the value of different monies and on exchange rates. Negotiations on these matters received legal expression in treaties of 1837 and 1838, later again in 1857.[30] Formally, two currency areas emerged from this, the Thaler and the Gulden areas, but the Thaler dominated (and became the standard). And the important point is that power politics–particularly the struggle between Prussia and Austria–and parochial fiscal interests seem to have dominated this institutional change, not the needs of "the economy"–which appears to have adjusted to the change readily.

The second part of the story builds on the first. It concerns the handling of paper money and note-issuing banks. As suggested elsewhere, in Prussia, fiscal needs and commitments supplemented aristocratic-agrarian suspicions about private business aims, leading to the decision (in 1846) in favor of a government monopoly of paper money and note issue. This was related to the Zollverein question, for negotiations about the metallic currency had not regulated paper money, while some German states permitted private banks of issue–which began to circulate their notes in Prussia. Prussia's answer to this was to impose penalties on the use of such money and expand her own note-issuing bank. This worked. By the 1860s the Prussian Bank's circulation dominated in most parts of Germany. By this time, also, a Zollverein treaty prohibited the issue of non-convertible paper money. This was one of the foundations for the development of German banking system as already discussed.

Reviewing this evidence, one is forced to conclude that in order to explain the evolution of banking structures and institutions in countries like Germany, one must devote considerable attention to political forces and institutions. Economic backwardness per se is certainly inadequate to the task, it seems to me. That is one important limitation of Gerschenkron's model as originally formulated.

3. *The standard of living*

Gerschenkron's model suggests that economically backward countries experience falling consumption levels and living standards during their early stages of industrialization–well into and possibly beyond their "big spurt" of growth. This is due to the stress in such countries on investment goods and possibly "forced savings" associated with credit creation and inflationary price increases for consumption goods. Such a characterization would seem to fit German experience as rendered by recent research. For a consensus has emerged that in that phase of Germany's development which most closely corresponds to the "big spurt"–from the 1840s to the 1870s–rapid industrial growth dominated by investment goods production went hand in hand with virtually stagnating living standards. The latter were also associated with food prices which tended to rise over the period and which accompanied a considerable rise in the terms of trade of agricultural vis-à-vis industrial prices. It is only in the 1880s that a clear and widespread rise in real incomes and living standards can be documented. Here, the consensus is equally strong: living standards rose. Our knowledge of these changes stems from general studies of wages and living standards,[31] from special sectoral and regional studies by Kirchhain on the cotton industry, by Holtfrerich and Tenfelde on Ruhr coal miners, by Fremdling on railroad workers, by Borscheid on Württemberg textile workers, or by Fischer and Noll on artisans ("Handwerker")–and from direct examination of consumption standards, themselves–e.g., by Teuteberg and Wiegelmann on food and nutritional standards or by Niethammer and Bruggemeier and also by Teuteberg and Wischermann on housing.[32] One must concede that huge gaps in our knowledge remain, especially for the considerable number of self-employed households in agriculture and (mainly rural) domestic industry. However, all recent work on this group points to average levels of income considerably below those enjoyed by employed labor in the 1840s and 1850s and to a noticeable decline in their number over the 1860s and 1870s.[33] That supports the periodization on living standards just outlined.

One needs to stress the distributional aspects involved, however. To reiterate: what we have is the disproportional growth and leading cyclical role of the heavy-investment goods sectors of the economy. For the German economy as a whole this was accompanied by a clear rise in per capita incomes. The estimated annual increase in per capita income of 1.6 per cent probably exceeded the growth rate in the previous period. However, it was not remarkable when compared with later periods and–most important for our present purposes–it was not matched by increasing real wage income, which rose hardly at all over the period (roughly zero from 1850 to 1873 and about 0.35 per cent per annum, 1844-80). Moreover, this stagnation in real wage income was accompanied–at least until 1873–by a clear rise in the degree of inequality of personal income distribution (as reflected in the Prussian data) and a probably related decline in the share of labor (or wages) in aggregate social product (from about 0.82 to 0.77. At the same time, we may note that both levels and growth rates of wages in the heavy industrial sectors stood well above the overall average and above the average of those in cotton textiles–a branch probably representative of the consumption goods sectors (see Table 3).

Given this pattern, and also the relatively constant and high share of foods and beverages in German consumption patterns over the period we have, I believe, some strong arguments for speaking of a "capital-oriented" growth process during the Take-off era.

It took place at the cost of virtually stagnating domestic wage incomes, whose distribution however made possible substantial gains not only to owners of property but also to sellers of agricultural products, and less substantial gains to the workers of the expanding heavy industrial sector. It was this kind of squeeze on overall living standards–with a hierarchy of exemptions running from "substantial" for property holders to "modest" for workers in the favored growth sectors–which lay behind the period of "big spurt." It was a squeeze which augmented agricultural and industrial profits and, in all probability, investment financed out of them. This is not to claim, I should add, that the German working class achieved no improvements in their standard of living during the Take-off period, for there were modest gains in real wages–supported, moreover, by some decline in hours of work and in the rate of mortality. Nevertheless, these gains did not lead to a pattern of growth which assigned a leading role to worker's purchasing power. This picture corresponds to Gerschenkron's typology, I believe.

Table 3

Wages in Selected Industries, 1840-80

Annual Growth Rates, Wages 1840-80 in Sectors (per cent)		Ratio of Wages Engineering/Cotton Textiles	
Cotton	1.3	1850-54	1.7
Coal (Prussian)	3.1	1860-64	1.5
Railways (German)	2.4	1870-74	1.8
Iron	1.5	1875-79	1.8
Engineering	1.4		

Source: Tilly, "Some Problems in the Measurement of Economic Growth."

However, German historiography has repeatedly drawn attention to a dimension of the question of living standards poorly handled in Gerschenkron's basic model: the role of demographic change. According to currently standard versions (Dickler, Harnisch, Hohorst, Knodel, Köllmann, Lee, etc.) Germany's population growth accelerated in the second half of the 18th century and remained high, though unstable, until the end of the nineteenth century. German experience seems to confirm the idea that children are the poor man's capital, for population growth resulted primarily from increased nuptiality and fertility among the rural poor. This is seen as a response to the demand for labor, e.g. in rural domestic industry as well as in agriculture. Shifts in labor demands and restrictions on child labor associated with urban living conditions weakened this nexus from the 1860s on. High and unstable child and infant mortality rates are also seen as determinants of fertility, with overshooting and externalities of reproduction decisions dominating population growth rhythms. When these rates fell, beginning in the 1870s, family planning became possible and fertility decline set in. There is some debate about which of these two forces weighed more heavily in explaining German population growth, but the importance of the question for my purposes here is that the latter produced what one may call a labor surplus economy in Germany for much of the 19th century. This view is implicitly supported by dozens of collections of contemporary descriptions of under-employment, living conditions, and work-seeking migration among the poor, by the gap, already noted, between growth of aggregate real income per capita and real wages, and by a number of studies suggesting that wage rates–especially for unskilled labor–tended to respond more to migration from rural areas and shifts in economic opportunities there than to productivity and demand shifts in the industrial center.[34]

There may thus be sense in viewing German industrialization as a case of rapid industrialization with abundant, elastic labor supplies, as well as the scarcity of capital stressed by Gerschenkron.

For an evaluation of this part of Gerschenkron's typology, however, it is useful to refer back to the periodization argument raised in the first section of the paper. In the period up to the 1870s, institutional change tended to strengthen the pressure on living standards just described. The capital accumulation and growing inequality which accompanied the early phase of industrialization reflected more than differential fertility among social strata: it reflected the highly unequal distribution of property and property incomes. This was largely a result of market forces in the nineteenth century. But these market forces operated on the basis of institutions which strengthened and supported owners of capital relative to laborers, e.g. the rules governing conversion of feudal claims and obligations upon peasant land and labor into private property, the laws weakening the gilds, the regressive taxes and the defences against labor organization. The situation changed in the 1870s, to be sure. From the 1870s onward, collective-official concern for workers' employment and living conditions mount. Due to urbanization and growing urban concentrations of working-class populations those problems become highly visible to public opinion and policymakers. Real wages and living standards begin to rise significantly in this period. This rise did not result solely or even primarily from mounting public concern, but a positive connection doubtless existed. Does this fit in with Gerschenkron's views? Perhaps one could argue that public concern for the workers resulted from the collective recognition that further German industrialization depended crucially on the country's human capital resources and hence, on related investments in health, education and welfare: Bismarck's social legislation as an essential part of the "second phase of industrialization." Certainly, Germany's industrialization pattern here did diverge significantly from that of Great Britain's, and there may be something to the idea of a link to the backwardness syndrome. But the issue is by no means clear-cut. It would seem panglossian to believe that such institutions came when the country's development "needed" them. And it would be ahistorical to ignore the fact that, e.g. Germany's educational institutions had powerful non-industrial features–and could thus be considered "dysfunctional"–or that social legislation was hotly contested and only gradually came to be a significant socio-economic factor in German daily life. Thus, any attempt to link living standards and the German "social question" to industrial backwardness alone would seem to be guilty of gross oversimplification.

4. *The role of ideology*

According to Gerschenkron the degree of economic backwardness is generally reflected in a country's dominant ideology. In the German case, that was the kind of economic nationalism articulated by Friedrich List. No doubt, economic nationalism–in the sense of giving priority to national over international or individual interests–is more visible as a professed creed among German political and business leaders in the 19th century than in

Great Britain. But on the whole, I do not find this interpretation too helpful. It has two related weaknesses: 1) it overlooks change over time; and 2) it is excessively aggregative, i.e., it neglects class differences and the structure of organized interests. These deserve brief discussion.

I begin by asserting that ruling élites are identifiable in 19th-century German history and that the ideologies these espoused shifted over time. If my assessment of the literature is correct, the middle decades of the 19th century are seen as witness to a shift in attitudes among Germany's ruling élite–particularly among the publicly articulate–in favor of capitalist industrial growth, and individualist, competitive economic activity. This shift accompanied a liberalization of economic policy–not least of all a remarkable shift in the sphere of commercial policy which had very important long-run political consequences–and was no doubt reinforced by the ongoing strength of the economic growth of the 1850s. But in the course of the depression of the 1870s a remarkable shift in this élite's ideology against economic and political liberalism took place. Most authors have seen the shift as a response to the threat posed to East Elbian owners of agricultural estates by intensified international and national competition in grain markets, to the desire of industrialists and their bankers to stabilize their shares in a shrinking national market, and to the interest of both and of the bureaucracy in combatting the growing socialist labor movement. It is important to note, however, that this shift was not complete. In many areas of public policy–involving housing, trade unions, voting rights, etc.–, the debate between liberals and conservatives–or reformists–went on. Moreover–and this takes us the second point referred to above–it goes without saying that the Marxist program of the socialist labor movement represented yet another and highly distinct position within the German ideological scene. In short, it seems unhelpful to look for a dominant ideology directly connected with a country's level of backwardness, not least of all since such a procedure begs the important question of whether or not ideologies are dysfunctional. I think there is a need here to tie up ideologies with analyses of interest group organization and political change on a more general level. That is, the links between ideologies and backwardness should be seen as *indirect* ones.

5. *Regional differences*

An obvious and frequently noted defect of Gerschenkron's "model" is its aggregative, national orientation, its failure to account for or even interpret regional differences. In the light of the previous discussion of ideology it is interesting to note a recent suggestion that the strength of nationalism as an ideology in Germany has to do with the strength of regional differences in development levels–seen as a political problem which must be overcome.[35] In fact, Gerschenkron himself pointed out this possibility–though he drew no consequences from it for his basic model. In any case, German economic historiography is clear on this issue. For one of the clearest results of recent work into regional dimensions of industrialization is the stress on its unevenness. There are leading regions–such as parts of Saxony or the Rhineland–as well as leading sectors. However, growth in one region need not and did not automatically contribute–through trade, factor movements or the diffusion of technical knowledge–to growth in another. This work shows that developmental differentials were *a*) a pre-industrial phenomenon which *b*) industrialization failed to remove, mainly because the East had remained largely agricultural.[36] It is thus possible that the temporal and sectoral shape of German industrialization could have varied substantially from region to region, so much so, in fact, that "German industrialization" could turn out to be a phrase denoting an average or artefact including highly diverse components. Explanation of German industrialization, in such a case, might require *a*) explanation of regional changes and *b*) explanation of how the weights of regions changed over time, as well as *c*) explanation of linkages between regions. Under extreme assumptions–factor immobility, internal trade barriers and impotent national policy–the development of national economies depends on these regional changes and not at all on policy. But most important, regional differences may have powerful consequences for the development of the national economy and polity. F. B. Tipton has used his own study of Germany's regional differences to raise this question, focusing particularly on the well-known East-West differential and on the response of agricultural protectionism and the problem of migration and labor supply. Ongoing industrialization shifted the economic center of gravity in Prussia–Germany's dominant state–westward and away from the agricultural regions. As Tipton puts it, "The eastern landlord class achieved and maintained an important political position, but the economic basis of that position eroded as their region began to drop behind the urban and industrial centers of central and western Germany." This erosion threatened to accelerate in the agrarian depression of the 1870s–which concentrated strongly upon the East Elbian grain producers. Tariff protectionism was the political response engineered by Bismarck, but this penalized labor-intensive mixed farming and taken together with the competition which continued despite the tariff, could not slow down and may have even encouraged the massive emigration out of the agrarian East which had begun in the 1860s. By the 1890s, this made recourse to seasonal immigrant labor a necessity for the East-Elbian grain producers and the result was that peculiar labor market which maintained, just as it reflected, the continued power of the eastern landlords. This further retarded any attempt to industrialize the East. Given the Prussian-German political structure and the disproportionate weight it gave to the eastern landlords (and the Conservative Party) in national questions, democratic change in the country as a whole proceeded more slowly than a regionally balanced industrialization would probably have allowed.[37]

6. *The role of the state and its political economy*

To some extent this concluding section is a residual. In the previous sections much of our topic has already been covered, for "the state" is virtually omnipresent in German economic historiography. A few remarks can suffice. To begin with, there is no question but that the state's direct role was greater in Germany's industrialization than in Great Britain's. A number of relevant areas can be cited: there is government's direct support for, even entrepreneurial involvement in the creation of social overhead capital (e.g. railroads); the agrarian reforms involved setting up government banks which facilitated the conversion of previously feudal claims into private property rights; there was commercial policy involving selective tariff protection and easing national economic integration; an ambitious program of social insurance; and so on. On the other hand, in the case of "moderate backwardness" (Germany's) Gerschenkron's model "predicts" the rapid development of private business entrepreneurship, and is thus consistent with a literature full of observations of inefficient government policies, e.g. such as those restraining entrepreneurs in banking, transportation, etc. This obviously differs from cases of extreme backwardness such as Tsarist Russia (where government policy might be inept but without alternatives). The trouble with Gerschenkron's formulations however, is that they do not offer an adequate explanation of changing levels of state action. His references to military priorities, while pertinent, certainly fall short of the mark, for the state is more than the collective national interest in military power. And the level of backwardness is not the only other conditioning factor. National governments or polities are coalitions of regional, sectoral and class interests which mobilize political resources and contend for control of the polity.

In fact, the economic history of German industrialization is full of government actions, economic policies, decisions having at most an indirect connection with economic backwardness and military considerations. I conclude this section by briefly mentioning one set of actions which illustrates an alternative approach: it concerns the Prussian and German customs unions: Ohnishi's well-documented study of the Prussian commercial union, 1818-33, demonstrates the strong, positive fiscal motivation and positive revenue effects of the union, but also surprisingly significant (and rising) protective effects as well.[38] R. Dumke's work extends the discussion. First, he shows that the Zollverein was an extension of the Prussian commercial union in the sense that *a*) the latter's relatively high rates became the norm for the former and, more significantly, that *b*) the revenue motive was of decisive importance, since Prussian statesmen, drawing on their own experience, could offer the rulers of small and poor states the prospects of revenues free of the necessity for democratic-parliamentary concessions as enticement to join the Zollverein. Parochial interests–including those of the civil servants in the several German states–explain commercial policy and not economic backwardness or military needs. The history of German state activity contains many chapters into which an approach stressing micro interests and political bargaining might be extended.

7. *Conclusions*

The main body of the paper offers quite a few generalizations about the appropriateness of Gerschenkron's basic model for the study of European industrialization. For this reason, these conclusions can be brief. They make only three points.

First, criticism and suggestions for revision are directed toward Gerschenkron's basic model (as developed in *Backwardness*), not toward his life's work as a scholar. That work is a treasure trove of differentiated knowledge about European history, even about German history, and contains numerous valuable insights into the process of economic development generally. And one must add that it also contains observations which qualify and even contradict the basic model. Nevertheless, it is the basic model–which necessarily suppressed much differentiated knowledge–which is on trial here, not Gerschenkron the scholar.

Second, Gerschenkron's fundamental insight–the importance of relative levels of backwardness as a determinant of industrialization patterns–automatically produces a significant deficiency in his model: the model's premise is that it makes sense to "tune in" on the industrialization drama of a country on the "eve of its initiation"–a point in time (or period?) determined by ex post knowledge of a great spurt which subsequently took place. But this way of formulating the problem begs the question of why a given country is backward in the first place, precludes discussion of factors present in "earlier developers" but not in their backward rivals. His model focuses on the successful followers, and neglects the failures or partial failures (with slow growth). Revision here implies at least three modifications of the basic model.

1) There is a need to distinguish between relative backwardness and absolute levels of development, a need which calls for an extension of what we wish to explain beyond the rate of industrial growth and the share of industry in total output and employment to more general measures of economic welfare.
2) A more general model should include factors generally–at least potentially–relevant for all countries; natural resources, labor supply, the levels of human and non-human capital, the level of technological knowledge, communications and mobility, the distribution of wealth and income and so on.
3) There is a need to relax (or forget) the assumption that modern economic growth takes the form of a "big spurt" discontinuity associated with rapid, dramatic industrial changes, a need to see modern economic growth as a more gradual, longer-run process extending well into the pre-industrial period.

Third, as suggested throughout the paper, Gerschenkron's basic model implicitly assigns great importance to the state but fails to develop an explanation of state behavior and changes in it. More generally, modern economic growth depends on the state, not only or even primarily as a source of specific services (financial and otherwise), but as the social arrangement through which the general rules and especially the legal framework of economic activity are negotiated, established and enforced. Explaining the role of the state means developing an appropriate model of political change. This is not the place to attempt to sketch out such a model, but in order to be useful as part of an explanation of the modern economic growth of nations it should probably cover at least the following needs:

1) the need to explain how regional, sectoral and socio-economic class interests are mobilized and organized and what determines their effectiveness at the national level. Is it the economic weight of their basic constituencies, their social homogeneity, the particular kinds of coalitions they enter into, ideology, or some combination of all of these?
2) the need to see the state as more than the collective expression of organized interest groups but as an autonomous actor (or collection of autonomous actors) pursuing independent aims. Autonomy may have a purely traditional base, which may correspond to no more than the inherited prerogatives of a small but highly cohesive minority. On the other hand, autonomy can be maintained by the adept exploitation of differences among contending socio-economic interest groups. What explains autonomy and with what kinds of state behavior is it associated?

That leads directly to

3) the obvious need to monitor the "output" of state action, to assess its efficiency in promoting or hindering modern economic growth in nations. This has been a traditional field of work for economic historians, so little more need be said. Nevertheless, if more attention is paid to the relationship between the political bases of government–the kinds of coalitions and bargaining its stability depends on–and the contribution of state activity to modern economic growth (net of resources absorbed by government), that work will doubtless yield higher returns.

Institut für Wirtschafts- und Sozialgeschichte,
Westfälische Wilhelms-Universität Münster.

1 A. Gerschenkron, *Bread and Democracy in Germany* (Berkeley, 1943).

2 In the following "industrialization," "modern economic growth" and "economic development" are used interchangeably; the same applies to the terms "typology" and "model" with respect to Gerschenkron.

3 A. Gerschenkron, *Continuity in History and Other Essays* (Cambridge, 1968), p. 137.

4 R. Tilly, "Das Wachstumsparadigma und die europäische Industrialisierungsgeschichte (Fontana Economic History of Europe)," *Geschichte und Gesellschaft*, 3 (1977), pp. 93-108.

5 H.-U. Wehler, *Deutsche Gesellschaftsgeschichte*, vol. 1: *Vom Feudalismus des alten Reiches bis zur Defensiven Modernisierung der Reformära, 1700-1815*, vol. 2: *Von der Reformära bis zur Industriellen und politischen "deutschen Doppelrevolution" 1815-1848/9* (Münich, 1987).

6 W. R. Lee, "Economic Development and the State in Nineteenth-Century Germany," *Economic History Review*, second series, 41 (1988), pp. 346-67.

7 R. Fremdling, *Eisenbahnen und deutsches Wirtschaftswachstum, 1840-1879* (Dortmund, 1975); R. Fremdling, "Railroads and German Economic Growth. A Leading Sector Analysis with a Comparison to the United States and Great Britain," *Journal of Economic History*, 37 (1977), pp. 583-604; R. Spree, *Die Wachstumszyklen der deutschen Wirtschaft von 1840 bis 1880* (Berlin, 1977).

8 Fremdling, "Railroads"; R. Fremdling, *Technologischer Wandel und internationaler Handel im 18. und 19. Jahrhundert. Die Eisenindustrien in Großbritannien, Belgien, Frankreich und Deutschland* (Berlin, 1986); F.-W. Henning, "Eisenbahnen und Entwicklung der Eisenindustrie in Deutschland," *Archiv und Wirtschaft*, 6 (1973), pp. 1-20; C.-L. Holtfrerich, *Quantitative Wirtschaftsgeschichte des Ruhrkohlenbergbaus im 19. Jahrhundert* (Dortmund, 1973).

9 H. Wagenblass, *Der Eisenbahnbau und das Wachstum der deutschen Eisen- und Maschinenbauindustrie, 1835 bis 1860* (Stuttgart, 1973); J. Krengel, "Zur Berechnung von Wachstumswirkungen konjunkturell bedingter Nachfrageschwankungen nachgelagerter Industrien auf die Produktionsentwicklung der deutschen Roheisenindustrie während der Jahre 1871-1882," in W. H. Schröder and R. Spree, eds., *Historische Konjunkturforschung* (Stuttgart, 1980).

10 Fremdling, *Eisenbahnen*, pp. 150-58; W. Steitz, *Die Entstehung der Köln-Mindener Eisenbahn*, Schriften zur rheinisch-westfälischen Wirtschaftsgeschichte, vol. 17 (Cologne, 1974); Spree, *Die Wachstumszyklen*, pp. 267-73.

11 H.-U. Wehler, *Deutsche Gesellschaftsgeschichte*, vol. 2: *Von der Reformära*, p. 614.

12 Spree, *Die Wachstumszyklen*; A. Gerschenkron, *Economic Backwardness in Historical Perspective* (Cambridge, 1962), pp. 77, 203.

13 Fremdling, "Railroads," pp. 132-63.

14 P. Kriedte, H. Medick and J. Schlumbohn, *Industrialisierung vor der Industrialisierung. Gewerbliche Warenproduktion auf dem Land in der Formationsperiode des Kapitalismus* (Göttingen, 1977); W. Mager, "Protoindustrialisierung und Protoindustrie. Vom Nutzen und Nachteil zweier Konzepte," *Geschichte und Gesellschaft*, 14 (1988), pp. 275-305; H. Kisch, *Die transindustriellen Textilgewerbe am Niederrhein von der industriellen Revolution* (Göttingen, 1981); H. Schulze, *Landhandwerk im Übergang vom Feudalismus zum Kapitalismus* (East Berlin, 1984).

15 H. W. Graf Finck von Finckenstein, *Die Entwicklung der Landwirtschaft in Preussen und Deutschland, 1800-1930* (Würzburg, 1960); G. Helling, "Berechnung eines Index der Agrarproduktion in Deutschland im 19. Jahrhundert," *Jahrbuch für Wirtschaftsgeschichte* (1965), part 4, pp. 125-43; G. Franz, "Landwirtschaft, 1800-1850," in W. Zorn, ed., *Handbuch der deutschen Wirtschafts-*

und Sozialgeschichte (Stuttgart, 1976) and M. Rolffs, "Landwirtschaft, 1850-1914," ibid.; C. Dipper, *Die Bauernbefreiung in Deutschland, 1790-1850* (Stuttgart, 1980).

[16] H. Harnisch, "Die Bedeutung der kapitalistischen Agrarreform für die Herausbildung des inneren Marktes und die Industrielle Revolution in den östlichen Provinzen Preussens in der ersten Hälfte des 19. Jahrhunderts," *Jahrbuch für Wirtschaftsgeschichte* (1977), part 4, pp. 63-82.

[17] Fremdling, *Technologischer Wandel*, pp. 335-37; H.-W. Müller, "Anfänge der deutschen Landmaschinenindustrie. Biespiele aus den 30er bzw. 50er Jahren des 19. Jahrhunderts," *Jahrbuch für Wirtschaftsgeschichte* (1987), part 3, pp. 169-94, especially p. 172.

[18] A. Kunz, "Die Verknüpfung von Märkten durch Transport, Verkehrsstatistik und Marktintegration in Agrarregionen," in T. Pierenkemper, ed., *Landwirtschaft und Industrielle Entwicklung – Zur ökonomischen Bedeutung von Bauernbefreiung, Agrarreform und Agrarrevolution* (Stuttgart, 1989).

[19] M. Kutz, *Deutschlands Aussenhandel, 1789-1834* (Wiesbaden, 1974); B. von Borries, *Deutschlands Aussenhandel 1836-1856* (Stuttgart, 1970); R. Dumke, "The Political Economy of Economic Integration: Tariffs, Trade and Politics of the Zollverein Era," Ph. D. Dissertation, University of Wisconsin, 1976.

[20] R. Dumke, "Anglo-deutscher Handel und Frühindustrialisierung in Deutschland, 1822-1865," *Geschichte und Gesellschaft*, 5 (1979), pp. 175-200; P. Beyer, *Leipzig und die Anfänge des deutschen Eisenbahnbaus* (Weimar, 1978).

[21] K. Borchardt, "Wirtschaftliches Wachstum und Wechellagen 1800-1914," in Zorn, ed., *Handbuch der deutschen Wirtschafts- und Sozialgeschichte*, vol. 2, pp. 208-9.

[22] R. Fremdling, "German National Accounts for the 19th and Early 20th Century. A Critical Assessment," *Vierteljahrschrift für Sozial- Wirtschaftsgeschichte*, 75 (1988), pp. 339-54.

[23] R. Tilly, "Das Wachstum industrieller Großunternehmen in Deutschland 1880-1911," in H. Kellenbenz, ed., *Wirtschaftliches Wachstum, Energie und Verkehr vom Mittelalter bis ins 19. Jahrhundert* (Stuttgart - New York, 1978).

[24] R. Tilly, "German Banking, 1850-1914: Development Assistance to the Strong," *Journal of European Economic History*, 15 (1986), pp. 113-52.

[25] W. P. Kennedy and R. Britton, "Portfolioverhalten und wirtschaftliche Entwicklung im späten 19. Jahrhundert. Ein Vergleich zwischen Großbritannien und Deutschland. Hypothesen und Spekulationen," in R. Tilly, ed., *Beiträge zur quantitativen vergleichenden Unternehmensgeschichte* (Stuttgart, 1985); R. Tilly, "Mergers, External Growth and Finance in the Development of Large-Scale Enterprise in Germany, 1880-1913," *Journal of Economic History*, 42 (1982), pp. 629-58; U. Wengenroth, *Unternehmensstrategien und technischer Fortschritt, Die deutsche und britische Statlindustrie 1865-1895* (Göttingen-Zürich, 1986).

[26] Kennedy and Britton, "Portfolioverhalten und wirtschaftliche Entwicklung"; Tilly, "German Banking."

[27] O. Jeidels, *Das Verhältnis der deutschen Großbanken zur Industrie, mit bes. Berücksichtigung der Eisenindustrie* (Leipzig, 1905); W. Feldenkirchen, *Die Eisen- und Stahlindustrie des Ruhrgebiets, 1879-1914* (Wiesbaden, 1982); Wengenroth, *Unternehmensstrategien*; P. C. Witt, *Die Finanzpolitik des Deutschen Reiches von 1903 bis 1913* (Lübeck-Hamburg, 1970); F. W. Pappi, P. Kappelhoff and C. Molbeck, "Die Struktur der Unternehmensverflechtung in der Bundesrepublik," *Kölner Zeitschrift für Soziologie und Sozialpsycologie*, 39 (1987).

[28] R. Rettig, "Das Investitions- und Finanzierungsverhalten deutscher Großunternehmen, 1880-1911," Dissertation, University of Munster, 1978; M. Pohl, *Form und Phasen der Industriefinanzierung bis zum 2. Weltkrieg, Bankhistorisches Archiv*, vol. 9, supplement (Frankfurt, 1986).

[29] M. Pohl, *Entstehung und Entwicklung des Universalbankensystems* (Frankfurt, 1986).

[30] C.-L. Holtfrerich, "The Monetary Unification Process in Nineteenth-Century Germany. Relevance and Lessons for Europe Today," in M. de Cecco and A. Giovannini, eds., *Monetary Regimes and Monetary Institutions: Issues and Perspectives in Europe* (Cambridge, forthcoming).

[31] R. Gömmel, "Wachstum und Konjunktur der Nürnberger Wirtschaft (1815-1914)," in H. Kellenbenz and J. Schneider, eds., *Beiträge zur Wirtschaftsgeschichte*, vol. 1 (Stuttgart, 1978) and *Realeinkommen in Deutschland. Ein internationaler Vergleich (1810-1914)* (Nürnberg, 1979); D. Saalfeld, "Lebensverhältnisse der Unterschichten Deutschlands im 19. Jahrhundert," *International Review of Social History*, 29 (1984), part 2, supplement, p. 219.

[32] G. Kirchhain, *Das Wachstum der deutschen Baumwollindustrie im 19. Jahrhundert* (New York, 1977); Holtfrerich, *Quantitative Wirtschaftsgeschichte*; C. Tenfelde, *Sozialgeschichte der Bergarbeiterschaft an der Ruhr im 19. Jahrhundert* (Bonn-Bad Godesberg, 1977); P. Borscheid, *Textilarbeiterschaft in der Industrialisierung* (Stuttgart, 1976); W. Fischer, "Die Rolle des Kleingewerbes im wirtschaftlichen Wachstumsprozess in Deutschland, 1850 bis 1914," in F. Lütge, ed., *Wirtschaftliche und soziale Probleme der gewerblichen Entwicklung im 15./16. und 19. Jahrhundert* (Stuttgart, 1968); A. Noll, *Sozioökonomischer Strukturwandel des Handwerks in der 2. Phase der Industrialisierung* (Göttingen, 1976); H.-J. Teuteberg and G. Wiegelmann, *Der Wandel der Nahrungsmittelgewohnheiten unter dem Einfluß der Industrialisierung* (Göttingen, 1971); L. Niethammer and F. Bruggemeier, "Wie wohnten Arbeiter im Kaiserreich?," *Archiv für Sozialgeschichte*, 16 (1976), pp. 61-134; D. Saalfeld, "Lebensstandard in Deutschland, 1750-1860," in I. Bog et al., eds., *Wirtschaftliche und soziale Strukturen im säkularen Wandel. Festschrift für W. Abel zum 70. Geburtstag*, 3 vols. (Hannover, 1974), vol. 2, pp. 417-43; Gömmel, "Wachstum," and *Realeinkommen in Deutschland.*

[33] F.-W. Henning, "Industrialisierung und dörfliche Einkommensmöglichkeiten. Der Einfluß der Industrialisierung des Textilgewerbes in Deutschland im 19. Jht. auf Einkommensmöglichkeiten in den ländlichen Gebieten," in H. Kellenbenz, ed., *Agrarisches Neben-Spätmittelalter und 19./20. Jahrhundert* (Stuttgart, 1975).

[34] Lee, "Economic Development"; Gömmel, "Wachstum," and *Realeinkommen in Deutschland*; W. Kollmann, *Bevölkerung in der Industriellen Revolution* (Göttingen, 1974); W. Fischer, J. Krengel and J. Wietog, eds., *Sozialgeschichtliches Arbeitsbuch*, vol. 1 (Munich, 1982); Fremdling, *Eisenbahnen*; Kirchhain, *Das Wachstum*; Saalfeld, "Lebensstandard"; G. Hohorst, J. Kocka and G. Ritter, eds., *Sozialgeschichtliches Arbeitsbuch. Materialien zur Statistik des Kaiserreichs 1870-1914* (Munich, 1979); R. A. Dickler, "Organization and Change in Productivity in Eastern Prussia," in W. N. Parker and E. Jones, eds., *European Peasants and their Markets* (Princeton, 1975), pp. 269-92.

[35] Lee, "Economic Development."

[36] R. Fremdling and R. Tilly, eds., *Industrialisierung und Raum. Studien zur regionalen Differenzierung im Deutschland des 19. Jahrhunderts* (Stuttgart, 1979), pp. 27-53; K. Borchardt, "Regionale Wachstumsdifferenzierung in Deutschland im 19. Jahrhundert unter besonderer Berücksichtigung des West-Ost Gefälles," in W. Abel et al., eds., *Wirtschaft, Geschichte und Wirtschaftsgeschichte. Festschrift zum 65. Geburstag von Fr. Lütge* (Stuttgart, 1966); H. Hesse, "Die Entwicklung der regionalen Einkommensdifferenzen im Wachstumsprozess der deutschen Wirtschaft vor 1913," in W. Fischer, ed., *Beiträge zu Wirtschaftswachstum und Wirtschaftsstruktur im 16. und 19. Jahrhundert, Schriften des Vereins für Sozialpolitik*, new series, vol. 63 (Berlin, 1971); G. Hohorst, *Wirtschaftswachstum und Bevölkerungsentwicklung in Preussen von 1816 bis 1914* (New York, 1977); Lee, "Economic Development."

[17] F. B. Tipton, "Farm Labor and Power Politics: Germany 1850 to 1914," *Journal of Economic History*, 34 (1974), pp. 951-79 and *Regional Variations in the Economic Development of Germany during the Nineteenth Century* (Middletown, 1976); K. Bade, "Transnationale Migration und Arbeitsmarkt im Kaiserreich: vom Agrarstaat mit starker Industrie zum Industriestaat mit starker Agrarbasis," in T. Pierenkemper und R. Tilly, eds., *Historische Arbeitsmarktforschung. Entstehung, Entwicklung und Probleme der Vermarktung von Arbeitskraft* (Göttingen, 1982).

[18] T. Ohnishi, "Zolltarifspolitik Preussens bis zur Gründung des Deutschen Zollvereins," Dissertation, University of Göttingen, 1973.

F. Crouzet

1 England and France in the eighteenth century: a comparative analysis of two economic growths*

The economic historian interested in the key problem of growth is bound to find the comparative approach particularly fruitful. A systematic comparison of the eighteenth-century English economy with that of another country – and France as the leading Continental power at that time seems the obvious choice – should bring out more clearly what factors were peculiar to England and might therefore have determined what is a unique phenomenon, the English Industrial Revolution of the eighteenth century.

The first point to be made is that the forwardness of England and the backwardness of France, clearly noticeable on the eve of the French Revolution, were not of sudden or recent appearance; this disparity between the two countries had already been evident early in the century, say at the death of Louis XIV. Moreover, to find an explanation for the contrasts between the social and economic structures of France and England we should have to look at the problem over a long period (going well back into the Middle Ages; one would stress, for example, the importance of wool in English history: wool production, the wool trade and woollen manufacture early made England part of the international economy and stimulated the rise of capitalism). That task cannot be attempted here, but it must at least be stressed that developments during the seventeenth century had a different effect on the two economies, which partly accounts for the time-lag between them at the beginning of the eighteenth century.

As is well known, a similar point was made some fifty years ago by J. U. Nef, who pressed it very hard with his assertion that the reason

* This article was edited for publication in R. M. Hartwell (ed.), *The Causes of the Industrial Revolution in England* (London, 1967). It had been first published in French in *Annales ESC*, XXI, 2, March–April 1966, pp. 254–91.

why England was the first country to undergo the eighteenth-century Industrial Revolution was simply that, unlike other countries and particularly France, she had already undergone a first industrial revolution in the late sixteenth and early seventeenth centuries, which meant that her lead had been established some two centuries earlier than was generally thought.[1] But this thesis has been strongly criticised from the British side, and little of it remains intact today.[2] The 'first English industrial revolution' existed chiefly in Nef's imagination: he attached far too much importance to a few technical innovations, to the understandably fast growth of the handful of new industries (coal excepted, their role was a very minor one – yet he ignored the slow growth in the major woollen industry), and lastly to a few non-representative cases of entrepreneurial giantism.[3]

Nevertheless, Nef deserves our thanks for having stressed the importance of the fast rise in English coal production and consumption during the century after 1540, and its influence on technological progress through stimulating the development and adoption of some completely new techniques, such as furnaces fired by coal or coke, which were quite unknown on the Continent. This early adoption of mineral fuels gave English inventiveness an impetus which proved to be lasting and which in France was lacking.[4] Again, it is possible that Nef was right in maintaining that between 1540 and 1640 England had an absolute lead over France in mining and metal production and also

[1] J. U. Nef, 'A Comparison of Industrial Growth in France and England from 1540 to 1640', *Journal of Political Economy*, XLIV, 3, 4, 5, June, August, Oct. 1936, pp. 289–317, 505–33, 643–66; see also by the same author: 'The Progress of Technology and the Growth of Large-Scale Industry in Great Britain, 1540–1640', *The Economic History Review*, V, 1, Oct. 1934, pp. 3–21; 'Prices and Industrial Capitalism in France and England, 1540–1640', *ibid.*, VII, 2, May 1937, pp. 155–85; 'L'Industrie et l'État en France et en Angleterre (1540–1640)', *Revue Historique*, CXCI, 1941, nos 1 and 2, pp. 21–53, 193–231.

[2] See particularly D. C. Coleman, 'Industrial Growth and Industrial Revolutions', *Economica*, n.s., XXIII, 89, Feb. 1956, reprinted in E. M. Carus-Wilson (ed.), *Essays in Economic History*, III (London, 1962), pp. 345–7; D. C. Coleman, 'Technology and Economic History, 1500–1750', *The Economic History Review*, 2nd series, XI, 3, April 1959, pp. 506–7, 509–10, 512; B. E. Supple, *Commercial Crisis and Change in England, 1600–1642* (Cambridge, 1959), pp. 2–8; F. J. Fisher (ed.), *Essays in the Economic and Social History of Tudor and Stuart England in Honour of R. H. Tawney* (Cambridge, 1961), pp. 6–7; Y. S. Brenner, 'The Inflation of Prices in England, 1551–1650', *The Economic History Review*, 2nd series, XV, 2, Dec. 1962, pp. 271–3.

[3] On the other hand, Nef overstated the stagnation of French industry during the same period. He also went much too far when he spoke of a 'phenomenal acceleration' in English industrial output between 1540 and 1640.

[4] W. H. Chaloner and A. E. Musson, *Industry and Technology* (London, 1963), pp. 20–1, 24–5, 29; 'The Origins of the Industrial Revolution', *Past and Present*, XVII, April 1960, p. 76; below, pp. 95–9.

had a higher output per head for products such as woollen goods and glass, and possibly for industrial output as a whole.[5]

However, it was after 1640 rather than before that significant disparities between the economic evolution of the two countries appeared. As is well known, recent research has produced a black picture of French economic history in the 'tragic seventeenth century' (admittedly, an overstated expression). This work has shown that after a relatively prosperous start to the century, the 1630s saw the beginning of a long period of stagnation and decline – turning in mid-century during the Fronde into collapse – which lasted up to the 1720s. Prices – and especially food prices – over the century show a downward trend, and more significantly, extreme instability; there were frequent and violent economic and demographic crises, attended by terrible mortalities; there was monetary famine and 'tightness of money', which crippled business activity and brought down rents and industrial and commercial profits; there was unemployment and pauperism. In view of all this, it is no wonder that industrial production stagnated and even declined.[6] Colbert's policy of industrialisation was really nothing but a desperate effort to counteract this declining trend; it was undertaken in the highly unfavourable conditions of deflation, falling prices, incomes, and consumption, and ended in semi-failure. Although some of his new industrial undertakings survived and developed later, many others declined and quickly disappeared; and on the whole, if there was any increase in total industrial output under Colbert's government, which is not certain, it was not very pronounced. It is true that recent research has shown that despite two great wars, several famines and protestant emigration, there were signs of recovery in the French economy during the latter part of the reign of Louis XIV, long thought to have been catastrophic. This may well reveal a rising underlying trend, a sign heralding the post-1715 growth, but any such recovery was largely counteracted by the bad effects of wars and

[5] Nef, 'A Comparison', p. 663.

[6] See P. Goubert, *Beauvais et le Beauvaisis de 1600 à 1730. Contribution à l'histoire sociale de la France du XVIIe siècle* (Paris, 1960), pp. 585–91, 595–6; graphs, 126–30, pp. 116–19 of the supplement; P. Deyon, 'Variations de la production textile aux XVIe et XVIIe siècles: Sources et premiers résultats', *Annales ESC*, XVIII, 5, Sept.–Oct. 1963, pp. 947, 950–3. See also P. Goubert, in F. Braudel and E. Labrousse (eds.), *Histoire économique et sociale de la France*, vol. II: *Des derniers temps de l'âge seigneurial aux préludes de l'âge industriel (1660–1789)* (Paris, 1970), pp. 329–48 (later abbreviated *HESF*, II); and vol. 2 of P. Léon, *Histoire économique et sociale du monde*, P. Deyon and J. Jacquart (eds.), *Les Hésitations de la croissance, 1580–1740* (Paris, 1978).

famines.[7] Therefore, even if there are a few positive items to enter on the French balance sheet for the seventeenth century, the general picture remains most unfavourable, and the population of France in 1715 or 1720 was hardly higher than it had been in 1600.

On the English side of the Channel the picture is certainly different, and English historians are on the whole optimistic in their estimation of England's economic performance during the seventeenth century. England was not, of course, totally unaffected by the adverse economic climate of the seventeenth century; there were periods of difficulty and stagnation, for example in the 1620s, during the Civil War and at moments in the wars against Louis XIV; England's main industry, woollen manufacture, went through a series of crises and its growth over the century as a whole was relatively modest; chronic poverty and underemployment were serious problems. But there is nothing to compare with the recurrent violent crises or the deep and lasting depression to be seen in France after 1630. The reversal in the trend of price movements and of the economy in general reached England later than France, about 1650, and in England the subsequent fall in prices was smaller, while short-term price fluctuations seem to have been – at least after 1661 – less violent, less irregular, and hence less injurious to economic activity.[8] F. J. Fisher was able to conclude that the global output of agriculture and industry increased slowly but appreciably during the century, and that despite the rise in population, average income per head probably rose.[9]

There are also other signs of economic progress, especially after 1660, for which there is no parallel in France: for example, the expansion of the English home market, due principally to the growth of London, which far exceeded that of Paris; the 'frontier-like' development of the northern and western counties;[10] and the fast and prolonged growth during the seventeenth century of English foreign

[7] P. Léon, 'La Crise de l'économie française à la fin du règne de Louis XIV (1685–1715)', *Information historique*, XVIIII, 4, Sept.–Oct. 1956, pp. 132–7; Deyon, pp. 953–5; P. Goubert, *Louis XIV et vingt millions de Français* (Paris, 1966).

[8] J. Meuvret, *Etudes d'histoire économique. Recueil d'articles* (Paris, 1971), pp. 103–4, 117–19, 123; A. B. Appleby, 'Grain Prices and Subsistence Crises in England and France, 1590–1740', *Journal of Economic History*, XXXIX, 4, Dec. 1979, pp. 865–87; also Brenner, pp. 276, 281–2; G. N. Clark, *The Wealth of England from 1496 to 1760* (London, 1946), pp. 108–10; E. H. Phelps Brown and S. V. Hopkins, 'Seven Centuries of the Prices of Consumables, Compared with Builders' Wage-Rates', *Economica*, n.s., XXIII, 92, Nov. 1956, pp. 299–301, 305, 312–13.

[9] Fisher, *Essays*, p. 3; F. J. Fisher, 'The Sixteenth and Seventeenth Centuries. The Dark Ages in English Economic History?', *Economica*, n.s., XXIV, 93, Feb. 1957, pp. 6–9, 12, 15, 16.

[10] Fisher, in *Economica*, pp. 10–11.

trade, largely thanks to early colonial expansion, which meant that from the 1660s England had an important trade in the re-export of exotic products, a trade which continued to rise and was the chief factor in the growth of English trade as a whole, at a time when French colonies and colonial trade were still relatively insignificant.[11]

Thanks to this slow but quite steady growth, at the beginning of the eighteenth century England already had a definite lead over France in several important fields. First, in agriculture: various technical improvements were now in use over a large part of the country, productivity was higher and more regular than in France, a fact which, together with England's geographical advantages, helps to explain why bad years were less disastrous in their effects. Secondly, in industrial technology: the earlier innovations which resulted from the use of coal as a fuel had been supplemented by the great burst of inventive activity of the late seventeenth and early eighteenth centuries, which threw up among others the major inventions of Savery, Newcomen, and Darby. Thirdly, England was also leading in the commercial field, with a higher volume of trade, both in relation to the population and in absolute terms,[12] with a much larger merchant navy, and with a faster accumulation of merchant capital. And she had won the first battles in the struggle for command of the seas and dominion over America. Lastly, England's superiority was also very evident in the financial field, where the establishment of the National Debt and the Bank of England were signs of a political and economic structure well in advance of France: attempts by Louis XIV to set up a National Bank foundered on the objections of the bankers who thought it incompatible with a 'pure monarchy', and John Law's attempt a few years later to transplant English financial institutions to France ended in disaster.[13]

In 1688 Gregory King asserted that England had reached a higher level of wealth than any other country, Holland alone excepted, and

[11] R. Davis, 'English Foreign Trade, 1660–1700', *The Economic History Review*, 2nd series, VII, 2, Dec. 1954, pp. 150–4, 159–63, and 'English Foreign Trade 1700–1774', *ibid.*, XV, 2, Dec. 1962, p. 285; D. A. Farnie, 'The Commercial Empire of the Atlantic, 1607–1783', *ibid.*, XV, 2, Dec. 1962, p. 206.

[12] E. Levasseur, *Histoire du commerce de la France, Première partie: avant 1789* (Paris, 1911), p. 405.

[13] H. Lüthy, *La Banque protestante en France, de la révocation de l'édit de Nantes à la Révolution* (Paris, 1959 and 1961, 2 vols.), vol. I, pp. 94–7, 290–1, 414; R. Mousnier, *Les XVIe et XVIIe siècles* (Paris, 1954), p. 299; C. P. Kindleberger, 'Financial Institutions and Economic Development: A Comparison of Great Britain and France in the Eighteenth and Nineteenth Centuries', *Explorations in Economic History*, XXI, 2, April 1984, pp. 117–19.

estimated the French average income per head at 20 per cent lower than the English; a generation later, just after the Treaty of Utrecht, Daniel Defoe was writing of England as 'the most flourishing and opulent country in the world'.[14] These opinions were certainly not unjustified, in view of England's modest but growing prosperity in the seventeenth century, when France was stagnant and even declining. Whatever the causes of this disparity – different socio-economic structures, political circumstances – once the unfavourable economic climate of the seventeenth and early eighteenth century had lifted and the French economy started to grow, it was already behind the English and began with a serious handicap. If it is accepted that the Industrial Revolution was merely the crown stage of a long process of growth and change, it is important to our understanding of the British 'take-off' in the last third of the eighteenth century that it came after about two centuries of growth – admittedly broken at times, but never for very long – whereas the growth of the French economy at the same period dated back less than half a century and followed a century of stagnation. It was in fact during this 'tragic' post-1630 seventeenth century that France was clearly outdistanced by England, and despite her relatively fast growth during the eighteenth century she was never able to catch up.

We have seen that the economic developments of France and England during the seventeenth century are in clear contrast; but for the three-quarters of a century between the end of the War of the Spanish Succession and the French Revolution the picture is quite different. In both countries this was a period of growth, and the available statistical data point to the somewhat surprising conclusion that the rates of growth were not at all dissimilar.[15]

Let us start with what is best known, foreign trade, which is also the sector where French growth was fastest – faster in fact than the British. The total foreign trade of England and Wales (imports + exports + re-exports) had an average yearly official value of £13 million for the five years 1716–20 and £31 million for the five years 1784–8, which

[14] Quoted by P. Deane and W. A. Cole, *British Economic Growth, 1688–1959. Trends and Structure* (Cambridge, 1962), p. 38, and M. Ashley, *England in the Seventeenth Century (1603–1714)* (London, 1952), p. 230.

[15] Naturally, statistics from the eighteenth century can give only rough orders of magnitude or very approximate indexes of growth, and they should be used with caution. A stronger warning applies to the calculations of national aggregates; the work of J. Marczewski and J. Toutain on France, P. Deane and W. A. Cole on Britain was received with reservations by many scholars. However, it would be a kind of intellectual Luddism to refuse to use their results.

means it multiplied by 2·4.[16] On the other hand, the average yearly value of France's foreign trade was 215 million livres tournois for 1716–20 and 1 billion 62 million for 1784–8, which means a fivefold increase.[17] However, the English official values were calculated at more or less constant prices, so that in fact they give a rough index of the volume of trade, while the French figures are at current prices. Since prices rose about 60 per cent between the 1730s and the 1780s, the comparison is only valid if we deflate the figures for 1784–8 to constant prices. But this would still leave an increase in the *volume* of French trade which at lowest is of the order of 1 to 3, and even so is higher than the English. True enough, it has been argued that available statistics overestimate the growth of French commerce,[18] but there is no doubt that in any case, the value of French foreign trade was by 1715 markedly lower than that of the English, and by the eve of the Revolution had reached about the same level (though the value per head of the population was naturally still much lower).

We must, of course, take into account the very low level to which French trade had fallen at the end of the War of the Spanish Succession, but it nevertheless grew faster than the English up to the time of the French Revolution. Moreover, there were several branches of international trade in which the French secured or maintained a dominant position; they continued to be the main suppliers of manufactured goods to Spain, and through Cadiz to Spain's American Empire, while the British had the monopoly of the smaller Portuguese and Brazilian markets; they dominated the markets of Italy and the Levant. Thanks to spectacular progress in sugar and coffee cultivation in San Domingo and to the low prices of these products, which competed successfully with those of the British West Indies where soils were becoming exhausted and production costs were higher, the French snatched most of the entrepôt trade in colonial produce from the English merchants and developed a large and fast-growing re-

[16] E. B. Schumpeter, *English Overseas Trade Statistics, 1697–1808* (Oxford, 1960), pp. 15–16, Tables I–IV.

[17] A. M. Arnould, *De la balance du commerce et des relations commerciales extérieures de la France dans toutes les parties du globe particulièrement à la fin du règne de Louis XIV et au moment de la Révolution* (Paris, 2nd edn, year III, 3 vols.), vol. III, Table 10; Levasseur, p. 512, n. 2.

[18] D. S. Landes, 'Statistics as a Source for the History of Economic Development in Western Europe: The Protostatistical Era', in V. R. Lorwin and J. M. Price (eds.), *The Dimensions of the Past. Materials, Problems and Opportunities for Quantitative Work in History* (New Haven and London, 1972), pp. 63–4. See, however, Riley (n. 62 below), p. 109 n. 1, who confirms my view.

export trade to Northern Europe. The British on the whole did not do well in European markets, where they came up against protective tariffs and French competition. In absolute terms, the increase in English exports to Europe was slow; they did not double between the beginning of the century and the 1780s; Europe's share in English total exports also diminished, falling from about four-fifths to under a half. English commercial expansion was due almost entirely to the fast growth of her colonial trade and especially of trade with the American possessions, which up to the revolt of the Thirteen colonies provided English industry with an almost fully protected market.[19] This 'Americanisation' of foreign trade is also noticeable, though less marked, in France; French colonial trade, despite losses of territory due to the Seven Years War, grew tenfold between 1716–20 and 1784–8, when trade with countries outside Europe made 38 per cent of total trade. French trade, then, was still more oriented on Europe than the English, and this European trade was growing almost as fast as total trade, certainly faster than that of England with the Continent.[20]

This optimistic picture of French foreign trade needs, however, to be qualified on several points. In particular, its relative position seems to have been more favourable during the first half of the period under review, that is before the Seven Years War. Up to the end of the forties, in fact, the development of English trade was 'painfully slow', because of the stagnation in the export of woollen goods and in re-exports; fast growth came only after 1748, with an annual rate of growth of 3·9 per cent between 1745 and 1760. At first, French trade, despite the whiplash of the '*Système*', grew at a rate similar to the English, but after about 1735 it accelerated sooner and faster, doubling in value in under twenty years (between 1736–9 and 1749–55). But this spurt was cut short by the Seven Years War, during which French trade was driven off the seas by the Royal Navy and fell 50 per cent, while English trade continued to grow. Between 1763 and 1771 there was a sharp recovery, whose chief effect, however, was to recover lost ground, and this was followed by distinct stagnation in the 1770s and then an outright recession during the American war. It was only on the very eve of the Revolution that French trade again became buoyant and reached a

[19] Davis, 'English Foreign Trade, 1700–1774', pp. 294, 298; Farnie, p. 214; Deane and Cole, pp. 34, 86, 88; Schumpeter, pp. 10–11; also pp. 216–18, 234–5.

[20] It is true that the main growth in French trade with Europe was made by re-exports (of colonial produce) which increased eightfold, while exports of French products barely trebled. Cf. Levasseur, pp. 487, 512, n. 2; Arnould, vol. I, pp. 326–7; II, Table 2; III, Table 12.

record level, before collapsing in 1793. English trade on the other hand prospered during the Seven Years War, suffered some slowing down during the 1770s and a fall during the American war; after the peace of 1783 it picked up very sharply, to make a rapid and sustained growth until the end of the century. There are also signs that the competitive position of French products deteriorated after 1770 and that they were losing ground in the Levant, in Spain (where the protectionist policy of Charles III was a further obstacle) and in Spanish America.[21] In addition, the course of British trade was much less irregular, because it did not suffer as much from the wars.[22]

Symptoms of an incoming crisis have been observed under the apparent prosperity of French colonial trade on the eve of the Revolution;[23] though the danger must not be overstated, an undoubted weakness of French trade was its great dependence on San Domingo, which in the 1780s was responsible for three-quarters of the trade with the French colonies and also supplied most of the re-exports, which themselves made about one-third of total exports. The proportion of re-exports was almost as great in English trade, but the British colonial empire was larger and more various than the French so that British colonial trade was more diversified; moreover, until 1776 it included as one of its chief markets the Thirteen colonies, with their fast-rising European population, their relatively high standard of living and expanding demand for manufactured goods of all sorts. Further, the percentage of manufactured goods to total exports was higher in British than in French trade – about two-thirds as against two-fifths in the 1780s; the leading French exports were coffee, sugar, and wines, linen and silk goods taking second place.[24]

However, in France, as in England, commercial expansion was a strategic factor in the growth of industry. This is a more difficult

[21] Davis, 'English Foreign Trade, 1700–1774', pp. 285–8, 294–3; Schumpeter, pp. 13–14; Deane and Cole, pp. 29–30, 42–9, 310–11; Levasseur, pp. 511, 521, 523; E. Labrousse, *La Crise de l'économie française à la fin de l'Ancien Régime et au début de la Révolution* (Paris, 1943), pp. xxxvi–xxxix.

[22] But there are also some striking parallels in its development, with alternating phases of fast growth (between the War of the Austrian Succession and the Seven Years War) and of slow growth or stagnation (after 1713, during the 1770s).

[23] F. Crouzet, in F. G. Pariset (ed.), *Bordeaux au XVIIIe siècle* (Bordeaux, 1968), pp. 316–23; Paul Butel, *Les Négociants bordelais, l'Europe et les iles au XVIIIe siècle* (Paris, 1974), pp. 382, 385; R. Stein, 'The State of French Colonial Commerce on the Eve of the Revolution', *Journal of European Economic History*, 12, 1, Spring 1983, pp. 105–17.

[24] E. Levasseur, *Histoire des classes ouvrières et de l'industrie en France avant 1789* (Paris, 2nd edn 1900, 2 vols.), vol. II, pp. 556–7; Levasseur, *Histoire du commerce*, pp. 515, 518 n., 521, n. 2.

question to discuss, because of the shortage of really satisfactory quantitative materials and the highly conjectural character of the structures which have been erected on the shaky foundations of those we possess. For England, W. Hoffmann's index shows that industrial production just trebled between 1700 and 1799; the later index of real output in commerce and industry established by Phyllis Deane and W. A. Cole goes up from 100 in 1700 to 285 in 1790, which means an average rate of growth of 1·17 per cent per year; more recent research suggests lower figures, at least for the pre-1760 period.[25] For French industry, Jean Marczewski published only some provisional results; according to these, the gross physical product of French industry and handicraft at current prices rose from a yearly average of 385 million livres for the decade 1701–10 to 1,573 million for 1781–90, which means its volume increased fourfold, in fact at a mean average rate of growth of 1·91 per cent per year.[26] But it seems that fuller and more precise calculations would modify these conclusions, reducing the rate of growth to little more than 1 per cent per year, which would be very close to the figure Deane and Cole obtained for English production. The idea that English and French industrial production increased at much the same pace between the early eighteenth century and the French Revolution may appear surprising, but comparison of the evolution of the main industries in each country appears to bear it out.

As regards the woollen industry, Deane considers that in England its output rose about 150 per cent in the eighteenth century taken as a whole. For France, Pierre Léon puts the increase between the early eighteenth century and 1789 at a mere 60 per cent, i.e. a mean rate of growth of 0·9 per cent per year. Later, T. J. Markovitch calculated that the volume of French wool textiles production rose by 144 per cent from 1702/4 to 1784/6, i.e. at a mean yearly rate of 1·1 per cent – roughly in his view at the same pace as the English wool industry. In France the linen industry was very important, in terms of labour force and value of output perhaps bigger than the woollen industry;

[25] W. G. Hoffmann, *British Industry, 1700–1950* (Oxford, 1955), Table 54, Part A; Deane and Cole, p. 78, Table 19. The index given by these authors refers to ten-year averages centred on the year quoted. See also N. F. R. Crafts, 'British Economic Growth, 1700–1831: A Review of the Evidence', *The Economic History Review*, XXXVI, 2, May 1983, pp. 182–4; C. K. Harley, 'British Industrialization before 1841: Evidence of Slower Growth during the Industrial Revolution', *Journal of Economic History*, XLII, 2, June 1982, pp. 267–89 (especially 272, 276, Table 5, 284, Table 8, 285).

[26] J. Marczewski, 'Some Aspects of the Economic Growth of France, 1660–1958', *Economic Development and Social Change*, IX, 3, April 1961, Tables 1 and 3; P. Léon, in *HESF*, II, p. 521.

according to Pierre Léon, its growth during the eighteenth century was faster than that of wool. In Britain, linen manufacture was of little importance at the beginning of the eighteenth century and though it later grew fast in Scotland, particularly at mid-century, by national standards it remained a relatively minor activity, and its growth, though sharp, can have had little consequence for industrial output as a whole. As for the silk industry, imports of raw silk into England slightly more than doubled between the beginning of the century and the 1780s, but the English silk industry remained small in comparison with the French, which was growing altogether faster: according to Léon, the number of looms at work in Lyons increased 185 per cent between 1720 and 1788.[27]

For cotton, the picture is of course different – but not as different as one would think. There is no need to dwell on its phenomenal growth in Britain. Net imports of raw cotton rose from just over 1 million pounds weight in the early eighteenth century to a yearly figure of over 15 million in 1780–9 (and in this last year to 30 million). However, in France the cotton industry was also growing very fast; J. Marczewski has calculated that its rate of growth between the first and the ninth decades averaged 3·8 per cent per annum; this was similar to the English growth rate but starting initially at a lower level. Thus in 1786 English net imports of cotton were 18 million pounds weight, French only 11 million. This shows the English cotton industry with a definite superiority, but the margin is not so great as it became later, and France had then the advantage in the production of printed cottons. It must also be remembered that until the very last years of the century cotton's contribution to British national income remained relatively small.[28]

For the mining and metal industries the quantitative data are very unsatisfactory. French coal production was of course much lower than

[27] Deane and Cole, pp. 52–3, 205, 207; P. Léon, 'L'Industrialisation en France en tant que facteur de croissance économique du début du XVIIe siècle à nos jours', *Première conférence internationale d'histoire économique, Stockholm, 1960* (Paris and The Hague, 1960), pp. 175–6, 178–9; *id.*, in *HESF*, II, p. 519; T. J. Markovitch, *Histoire des industries françaises. Les industries lainières de Colbert à la Révolution* (Genève, 1976), especially pp. 458, 460, 484–6; this last work is based upon archival sources, very thorough and rather reliable.

[28] Deane and Cole, pp. 50–2, 163; Léon, 'L'Industrialisation', p. 178; Levasseur, *Histoire des classes ouvrières*, p. 524–6, 545, 690; Marczewski, Table 7; P. Bairoch, *Révolution industrielle et sous-développement* (Paris, 1963), pp. 235–6, 305–7; C. Ballot, *L'Introduction du machinisme dans l'industrie française* (Paris, 1923), p. 120; Lüthy, vol. II, pp. 663–5; S. D. Chapman and S. Chassagne, *European Textile Printers in the Eighteenth Century. A Study of Peel and Oberkampf* (London, 1981), p. 8, Table 1.

British, but in the eighteenth century it started to grow very fast and by the time of the Revolution had some national importance. As for the iron industry, although there is still much that is obscure and controversial about its evolution in England during the first part of the century, its growth up to about 1760, in fact up to the time of Henry Cort's inventions, was certainly slow; it was only as a result of Cort's inventions that iron production showed dramatic increase, doubling in ten years. Some figures suggest that until the 1780s output in France grew faster. At all events, on the eve of the Revolution, France probably had an output of from 130,000 to 140,000 tons of pig-iron, as against only 60,000 in England.[29]

This brief survey thus does not contradict the impression left by estimates of global industrial output, and as regards *volume of output*, French industrial performance up to the Revolution compares not unfavourably with the English. France was producing less coal, non-ferrous metals, ships and cotton goods than Britain, but more woollens, linens, and silks, as well as more iron. French total industrial production was appreciably higher than the English, but production per head remained smaller, as it had already been in the seventeenth century.

It is true that in industry, as in foreign trade, French performance seems to have been relatively better in the first half of the period under review than in the second. Recovery in output after the depression of the later years of Louis XIV occurred before 1720 in some sectors (Dauphiné, Amiens), after this date in others (Beauvais for example). The recovery was sharp: output of woollens seems to have doubled at Amiens between 1715 and 1750 and trebled at Beauvais between 1724 and 1735. In any case, after 1730 growth was

[29] Léon, 'L'Industrialisation', pp. 177, 179; P. Léon, 'Techniques et civilisations du fer dans l'Europe du XVIIIe siècle', *Le Fer à travers les âges. Hommes et Techniques* (Nancy, 1956), p. 233; B. Gille, 'La Métallurgie française d'Ancien Régime', *Revue d'histoire de la sidérurgie*, V, 2, April–June 1964, p. 156; H. d'Herouville, 'Réflexions sur la croissance', *Etudes et conjoncture*, XIII, 11, Nov. 1958, p. 997; E. F. Heckscher, 'Un grand chapitre de l'histoire du fer: le monopole suédois', *Annales d'Histoire économique et sociale*, IV, 14, March 1932, p. 132; Levasseur, *Histoire des classes ouvrières*, vol. II, p. 675; Marczewski, Table 8; Bairoch, pp. 238, 247–8, 313–14; Deane and Cole, pp. 55, 221; T. S. Ashton, *An Economic History of England. The 18th Century* (London, 1955), p. 124; B. R. Mitchell, with the collaboration of P. Deane, *Abstract of British Historical Statistics* (Cambridge, 1962), p. 221; M. W. Flinn, 'The Growth of the English Iron Industry, 1660–1760', *The Economic History Review*, 2nd series, XI, 1, August 1958, pp. 151–2; Léon, in *HESF*, II, pp. 321, 527; C. K. Hyde, 'The Adoption of Coke-Smelting by the British Iron Industry, 1709–1790', *Explorations in Economic History*, X, 4, Summer 1973, p. 408.

general and quite fast.[30] English industrial growth at this period was somewhat sluggish: Deane and Cole observed a faster pace only in the 1740s and 1750s, and even this movement slackened after 1760. Not everyone will agree with this interpretation, but there is broad agreement about an acceleration during the 1780s in all manufacturing output, which signifies that England had truly entered the Industrial Revolution.[31] In France, according to Ernest Labrousse, there was a slackening in industrial growth from mid-century and after 1770 prolonged stagnation. The old textile industries stopped growing and in some districts – for example Amiens, Normandy, Brittany, and Languedoc – even declined. The agricultural depression around 1780, the worsening lot of the poorer classes, made the difficulties of industry more acute and were a preface to its collapse in 1788–9. These difficulties should not, however, be overestimated, since the stagnation of the old industries (also noticeable in England, where competition from cotton damaged wool and linen) was in part compensated by the fast rise of new industries – cotton in particular, but also coal, iron, glass, and chemical products. However this may be, during the eighteenth century, French industry certainly displayed considerable vitality and buoyancy.[32]

There remains agriculture, a thorny question which will be considered in the next chapter. According to J. Toutain, France had a 60 per cent increase of the deflated agricultural product between 1701–10 and 1781–90; this figure is undoubtedly too high. On the other hand, Deane and Cole have estimated the growth in real output of British agriculture between 1700 and 1790 at 35 per cent. As in the preceding cases, some roughly parallel growth is not to be excluded. Admittedly,

[30] Goubert, *Beauvais*, pp. 586, 589; P. Deyon, 'Le Mouvement de la production textile à Amiens au XVIIIe siècle', *Revue du Nord*, XLIV, 174, April–June 1962, pp. 204, 207; P. Léon, *La Naissance de la grande industrie en Dauphiné (fin du XVIIe siècle–1869)* (Paris, 1954, 2 vols), vol. I, pp. 118 ff.

[31] Deane and Cole, p. 78, Table 19; Ashton, *The 18th Century*, p. 125; A. H. John, 'Aspects of English Economic Growth in the First Half of the Eighteenth Century', *Economica*, n.s., XXVIII, 110, May 1961, pp. 176–90. According to Crafts and Harley (see n. 25), the acceleration of the 1780s has been overestimated by most writers; still they do not deny it existed.

[32] C. E. Labrousse, *Esquisse du mouvement des prix et des revenus en France au XVIIIe siècle* (Paris, 1933, 2 vols), pp. 506–8, 548, 557; *La crise*, pp. viii, x, xxii–xxiii, xxxii–xxxvii, 177; Deyon, 'Le Mouvement de la production', pp. 207–8; Levasseur, *Histoire des classes ouvrières*, pp. 527, 682, notes 6 and 7; Lüthy, vol. II, p. 594; Léon, 'L'Industrialisation', p. 179; 'Structure du commerce extérieur et évolution industrielle de la France à la fin du XVIIIe siècle', in *Conjoncture économique, structures sociales. Hommage à Ernest Labrousse* (Paris – The Hague, 1974), pp. 415–21; Markovitch, p. 486.

English agricultural techniques underwent great improvements during the eighteenth century, for which there is no equivalent in France, but French agriculture was not as conservative as is often thought and eighteenth-century France did not suffer the famines and mass mortalities which had been so frequent in the seventeenth.[33]

As regards the global growth of the two economies, for Britain the Deane and Cole index of total output rises from 100 in 1700 to 190 in 1790 (and Crafts has a doubling from 1700 to 1780), for France an index of the gross physical product at constant prices, as computed by Marczewski, rises from 100 to 260 between the first and ninth decades of the century; the last figure is undoubtedly too high.[34] But between 1700 and 1781 the population of France apparently increased 28 per cent (from 21·5 to 26 millions), that of England and Wales 39 per cent (from 5·1 to 7·0 millions).[35] The growth in average real output and income per head might therefore have been roughly of the same order of magnitude in both countries – and possibly faster in France.[36]

As we have seen, however, at the beginning of the eighteenth century France was behind England, especially as regards the level of average industrial output, foreign trade, and incomes; as both economies grew at about the same rate in the next three-quarters of a century, France could not overtake her rival (although some ground was no doubt made up during the second third of the century), and the close similarity in the rates of growth, which in this paper is deliberately stressed because it has often not been recognised, ought not to disguise the differences between them. In the 1770s, though the economic power of France was, like in the seventeenth century,

[33] J. C. Toutain, *Le Produit de l'agriculture française de 1700 à 1958* (Paris, 1961, 2 vols, *Histoire quantitative de l'économie française*, vols I and II), vol. I, pp. 213, 215; vol. II, pp. 128 (diagram 110), 133, 136, 139, 204, 276; Deane and Cole, p. 78, Table 19; Chapter 2, pp. 52–4.

[34] Deane and Cole, pp. 78–9; Marczewski, 'Some Aspects', Table 4; his figures are tainted because of Toutain's mistakes. The last estimates – by Crafts and Harley – of British national product give a somewhat slower growth than Deane and Cole.

[35] E. A. Wrigley and R. S. Schofield, *The Population History of England, 1541–1871. A Reconstruction* (Cambridge, 1981), pp. 528–9, Table A3. 1; J. Dupâquier, *La Population française aux XVIIe et XVIIIe siècles* (Paris, 1979), pp. 34, 37, 81–2, 87; 'La Non-Révolution agricole du XVIIIe siècle', *Annales ESC*, XVII, 1, Jan.–Feb. 1972, p. 83; E. Le Roy Ladurie, in *Seventh International Economic History Congress. Edinburgh 1978. Four 'A' Themes* (Tonbridge, 1978), p. 122.

[36] A rise of the French income per capita is undoubted, but no figure can be presently given. For England, there was an increase of 20–30 per cent from 1700 to 1760 and then stagnation; Deane and Cole, p. 78, Table 19; Crafts, p. 183, Table 5.

clearly superior in mercantilist terms,[37] England was still the more 'developed', the richer country, with certainly a higher average income per head. England was more urbanised, more industrialised,[38] more involved in international trade; industry employed a higher percentage of the active population and contributed more to national income, roughly over one-quarter, as against one-fifth in France.[39] Moreover, the subsistence sector had virtually disappeared from England as early as the seventeenth century, ousted everywhere by the mercantile and monetary economy; in France the quasi-autarchic subsistence sector was still important and prevailed in extensive areas, acting as a brake on the growth of the economy as a whole (mostly because of obvious geographical reasons).

The fundamental difference between the two economies, however, is in the technological field. The relatively fast (but extensive) expansion of French industry during the eighteenth century took place within a framework which was still, as regards organisation and methods, largely the traditional one. Some changes can be observed, for example the tendency towards concentration and growth of commercial capitalism, the migration of industry from the towns to the countryside, the appearance of new products or even of new industries such as cotton, but in general the traditional structures held. On the eve of the Revolution, the French economy was not basically different from what it had been under Louis XIV: it was merely producing much more. In England, on the other hand, where growth during the first half of the century was of the 'French' type, that is, within the traditional framework, after 1760 some revolutionary changes appeared and economic structures were modified in depth by a series of technical inventions which heralded a general revolution.

This is the heart of the matter: Britain was the place where all the basic inventions which created modern industry – the spinning machine, the flying shuttle, the power-loom, the printing drum, the coke furnace, puddling, and most revolutionary of all, the steam engine – were made, perfected and introduced into industry. In France inventions were far fewer and restricted to improvements for the silk

[37] P. O'Brien and C. Keyder, *Economic Growth in Britain and France, 1780–1914. Two Paths to the Twentieth Century* (London, 1978), pp. 57, 60.

[38] Possibly the difference in this respect between the two countries was not large; cf. O'Brien and Keyder, pp. 92–3; P. Bairoch, 'International Industrialization Levels from 1750 to 1980', *Journal of European Economic History*, XI, 2, Autumn 1982, p. 281, Table 4.

[39] These estimates are possibly wrong, but nothing better can be suggested.

industry (Vaucanson and later Jacquard) and the chemical industry (Berthollet and Leblanc). Eighteenth-century France had some excellent technicians in fields like shipbuilding, ordnance (Gribeauval), and public works, not to mention many marvellous craftsmen, but their talent was not applied to the improvement of industrial techniques. French industry only developed on the technical side through taking up foreign machines and techniques, and to assimilate them usually required the help of foreign technicians, most of them British, though some were Swiss or German.[40]

England experienced a real outburst of inventiveness, which in France was almost completely lacking, and this decisive British superiority in ingenuity and willingness to innovate is the basic fact which accentuated the structural discrepancy between the two economies during the second part of the eighteenth century, and which we must now try to explain.

This is a difficult task, all the more so because the foregoing analysis has removed one convenient explanation: if, as was long believed, British output had increased much faster than the French, British inventiveness could then very easily be explained as a function, a by-product, of growth. But as things are, the parallel growth in output in the two countries makes the British superiority harder to understand, since there seems nothing specific or unique about the development of the English economy, or at least of a large part of it. And in so far as individual genius remains something of a mystery, it will never be fully explained why England had so many great inventors in the eighteenth century. All the historian can hope for is to understand the environment which favoured the making of inventions and which made manufacturers eager to take them up.

The earliest economic historians, when they considered this question, attached great importance to the English institutional framework. Efforts by the State to control and regulate industry had been abandoned as early as the seventeenth century, and though the relevant laws were still theoretically in force, they were never applied; at the same time, the guild system had fallen into decay.[41] Eighteenth-century England was thus a country of *laissez-faire*, in which the field was left free for individual initiative. In France, on the other hand, the

[40] See particularly P. Léon, 'Tradition et machinisme dans la France du XVIIIe siècle', *Information Historique*, XVII, 1, Jan.–Feb. 1955, pp. 5–14.
[41] T. S. Ashton, *The Industrial Revolution, 1760–1830* (London, 1948), pp. 11–12.

guilds survived and their members resisted the development of large-scale enterprises and the introduction of new techniques. Moreover, Colbertist regulations, by prescribing detailed standards of workmanship, discouraged innovations. And in England, such vestiges of corporate regulation and public control as did survive, as for example in the woollen industry, had precisely the effect of encouraging routinism.

Historians like Levasseur, however, went too far in envisaging eighteenth-century French industry as imprisoned in the double strait-jacket of guilds and regulations, which they saw as an insuperable barrier to progress and the reason why French inventions were so few.[42] Nowadays, it is accepted that the guild system was never general in France and in fact did not apply to the greater part of industry; it was unknown in the economic life of many towns, for example Lyons, in royal manufactures, in rural domestic industries, and in new industries such as cotton.[43] Moreover, the system went into rapid decline in the second half of the century, as did Colbertist regulations, which had always been evaded and which from mid-century were less and less strictly enforced, as governmental circles were converted to *laissez-faire* policies. The contrast betweeen French 'dirigism' and English liberalism should thus not be overestimated and it seems that the influence of the institutional framework, though real, was limited.

Some recent historians have stressed the importance of socio-psychological factors, involving a distinct contrast between the English and French social structures and 'scales of value' to explain the British superiority in inventiveness and the entrepreneurial spirit.[44] These writers have shown, with justice, that there was greater vertical social mobility in eighteenth-century England than in France, but it is odd that they should emphasise an aspect of the problem which seems both disputable and of secondary importance. They maintain that in England the social barrier separating the landed classes from industry and commerce, the barrier, that is, between the aristocracy and the business world, was relatively low. But this can have very little bearing

[42] Levasseur, *Histoire des classes ouvrières*, vol. II, p. 305. A. Toynbee and W. Cunningham also overestimated the influence of *laissez-faire* on the English side.

[43] The rise of this industry was possibly delayed by the prohibitions against printed cottons which remained in force until 1759, though Chapman and Chassagne, p. 190, doubt it.

[44] See D. S. Landes, *The Unbound Prometheus* (Cambridge, 1969), pp. 49–52, 66–74; H. Perkin, *The Origins of Modern English Society, 1780–1880* (London, 1969), pp. 63 ff.; A. Milward and S. B. Saul, *The Economic Development of Continental Europe, 1780–1870* (London, 1973), pp. 34–5.

on the problem, since the Industrial Revolution was certainly not the work of the British aristocracy nor even of the gentry. As we know, England had the Duke of Bridgewater and his canal and some great landowners who took an interest in business, but on the whole active participation and investment in industry by the nobility and gentry seems to have actually declined during the eighteenth century, notably in metallurgy (it has been maintained, however, that it gave legitimacy to the pursuit of profit). Moreover, despite the contrary assertions of some English-speaking sociologists, the French nobility was not a close caste, nor was it totally uninterested in business. It controlled a good part of the metallurgical and glass industries and the coal-mines; it invested in foreign trading ventures and West Indian plantations, and on the eve of the Revolution a number of *grands seigneurs* like the Duke of Orleans and the Count of Artois were active in enterprises concerned to introduce English technology into France.[45]

No one, however, will deny that the social prestige of business was lower in France than in England, that the ideal of *vivre noblement* (that is of doing nothing) was stronger there, contempt for work more widespread. But here, too, caution is necessary. For example, writers have often stressed that the French bourgeois who got rich in business retired from it as soon as possible to buy land or public office. And the sale of offices has been rightly criticised as an important factor in diverting capital and talents away from productive activities (tax-farming and financial deals with a government perpetually hard-up played the same role). But this factor was much less important in the eighteenth than in the seventeenth century, as the State was no longer resorting to mass creations and sales of offices. As for the diversion of capital to land, this was also a marked feature in England, where it was the dream of every merchant who got rich to become a country gentleman: and though he himself fairly often retained an interest in the business which had made his fortune, it very rarely survived into the next generation.

As regards our present problem, vertical mobility and scales of values may well be less important than differences between the two countries in the mentality of the men actually engaged in business. In English society there was a more 'capitalist', a more commercial, a more acquisitive spirit; and according to contemporary accounts there was in England a harshness, a ruthlessness, a concentration on the

[45] There has been a good deal of discussion about the French nobility; see particularly G. Chaussinand-Nogaret, *La Noblesse au XVIIIe siècle: de la féodalité aux Lumières* (Paris, 1976).

pursuit of gain, which was absent in the more easy-going France of the *Ancien Régime*.[46] Yet the French social and psychological environment was not basically hostile to innovation, and many manufacturers were quite ready to take up foreign inventions (although they often had difficulty in making them work). French merchants too displayed plenty of initiative and daring in overseas ventures. In any case, such observations are primarily descriptive and quite superficial, leaving the differences in the two mentalities unaccounted for.

Basically, the problem is still obscure and new research is needed into the relationship between the English and French social structures on the one hand and the entrepreneurial mentalities and attitudes on the other. What is most needed is a close analysis at the local or regional level, since to speak of English society and French society as a whole only leaves one with dubious generalities. After all, the Industrial Revolution was not made in England but in a few small districts of England – south Lancashire, some sectors of the East Midlands and Yorkshire, Birmingham, and the Black Country.[47] The inventors and innovators were recruited from the middle class of these industrial districts, from the merchant-manufacturers and the skilled, well-to-do artisans, and England's wealth in qualified cadres of this type was certainly an important factor. A detailed comparison between these English nurseries of the Industrial Revolution and some French industrial centres would be very instructive. Another pointer is the fact that most of the English discoveries were the product of joint effort, as for example with the notable collaboration of Black, Watt, Boulton and Wilkinson over the steam engine; in France, a larger and more rural country, where manufacturing was more dispersed, an element of isolation may have hampered this type of co-operation and cross-stimulus.

Another difficult question is the influence of the Enlightenment. The relationship between the Industrial Revolution and the English philosophical and scientific movement of the late seventeenth century and the eighteenth century is well known. Recent research has stressed the importance of the belief, which goes back to Newton, that industrial progress was possible through observation and experiment; it has shown that most of the inventions (although largely the work of

[46] See Landes, pp. 71–2.

[47] The idea of the Industrial Revolution as a 'regional phenomenon' has been developed by S. Pollard, *Peaceful Conquest: The Industrialization of Europe, 1760–1970* (Oxford, 1981), especially pp. 3, 14, 20, 40–1.

practical men with little education) were backed by systematic thinking, that British scientists and manufacturers were often in close contact, for example in bodies such as the Birmingham Lunar Society, and that scientific knowledge had penetrated industrial society down to a very modest level. But the philosophical and scientific movement was at least as powerful in France, which had plenty of scientists, some of whom, like Buffon and Réaumur, tried their hand at industrial research; scientific societies and academies existed there in profusion and numbered business men among their members; belief in progress and the wish to improve man's material condition were at least as widespread as in England, and Diderot's *Encyclopédie* is striking proof of the interest cultivated people took in technology. But unlike what happened in England, all this intellectual activity had very little practical result; discussions in learned societies remained theoretical in character and those who took part lacked a sense of the concrete.[48] A study of this question which also took education into account would again be very useful. In Britain, dissenters and Scotsmen played a major role in the Industrial Revolution because, thanks to the dissenting academies and the much more advanced Scottish educational system, they were the best-educated section of the middle class. French businessmen had either a very elementary or a purely classical education. This difference may have had some consequences.[49]

To sum up, although neither the differences in social structure and mentality between the two countries in the eighteenth century, nor the features of the English social environment which favoured the entrepreneurial spirit should be minimised, the contrast does not appear to be as sharp as is often said. Moreover, the real influence of such concrete differences as did exist is still obscure, and though they may in some respects account for British technical superiority, they can only

48 There is a large literature on this problem; see particularly Ashton, *The Industrial Revolution*, pp. 14, 16; A. E. Musson and E. Robinson, 'Science and Industry in the Late Eighteenth Century', *The Economic History Review*, XIII, 2, December 1960, pp. 222–44; Musson and Robinson, *Science and Technology in the Industrial Revolution* (Manchester, 1969); Musson, *Science, Technology and Economic Growth in the Eighteenth Century* (London, 1972); P. Mathias, 'Who Unbound Prometheus? Science and Technical Change, 1600–1800', article of 1972, reprinted in *The Transformation of England* (London, 1979), pp. 54–5; also 'Science and Technology During the Industrial Revolution: Some General Problems', *ibid.*, pp. 72–87; J. R. Harris, *Industry and Technology in the Eighteenth Century: Britain and France. An Inaugural Lecture delivered in the University of Birmingham on 27th May 1971* (Birmingham, 1972), pp. 15–16; 'Skills, Coal and British Industry in the Eighteenth Century', *History*, LXI, no. 202, June 1976, pp. 167–9.

49 On the whole the level of literacy was higher in the eighteenth century in Britain than in France.

be a secondary factor in comparison with strictly economic forces. After all, technical progress is closely bound up with economic phenomena, and the explanation we are looking for must be first and foremost an economic one. The contention is that the French did not innovate because, unlike the British, they were not made to do so by the strong pressure of economic forces.

Did this pressure arise from demand? Several writers have maintained that a fast growth in the demand for manufactured goods, from both the home and the export markets, presented British industry with a challenge it could meet only by a technical revolution. This is an attractive hypothesis, but its major premise, that there was an abnormally fast growth in demand in eighteenth-century England is still unproved, and in any case there seems no reason why effective global demand should have grown much faster in England than in France.

There was population growth in the two countries, but at a faster pace in England than in France: this increase in population from 1700 to 1781 was respectively 39 per cent and 28 per cent, as already mentioned (but in absolute numbers, France gained three times more people than England). There has been much debate about the relationship between the industrial and demographic revolutions, but some of the contributors are surely right in stressing that while population growth was doubtless a necessary condition for the Industrial Revolution, it was by no means a sufficient or decisive one, since it was general throughout Western Europe, in particular the upsurge which occurred around mid-century.[50]

Turning to average real incomes per head, they certainly rose in England during the first half of the eighteenth century, when the population was almost stationary, food prices falling, and money wages rising; the internal market therefore expanded, consumption of many articles increased, there was a broadening in range and improvement in quality; A. H. John has pointed out that this permanently affected the level of consumption in most classes and even aroused an appetite for mass consumption. This progress, however, did not last, for from mid-century the expansion of the economy was absorbed by the upsurge in population, and average output per head hardly

[50] This problem would need much fuller treatment in the light of recent research.

increased afterwards.[51] Development in France seems to have been roughly parallel, with a rise in average incomes during the first half of the century followed by a degree of stagnation. Labrousse has shown that, from the 1760s, food prices and rents were rising faster than money wages, which means a fall in real wages, affecting not only the urban wage-earner and the landless rural proletariat, but also many poor peasants with some land of their own, who derived part of their income from wages; a large section of the population therefore became poorer, and this must have meant a fall in its average consumption of manufactured goods. However, the rise in rents and agricultural profits was at the same time enriching a not insignificant part of the population, not only the 'feudal' landlords but also the well-to-do peasants. This was a stimulus to trade, and in particular trade in colonial products, and to industry and urban development, and produced a rise in non-agricultural incomes. There is therefore no reason why average effective demand for manufactured goods should have decreased, except possibly during the late 1770s and the 1780s when, according to Labrousse, the fall in food prices brought about a general stagnation.[52] But, and this is the important point, the slowing down in French demand would have come only after the English had made their main innovations, which were achieved during a period of prosperity on both sides of the Channel.

There is also the demand from foreign markets to consider. In both countries it was more buoyant and increased much faster than internal consumption (although it absorbed a much smaller proportion of output); but as we have already seen, British exports of manufactures increased no faster than French, with the important exception of cotton goods.[53] In addition it must be said, against a currently held opinion, that French losses of colonial territories as a result of the Seven Years War had little effect on exports, since few French goods were taken up by the Canadian and Indian markets, which were only

[51] Ashton, *The 18th Century*, p. 232; John, pp. 180–3, 190; Deane and Cole, pp. 18–19, 21, 41, 80–1; also note 36 above and p. 89 below. There is presently a broad consensus that real wages stagnated or even fell during the second half of the eighteenth century.

[52] Labrousse, *Esquisse*, vol. II, pp. 379, 382–3, 444, 497, 598–9, 610–11, 617; *La Crise*, pp. xxv–xxviii, xxxi.

[53] Schumpeter, p. 12; Deane and Cole, p. 59; see also *Past and Present*, XVII, April 1960, p. 74, and K. Berrill, 'International Trade and the Rate of Economic Growth', *The Economic History Review*, 2nd series, XII, 3, April 1960, p. 358, which stresses the importance of England's capture of the expanding market for this article of mass consumption.

of potential value; nor did England derive much immediate advantage from her conquests.[54]

In fact, the important differences between the English and French markets for manufactured goods are to be observed in their structure rather than their evolution.

In the first place, the British home market – unlike the French – was a truly national market, since the country was an economic unit and there were no internal customs or tolls. This factor should not be overstressed, since the principal French customs unit – the 'Five big Farms' – covered a population about as large as Britain. Nevertheless, as K. Berrill rightly stressed, during the eighteenth century Britain and her colonies formed the largest 'free trade area' in the world. Furthermore, the small size of the country, its configuration, and the early improvement to its transports, contributed to making this national market a reality and internal circulation much more active than in France.[55]

Again, demand within this market was more intense, because as we have already seen, from the beginning of the eighteenth century Britain had a higher average income per head and a higher standard of living. There is no doubt that the British masses were better fed, better clothed, better shod than their French counterparts; the percentage of really poor in the total population may have been smaller and that of well-to-do middle-class people greater. In consequence, if average demand for manufactured goods did not grow faster than in France, it still kept to a higher level for the whole of the period under review.[56]

Lastly, the pattern of consumption was also different, though not in the sense often claimed. Some English-speaking writers maintain that France had only luxury industries such as silk, crystals, porcelain, cabinet-making, which were unsuited to machine production, while England specialised early in making cheap goods for a mass market, which could be mass-produced by machinery in factories.[57] France certainly had luxury or semi-luxury industries which catered for the upper classes; their relative importance was greater than in England and they were little suited to mechanisation. But, in addition, France had large industries which catered for popular consumption, turning out coarse woollens and linens for the peasantry, for West Indian

[54] Deane and Cole, p. 85, n. 2. [55] Berrill, p. 358; John, pp. 185–8.
[56] The 'high' standard of living of the bulk of the British people must not be overestimated.
[57] *Past and Present*, XVII, April 1960, pp. 79–80; Clark, pp. 169–70.

slaves and for the Spanish colonies. Now this production for the masses was not large-scale mass production, since the market was fragmented and the places of manufacture dispersed; moreover, it catered for an unfastidious clientèle and used very cheap rural labour, so there was no pressing need to make technical improvements. On the other hand, it appears that British superiority in the eighteenth century was in the manufacture of good quality products suitable for middle-class consumption; and when Frenchmen complained (at any rate before the 1780s) about British competition, they stressed the superior quality, finish and appearance of the goods manufactured in Britain. In part the contrast was a reflection of the differing social structures of the two countries and their colonies: England, relatively speaking, had a larger middle class and the fast-growing population of the Thirteen colonies had a standard of living close to that of the middle classes in Europe. English industry was therefore not catering for a true 'mass market' – this appeared only in the nineteenth century – but was in a position to make longer production runs than the French.[58] In the result, output had to be both somewhat standardised and of good quality, which undoubtedly encouraged increased division of labour and technical improvements in order to keep prices low, and also the concentration of work in factories, to maintain quality by better supervision.

We may also wonder whether the growth in demand for manufactured goods was not more regular in England than in France, where recurrent crises and wars were more serious interruptions. T. S. Ashton has pointed out that in England the number of patents rose during periods of expansion and optimism and decreased during depressions or wars. Therefore if economic fluctuations were much more violent in France, this could have had a discouraging effect on innovation. Since it is generally agreed that short-term fluctuations in eighteenth-century industrial output were due mainly to variation in harvest yields, a comparison of grain prices on both sides of the Channel is of interest. In this respect, J. Meuvret observed that the amplitude of fluctuations in wheat prices markedly decreased in both countries after 1741; none the less, fluctuations remained sharper,

58 One wonders whether writers who have maintained that mass consumption was encouraged in Britain by a relatively equalitarian distribution of income have ever visited a large country house... Chapman and Chassagne, pp. 214–15.

more violent in France (she, however, did not suffer any more from country-wide famines after 1709/10).[59]

As for fluctuations in output, it is known that in some centres in France production fell about 50 per cent in the course of a business cycle, but such cases are exceptional. In England two reasonably well-founded manufacturing series, one for printed goods and the other for West Riding cloth, show that in several cases output fell 20 per cent or even 25 per cent in the course of a given cycle.[60] As a hypothesis, we might conclude that fluctuations were less violent than in France, probably because a larger proportion of output was intended for export, so that in a year of bad harvest the foreign, and especially the non-European, demand was more likely to compensate for the fall on the home market.

The impact of the wars of the eighteenth century has stimulated some interesting debates among British historians: for them the problem centres on whether their favourable effects – such as encouraging output and technical innovation in the metal industries – were outweighed by their depressive effects on other branches of the economy.[61] For France the question hardly arises, since there is no doubt that the balance was unfavourable in view of the very serious interruption to seaborne trade during hostilities. Admittedly, out of the three wars the only really disastrous one for France was the Seven Years War. But whatever the outcome of particular battles and the conditions of peace, in time of war the Royal Navy was always in command of the seas, leaving France, despite the use of neutral ships, to a large extent cut off from overseas countries and particularly her colonies. The British sea lanes, on the other hand, were kept open and British commerce, despite French privateering, had real protection. French trade thus fell sharply during each of the wars, though the worst collapse was naturally during the Seven Years War, when its annual average was barely half that of the preceding years; even so, the situation was little better during the American war; only the War of the

[59] Meuvret, p. 119; Appleby, pp. 866, 870, 882.

[60] Labrousse, *Esquisse*, vol. II, p. 549; *La Crise*, p. xl; Deyon, 'Le Mouvement de la production', p. 211; Ashton, *The 18th Century*, pp. 248–9.

[61] A. H. John, 'Wars and the English Economy, 1700–1763', *The Economic History Review*, 2nd series, VII, 3, April 1955, pp. 329–44; Ashton, *The Industrial Revolution*, pp. 90–1; *The 18th Century*, pp. 126–7; *Economic Fluctuations in England, 1700–1800* (Oxford, 1959), pp. 69–83; L. D. Neal, 'Interpreting Power and Profit in Economic History: A Case Study of the Seven Years War', *Journal of Economic History*, XXXVII, 1, March 1977, pp. 20–35; W. W. Rostow, *How It All Began. Origins of the Modern Economy* (New York, 1975), pp. 42–4, 235.

Austrian Succession was without really adverse effects. The fall in French exports caused a decline in industrial activity but from Labrousse's figures, the recession was not so great as in foreign trade, which is normal since the greater part of industrial output was distributed on the home market. The adverse influence of the wars should therefore not be overstressed, and indeed H. Lüthy and A. Rémond have maintained that on the whole France was fairly prosperous during the Seven Years War.[62] On the other hand, while English foreign trade rose significantly during the Seven Years War, it fell during the War of the Austrian Succession and fell badly (although less than the French) during the American war. A relationship might be seen between the innovations of the 1760s and the optimism, the confidence in expansion, aroused by England's crushing victories; yet the disaster of American independence was immediately followed by acceleration in the rate of economic growth. On the whole, British maritime and colonial victories do not appear to have been important factors in British technical superiority; it was only during the Revolutionary and Napoleonic wars that complete command of the seas gave Britain the monopoly of overseas markets and thus a stimulus to growth.[63]

There are, then, a number of interesting divergences to be noted between the character and evolution of demand for manufactures in France and England; but these appear less important than those to be observed on the supply side of industry, in the analysis of productive factors.[64]

The Industrial Revolution was not something gratuitous, a triumph of technical progress for its own sake, but a determined effort to solve some concrete problems facing British industry; the character of the

62 Labrousse, *Esquisse*, vol. II, p. 548; Levasseur, *Histoire des classes ouvrières*, vol. II, p. 551, n. 2; Lüthy, vol. II, pp. 12, 42–4, 357–8; A. Rémond, 'Trois bilans de l'économie française au temps des théories physiocratiques', *Revue d'histoire économique et sociale*, XXXV, 4, 1957, pp. 420–1. These views are confirmed by recent research: J. C. Riley, *The Seven Years War and the Old Regime in France. The Economic and Financial Toll* (Princeton, 1986).

63 H. J. Habakkuk, *American and British Technology in the Nineteenth Century. The Search for Labour-Saving Inventions* (Cambridge, 1962), pp. 185–6. Also O'Brien and Keyder, pp. 75–6, 98–9, 197. The dominating position in the international economy which Britain eventually achieved was largely the product of two centuries of aggressive and successful mercantilism, but the connection of this policy (the dividends of which came only gradually) with the Industrial Revolution is tenuous.

64 Some neo-classical writers have denied that demand played a major part in the coming of the Industrial Revolution.

most significant inventions reveals what those problems were. The inventions were designed to make possible the replacement of relatively scarce and expensive resources, such as wood, water power and labour, by others which were relatively plentiful and cheap, such as coal, steam power, and capital, the last in the form of labour-saving machines and processes.[65] This underlying character reflects the fact that English industry in the eighteenth century suffered from relative shortages, particularly of fuel in the iron industry and of labour in textile manufactures, and that such bottlenecks were hampering expansion in output. In other words, unless a technical breakthrough was achieved, the available productive factors were inadequate (except through an excessive rise in prices) to meet the relatively fast increase in demand which started in mid-century when the population growth made itself felt. The contention is that these shortages and bottlenecks, which in England exerted strong pressure in favour of innovation, did not exist in France.[66]

In the primary iron industry the contrast is quite evident. Admittedly, we must not overestimate the wood 'famine' which some writers say existed in England from the early seventeenth century; there was no exhaustion of resources nationally, but at worst local shortages. Michael Flinn has shown that between 1550 and 1750 charcoal prices rose little higher than prices in general, and that ironmasters could secure a fuel supply either by systematic copse 'cultivation' or by moving their works to remote and wooded districts. Even so, in any region, every attempt to increase output meant a fast rise in marginal costs, so that prospects for expansion were limited, and fuel costs were markedly higher in England than in Sweden. On the other hand, from the sixteenth century coal had been cheaper than charcoal in terms of heat output per unit. The ironmasters were aware of the economic advantages of coal, and they also knew that vast reserves of it lay hidden underground. There was therefore a strong incentive to look for methods allowing the substitution of coal for charcoal in blast furnaces, a search which began in the sixteenth century. When it had been successful, by the early eighteenth century, and especially when, from *c.*1750 onwards, coke pig-iron had become less expensive than charcoal pig-iron, ironmasters who wished to take advantage of the growing demand for iron, were induced increasingly

65 Ashton, *The 18th Century*, pp. 108–9; *The Industrial Revolution*, p. 91; Rostow, p. 130.
66 This interpretation will be discussed in more detail in Chapter 2.

to replace charcoal, which was expensive and limited in supply, by coke, which was a cheaper and much more elastic source; eventually, in the last quarter of the century, ironmasters found they either had to go over to coke smelting or close down. Technical progress therefore resulted from the pressure of growing demand on inadequate wood resources, and this explains why coke smelting was discovered and taken up in England rather than in countries where supplies of wood were cheaper and more elastic.[67] In France, where large forests were still extant, the situation was quite different, and though there was some pressure on wood resources during the eighteenth century and prices rose sharply, such shortages as developed remained purely local and French ironmasters were never faced with the choice between innovation or extinction; and in France coal was scarce and expensive.[68]

Moreover, in England the growing demand for coal for both industrial and domestic purposes during the eighteenth century led to deeper working of the mines, which created the need for more efficient pumping devices; and this of course was a powerful incentive in the development of the steam engines, which was in fact a by-product of the rise of the British coal industry. The same incentive was lacking in France, because of the different balance between wood and coal resources.

A contrast between France and England is also evident in the textile industry, but here it is in the supply of labour. The putting out system (domestic labour allied with commercial capitalism) which prevailed in the industry had many advantages, but also the disadvantage that expansion of output was hampered by rising marginal costs. Once the development of an industrial centre had passed a certain limit, to increase production work had to be put out over an ever-widening

[67] M. W. Flinn, 'Timber and the Advance of Technology: a Reconsideration', *Annals of Science*, XV, 12, June 1959, pp. 109–20; Habakkuk, pp. 157–8; G. Hammersley, 'The Charcoal Iron Industry and its Fuel, 1540–1750', *The Economic History Review*, XXVI, 4, November 1973, pp. 593–613; C. K. Hyde, 'The Adoption' (note 9), pp. 397–418, and *Technological Change and the British Iron Industry 1700–1870* (Princeton, 1977); Brinley Thomas, 'Was There an Energy Crisis in Great Britain in the 17th Century?', *Explorations in Economic History*, XIII, 2, April 1986, pp. 124–52.

[68] See Labrousse, *Esquisse*, vol. II, pp. 346–7; Andrée Corvol, *L'Homme et l'arbre sous l'Ancien Régime* (Paris, 1984), pp. vi, 631; Institut d'histoire moderne et contemporaine, *Aspects de la recherche sur l'histoire des forêts françaises. Table-Ronde, Paris. 2 février 1980* (Paris, 1980); Groupe d'histoire des forêts françaises, *Histoire des forêts françaises, Guide de recherche* (Paris, 1982), pp. 60, 76, 95, 132; D. B. Young, 'A Wood Famine? The Question of Deforestation in Old Régime France', *Forestry*, XLIX, 1, 1976.

area, and at a period when communications were slow this meant increased distribution costs, which were soon exceeding profits; the worker spent too much time going to and fro; supervision of the work was becoming impossible and the risks of bad workmanship and of embezzlement of the raw materials were growing. Putting out could, of course, be concentrated on a restricted area where the workers could be made to work longer and faster, but such attempts were thwarted by a regressive labour supply, since the bait of increased earnings was not enough to lure the worker away from his traditional way of life, with its leisure hours and convivial drinking sessions.

There is plentiful evidence of the appearance of this kind of situation during the first half of the eighteenth century in a number of English industrial districts where the putting-out system had developed early and was widespread, with the result that all available labour within reasonable reach of each centre was used up. Nor was there an unlimited supply of mobile labour: the innovations in agriculture, far from creating unemployment, were stimulating demand for labour, which contradicts the older view that England was at this period suffering from large-scale rural underemployment and that there was a mass exodus from the countryside to the industrial centres. We must also remember that up to mid-century the population grew only very slowly. Although it then started growing fast, a number of years had to pass before a large supply of new hands was available for the labour market, and by this time demand was again increasing. There was therefore a relative shortage of labour in industrial districts, as is proved by the quite sharp rise in money wages there (not found in the south of England) during the first half of the eighteenth century.

Manufacturers were therefore faced with high and rising labour costs, which was particularly embarrassing in a young industry like cotton, which in practice had to build up its labour force at the expense of the older industries. There must therefore have been great difficulty during the 1750s in meeting the fast-growing demand for cottons, particularly for export to the colonies. But in the 1760s and 1770s, when there was some slackening in foreign demand, the rise in manufacturing costs was really dangerous. It was now imperative to reduce labour costs and therefore to invent and take up labour-saving machinery. The relative shortage of labour which affected English industry seems therefore to have been one of the most powerful incentives to innovation, not only in the cotton industry but in several others as well.[69]

In France the situation is again obviously different. The countryside was relatively less industrialised, the relaxation of the regulations against rural manufactures (especially after the decree of 1762) freed vast reserves of labour, and this meant it was possible to put work out in a wider area around the industrial centres without unduly increasing labour costs. Moreover, since the rise in the French population was due solely to a fall in deaths (whereas in England there was also an increase in births), in France the number of people of working age may have grown faster. Lastly, and most important, there was in the French countryside a large pool of semi-employed proletarians which putters-out could use without bringing pressure on wages. Labrousse has shown that this proletariat increased fast during the eighteenth century, outpacing the agricultural demand for labour and glutting the labour market.[70]

In sum, during the first half of the eighteenth century, industrial growth in England had reached limits not to be crossed without a technological breakthrough, which the increase in population and in demand from the 1740s made imperative. In France, on the other hand, there was no shortage of labour, output could be increased to meet demand without looking for drastic innovations. This may rank as the most important of the differences to be observed between England and France, but it was in part due to England's earlier and more intense industrialisation, and hence to the situation in the seventeenth century.

There remains the question of capital resources. In eighteenth-century England capital was relatively abundant, as is shown by the fall in interest rates over a long period; Ashton saw this as the deciding factor in the Industrial Revolution, but most British historians have not followed him.[71] In any case, there was the same fall in interest rates in France. From what little we know about the financing of French industry in the eighteenth century there is nothing to suggest that its

[69] Habakkuk, pp. 132–6, 144–7, 161–3; Landes, pp. 57–60; Ashton, *The Industrial Revolution*, pp. 109 sqq., 125; *The 18th Century*, p. 47; Deane and Cole, pp. 59, 96–7; John, 'Aspects', pp. 178, 189; Schumpeter, p. 12.

[70] Labrousse, *Esquisse*, vol. II, pp. 491–2, 598–9; *La Crise*, pp. xxix–xxxi.

[71] Ashton, *The Industrial Revolution*, pp. 11, 58, 94.

development was hampered by a shortage of capital resources[72] – especially at a time when industrial investment had a very low threshold and the earliest spinning jennies, for example, could be got for a few pounds sterling. We know, of course, that the English banking system was much more advanced than the French, but banks played only a minor and indirect part in financing the Industrial Revolution: industries for the most part financed themselves, through ploughing back profits. However, in England capital accumulation was at a faster rate than the growth of other factors of production, especially labour supplies, and this was a powerful incentive to innovation.[73] Things may have been different in France, but this could only be proved by research into the problem, which has not yet been done.

This analysis may appear inconclusive because in many cases it has tried to play down the 'sharp contrasts' between the French and English economies dear to the textbooks, contrasts which have been given too readily as the explanation for Britain's superior inventiveness and economic lead. There is no wish to deny that differences existed between the two countries, though many seem to have been less pronounced than is generally thought, and a matter of degree rather than kind. What is important is that these differences nearly all point in the same direction: in all the various fields investigated, in England the conditions for innovation seem to have been more favourable than in France.[74] The accumulation of these relative differences, many of them small but important as regards factors such as fuel and labour supplies, seems to have been sufficient to set in motion a cumulative and self-sustaining process of technical advance; in France there was no such movement, because there was no need for it. In a paper read in Paris in 1961, David Landes used an illuminating metaphor borrowed from nuclear physics: he spoke of a 'critical mass', a piling up of various factors favouring England's growth which triggered off a chain reaction – the Industrial Revolution.[75] In France, on the other hand,

[72] For a different view see Lüthy, vol. II, p. 41, and C. Fohlen in his Introduction to the French edition of T. S. Ashton, *The Industrial Revolution* (Paris, 1955), pp. xvi–xviii.

[73] Habakkuk, pp. 1–3, 132; and Chap. 5 below.

[74] Of course, many of these differences had deep historical roots, and for example, the contrast between the English and French colonial markets has its origin in the settlement of the English in Virginia and Massachusetts and of the French in Canada and San Domingo.

[75] D. S. Landes, 'Encore le problème de la Révolution industrielle en Angleterre', *Bulletin de la société d'histoire moderne*, 12th series, XVIII, p. 6. This expression was also used in 1962 by P. Lebrun; see P. Lebrun *et al.*, *Essai sur la Révolution industrielle en Belgique, 1770–1847* (Brussels, 2nd edn, 1981), pp. 33–41; particularly p. 36, note 1.

there was no such critical mass, which is why France did not start spontaneously an Industrial Revolution. The external stimulus of serious competition from cheaper English goods, first in foreign markets and after the treaty of 1786 in France itself, was needed to set in train a number of French efforts, intensified during the 1780s, which aimed at introducing the new English technology into France. At this date English industry had a clear superiority, but was only just entering the stage of fast growth and widespread revolutionary changes; France was not disastrously behind, and the Industrial Revolution might have taken off there with only a few years' delay in relation to England. But the 'national catastrophe' which the French Revolution and the twenty years war meant to the French economy[76] would intensify the discrepancy and make it irremediable. In 1815 it would be more pronounced than in 1789, because during this quarter of a century, despite a delaying effect due to the wars, the British economy had continued to change and to make rapid growth.[77]

[76] M. Lévy-Leboyer, *Les Banques européennes et l'industrialisation internationale dans la première moitié du XIXe siècle* (Paris, 1964), p. 29.

[77] See F. Crouzet, *L'Economie britannique et le Blocus continental (1806–1813)* (Paris, 1958, 2 vols., 2nd edn., 1987), vol. II, pp. 863–72, and the introduction to the 2nd edition (Paris, 1987); 'Les conséquences économiques de la Révolution. A propos d'un inédit de Sir Francis d'Ivernois', *Annales historiques de la Révolution française*, 34th year, nos. 168 and 169, pp. 182–217, 336–62; reprinted as Chapter 10 of the French edition of the present book.

2 Criticisms and self-criticisms of a comparison*

The previous chapter derives from two lectures delivered at the London School of Economics in January 1964; the text was revised and published in *Annales ESC* in 1966. This was a period when French historians tended to exaggerate England's 'modernity' and the lead it enjoyed with respect to France, whereas English-speaking historians were inclined to gloat over the stagnation and archaism of France. When I began to draft these lectures, I believed that growth in the eighteenth century must have been much more vigorous in Britain than in France, but I soon realised that a number of preconceptions needed correcting. This article therefore belongs to the revisionist school as far as the economic history of France is concerned, a school which a number of French, British and American historians have joined in the course of the last twenty years and which has been mainly concerned with the nineteenth and twentieth centuries.[1] In order to combat the clichés which were then current, I had at times opted for somewhat extreme formulations but I had tried to avoid being provocative. I had not wanted to shock or offend my English audience, a large part of which was very taken aback by my hypotheses, nor had I wished to lay myself open to the charge of chauvinism. It is a little ironical, therefore, to find myself accused by an

* The 1966 article – which is reprinted in Chapter 1 – has provoked a certain amount of debate, and it has therefore seemed useful to review the question some twenty years later. I have not attempted, however, to survey all of the studies published in the last two decades on eighteenth-century France and England, and I have referred mainly to those which mention my original article or which are directly concerned with a comparison between the two countries. I have no wish to accord this article more importance than it merits or to deck myself with laurels.

[1] This effort culminated in François Caron's remarkable synthesis, *An Economic History of Modern France* (New York, 1979); French edition: *Histoire économique de la France* (Paris, 1981); an extreme version of it may be found in an article by R. Cameron and C. E. Freedeman, 'French Economic Growth: A Radical Revision', *Social Science History*, VII, 1, Winter 1983, pp. 3–30.

English author of having unfairly denigrated the eighteenth-century French economy![2]

An eminent historian described my original article as 'stimulating', which, in academic circles, means that it is of some interest to read but does not stand up.[3] Others have criticised the article more directly (but without a detailed discussion of it), on the absurd pretext that it implies a deliberate attack upon the French Revolution, as if I had charged it with having shattered the impetus of French growth.[4] In general, however, my conclusions have been accepted in France. Thus, Pierre Léon, in his authoritative contribution to the second volume of the *Histoire économique et sociale de la France*, suggested a growth rate for French industrial production, from the beginning of the eighteenth century to 1789, of between 1·5 per cent and 1·9 per cent per year, which, he observed, comes very close to that suggested by me, on the basis of Jean Marczewski's findings. Pierre Léon also accepted my conclusions regarding the growth rates of national product in Britain and in France, and as regards a similar increase in the *per capita* product of the two countries.[5] In 1974, he added:

> In the light of the very detailed studies conducted by Jean Marczewski and François Crouzet, it really does seem as though, from around 1700 to around 1780, France and England advanced at the same pace towards modern industrialisation, with a relatively high annual growth rate of over 1 per cent. The two dominant powers in Europe thus achieved a complete break with the slow, almost imperceptible evolution which industry had known in the past.[6]

In a more sober vein, François Caron wrote: 'During a good part of the eighteenth century, France underwent a demographic, agricultural

[2] N. F. R. Crafts, 'Industrial Revolution in England and France: Some Thoughts on the Question, "Why was England First?"', *The Economic History Review*, XXX, 3, August 1977, pp. 438–41. Some authors have taken the rehabilitation of the French economy still further, as the reader will discover below.

[3] E. Le Roy Ladurie, 'The Chief Defects of Gregory King', in *The Territory of the Historian*, translated by Ben and Sian Reynolds (Hassocks, 1979), p. 185. This remark was made in the context of criticisms of the work of J. Toutain (see below, p. 52), but actually I had only referred to his conclusions in passing and had presented them as hypothetical.

[4] See below, Chapter 3, pp. 105–6.

[5] In F. Braudel and E. Labrousse (eds.), *Histoire économique et sociale de la France*, vol. 2, *Des derniers temps de l'âge seigneurial aux préludes de l'âge industriel (1660–1789)* (Paris, 1970; referred to from here on as *HESF*), pp. 521, 527. In the conclusion to this work (pp. 699, 708), Ernest Labrousse himself expressed his agreement with Léon's perspective on the rapid growth of several French industries during the eighteenth century. My article was heavily indebted to earlier studies by Pierre Léon.

[6] In *Conjoncture économique, structures sociales. Hommage à Ernest Labrousse* (Paris–The Hague, 1974), p. 407.

and industrial growth parallel to that of England',[7] but, this being so, the discrepancy between the two countries which existed at the beginning of the eighteenth century was maintained.

This point of view has even been endorsed by the late Albert Soboul, the advocate of the most traditional, jacobino-leninist interpretation of the French Revolution. In his *La Civilisation et la Révolution Française*, Soboul did draw quite heavily on my article, although he was not overgenerous with quotation marks. He wrote as follows:

> Statistical data, although their value is obviously only approximate for the eighteenth century, surprise us somewhat by revealing that the rhythms of growth [of industry] in England and France were remarkably close to each other ... Taking population increase into account, François Crouzet, whose study I follow here, reckons that the growth in average real production and *per capita* income was approximately the same for both countries. It is only fitting that the parallelism in rhythms of development, at any rate up until the 1780s, be emphasised . . .[8]

More recently, Jean-Charles Asselain, an economist, has also expressed his agreement with most of my suggestions. Even if, Asselain observes, one allows that Marczewski has exaggerated French growth somewhat and that Deane and Cole have by the same token under-estimated English growth, 'one must accept that, in quantitative terms, at any rate, France's industrial growth was not inferior to that of England during the greater part of the eighteenth century'. Asselain continues: 'At first sight, a comparison of growth rates also belies the thesis of Great Britain's superiority, since the growth of foreign trade and of industrial production ran roughly parallel in both countries throughout the eighteenth century.' However, since England was more developed at the beginning of the century, this parallelism implies that the gap was not closed but simply stabilised. It is true that Asselain placed greater emphasis than I had done upon 'the different nature' of the two economies, which 'the similarity in rhythms of growth' should not be allowed to obscure. French growth remained typically 'extensive' in character, whereas the English economy displayed a 'tendency to give rise to an increasing flood of inventions, especially in a group of industrial branches which were engines of growth, as they opened up outlets for each other and drew each other on', to such an extent that the gap between the two countries was becoming wider by around

[7] Caron, pp. 12, 17, rightly observes that 'France was . . . a great industrial country rather than a great industrialised country.'

[8] A. Soboul, *La Civilisation et la Révolution Française. 1: La Crise de l'Ancien Régime* (Paris, 1970), pp. 342–7 and 351–2, particularly 343 and 346.

1770. Asselain also regards the timelag between the beginning of the growth in agricultural production and the demographic spurt as highly significant, a timelag which was one generation in England but which did not exist in France. This occasioned a rise in nominal and real wages in England, which affected both supply and demand, by widening available markets and by making labour more costly. Finally, Asselain recalls the 'vulnerability of the French economy', and of agriculture in particular; the growth of French industry was therefore precarious and the catastrophe which followed the Treaty of 1786 demonstrated its fragility, along with England's superiority.[9]

The *Annales* article, which was soon translated and published in a widely available collection,[10] aroused more interest in Britain and the United States than in France;[11] it was the object of serious debate on the part of several economic historians and the starting-point for a new analysis of the problems which it had tackled.

Most commentators were prepared to accept the conclusion which I had advanced, namely, that the growth rates of the British and French economies had been very similar, from the beginning of the eighteenth century up until the 1780s, although the unreliability of the quantitative data upon which it was based had led me to present it as conjectural. Thus, David Landes, who mistrusts 'guessestimates' of France's national income in the eighteenth century and who often adopted a pessimistic approach to the French economy, has nevertheless written: '... it seems reasonable to suppose that income per head was moderately lower than in Britain at the beginning and that it kept pace fairly well until the last quarter of the century'.[12] Emmanuel Le Roy Ladurie's later criticisms of Toutain has, admittedly, caused Landes to declare that my work ought in fact to be reassessed, given Toutain and Marczewski's overestimation of French growth, and also Michel Morineau's writings.[13] Milward and Saul, on the other hand,

9 J.-C. Asselain, *Histoire économique de la France du XVIIIe siècle à nos jours. 1. De l'Ancien Régime à la Première Guerre Mondiale* (Paris, 1984), pp. 88, 93, 97–8, 102–8.

10 F. Crouzet, 'England and France in the Eighteenth Century: A Comparative Analysis of Two Economic Growths', in R. M. Hartwell (ed.), *The Causes of the Industrial Revolution in England* (London, 1967), pp. 139–74. This translation was slightly abridged.

11 See A. Milward and S. B. Saul, *The Economic Development of Continental Europe, 1780–1870* (London, 1973), p. 117.

12 D. S. Landes, *The Unbound Prometheus* (Cambridge, 1969), p. 13.

13 D. S. Landes, 'Statistics as a Source for the History of Economic Development in Western Europe: The Protostatistical Era', in V. R. Lorwin and J. M. Price (eds.), *The Dimensions of the Past. Materials, Problems and Opportunities for Quantitative Work in History* (New Haven and London, 1972), pp. 73–4. In fact, if one takes Toutain and Marczewski's figures (together with

claim that '. . . on the whole it does not seem unreasonable to accept that the overall rate of growth of the French economy in the eighteenth century was about, or slightly above, 1 per cent per annum. It was, therefore, growing at about the same rate as, or perhaps slightly faster than, the British economy';[14] likewise the industrial product *per capita*. Ralph Davis, in spite of his doubts about several points in my article (see below), nevertheless asserted that the national incomes of France and England had more than doubled during the first three-quarters of the eighteenth century, with France probably enjoying a slight advantage between 1730 and 1770.[15] As for W. W. Rostow, after having carefully examined the quantitative data, he recognised that aggregate indices should be used with caution and admitted that Toutain had exaggerated the growth of French agriculture, he none the less concluded that 'from the 1720s, French production and trade moved up more rapidly than that of Britain'; thus, 'from the 1720s until 1783, France, from a depressed base, gained relatively on Britain. Both moved forward thereafter, Britain with greater momentum, but the period of revolution and war set back France severely'.[16]

Peter Mathias and Patrick O'Brien, and the latter in collaboration with Caglar Keyder, have started out again from scratch, producing new calculations of France's and England's production in the eighteenth century, although again relying, it must be admitted, on the same substratum (the figures of Marczewski and his team, on the one hand, and those of Deane and Cole on the other), but with significant developments, including backward extrapolations in particular. They have thus constructed two 'hypothetical growth curves' of total *physical* product, at constant prices, whose aspect is fairly striking: on a basis of 100 for 1715, the index for 1785 is 199 for France and 145 for Great Britain.[17] These authors are mindful of the criticisms levelled at the work of Toutain and at the *Histoire quantitative de l'économie française* in

those of Deane and Cole), one arrives at a faster growth for France than for England (up until the Revolution, of course); this is the conclusion that D. R. Leet and J. A. Shaw, for example, come to, in 'French Economic Stagnation, 1700–1960: Old Economic History Revisited', in *Journal of Interdisciplinary History*, VIII, 3, Winter 1978, p. 538. It was in assuming that these figures were too high that I proposed roughly similar rates of growth.

14 Milward and Saul, pp. 29, 31, 38; they identify a 'very small' and shrinking development gap between the two countries.

15 R. Davis, *The Rise of the Atlantic Economies* (London, 1973), pp. 288, 301.

16 W. W. Rostow, *How It All Began. Origins of the Modern Economy* (New York, 1975), pp. 94, 170–9. C. Trebilcock has arrived at the same conclusion, in *The Industrialization of the Continental Powers, 1780–1914* (London, 1981), pp. 112–13.

17 For 1790–1, the indices are 210 and 157.

general, but they maintain that their curves fit, 'by and large', with the 'qualitative' conclusions of French historians (and with the periodisation advanced in the *Histoire économique et sociale de la France* in particular). They have also calculated the per capita commodity output at constant prices, which, between 1715 and 1785, rose by 41 per cent in France[18] and by only 5 per cent in Great Britain; they admit that these latter figures underestimate the growth of the British economy, since services played a more important role there than in France, but the gap remains a considerable one, even if one takes this tertiary sector into account.[19]

O'Brien and Keyder have calculated, on the basis of the above series of production at constant prices, that the average annual growth rates for the period 1701–10 to 1781–90 were:[20]

	Agricultural production (%)	Industrial production (%)	Total Commodity production (%)
Great Britain	0·4	1·1	0·7
France	0·6	1·9	1·0

These authors have also calculated that, in 1701–10, the commodity production of France could have been two and a half times that of Britain, which is compatible (the French population being three times the size of the British one) with G. King's idea that the income per capita of the English in 1700 was 10 to 30 per cent higher than that of the French. Since total physical production later increased more rapidly in France than in Britain, the relative gap between the two aggregate products, on the eve of the Revolution, had increased. As far as per capita product was concerned, France had very probably caught

18 This means an average rate of growth of 0·5 per cent per year. J. C. Riley and J. J. McCusker, 'Money Supply, Economic Growth and the Quantity Theory of Money: France, 1650–1788', *Explorations in Economic History*, XX, 3, July 1983, pp. 280, Table 2, 282, have calculated a rate of 0·3 per cent per year for the growth of output per capita and consider that a higher figure, like 0·5 per cent, is possible; see below, note 27.

19 P. Mathias and P. O'Brien, 'Taxation in Britain and France, 1715–1810. A Comparison of the Social and Economic Incidence of Taxes Collected for the Central Governments', *Journal of European Economic History*, V, 3, Winter 1976, pp. 607–9, 611, 613, Tables 3, 4, 5, 6.

20 P. O'Brien and C. Keyder, *Economic Growth in Britain and France, 1780–1914. Two Paths to the Twentieth Century* (London, 1978), p. 57, Table 3.1; *id.*, 'Les Voies de passage vers la société industrielle en Grande-Bretagne et en France (1780–1914)', *Annales ESC*, XXXIV, 6, November–December 1979, p. 1301.

up and, in the 1780s, its per capita product 'seems to have been' above the British level: O'Brien and Keyder's calculations present it as being, for 1781–90, 14 or 37 per cent (depending upon the exchange rates adopted) higher than that of Britain. The latter enjoyed, however, thanks to the sale of services abroad and to income accruing from foreign investments, an imports surplus amounting to 10 per cent of domestic product (while this surplus was very small in the case of France), which increased the quantity of goods which were available in England. Therefore, the per capita incomes of the two countries 'could not have been very far apart', but it is possible that France had a slight advantage.[21] It is worth noting that O'Brien and Keyder are very prudent and make no attempt to conceal their doubts when they formulate their conclusions. This is because they have used Toutain's figures in their calculations, and they mention that they are 'controversial' and regarded with scepticism by numerous French specialists (which is something of an understatement), so that the results obtained for France are probably too high.[22]

These authors are just as cautious in their presentation of another of their discoveries; in the years 1781 to 1790, the productivity of the workforce employed in commodity production was higher in France than in Britain (two or three times higher in industry, which more than compensated for the much lower productivity of French agriculture). They acknowledge that this is a somewhat difficult view to admit.[23] This is not the place to embark upon a technical discussion of this point and I will simply quote some sensible remarks made on this subject by J.-C. Asselain:

> Quantitative history has recently produced one of the 'shocking' results we have come to expect of it . . . a figure [for the productivity of French industry] which is genuinely hard to believe, and whose exaggerated size can be explained by a whole series of statistical biasses . . . This undoubtedly suspect assessment has nevertheless the real virtue of reminding us just how little we really know and also, more precisely, of acting as a counterbalance to the opposite error, which is traditionally committed, of judging industry as a whole, in a far too exclusive manner, in terms of the nucleus of 'modern' industry.[24]

[21] Although real wages were higher in Britain, by a full third, if not more; O'Brien and Keyder, pp. 62, 70, Table 3.7.

[22] O'Brien and Keyder, *Economic Growth*, pp. 51, notes 9 and 10, 57, 60, 61, Table 3.3, 77, note 4, 194; 'Les Voies de passage', p. 1287; P. O'Brien, in D. Johnson, F. Bédarida, F. Crouzet, (eds.), *Britain and France. Ten Centuries*, (Folkestone, 1980), pp. 176–7.

[23] O'Brien and Keyder, pp. 87, 91, Table 4.3, 103, 146.

[24] Asselain, p. 96.

Rondo Cameron and C. E. Freedeman have written as follows of the work of O'Brien and Keyder: 'Neither their methods nor their sources are immune from criticism, but, as far as that is concerned, the same thing could well be said of virtually every quantitative assertion in history.'[25]

Is it therefore possible to conclude that I had been quite cautious in suggesting that '... the growth in average real output and income per head might therefore have been roughly of the same order of magnitude in both countries'?[26]

It is worth adding that the most recent quantitative studies on the growth of the British economy in the eighteenth century arrive at growth rates which are lower than those posited by Deane and Cole, which had been used in my 1966 article (and also by O'Brien). N. F. R. Crafts, for example, gives the following figures:[27]

Rates of average growth per year of real product (%).

	Agriculture	Industry	Total product	Per capita product
1700–1760	0·60	0·71	0·69	0·31
1760–1780	0·13	1·51	0·70	0·01
1780–1801	0·75	2·11	1·32	0·35

The main novelty here is that growth (particularly that of industry and of trade) was clearly less rapid between 1780 and 1800 than Deane and

[25] Cameron and Freedeman, p. 8.

[26] Crouzet, 'England and France in the Eighteenth Century', p. 154; see p. 25.

[27] N. F. R. Crafts, 'British Economic Growth, 1700–1831: A Review of the Evidence', *The Economic History Review*, XXXVI, 2, May 1983, pp. 183, Table 3, 187, Table 5, also 182–4, 186, 194. Crafts had presented somewhat different rates earlier, in R. Floud and D. McCloskey (eds.), *The Economic History of Britain since 1700, Volume 1: 1700–1860* (Cambridge, 1981); see p. 2, Tables 1.1 and 1.2, also p. 5, where he wrote: 'The growth of the [English] economy prior to 1780 ... was similar to that of other areas of Europe – France, for example' (he cites Rostow, O'Brien and Roehl as his authorities). In the same work (p. 64, Table 3.2), W. A. Cole gave new estimates for the English national product at constant prices, according to which rates of growth would be calculated as follows: 0·77 per cent per year for 1700–80, 0·82 per cent for 1700–90. C. K. Harley, however, in 'British Industrialization Before 1841: Evidence of Slower Growth During the Industrial Revolution', *Journal of Economic History*, XLII, 2, June 1982, p. 286, Table 9, argues for a slower total growth than does Crafts, but the difference is very slight as regards per capita income:

	Rate of growth per year (%)	
	National income	Per capita income
1700–1770	0·56	0·27
1770–1815	1·31	0·33

Cole had supposed. On the other hand, these writers would have underestimated the growth of agriculture prior to 1760 and overestimated it after that date. As for the growth of per capita product, it was markedly slower than Deane and Cole had assumed.

It is not in a spirit of self-congratulation that I have reviewed the conclusions several authors have reached; I merely wished to show that broad agreement had been reached as to the correctness of the hypothesis which I had originally advanced in 1964. I am only too ready to allow that my demonstration is far from being a conclusive one. In all studies, the same materials (Deane and Cole's statistical series, on the one hand, those of the ISEA on the other) were, by and large, used, and subjected to more or less refined or barbarous handling, so as to be completed or corrected. No amount of sophistication in the use of statistics would enable one to compensate for their shortcomings.[28]

The real bone of contention is France's agricultural production (one cannot be certain that English agricultural production is better known, but it has not aroused such fierce passions). In my 1966 article, I had quoted, not without some reservations, a recently published work by J. Toutain. However, in 1968, this study was subjected to a devastating critique by Emmanuel Le Roy Ladurie, in an article whose title, 'The Fantastic Accounts of Gregory King', showed more charity and courtesy towards the living than the dead, for really it was Toutain whose shortcomings were under fire, whereas honest Gregory King had done the best he could with the means at his disposal at the time. Toutain's study is described as a colossus with feet of clay, resting upon 'extremely fragile foundations', which are 'sawn off' by a critical examination and turn out to be evanescent; more particularly, he accepts figures for the production of 1701–10 which are far too low, so that the growth during the rest of the eighteenth century is overestimated. Le Roy Ladurie's critique is a wholly convincing one.[29] Nevertheless, British and American experts have gone on using

[28] Note also S. Pollard's critique, in *Peaceful Conquest: The Industrialization of Europe, 1760–1970* (Oxford, 1981), p. 39, namely, that in a period when industrialisation is regional, it is misleading to use national statistics.

[29] E. Le Roy Ladurie, *The Territory of the Historian*, p. 190. See also, M. Morineau, 'Y a-t-il eu une révolution agricole en France au XVIIIe siècle?', *Revue historique*, CCXXXIX, 486, April–June 1968, pp. 303–5; M. Garden, in P. Léon, *Histoire économique et sociale du monde*, vol. 3, L. Bergeron (ed.), *Inerties et Révolutions, 1730–1840* (Paris, 1978), pp. 173, 190, 192–3.

Toutain's study, as we have seen, merely emphasising that his estimates were on the high side.[30]

It is true that researches into the yield of tithes, conducted under the supervision of Le Roy Ladurie, have provided a more solid foundation, namely, archival sources, than have the phantasies of earlier authors. The first overall conclusions to be drawn were that the real agricultural product of France rose between 1700–9 and 1780–9, by at least 25 per cent and at the most 40 per cent, thus clearly less than Toutain had maintained. Subsequently, Le Roy Ladurie has gone further: 'The undeniable growth of French agricultural product must have come close to 40 per cent between 1715 and 1789, but it can hardly have been higher than that.'[31] It has, admittedly, been emphasised that the production levels at the end of Louis XIV's reign were abnormally low, lower, indeed, than those achieved in Mazarin and Colbert's days; therefore a part of the post-1715 growth – up until 1750, and possibly even beyond – was just 'recovery'. This is a plausible line of argument, but I am not concerned here with the very long term, only with the eighteenth century. Sources on tithes, their use and their interpretation have also given rise to controversies which it would be pointless to linger over.[32] In any case, the increase in French agricultural production implied by the study of tithes would seem to be of the order of that proposed for Great Britain by Deane and Cole (35 per cent from 1700 to 1790). However, more recently, Eric Jones suggested that the increase in *English* agricultural production was 61 per cent from 1700 to *1800*.[33]

Michel Morineau's theses would seem, however, to undermine this

[30] O'Brien and Keyder, p. 27, remark, with regard to the estimates of the *Histoire quantitative* and of Deane and Cole, that, for all their faults, 'the estimates have been properly constructed and are the best available for the time being'.

[31] E. Le Roy Ladurie, 'Tithes and Net Agricultural Output (Fifteenth to Eighteenth Century)', in *The Territory of the Historian*, p. 202; Le Roy Ladurie, in G. Duby and A. Wallon (eds.), *Histoire de la France rurale*, vol. 2, *L'Age classique des paysans, 1340–1789* (Paris, 1975), pp. 394–6, 417, 580; E. Le Roy Ladurie and J. Goy, *Les Fluctuations du produit de la dîme. Conjoncture décimale et domaniale de la fin du Moyen Age au XVIIIe siècle* (Paris–The Hague, 1972), pp. 23–4, 369 ff.; see also Garden, in Léon, vol. 3, p. 193; Labrousse, in *HESF*, II, pp. 460–2, 696. H. Neveux has suggested that, as a consequence of increasing specialisation, total production had grown more than that of cereals.

[32] See J. L. Goldsmith, 'The Agrarian History of Preindustrial France. Where do we go from here?', *Journal of European Economic History*, XIII, 1, Spring 1984, pp. 175–99, who has some harsh criticisms of French works on rural history and the 'ideological blockages' from which their authors suffer.

[33] In Floud and McCloskey, pp. 67–9; Jones adds that this estimate is 'purely illustrative'. According to recent growth rates calculated by Crafts (p. 55), one would arrive at an increase slightly under 50 per cent between 1700 and 1780.

apparent parallelism. He has repeatedly denied that there was an agricultural revolution in France in the eighteenth century (I would have no hesitation in granting this point), or even prior to 1840; but he did not maintain that there was a complete standstill in French agriculture. His basic idea is that, between the fifteenth and the middle of the nineteenth century, the average yields of corn underwent fluctuations, in the course of which they attained the same maximum on several different occasions, without ever passing beyond it. One of these occurred after 1750 but, since yields had fallen to a low level at the beginning of the century, the subsequent recovery appeared all the more vigorous, at any rate in the two villages of Hainaut, which Morineau studied with great care, and which he considered typical of the whole of North-West France. A 'recovery' was undoubtedly involved, but this again is not relevant to the present case (it would merely illustrate the notion that, at the beginning of the eighteenth century, the French economy was in a trough, while the English economy was expanding).[34] Altogether, there has been a consensus among most specialists of eighteenth-century French agriculture that both a technical revolution and immobilism can be ruled out, in favour of an increase in production, through a large number of 'small steps', of local and scattered efforts, with strong regional variations. Harvests and prices no longer oscillated as wildly as in the seventeenth century and ceased to have such catastrophic consequences; 'scarcity eased off', in spite of a population increase of six or seven million.[35] Le Roy Ladurie has even written of the end of the 'cold economy', which prevailed since the fourteenth century, and of a 'genuine' but 'hesitant' growth, with a slight increase of corn yields in the zone of open fields.[36] From my point of view, this is rather reassuring!

I want now to pass from quantitative questions to problems of

[34] Morineau, 'Y a-t-il eu une révolution agricole', p. 326; Morineau, *Les Faux-semblants d'un démarrage économique: agriculture et démographie en France au XVIIIe siècle* (Paris, 1971), pp. 31, 35, 93–5, 104–6 and the graphs between pp. 160 and 161; Morineau, 'Cambrésis et Hainaut: des frères ennemis?', *Revue historique*, CCLVII, 522, April–June 1977, pp. 332, note 2, 341. Between 1701–10 and 1781–9, the increase in tithes was 48 per cent in one village and 61 per cent in the other.

[35] Garden, in Léon, vol. 3, pp. 174–5, 183, 193–5; Asselain, pp. 36, 40, 47, 49. Morineau also accepts the view that harvests were more regular in the eighteenth century. He writes: 'There was therefore some progress, as I would take some care in emphasising', but he deems it peripheral, fragmentary, minor, ambiguous, costly and impoverishing.

[36] Le Roy Ladurie, in Duby and Wallon, vol. 2, pp. 395–6; this contribution earned him savage criticisms from M. Morineau, 'History and Tithes', *Journal of European Economic History*, X, 2, Autumn 1981, p. 446.

interpretation. One of the aims of my 1966 article had been to consider the origins of the Industrial Revolution in a new light. However, my approach left me vulnerable to criticism. Peter Mathias, for example, judged that the search for an operative variable which would allow one to differentiate between the English experience and that of all the other countries, for a factor present in England but absent elsewhere, was a methodological error, linked to the vain quest for a monocausal explanation of the Industrial Revolution; he believed that there were too many factors which could have had an influence and that their mutual interactions were too complex. Comparative history is not genuinely scientific, although it can supply, if not proofs, at any rate indications as to the relative importance of the various forces which were involved. Furthermore, in the case which concerns us here, 'many aspects of Britain's economic development in the eighteenth century were certainly not unique to herself' and, in a comparison with France, 'common features ... stand out as prominently as so-called unique ones'. Mathias concludes, therefore: 'If England's growth was not unique in most, if not all, particulars its uniqueness ... lay in the combination of relationships that focused within England during the eighteenth century but in no other European power.'[37]

Milward and Saul, however, have criticised such a position for being tautological; they have also expressed some regret that historians have been too much preoccupied with why it was in Britain that the Industrial Revolution arose first. This is to measure other societies too much in terms of the British yardstick, and to regard them as aberrations with respect to the British norm (this is a valid criticism of my original article, which revolved around this question). Be this as it may, Milward and Saul also reckoned that it was very hard to isolate *one* single economic factor which was operative in eighteenth-century Britain, but not in a number of other European countries, whose economies were in fact very diverse (England, they noted in passing, more closely resembled France than France did Romania). It is even difficult to give a precise definition of the *economic* differences between England and the other countries; all one can say is that it was more 'developed', and that it had attained to a greater 'maturity'. But this situation can only be understood if one takes account of the global

[37] P. Mathias, 'British Industrialisation: Unique or not?', in P. Léon, F. Crouzet, R. Gascon (eds.), *L'Industrialisation en Europe au XIXᵉ siècle* (Paris, 1972), pp. 497–9, 502–3; 'La Révolution industrielle en Angleterre: un cas unique?', *Annales ESC*, XXVII, 1, January–February 1972, pp. 35–7; *The Transformation of England* (London, 1979), pp. 12–16.

historical experience during the centuries prior to the Industrial Revolution; if the latter did not take place in the most productive, the richest and the most powerful country, i.e. France, it was because of the nature of this long-term process of development. Milward and Saul reckon that the problem was fundamentally a social one; but it was not so much a question of 'social attitudes', which they feel have been too much discussed, as of social structures, which are themselves the outcome of a long-term process in economic history.[38]

N. F. R. Crafts advanced a more radical and more detailed methodological criticism of my article in 1977. In his view, the question 'Why was England first?', raised by several authors, myself among them, was a badly formulated one, which cannot be properly answered and ought therefore to be discarded (it may even have distorted our understanding of the economic history of the eighteenth century).[39] Crafts believes that economic development in general, and technological progress in particular, are stochastic processes, highly uncertain in their nature and as regards the dates at which they produce results, and involving, furthermore, a random variable (which is not to say that they are lotteries or that no causal relationships are involved); specialists in technological innovation tend to agree on this point. England was, admittedly, the first to realise the truly decisive innovations, but that does not mean that it had *ex ante* greater chances than France of doing so.

One ought not to perpetuate the common mistake of *post hoc ergo propter hoc*, especially if one only has a single set of observations to hand.[40] Just because England was the first country to carry through an Industrial Revolution, one cannot therefore claim that conditions there were especially propitious (although this may have been the case), that the probability of it making the decisive inventions was higher than elsewhere, and that its economy was superior to that of the other countries. Even where there is little probability *ex ante* of an

[38] Milward and Saul, pp. 30–4, 38–40, 114.

[39] Crafts, 'Industrial Revolution in England and France', pp. 429–31, and, for what follows, 433–4, 436–41. Crafts has also rebuked me for having stated at the beginning of my article that a comparison between England and France can help to understand economic growth in general, a notion which he described as 'otiose'. No doubt I deserved this rebuke, for I had clearly fallen for the fashion which prevailed at the time that every empirical study needed to have theoretical implications.

[40] Crafts sums up (p. 441) by noting that one ought not to expect too much from comparative economic history, for 'some of the uncontrolled experiments that history performed were unique, non-repeatable events'.

economy making decisive progress, it may nevertheless, if it has a good opportunity early on, so position itself as to enable it, eventually, to win the race against another country, a race which is only contested once. Thus, the fact that in 1790 Britain was more 'advanced' and had every chance of making further progress in the industries that were then the leading ones, does not necessarily mean that, around 1740, there was a greater probability of its achieving the Industrial Revolution; its situation in 1790 does not oblige us to find reasons, extending far into its history, for its inevitable supremacy. Britain's ultimate ascendancy does not justify the postulate that it possessed the elements of that superiority long before. To the question as to whether the English economy was visibly superior to that of France in the middle of the eighteenth century, one must answer, Crafts maintains, with a 'resounding no'; considered as a whole, the French economy of this period has been unjustly and prematurely set aside as inferior to the English. Crafts borrows from Rostow various figures, which refer to the respective situations of the two countries in 1700 and in 1780, and which demonstrate that France enjoyed a superiority in absolute terms and a faster growth in almost all domains. He quotes Peter Mathias on the impressive French achievement in science and innovation, and he recalls that, according to S. T. McCloy,[41] the French twice came close to inventing the spinning jenny, before Hargreaves had managed to do so. Crafts also rejects the point of view of those such as David Landes, who have come to the conclusion, on the basis of the situation at the end of the eighteenth century and the beginning of the nineteenth, that the French were less open to innovation than were the British. The French were, after all, at war in this period, the English had taken the lead, but this situation tells us nothing in particular about the ex ante potential of France. The 'unjust treatment' which France has suffered at the hands of economic historians derives in part from the incorrect formulation of the question 'Why was England first?', for France's inferiority is here simply a corollary of England's chronological priority. However, Crafts does, it must be admitted, insist that he never claimed that the Industrial Revolution was a wholly fortuitous event and that the likelihood of its arising in France was greater than that of its occurring in England.

In response, W. W. Rostow chided Crafts for having neglected the acceleration in the rhythm of invention which took place in Britain, by

[41] S. T. McCloy, *French Inventions of the Eighteenth Century* (Lexington, Kentucky, 1952).

contrast with France, from the 1760s onwards; French observers of the period were the first to realise that they were losing ground. Furthermore, one ought also to distinguish between invention and innovation; the stochastic hypothesis cannot be applied to the latter, on account of the 'innovatory zeal' which was widespread in England after 1760 – to a far greater degree than in France – and which covered much more ground than a small number of 'decisive' inventions. Rostow admits that it was not inevitable that the Industrial Revolution should have taken place in England, but he maintains that various factors made it more probable that it should have occurred there rather than in France. There was not one unique match, that might have been won by a weaker team, but rather a whole series of challenges. Rostow thus concludes that the Industrial Revolution was not 'a random walk'.[42] Crafts answered that Rostow had accepted his main thesis, namely, that the Industrial Revolution should be considered in terms of probabilities and that, as a consequence, it had still to be demonstrated that the English economy was 'superior' to the French. Crafts reproached Rostow with having overestimated the inventiveness and innovatory zeal at work in this period in England and he noted that their impact upon growth, such as is shown by a calculation of total factor productivity, had anyway been insignificant before 1780. Crafts maintained that it was impossible to formulate an answer to the question 'Why was England first?', and likewise to discover the causes of the Industrial Revolution by means of the comparative method.[43]

I find Crafts' theoretical or philosophical arguments perfectly valid. There is good reason to react against the frequent indulgence in explanations of the *post hoc ergo propter hoc* kind, and some of Crafts' ideas are very appealing, namely, the notion that one should think of the Industrial Revolution in terms of probabilities, that there is a random element to the date at which particular inventions occur and that the response to economic incentives is an uncertain one. Conversely, flattering to French *amour propre* though Crafts' perspectives on France's economy in the eighteenth century may be, it is nevertheless the case that everything that we know of France and England during the Enlightenment suggests that France was qualitatively inferior to

[42] W. W. Rostow, 'No Random Walk: A Comment on "Why was England First?"', *The Economic History Review*, XXXI, 4, November 1978, pp. 610–12.

[43] N. F. R. Crafts, 'Entrepreneurship and a Probabilistic View of the British Industrial Revolution', *The Economic History Review*, XXXI, 4, November 1978, pp. 613–14; its author acknowledged that his article of 1977 had been deliberately provocative.

England. My 1966 article was meant to show that the gap was not so yawning a one as the 'Vulgate' had traditionally argued, but I cannot subscribe to Crafts' claim that such a gap did not exist at all. Furthermore, the comparative method may not be a panacea but it is still, to my mind, a fruitful approach to problems.

As far as the basic issues are concerned, two economic historians have discussed my proposals in some detail, agreeing with some and rejecting a number of others. Thus, the late Ralph Davis favoured a more traditional conception of the comparison between France and England;[44] he held that real and important contrasts existed between the social structures of the two countries (which I would be the last to deny), so that English society provided a climate that was more congenial to innovation than French society. The former was more 'open', it included the dissenters (for which there was no equivalent in France) and it was less subject to state regulation. Thus, industrial innovation was pursued with more vigour and accepted with greater eagerness in England than in France. Davis has also emphasised the differences between the agrarian structures of the two countries, the brake to economic expansion constituted by the French peasantry, and he wrote: 'The agricultural base of the French economy once more revealed in the 1770s and 1780s, its inability to sustain prolonged growth', whereas, by the second half of the seventeenth century, England had to all intents and purposes solved the problem of its food supplies. It was in fact the attempts to modernise agriculture which triggered off the French Revolution: 'Perhaps it was now impossible for France to undergo its industrial before its political revolution.' Davis went on to observe that the demographic spurt of the eighteenth century in England had occurred on the basis of a historical maximum, since the population had not fallen during the second half of the seventeenth century, whereas in France there had first been a recovery from the losses of the late seventeenth and early eighteenth century; in addition, after 1770, a Malthusian trap began to make itself felt (although symptoms of this were also evident in England[45]). Finally, Davis reckoned that France suffered from the lack of a large colonial

[44] The final two chapters of *The Rise of the Atlantic Economies* are concerned with this comparison.
[45] It is not wholly accurate to claim that English population had not diminished after the middle of the seventeenth century. See below, pp. 70 and 71. I will return to the question of the different agricultures of the two countries. Milward and Saul, p. 41, also reckon that the Revolution had its roots in the *Ancien Régime*'s incapacity to resolve the problem of agricultural production.

market for its industry, of the kind furnished by the Thirteen Colonies.[46]

Admittedly, Davis also believed that many of these factors had nothing to do with the triggering of the Industrial Revolution, which in his view amounted to a particular innovation, namely, the transformation of the technology of cotton spinning, which set in motion some remarkable changes.[47] But why were spinning machines invented? Davis reckoned that it was because England needed to economise on labour (although he rejected the notion that there was a relative shortage there by comparison with France), and because the atmosphere conducive to innovation, which was mentioned above, 'brought forth the inventors'.[48] For all the respect one owes to the memory of an excellent historian, whose early death is a cause for deep regret, his contribution to the debate strikes as a limited one.

W. W. Rostow, on the basis of the statistics concerning the issuing of patents (whose number rose much more sharply in England than in France, in the course of the eighteenth century), has concluded that the operational demand for inventions was lower in France than on the other side of the Channel, and therefore that the French entrepreneurs had not displayed the same innovatory zeal as their British rivals.[49] Rostow has accepted, for a number of sectors,[50] my hypothesis that the English ran up against difficulties which necessitated, far more urgently than elsewhere, the introduction of new technologies: 'the economic pressure to innovate was less in France than in Britain'. This passage continues, however, as follows:

> it is the responsiveness of British society to the economic challenges and opportunities that emerged in the eighteenth century that, I believe, made the difference ... What distinguished Britain from France, as the eighteenth century wore on, was the greater scale of the inventive effort that went into the breaking of crucial technological bottle-necks, and the greater scale of the entrepreneurial corps that introduced them as the century moved towards its close.

46 Davis, pp. 289, 293, 296, 298, 305–10, 313.

47 This view of the problem has been accepted by Crafts, 'Industrial Revolution in England and France', pp. 431–2.

48 Davis, pp. 311–13; see below, p. 92.

49 Rostow, *How It All Began*, pp. 175–8. It is true that statistics for Great Britain are for the number of patents granted, whereas those for France refer to the number approved by the Academy of Sciences (according to McCloy); they are therefore not exactly comparable and probably involve a bias against France (although it is the broad trend which is important). R. Roehl, 'French Industrialization: A Reconsideration', *Explorations in Economic History*, XIII, 3, July 1976, p. 250, had drawn almost the opposite conclusions.

50 In particular, the industrial use of coal, the development of 'coal fuel technology', with its fall-out, but not the cotton industry (see below, p. 92, note 143).

This difference was too great, in Rostow's mind, to be explained in terms of discrepancies in per capita income, percentages of the population that were living in cities, and so forth: 'Crouzet is inclined to explain the difference in terms of the sharper challenges confronted by the British, challenges which more urgently required technical innovation than in France. I am inclined to seek the answer in another direction'.[51]

The other direction to which Rostow refers is that of English Nonconformists and Scottish Presbyterians. Drawing his inspiration from T. S. Ashton and from E. Hagen's attempt at quantification,[52] Rostow has emphasised that there is 'a formidable disproportion' between the presence of dissenters among inventors and innovators and their numbers as regards the population as a whole. At the end of the seventeenth century, Rostow adds, the puritans' ardour shifted from theological debates to 'the market place', although it was the dissenters' minority position, a minority excluded from public life, which directed them towards 'modernising activities'. It was the dissenters who showed the necessary patience, determination and, indeed, obstinacy for the long, slow and painful task which would turn an intuition into an invention that could actually be put to use. However, under these circumstances, there was a direct relationship between the religious problem of the seventeenth century and the locus of the Industrial Revolution. After a series of bitter conflicts, England had managed to resolve this problem (at home and in Scotland) by means of toleration, after the Revolution of 1688. This was for non-economic reasons but it occurred in such a way as to strengthen England's economy, which, in spite of having a much smaller population than France, was fortunate enough to be able to call upon 'big battalions' of potential innovators among the dissenters.

On the other hand, the Revocation of the Edict of Nantes was doubly damaging to France, for it both deprived it of entrepreneurs and further reinforced the creative capacities of its rivals, and of England in particular. Admittedly, French eighteenth-century society, Rostow observes, was also on the move, with considerable opportunities for

[51] Rostow, *How It All Began*, pp. 177–80, 183–4, 253, no. 48.

[52] E. E. Hagen, *On the Theory of Social Change. How Economic Growth Begins* (London, 1964), pp. 296–7; using a sample of some 92 'innovators' for the period of the Industrial Revolution, Hagen discovered that 41 per cent of those English and Welsh innovators whose religious denominations were known, were Nonconformist, although the latter category represented only 7 per cent of the population of England.

vertical-mobility and with some businessmen who were vigorous entrepreneurs; but the reserves of entrepreneurial talent had been reduced and so much so that, to take one example, the French were quite unable to profit from the potential mass market in cotton fabrics which was very much within their grasp. Rostow therefore concludes that: '. . . the British Nonconformists contributed a scale and a thrust to the generation and application of new technology that larger France could not match. They were, in my view, the critical margin France denied herself in 1685.' One could admittedly, Rostow adds, invoke the examples of Holland and of the Thirteen Colonies, which had not had the same 'political and social inhibitions' as France but did not achieve an industrial breakthrough either; however, the profitability of industrial enterprise in these places was insufficient.[53]

Rostow's account implies that development be considered in the very long term and that a broad framework be used. If England's economic superiority is to be understood, the whole of the country's history in the modern period (together with that of France), and its political and religious history in particular, must be taken into account. Furthermore, David Landes' recent book on watch-making is grist to Rostow's mill; the persecution of the Huguenots and the Revocation of the Edict of Nantes, 'devastated' the French watch industry and helped to transfer to England and to Geneva predominance in this branch.[54] However, on the one hand the haemorrhage of talent occasioned by the emigration of the Huguenots had begun long before the Revocation and, on the other hand, contrary to what the casual reader of Rostow's book might suppose, not all the protestants did leave France. Besides, though Peter Mathias has written ironically of the notion that Nonconformists were the 'secret weapon' of economic growth, he places much emphasis, in his *First Industrial Nation*, upon the role that dissenters played in the English economy of the eighteenth century, upon the effects of their value system, of their minority status and of the solidarity networks binding them one to the other.[55] However, if the 'historical compromise' of

[53] Rostow, *How It All Began*, p. 175, 184–9.

[54] D. S. Landes, *Revolution in Time. Clocks and the Making of the Modern World* (Cambridge, Mass., and London, 1983), pp. 92–3, 412–13, notes 19 and 21. Landes rejects the thesis of W. Scoville, in *The Persecution of Huguenots and French Economic Development, 1680–1720* (Berkeley, 1960), who had denied that the Revocation had damaging effects upon the French economy.

[55] P. Mathias, *The First Industrial Nation. An Economic History of Britain 1700–1914. Second Edition* (London, 1983), pp. 7, 142–5.

1688–9 really played the crucial role that Rostow claims, one cannot help but wonder why three-quarters of a century had to pass before English inventiveness manifested itself on a large scale. Conversely, the 'prosperity' enjoyed by England at the end of the seventeenth century had begun under the Restoration, and the Glorious Revolution did not increase it overmuch. Everything suggests that the presence of Nonconformists in England constituted *a* difference with respect to France; I do not believe that it constituted *the* difference.[56]

After this reminder of the criticisms and observations to which my 1966 article gave rise, it seems appropriate to return to a number of problems touched upon there which have been further illuminated by research in the last twenty years.

At the beginning of the article I had emphasised the initial advantage enjoyed by England at the beginning of the eighteenth century,[57] as a consequence of the more favourable conjuncture from which it had benefited, by contrast with France, in the course of the seventeenth century ('conjuncture' being understood here in a purely descriptive sense, namely, the more or less prosperous or depressed situation of an economy, and not, *à la Braudel*, as a *deus ex machina*).

Several historians have confirmed the hypothesis that there was a perceptible difference in the economic development of the two countries in the seventeenth century and in their position at the beginning of the eighteenth. Pierre Goubert, for example, has produced a strong statement of this point of view, observing that, during the second half of the seventeenth century, 'recession, depression, or at any rate stagnation, were glaringly apparent' in France, whereas Holland and England had more or less escaped this great depression.[58] However, one should not push the contrast between France and England to the point of caricature, as Robert Mandrou, both because of his complete ignorance of conditions across the Channel and because of his ideological bias, has done.[59] It is going too far to claim that, after a 'total'

[56] Recent research has not confirmed the notion that there was a clear contrast between Anglican and dissenting schools, although some authors had represented the latter as being like business schools!

[57] A line of argument which Caron, *Histoire économique*, p. 17, accepts. See also O'Brien and Keyder, *Economic Growth*, pp. 57, 194; P. Chaunu, *La Civilisation de l'Europe des Lumières* (Paris, 1971), p. 351.

[58] Goubert, in *HESF*, vol. II, pp. 329–30, 347. A. Corvisier, *La France de Louis XIV, 1643–1715. Ordre intérieur et place en Europe* (Paris, 1979), pp. 139, 146, 153, criticises this excessively sombre balance sheet.

[59] R. Mandrou, *Louis XIV en son temps* (Paris, 1973), especially pp. 21–9, 309–11, 376–81, 386, 403 and Chapter V, pp. 404 ff. According to Mandrou, around 1660, the whole of the French

triumph in 1713, England found herself 'at the head of Europe if not of the world'[60] and that she was 'ready for the great industrial adventure',[61] which in fact only began half a century later.

Nevertheless, when I wrote my original article, I was still too much influenced – like most French historians of my generation – by F. Simiand and Ernest Labrousse, especially as far as the meaning of price movements and their relationship with production were concerned, and by the concept of the 'general crisis' of the seventeenth century. That century had, admittedly, seen several grave crises, but most scholars would now question the notion that it had ever known *one* generalised and continuous depression.[62]

Indeed one ought to add a little light and shade to the sombre picture that had been drawn, up until around twenty years ago, of the seventeenth-century French economy, and in particular of the state of things at the end of Louis XIV's reign. Admittedly, new research into agriculture, and into tithes in particular, has confirmed that there was indeed long-term stagnation – up until 1720 – in cereal production, and that the latter suffered several collapses, in the short term, during wartime.[63] However, alongside this rural stagnation, there was some progress in trade, industry (in rural manufactures in particular) and capitalist organisation. As far as manufactures are concerned, Pierre Deyon, who had emphasised their archaic condition when writing in 1966, took a much less sombre view in a subsequent book; he even noted that those of France and England had undergone virtually

economic machine was jammed, industry and trade were blocked and so doomed to relatively low levels of development, the country was hit by a 'monetary malediction'(?); as for the second part of Louis XIV's reign, it was a time of protracted crisis, whereas England's economic advance is presented in glowing colours, although not without many errors.

60 *Ibid.*, p. 376. Corvisier, p. 340, writes that the treaties of 1713–14 were not disastrous for France, and that it is premature to speak of England as being already dominant; this term is also deemed anachronistic by P. Deyon and J. Jacquart, in the volume they edit of P. Léon, *Histoire économique et sociale du monde*, vol. 2, *Les Hésitations de la croissance, 1580–1740* (Paris, 1978), p. 505; they, like Chaunu (p. 20), prefer to speak of a shared ascendancy, between France and England, although they recognise that the latter held some trump cards, both economic and socio-political, which account for her successes in the War of Spanish Succession.

61 Deyon, in the work mentioned in note 60, pp. 204–8.

62 See Morineau, in Léon, *Histoire économique et sociale*, vol. 2, pp. 86–9, 98–101, 104–5, 193–4; Deyon, in the same work (p. 578), emphasises that the concept of a 'general crisis' has been dealt some hard blows.

63 E. Le Roy Ladurie, in *Seventh International Economic History Congress. Edinburgh 1978. Four 'A' Themes* (Tonbridge, 1978), p. 11; Jacquart, in Léon, vol. 2, pp. 350–3; H. Neveux, 'Les Mouvements longs de la production céréalière en France du XIVe au XVIIIe siècle', *Bulletin du département d'histoire économique, Faculté SES. Université de Genève*, no. 8, 1977–8, pp. 20–4. The latter argues for a depression of around fifty years at the end of the seventeenth century and the beginning of the eighteenth.

simultaneous ups-and-downs in the course of the seventeenth century. It is admittedly true that England enjoyed better results after 1688, but the overall picture of French industry at the end of Louis XIV's reign contains both positive and negative components. As for foreign trade, Jean Delumeau has described its development after 1660, which, though slow and irregular, was evident and characterised by the crucial fact of the emergence of French colonial trade. It would therefore be an error to suppose that in 1715 France was ruined.[64]

It is, nevertheless, the case that the English economy was ahead of the French at the beginning of the eighteenth century, as recent studies have clearly confirmed. England, together with the United Provinces, was in a category apart; one could describe it as 'an advanced pre-industrial economy'.[65]

There can be no doubt that England in the seventeenth century, especially in its latter half, displayed many signs of economic progress, on a broad front and in almost all sectors (whereas in France, innovation, where it existed, was localised). There would be little point in reviewing the increasing prosperity and the innovations of the period; to be brief, agriculture was changing, industry and foreign trade were expanding and diversifying, a new system of public finance and an embryonic banking system were established. D. C. Coleman has therefore emphasised what he calls 'the divergence of England', from the middle of the seventeenth century on, which consisted in the main in a rise in the standard of living, in the gradual accumulation of a reserve of disposable income that was larger and more widely distributed than in any other part of Europe; all of which served to guarantee a particular extension and depth to the domestic market. Coleman has identified the causes of these developments as being the rise in agricultural productivity after 1650, the advance of the 'frontier' in the North of England and the increase in the percentage of the population that was living in towns and engaged in non-agricultural

[64] See in particular, D. Richet, 'Croissance et blocages en France du XVe au XVIIIe siècle', *Annales ESC*, XXIII, 4, July–August 1968, pp. 781–2, 785–6; J. Delumeau, 'Le Commerce extérieur français au XVIIe siècle', *XVIIe siècle*, no. 70–1, 1966, pp. 90, 94 ff., 104–5; P. Deyon, 'La Production manufacturière en France au XVIIe siècle et ses problèmes', *ibid.*, pp. 47–63; Deyon, in Léon, vol. 2, pp. 287–8; Corvisier, pp. 12, 151, 346.

[65] For A. Maddison, *Les Phases du développement capitaliste* (Paris, 1981), pp. 35, 39–40, the United Provinces having a more advanced economic structure and England having borrowed large sums from them, the latter was a follower country rather than a leader. As for Morineau (in Léon, vol. 2, pp. 189–91), he observes that England had not had to bear the burden of the Thirty Years' War, so that revolution broke out in the country which was the most exempt from the misfortunes of the early seventeenth century and which was the least hurt by taxation.

activities. In France, Coleman added, domestic demand continued for a longer time to be depressed, the standard of living of the masses was lower, because of an inadequate agricultural base, and of other causes, like the absence of an integrated national market such as England was fortunate enough to possess.[66]

The same notion of 'divergence' features in the work of Eric Jones, who maintains that England, during the seventeenth and the beginning of the eighteenth century, had an exceptionally vigorous performance, by comparison with that of the Continent (including, when all is said and done, Holland), through the improvement of the volume, methods and organisation of production. Jones is careful not to attribute everything to agriculture, but he reckons that the main divergence arose in that sector, a judgement that accords with Denis Richet's views on France. Richet has in fact highlighted the French contrast between progress in trade and industry and stagnation of the rural areas, the distortion, after 1680–90, between the depressed cereal economy and the expanding capitalist sector.[67] Jones believes that the depression in Continental Europe was caused by the fall in cereal prices, which was itself triggered by decline in population.[68] England did not escape this fall in cereal prices either, but it was late in coming, less severe and less protracted. However, what made England so distinctive was her atypical response, through innovation, to the difficulties occasioned by the fall in prices. The innovation that was pioneered in these years was mixed farming, based upon the introduction of fodder crops, which created a 'virtuous circle' and guaranteed increases in production (J. Jacquart observes that, thanks to the new methods, farming remained profitable, well-to-do cultivators did not suffer from the same difficulties as in France). However, we still need to know how it was that the English had managed to develop this system, when the Continentals had failed to do so. Jones lays particular emphasis upon the system of land ownership and of cultivation which prevailed in England: the predominance of large estates, the peculiarly British relationship between landlord and tenant, which favoured investment in agriculture. On the Continent, on the other

66 D. C. Coleman, *The Economy of England, 1450–1750* (Oxford, 1977), pp. 197–200 (note that his analysis extends as far as 1750). J. Jacquart, in vol. 2 of Léon, p. 462 and elsewhere, emphasises the pauperisation of rural France.

67 Richet, 'Croissance et blocages'; see note 64.

68 This view of a depression resulting from a fall in agricultural prices has been challenged by Morineau, in Léon, vol. 2, pp. 86–9, 98.

hand, social structures tended to deprive agriculture of capital (the French upper nobility was an absentee class, which lived off its seigneurial dues and, Jones has observed, made no improvements in its estates, whereas English magnates were beginning to take an interest in agronomy).[69] The 'Agricultural Revolution' was not, as the reader will discover below, a crucial factor in the Industrial Revolution, but what concerns us here is the importance of agriculture in 'the divergence of England' from mainland Europe.

England enjoyed the immense advantage of having overcome famine or at least of having reduced the deleterious effects of bad harvests, so that they no longer resulted in demographic disasters.[70] It is worth recalling that the last famine to affect England *on a national scale*, though, admittedly, a truly murderous one, was that of 1596–7; at the regional level, subsistence crises continued to occur for longer, in poor and isolated zones, such as present-day Cumbria, but the last real famine there dates from 1623. After 1650, the classic subsistence crisis had disappeared from England, and A. A. Appleby was thus prepared to assert that the latter had solved the problem of feeding its population sixty years before France (where famines had come to a definite end, according to Emmanuel Le Roy Ladurie, by 1715–20), although R. B. Outhwaite has qualified this view, observing that demographic immunity to bad harvests was by no means complete.[71]

B. A. Holderness has considered this question of crises in a long-term perspective, and, in a very evocative passage, recalls the resemblances between southern England and northern France which together 'formed perhaps the most productive, and naturally most

[69] E. Jones, 'English and European Agricultural Development. 1650–1750', in R. M. Hartwell (ed.), *The Industrial Revolution* (Oxford, 1970), pp. 42–3, 52–3, 58, 62, 65–9, 71. J. Jacquart, in Léon, pp. 412, 468, reckons that the ruling classes' offensive against the peasantry was more successful in England than on the Continent; hence the progress of agriculture in that country.

[70] J.-P. Poussou, in Léon, vol. 2, p. 49.

[71] A. B. Appleby, *Famine in Tudor and Stuart England* (Stanford, 1978); 'Grain Prices and Subsistence Crises in England and France, 1590–1740', *Journal of Economic History*, XXXIX, 4, December 1979, pp. 865–7, 883; R. B. Outhwaite, 'Dearth and Government Intervention in English Grain Markets, 1590–1700', *The Economic History Review*, XXXIV, 3, August 1981, pp. 393, 397, 401–2; Le Roy Ladurie, in Duby and Wallon, vol. 2, p. 386. According to Appleby, from the sixteenth century on, subsistence crises were less grave in England than in France. Subsequently, according to E. A. Wrigley and R. S. Schofield, *The Population History of England, 1541–1871. A Reconstruction* (Cambridge, 1981), pp. 325, 328, 354, 372, in England, bad harvests affected nuptiality and fertility much more than mortality (although they could result in epidemics). V. H. T. Skipp, *Crisis and Development: An Ecological Case Study of the Forest of Arden, 1570–1674* (Cambridge, 1978), has studied a small region in the Midlands in which the link between bad harvests and mortality was probably broken as early as 1538. In Scotland, admittedly, the connection persisted up until the 1690s.

fertile, cereal-growing region in agrarian Europe, and the different experience of the two countries is therefore doubly significant'. The 'most obvious contrast' between these two large regions, 'at the simplest, most fundamental and most poignant level' lies in 'the infrequency and comparative mildness of hunger crises in England after 1350' compared with the frequency and gravity of such episodes in France. So it was that England had a 'secure basis' upon which to diversify its economy, in both the industrial and the commercial domains (which depended largely on the capacity of the agricultural sector to produce surpluses of foodstuffs and raw materials), without running up against the blockages and setbacks suffered by France. Holderness reckoned that medieval England, although it was technically backward in many respects, 'was still one of the best favoured and least precarious economies in Europe'. In subsequent centuries, England maintained and even increased its initial advantage and, from 1350 onwards, the contrast between English and French society grew sharper (although their main demographic fluctuations ran parallel to each other and, as Holderness observed, there was an element of chance in the outcome, for England had not undergone a series of devastating wars at home and had a better endowment of natural resources). Holderness also made the interesting observation that the contrast between the two countries would seem to be reflected in their respective historiographies, with French historians having paid most attention to cycles, fluctuations and the circular flow of economic change (which ends up in zero growth in the long term), to what Emmanuel Le Roy Ladurie has called 'immobile history', whereas British historians have been concerned above all with '(the) long period of hard training in the exercise yard' of economic development, which their country had undergone during three or four centuries, and which had given it an invaluable advantage even before the start of the eighteenth century.[72] One must add, however, that few British historians share the miserabilist obsession of the majority of their French colleagues!

Indeed, several authors have emphasised the 'very long term growth' (R. M. Hartwell) from which England benefited during the two centuries prior to the Industrial Revolution and which the other countries of Europe did not experience. S. Kuznets has arrived at the

[72] B. A. Holderness, *Pre-Industrial England. Economy and Society from 1500 to 1750* (London, 1976), pp. 232–4. The passage from which I am quoting constitutes the conclusion to his book.

figure of 0·4 to 0·5 per cent per year for the total product and 0·2 per cent for the per capita product, between 1500 and 1750. These are quite modest rates (growth was, furthermore, fairly irregular), but by no means negligible.[73] It is of course tempting to emphasise this long-term growth and the deep roots of the Industrial Revolution (such as the 'commercialisation' of the economy, its market orientation, features which developed early and were quite pronounced). If one subscribes to this line of argument, one would see the Industrial Revolution as having emerged quite naturally out of the most advanced of the *Ancien Régime* economies, which had attained to such a degree of maturity that it effected the transition to a higher stage without any particular force, whether exogenous or endogenous, impelling it.[74] However, to invoke an argument of this kind is merely to sidestep the difficulty.

If we return now to the seventeenth century, which is a kind of statistical desert as far as aggregates are concerned, the only acid test of economic performance has probably to be a demographic one. As far as France is concerned, the very bleak hypotheses, which were at one time current, have now been abandoned. According to Emmanuel Le Roy Ladurie, between 1570 and 1730 the French population curve was a sinusoid between a ceiling of twenty millions and a floor of eighteen millions, with linear adjustment producing a horizontal straight line for a hundred and sixty years. J. Dupâquier has arrived at a more cheerful assessment, arguing that population was more or less stationary during Louis XIV's reign, and that upward movements in the birth-rate compensated rapidly for peaks in mortality; the total that he produces for France within its present borders at the beginning of the eighteenth century is a little higher than the figure for the beginning of the seventeenth century. J.-P. Poussou holds an intermediate position, but he emphasises the demographic disasters at the close of the Sun King's reign. From my own perspective, the important point is that, even if one opts for the most optimistic estimate, France's population only increased by a modest amount (perhaps 10 per cent at

[73] R. M. Hartwell, *The Industrial Revolution and Economic Growth* (London, 1971), p. 31; S. Kuznets, 'Capital Formation in Modern Economic Growth (and Some Implications for the Past)', *Troisième Conférence internationale d'Histoire économique. Munich, 1965* (Paris–The Hague, 1968), vol. I, pp. 30–1, 33–4.

[74] All the more so given the difficulty (see below, p. 102) of identifying any direct, immediate causes of the 'explosion' of innovation at the end of the eighteenth century. Conversely, it is possible to see no necessary connection between the Industrial Revolution and the various changes occurring during the two centuries prior to it.

the utmost) in the course of a century, with sharp fluctuations in the meantime.[75]

England's demographic history, now well known, thanks to E. A. Wrigley and R. S. Schofield's book',[76] was undoubtedly less turbulent and tragic than that of France. Admittedly, after a steep rise during the sixteenth century and the first half of the seventeenth,[77] the population of England levelled out and even fell, from a maximum of 5,281,000 inhabitants in 1656 to a minimum of 4,865,000 in 1686, but the main reason for this fall (accounting for 58 per cent of it) was large-scale emigration, which probably involved the loss of some 500,000 persons between 1630 and 1680; furthermore, demographic indicators can hardly be said to be favourable during the seventeenth century, particularly between the 1640s and the 1680s.[78] After 1686, population began to rise again, but only very slowly, and it was a full eighty years before the figure that had been reached at the mid seventeenth century was definitively exceeded. However, whereas the figure for 1701 was lower than that for 1641, it was nevertheless 23 per cent higher than that for 1601, which implies that the best part of the growth of the first half of the century (28 per cent between 1601 and 1656) had been preserved. Furthermore, seventeenth-century England did not experience the violent fluctuations suffered by the population of France; it was, as we have seen, virtually immune to the disastrous effects of poor harvests; high mortality peaks were rarer and less steep.[79]

Wrigley and Schofield account for these contrasts by invoking the original demographic system that was operative in England. This was a 'low pressure' system, whose regulator consisted of variations in

[75] Le Roy Ladurie, in Duby and Wallon, vol. 2, pp. 359–61, 363, 376; he points out, however (p. 386), that the last third of the seventeenth century saw a diminution in the kind of large-scale disasters which cause spectacularly high mortality rates, and, more especially, the end of plagues (the one in Marseilles in 1720 being merely an episode, tragic, but short-lived and localised); J. Dupâquier, *La Population française aux XVIIe et XVIIIe siècles* (Paris, 1979), pp. 11, 43, 45, 67–8, 87; J.-P. Poussou, in Léon, vol. 2, pp. 42, 47–8, 50. According to Poussou, there was a population peak in 1705, higher than that of the beginning of the seventeenth century; Dupâquier has underestimated the effects of the crisis of 1709–10, and the 1715 figure was lower than that for 1705.

[76] For what follows, see Wrigley and Schofield, pp. 208–11, 228, 441, 533.

[77] D. M. Palliser, 'Tawney's Century: Brave New World or Malthusian Trap?', *The Economic History Review*, XXXV, 3, August 1982, pp. 339–53; the strong demographic growth between 1540 and 1640 did not create any Malthusian trap, but it stimulated the economy; Wrigley and Schofield, p. 440: the drop in fertility at the beginning of the seventeenth century meant that England avoided the dangers of the 'positive check'.

[78] A fall in nuptiality and in life expectancy, a rise in mortality: Wrigley and Schofield, pp. 161–2, 236, 413–14, 417.

[79] *Ibid.*, pp. 325, 328, 335, 341, 354, 372, 414.

fertility (or, rather, in the last resort, in nuptiality and age at first marriage), which guaranteed an equilibrium between population and resources (an equilibrium which actually improved between 1650 and 1750). On the other hand, France and the other European countries had 'high pressure' systems, in which mortality (or, to be more precise, high mortality peaks) was the regulator. Thus, the positive checks had disappeared from England by the seventeenth century, and perhaps even earlier (although England experienced several years of high mortality resulting from epidemics, in the second quarter of the eighteenth century), whereas it was still a crucial fact in France at the beginning of the eighteenth century. Since its standard of living was low, French population was vulnerable to short-term shocks and both the death-rate and the birth-rate were higher than in England.[80]

Wrigley and Schofield have also calculated an index of real wages which falls sharply from 1571 to 1606 and then rises almost uninterruptedly from 1626 to 1741.[81] From the 1690s, and even earlier, this index also becomes more stable, which suggests that there was an improvement in the production and distribution of foodstuffs. England therefore enjoyed an exceptionally long 'benign' period; for a whole century she benefited from the advantages which derive from a rising level of real incomes and she underwent as a consequence certain structural changes (in the pattern of demand, in particular, which operated to the advantage of goods other than basic necessities),[82] without this rise having provoked an increase in population such as to prevent further progress. Wrigley and Schofield regard this as the explanation for 'the divergence of England' mentioned by Coleman, a divergence which raised the country to levels of prosperity unknown in

80 *Ibid.*, pp. 1, 9, 162, 268–9, 328, 451–3, 467, 469, 471–3, 478–9. However, J. Dupâquier, pp. 67–9, denies that mortality was the main regulator of French population under Louis XIV; more crucial determinants, according to him, would be variations in the birthrate in response to variations in mortality: young unmarried people only had to increase their nuptiality in order to compensate for the devastating effects of high mortality rates. D. R. Weir, 'Life Under Pressure: France and England, 1670–1870', *Journal of Economic History*, XLIV, 1, March 1984, pp. 40, 42–4, admits that between 1670 and 1739, the demographic history of France was more 'turbulent' than that of England (French mortality rates were more sensitive to variations in the price of grain); but he denies that France, in relation to England, had a 'high pressure' system; nuptiality was sensitive to economic jolts in the two countries.

81 I am quoting here their moving average of twenty five years.

82 The ages pyramid of the period 1626–96 had the same favourable effect, with the dependency ratio (percentage of children and old people) being at its minimum (it was higher in the early eighteenth century, and clearly much higher at its end); Wrigley and Schofield, pp. 443–4, 448–9.

the pre-industrial world. This 'good fortune', namely, the unusual length of this benign period, was partly the 'gift', as it were, of the economic and demographic régime of early modern England. On the one hand, age at marriage and fertility tended to adjust themselves *slowly* to economic changes; this is why there was a long period of late marriages and therefore of somewhat low fertility and reproduction rates, which brought significant economic advantages, thanks in particular to a relatively low rate of dependence. On the other hand, the English economy was capable either of supporting a population growth at a rate of 0·5 per cent per year without inducing a fall in real wages, or of guaranteeing a rise in real wages of 0·5 per cent per year if there was nil demographic growth.[83] Admittedly, Wrigley and Schofield's index of real wages has been criticised by Michael Flinn, who felt that it rested upon a narrow and 'terribly frail foundation'.[84] It is worth noting, anyhow, that as early as 1971 Pierre Chaunu had observed that the Malthusian blockage experienced by England after 1640, through the rise in the age at first marriage and the fall in the birth-rate, had prepared the way for the explosive 'growth' of the eighteenth century.[85]

Pierre Chaunu had also observed that it was during the seventeenth century that England and Scotland had acquired a strong lead in literacy, so that, by 1700, the percentage of persons able to sign their names on documents was roughly twice as high in England as in France (although some regions of France were as literate as the most advanced areas of Britain). Since literacy advanced at the same pace on both sides of the Channel during the eighteenth century, the British held on to their lead. This is a fact of considerable importance, even if one does not wholeheartedly endorse Pierre Chaunu's conclusion: 'It was in the seventeenth century that England gained the trump cards which ensured its take-off at the end of the eighteenth century. Seventeenth-century England had invested in brains.'[86] It had also invested in colonies. Thus, in 1700, New France, with its 15,000 European inhabitants, was twenty times less populated than were the

[83] Wrigley and Schofield, pp. 319, 408–9, 439–40, 481–2. These authors reckon that the peculiar features of the English marriage system helped to prepare for the Industrial Revolution but were not themselves capable of triggering it.

[84] M. W. Flinn, 'The Population History of England, 1541–1871', *The Economic History Review*, XXXV, 3, August 1982, pp. 455–6. This index was calculated on the basis of that of Phelps Brown and Hopkins, completed, for the period after 1712, with that of E. Gilboy.

[85] Chaunu, pp. 136–7. [86] *Ibid.*, pp. 145–8.

English colonies of North America, whose natural resources and economic possibilities were moreover far superior.[87]

I shall end this retrospective account with one question: were the causes for England being ahead of France at the beginning of the eighteenth century mainly structural or, conversely, were circumstances partly responsible?

The former position is held by historians who are violently opposed to the *Ancien Régime*,[88] and by American neo-institutionalists. Thus, according to North and Thomas, economic competition at the end of the seventeenth century had produced some winners (England and the United Provinces), some 'also rans' (such as France) and clear losers (Spain, Italy and Germany). 'For the first time in history, some regions and nations had been able to escape the iron teeth of the Malthusian trap'; in England, real wages and real per capita income had progressed, in spite of the increase in population; in France, on the other hand, they had stagnated, even though the population had not risen.[89] For North and Thomas, the lack of sustained growth in France was to be attributed to the fact that the State had not managed to develop 'an efficient set of property rights'; conversely, England had imitated the property rights and institutions of Holland, indeed so successfully that after 1700 she overtook the latter from the point of view both of efficiency and of rapidity of growth. England had established an institutional framework which 'provided a hospitable environment for growth' and 'the stage was now set for the industrial revolution' which, nevertheless, as I ought yet again to remind the reader, only finally arrived half a century later.[90]

D. R. Leet and J. A. Shaw have countered this interpretation with a very simple but nevertheless pertinent objection, namely, that France's 'institutions', reckoned by North and Thomas to be 'medieval', remained largely the same in the eighteenth century as they had been in the seventeenth, and yet the French economy underwent a relatively rapid growth (more rapid, in the view of these authors, than that of England). This implies that France's economic misfortunes in the seventeenth century should be attributed to non-institutional causes, above all the handicap of virtually continuous

87 Deyon, in Léon, vol. 2, pp. 237, 241, 243.
88 For example, Mandrou, p. 21.
89 According to Morineau, in Léon, vol. 2, p. 101, France also experienced a rise in real wages.
90 D. C. North and R. P. Thomas, *The Rise of the Western World. A New Economic History* (Cambridge, 1973), pp. 103, 105, 117–18, 127, 146, 155–6.

warfare.[91] The perspective of these two American historians thus converges with that of several French specialists. Emmanuel Le Roy Ladurie has emphasised the part played by war in France's demographic misfortunes: 'The tremendous wars of the classical age . . . had, on three separate occasions between 1560 and 1715, checked French demographic growth.'[92] J. Jacquart has emphasised the important role played by wars in causing agricultural depression and reducing the peasantry to poverty, both by their direct effects and by the voracious fiscalism to which they gave rise in France, while England was more or less spared such depradations.[93]

The case for the institutionalist hypothesis is strengthened, however, by the testimony of John Harris, whose researches in French archives have uncovered many cases of

> avoidance of commercial and industrial innovation by businessmen pleading the excellence of existing local or national regulations, of low social mobility in the industrial context, the workmen not looking to rise to an entrepreneurial role, the noble requiring a royal privilege to preserve his status before he would enter on anything so demeaning as industrial enterprise. The magnificent illogicality of the French tax, tariff and toll structure created a world of economic Kafkaism, with the entrepreneur as anti-hero.[94]

These observations refer to the eighteenth century, but they are valid *a fortiori* for the seventeenth also.

What I have written in the above pages makes it plain that there is a wide consensus among historians as to the very pronounced differences between the English and French economies at the end of the seventeenth and the beginning of the eighteenth century. One voice, however, has been raised in opposition to this consensus in recent years, that of Immanuel Wallerstein. This curious book is neo-marxist and in sympathy with the Third World, but it is also stuffed full of

[91] Leet and Shaw, pp. 532–5, 538–44. These authors maintain that, prior to the nineteenth century, political events determined economic progress; they attribute to these events a crucial role in the chronology of French growth, with wars (especially those waged on French soil) proving damaging for the latter. Had it not been for 'the Napoleonic holocaust', France could, they claim, have overtaken England. Rostow, *How It All Began*, p. 253, note 46, has also rejected North and Thomas's views on 'property rights' in France.

[92] Le Roy Ladurie, in Duby and Wallon, vol. 2, p. 389.

[93] Jacquart, in Léon, vol. 2, pp. 380–1, 385, 406: also, Dupâquier, pp. 18, 20–1; Corvisier, pp. 146, 151–2, 343, who notes that the French economy did nevertheless stand up to the ordeal of two long periods of war.

[94] J. R. Harris, *Industry and Technology in the Eighteenth Century: Britain and France. An Inaugural Lecture delivered in the University of Birmingham on 27th May 1971* (Birmingham, 1972), p. 9. In the seventeenth century England became by and large a *laissez-faire* country, as far as industry was concerned.

sophistries, half-truths and contradictions;[95] nevertheless, it contains a number of ideas which deserve to be taken seriously, particularly as regards the problem with which we are concerned here.

Wallerstein argues that the 'core' of the European world-economy in the period 1600–1750 was in north-west Europe, in Holland and Zeeland, in London, the Home Counties, and East Anglia, and in northern and western France (on the other hand, a good part of the French kingdom did not belong to this centre); 'economically these zones were more alike than different'. Wallerstein goes on to add: 'My strong suspicion is that there was far less of a difference between the real agro-industrial efficiencies of England and France in this period than we tend to assume. The small differences that emerged as of 1763 were magnified *politically* into the significant differences of a century later.'[96]

Wallerstein has argued that, even in agriculture, there were only 'relatively minor' differences between the two sides of the Channel prior to 1750, even as regards productivity. He doubts that French agriculture was as backward as is commonly supposed and maintains rather that there were parallelisms between, for example, the rise of stockbreeding in England and that of viticulture in France. England did, admittedly, become a large exporter of cereals, thanks to the bounties instituted by the Corn Law of 1688 (*sic*),[97] but northern France used also to 'export' to other less fertile French provinces, and the south of France exported to Mediterranean countries. In both countries, the development of industry was encouraged by a high degree of protection and there was a shift towards rural areas.[98] Wallerstein's views may at first seem surprising but, upon second thoughts, it is not at all absurd to maintain that England's economic evolution up until the start of the Industrial Revolution was not fundamentally different from that of the rest of Western Europe, if one

95 I. Wallerstein, *The Modern World System: II: Mercantilism and the Consolidation of the European World-Economy, 1600–1750* (New York, 1980).

96 *Ibid.*, pp. 37, 81; also 120, 288.

97 Curiously enough, Wallerstein (pp. 260–2, 267–8) is here reviving eighteenth-century views (p. 138), and those of W. Cunningham (1882), regarding the significance of this Corn Law, which is held to have stimulated exports and therefore agricultural production, and, indirectly, merchant shipping; it is therefore supposed to have led the English to concentrate on foreign trade, therefore on the navy, etc. This, in Wallerstein's view, accounts for their subsequent victories over France. In a review of this book, Eric Jones has written of a 'surprising explanation of Britain's eventual success in terms of state support for farm exports', *The Economic History Review*, XXXIV, 2, May 1981, p. 343.

98 Wallerstein, pp. 83–91.

excepts a higher degree of regularity which may be largely explained in terms of the country being an island and thus immune to invasion and devastation.

Wallerstein allows, however, that there were serious differences between France and England as regards colonies, trade and finance. Yet if important settler colonies were founded by the English in the course of the seventeenth century, it was because the development of their industries needed external outlets, whereas France, which had a huge (albeit poorly integrated) domestic market, had scarcely any need to increase its foreign trade. In this case, as in a number of others, Wallerstein (and also Fernand Braudel, as we shall see below) has attributed a determinant role to the difference in size between England which, being small, had no choice but to turn its attention towards the outside world, and France, a huge country.[99] This explanation is of course vulnerable to the usual criticisms levelled at a crude geographical determinism.

As far as public finance and banking were concerned, England was far more successful than France, because, appearances notwithstanding, the English state was just as 'strong' by 1689 as the French monarchy (which was absolute in theory but not in practice) and, from the beginning of the eighteenth century, it became manifestly the stronger of the two. Wallerstein's thought revolves around this crucial point; in a capitalist system, a state is strong if it enables the owners of the means of production to extract the highest possible surplus, by helping them in particular in their struggle against their competitors on the world market, by armed force if necessary, which presupposes a capacity to mobilise the country's resources for military purposes. But the key factor is the existence of a stable 'hegemonic bloc' (to use Gramsci's term), which balances out the interests of the owners of means of production.[100] The seventeenth century saw serious attempts on both sides of the Channel to strengthen the state, but they enjoyed more lasting success in England, where the two Revolutions resulted in a stable compromise, with the reconstitution of a unified ruling class, and with the adoption of a political strategy by means of which the

[99] *Ibid.*, pp. 101–4, 267. See also pp. 112 and 99–101, for other bizarre views, for instance about the replacement of wood with coal in England, which Wallerstein rightly attributes to a different 'ecology' from that of France, but he goes on to commit the shocking error of confusing 'timber', which was used for shipbuilding and was imported from the Baltic, and 'wood' for making charcoal, which never entered international trade.

[100] *Ibid.*, pp. 11–14, also 94, 102.

state devoted itself to stimulate economic development at the expense of other countries. France, because of its peculiar geography – its large size and its heterogeneity, did not achieve a compromise of this nature but merely an 'imposed truce' under Colbert; the reconciliation of the upper classes, which, after their divisions in the seventeenth century, accounted for England's stability in the eighteenth, was never achieved in France, where, in the end, the state's weakness was to bring about the Revolution of 1789.[101]

It is undoubtedly true that the making in England of a genuine ruling class (such as France has never had), together with its takeover of the state, were phenomena of the greatest importance, as Pierre Chaunu had already observed.[102] Furthermore, Wallerstein is quite right to point out that the strongest state is not the one which has the most developed apparatus or whose decisions are the most arbitrary; he is also justified in rejecting the old clichés about a constitutional and liberal England, which tended to favour private enterprise and mass production, by contrast with an absolutist and bureaucratic France, which was supposed to stifle such developments.[103] However, the curious thing about Wallerstein's position is that it leads him to put the emphasis upon political factors (he holds that the strength or weakness of a state depends upon politics rather than upon economics) and upon political decisions respecting the economy (he attaches great importance to the Navigation Acts and, quite unjustifiably, to the Corn Law of 1689).[104] Finally, Wallerstein has rehabilitated the military factor: 'Military strength is one key to efficacity in this regard' (the struggle against economic rivals); the 'crucial Battle of Barfleur' (which was in 1692, and not in 1694) gave the English at a single stroke mastery of the seas.[105] There is some justification for such positions but what is surprising is that they have been formulated by an author who, in the

[101] *Ibid.*, pp. 116–25, 246, 285–7. However, some other passages appear to minimise the differences between the compromises realised in the two countries, at the end of the seventeenth century, between the old aristocracy and the 'new capitalists', and between the monarchy and the nobility (pp. 142, 284–5). Differences between the social structures of the two countries are also described as 'minor'.

[102] Chaunu, pp. 202, 362; he sees also in the 1680s a significant divergence between France and England, when the 'reformations' of the nobility in France served to reinforce social rigidity. See also his suggestion, pp. 198–200, that one should not exaggerate the contrast between the English and French political systems of the eighteenth century.

[103] Wallerstein, pp. 94, 102, 113–14.

[104] *Ibid.*, pp. 102, 113, 179, 236; he also remarks, on p. 264, that the state's intervention (through the prohibition of imports) was the crucial factor in the growth of textiles production.

[105] *Ibid.*, pp. 113–14, 248.

United States, has been the high priest of the cult of Fernand Braudel, who had been fiercely opposed not only to old-fashioned military and political history, but also to any consideration of the military and the political dimension.

Be this as it may, Wallerstein has argued that the 'ultimate triumph' of England in the course of the second One Hundred Years War (which he sees as coming to an end, curiously enough, in 1763) was a victory not of liberalism, but of the strongest, least democratic and most mercantilist state in Europe.[106] However, this was also a state which enjoyed some strong support from abroad – from the United Provinces. Wallerstein is very insistent on this point. The reconciliation between England and Holland at the end of the seventeenth century enabled the English to borrow from Amsterdam and the international Huguenot community, to set up the Bank of England, to contain France militarily and to establish Walpole's stable, one-party (*sic*) state, whereas France's finances descended into chaos.[107] Wallerstein concluded his account as follows: 'It was this steady increase in the relative strength of the English state – rather than significant differences in how French and English production was organized in the period from 1600 to 1750 or in their value systems – that accounted for the ability of England to outdistance France decisively in the period from 1750 to 1815.'[108] Wallerstein's study on the Industrial Revolution itself will be eagerly expected.

At the end of my 1966 article, I had expressed some regret at not having examined the problem of the relations between the political revolutions which England underwent in the course of the seventeenth century and the Industrial Revolution of the following century.

It is worth going into this problem, for marxist or *marxisant* historians hold that, if England was the first country to carry out the Industrial Revolution (regarded as the blossoming of industrial capitalism), it was because, a century before, it had been the first country to make a bourgeois revolution, which had liquidated

[106] *Ibid.*, pp. 115, 258, 268, 286.

[107] *Ibid.*, pp. 115, 277–9, 281, 283, 285, 288. Wallerstein acknowledges, on pp. 249 and 277, that neither France nor England could hope to be strong on both land and sea at the same time, that it was natural for France to lean towards expansion on the Continent, and that in fact she had no choice. This is once again, with some justification, to favour the geographical factor.

[108] *Ibid.*, p. 288. He had noted, on p. 120, that the *small* differences existing prior to 1763 enabled the English subsequently to pull ahead in terms of productivity and economic dominance, an advance for which the rise of their foreign trade and their colonies had also prepared them (p. 104).

feudalism and eliminated all obstacles to the free advance of capitalism. Conversely, France, which had to wait until 1789 for its bourgeois revolution, was unable to rival England on the economic plane.

Christopher Hill has enlarged upon this theme in his book, *Reformation to Industrial Revolution*, which is now widely regarded as a classic. Hill argues that the English Revolution of the seventeenth century transformed the economic and social life of the country and gave rise, as early as the seventeenth century, to 'revolutions in trade and industry which had far-reaching effects on the whole of society'. Because of these changes, and because it had founded 'a political system within which commercial and industrial capital had freedom to develop', the Revolution 'created the conditions for that uniquely favourable balance between population and resources in England from which the Industrial Revolution was to result'.[109]

On the French side, Albert Soboul has likewise stressed that the path from feudalism to capitalism had as revolutionary a character in seventeenth-century England as in France at the end of the eighteenth. In both countries, feudal relations of production were abolished 'as categories or structurally'. The English Revolution 'played an equivalent role in the history of England to that of the French Revolution in the history of France'. To a very great extent it 'cleared a path for the development of capitalism', by sweeping away the last vestiges of the feudal order, by abolishing monopolies, by lessening the influence of the guilds etc.[110] G. Lemarchand has written that, during the sixty or eighty years following 1640, the last vestiges of the feudal mode of production (FMP) were liquidated, by revolutions that were bourgeois in the sense that they shattered the institutional barriers which impeded the free advance of capitalism and they enabled the merchant upper bourgeoisie to have a share in political power.[111]

The main objection that one can raise to the crucial role that these historians appear to be attributing to the revolutions of the seventeenth century is that by 1640 there was not much feudalism left to abolish. Actually Lemarchand admits that it had been steadily declining in England since the fourteenth century; the manor had disintegrated

[109] C. Hill, *Reformation to Industrial Revolution* (Harmondsworth, 1969), pp. 14–16, 22, and Chapters 3 to 5 of the third part of the book.

[110] Soboul, *La Civilisation*, vol. II, pp. 66, 68–9.

[111] G. Lemarchand, 'Un cas de transition du féodalisme au capitalisme: l'Angleterre', *Revue d'histoire moderne et contemporaine*, XXV, April–June 1978, p. 297. It is worth noting the somewhat embarrassed formula employed with respect to 'bourgeois' revolutions.

and had assumed the form of the large estate, which was rented out to tenant farmers and worked by wage-earning labourers. The FMP had ceased to be dominant as early as the sixteenth century; it had at first continued in symbiosis with the new capitalist mode of production (CMP), but this latter prevailed fairly quickly, without being 'coopted' by the feudal mode (as was to be the case in France in the modern period), so relentless was its advance and so weakened had feudalism been as early as the fifteenth century. Soboul himself has written that in England 'there had been a free and rapid development of capitalism, linked with the utter dissolution of rural communities through the mediation of rent in money'; in France, things proceeded at a slower pace, in spite of a more radical revolution, because the dissolution of peasant communities had never been completed and because rents in kind survived.[112]

Whatever the validity of this scheme, the major point is the idea that the transition from feudalism to capitalism in England was a process which took a very long time and that it was already very advanced well before the Revolution of 1640 to which, when all is said and done, Soboul and Lemarchand (whose marxist–leninist credentials can not be suspected) did not attribute a crucial importance.[113] Thus, English agriculture was in the main commercial and capitalist. The Long Parliament did, admittedly, abolish a number of 'feudal' survivals (such as the Court of Wards) in the field of property rights, together with a number of restrictions upon free enterprise, but it is hard to believe that these details were anything more than finishing touches. Christopher Hill has certainly exaggerated the influence of the political revolution upon the 'Agricultural Revolution' of the seventeenth century (a revolution that was still modest both in its nature and in its spread). His account becomes more convincing when it deals with the impact of the Navigation Act and of the Commonwealth's foreign policy upon the 'Commercial Revolution' and upon the emancipation of England from Dutch tutelage. Finally, is it really necessary to point out yet again that there was no social revolution in England in the seventeenth century and no significant transfer of power to the advantage of the 'bourgeois' and the 'capitalists'? Furthermore, the peculiarities of the evolution of agrarian society in England, of its

112 Soboul, vol. II, p. 66; Lemarchand, pp. 277–82, 286–7, 304.

113 Lemarchand is very eclectic and draws into his argument multiple factors in order to explain the transition from the FMP to the CMP. Of course, it is quite another matter to decide whether this transition is or is not a pseudo-problem.

passage from the FMP to the CMP (if this jargon has to be used), are beyond dispute, and proved to be a major factor in the Agricultural Revolution, but the causal relation linking this latter and the Industrial Revolution, as we shall see below, was rather weak.

A second objection to Christopher Hill's thesis concerns chronology. Lemarchand himself has observed that the development of capitalist industry, that is, the Industrial Revolution, occurred at least a century after the disappearance of the FMP, and he has shown some surprise (as have a number of other authors) at the fact that this revolution had not arisen during the 'period of rupture', between 1689 and 1715 (or 1720), when the FMP had all but disappeared and when there took place a 'genuine boom' (*sic*), together with a number of technical inventions.[114] We should recognise, nevertheless, that the 'divergence of England' did become apparent ... at the Restoration.

As for eighteenth-century France, Lemarchand has written that she 'experienced a significant economic expansion but [that] her basic structures did not greatly change, which gave rise to grave imbalances after 1770'. Yet feudalism does not feature among the three series of causes which, according to Lemarchand, account for the absence of an Industrial Revolution in France, namely, the relative abundance of wood, water and labour; the cast of mind of the businessmen which, formed by an education which was still humanist and religious, had not turned towards technical matters; and, finally, the attitude of certain merchants who, since they enjoyed a *de facto* monopoly within the framework of commercial capitalism, had no particular incentive to increase production.[115]

Curiously enough, it was an English marxist, Tom Kemp, who formulated the idea that the French economy had been blocked by a system that was still feudal, and that it was necessary to shatter that system if an Industrial Revolution were ever to be realised. Kemp accepts, albeit begrudgingly and with reservations, that growth may have pursued a parallel course in France and Britain: 'the curve of income per head and total output [of France] moved unmistakably upward; at a rate, it has been claimed, comparable with that in

[114] Lemarchand, pp. 295, 297–8, 301–2: Soboul, while he makes the liquidation of the last vestiges of feudalism a necessary condition for the development of industrial capitalism, does not suggest that this condition is a sufficient one; vol. II, pp. 64–5.

[115] Lemarchand, p. 298; also his earlier article, 'Bilan et perspectives de recherches: Le XVIIe et le XVIIIe siècles en France', *Annales historiques de la Révolution française*, 1969, no. 198, pp. 647–8.

England'; further on he admits that the French economy was not immobile and displayed 'a capacity for change and expansion which has enabled the claim to be made that the growth in production and in average income per head was about the same as that in England'. However, Kemp maintains that quantification is 'misleading', for profound changes were blocked by various factors which are not really susceptible to quantification: 'They depend upon an analysis ... of the totality of social relations.' Kemp's account continues:

> The fact that growth took place, in some spheres at an impressive rate, does not, therefore, seem to be the significant thing about French development in this period. Rather the emphasis must be upon the comparatively narrow scope for entrepreneurship of a modern kind in a society which remained overwhelmingly agricultural and traditional. What seems significant is not the fact that growth took place ... but rather the lack of conditions for carrying through a structural transformation. To explain this it is necessary to look outside the sphere of production ... at the character of the society. Stress has therefore been laid on the ways in which the general characteristics of the old régime limited the possibilities of economic transformation.

Given conditions of this sort, after 1780, the divergences between the French and English economies became more pronounced, with tensions and symptoms of crisis appearing in France and with industry making great strides in England. As far as Kemp is concerned, 'France appears basically as a feudal society in the eighteenth century'; even industrial and commercial undertakings had a 'feudal' character, with very little productive investment or entrepreneurial spirit. In such conditions, tendencies towards inertia continued to be 'overwhelmingly strong', especially in agriculture. Capitalist development could not intervene 'as long as the grip of the old social relations had not been broken or loosened'. Kemp heaps up all the most traditional clichés regarding the French economy of the *Ancien Régime*, and his refusal to resort to quantification is, it must be said, frankly obscurantist (an attitude which is to be found among a number of historians of the Left, when statistics do not agree with their ideology).[116]

The best response to Kemp must be the following perspicacious passage from Emmanuel Le Roy Ladurie:

> The contradictions between bourgeoisie and feudalism, and between capitalism and feudalism ... only play a fairly weak role, since in the countryside, it is actually the *seigneurie* (centred on the demesne) which is the matrix for the capitalism of the big

[116] T. Kemp, *Economic Forces in French History* (London, 1971), pp. 66, 69–72, 74–6. See also the whole of his first four chapters; note 1, on pp. 79–80, shows that my 1966 article was one of his targets.

farmers, when it exists. This capitalism, though not without some internal disagreements, is embedded in the *seigneurie*; it pits itself against the familial, plot-holding economy of the peasant masses (which was the real obstacle to change).[117]

This observation quite clearly breaks with the usual 'catechism' regarding the *Ancien Régime*.

I had indeed neglected in my 1966 article to survey the differences between the agrarian systems of England and France or to analyse their possible consequences for the industrialisation process.[118] I would have needed, admittedly, to extend the field of analysis, probably too much so. One would have, indeed, to take account of the geographical factor, which Patrick O'Brien and D. Heath reckon to be as important as the institutional one, for England has a higher percentage of fertile soils than France and natural conditions which are more suitable for stockbreeding, which is a fact of some significance. Furthermore, if one wished to understand differences in land ownership and in patterns of tenancy, one would have to go back a long way, to the medieval period itself, and to developments taking place, as Marc Bloch had long ago pointed out, between the eleventh and the thirteenth centuries; one would have to begin by comparing the French *seigneurie* and the English manor, and even the feudal systems of the two countries. At any rate, 'the disappearance of the peasantry' in England (by the end of the seventeenth century, less than a quarter of the land was still in the hands of owner-occupiers) and its survival in France, the concentration of both ownership and cultivation of land in the one country, and their fragmentation in the other,[119] were processes which took several centuries. One must likewise take a long-term view if one is to grasp why the density of rural population was, in the seventeenth

[117] Le Roy Ladurie, in Duby and Wallon, vol. 2, p. 591; see also pp. 431–2, where he shows that the nobility's fiscal privileges favoured the progress of large-scale farming: 'fiscal iniquity and modernisation went hand in hand'. R. Forster, 'Obstacles to Agricultural Growth in Eighteenth-Century France', *American Historical Review*, LXXXV, 6, October 1970, p. 1614, reckons that the retardative effects of the seigneurial régime have been much exaggerated.

[118] I did look at this problem in another article, 'Agriculture et Révolution industrielle. Quelques réflexions', *Cahiers d'Histoire*, XII, 1–2, 1967, pp. 67–86. S. Pollard, *Peaceful Conquest*, p. 40, gives a summary of the traditional view regarding England: 'No peasants, no feudal privileges, but a class of wage-earners accustomed to working for wages.'

[119] Forster, pp. 1601–3, provides a pertinent discussion of the consequences of this fragmentation. See also C. Heywood, 'The Role of the Peasantry in French Industrialization, 1815–1880', *The Economic History Review*, XXXIV, 3, August 1981, pp. 359–60, who condemns the thesis, which dates back to Arthur Young, that small family farms act as a brake upon industrialisation. It is worth noting that in England some of the cradles of the Industrial Revolution were areas of small farms and even division of landownership.

and eighteenth century, much lower in England than in France (according to Michel Morineau, the former had not been, at the beginning of the fourteenth century, a part of the *monde plein*). This was an important advantage, which may have helped to set the Agricultural Revolution in motion or at any rate to make it possible; whereas in France strip farming and concentration on cereal production still tended to prevail, in England the amount of cultivable land available per person employed in agriculture (and, consequently, the productivity of labour) and the capital/land and capital/labour ratios were higher and therefore more favourable. The animals, being both input (as supplying draught and manure) and output, are the key element in agricultural capital, so much so, indeed, that O'Brien, Heath and Keyder have placed the beast/land ratio at the very heart of their explanation of France's agricultural backwardness with respect to England; but they note that only a study concerned with the long term would explain how English cultivators had managed to accumulate a larger stock of productive assets, and of animals in particular, per unit of cultivated land; an advantage which had been won as early as the beginning of the eighteenth century and probably long before, and which increased as time went on, guaranteeing them increasing yields.[120] The specifically English landlord–tenant relationship was also something that had developed gradually over a period of time, and there is a general consensus (at any rate, among English-speaking historians) that it favoured a relatively high rate of reinvestment of rents and of profits. Conversely, French landowners, including the greatest of them, who were drawn to Paris and Versailles, took no interest in their land and in its improvement, did not invest in it, squeezed more heavily their tenant farmers than was the case in England (and the *gros fermiers* in particular, who were the equivalents

[120] For all of this, see M. Bloch, *Les Caractères originaux de l'histoire rurale française* (Paris, 1952, 2nd edn.), p. 132; P. O'Brien, 'Agriculture and the Industrial Revolution', *The Economic History Review*, XXX, 1, February 1977, pp. 167–9; in D. Johnson *et al.*, pp. 200, 202; P. O'Brien and D. Heath, with C. Keyder, 'Agricultural Efficiency in Britain and France, 1815–1914', *Journal of European Economic History*, VI, 2, Autumn 1977, pp. 349–53, 362, 374–77, 380–1, 390; O'Brien and Keyder, 'Les Voies de passage', p. 1294; O'Brien and Keyder, *Economic Growth*, pp. 129–30, 132, 138–9. See also M. Morineau, *Les Faux-semblants*, pp. 68–9, 76, 84, and 'History and Tithes', pp. 475–6; J. Mulliez, 'Du blé, "mal nécessaire". Réflexions sur le progrès de l'agriculture, de 1750 à 1850', *Revue d'histoire moderne et contemporaine*, XXVI, 1979, pp. 3–47; the animal/land ratio was higher in some French regions than in the openfields of the Paris Basin.

of the 'farmers' on the other side of the Channel, and the sole potential agents of agricultural progress).[121]

One cannot help but wonder, none the less, whether there has been a tendency, going back as far as the travellers and agronomists of the eighteenth century, Arthur Young in particular, to exaggerate the superiority of English agriculture. Michel Morineau has observed, half in jest, that 'it was the English farmers who gave French agronomists a feeling of guilt'; reviewing the various figures quoted by Arthur Young for agricultural yields, Morineau has found that the English traveller was heavily biassed against France and that, taken all in all, the French wheat yields *c.*1789 were not so bad by comparison as Young had asserted. After having made calculations which extrapolated backwards the English data for 1837, Morineau concludes that the average wheat yield in England in 1789 was from 40 to 50 per cent higher than that of France. However, if one restricts the comparison to north west France, where soil and climate resemble those of England, the latter's advantage in yield per hectare falls to 28 per cent. Morineau holds that this figure allows one to scale down the progress in efficiency which is habitually attributed to English agriculture during the eighteenth century, at the end of which the latter's output was not managing to keep up with the country's demographic growth, as England, which had been an exporter of cereals on a large scale at mid-century, was becoming a net importer.[122]

P. O'Brien has pursued a similar line of argument, expressing some doubt 'about the *uniquely* progressive state of British agriculture from 1660 to the mid eighteenth century'; 'there appears to be nothing extraordinary about the British achievement from 1700 to 1850'. The post-1750 period saw a rapid rise in agricultural prices and a very clear and continuous shift in the intersectorial terms of trade to the detriment of industry, which means that agricultural production did not increase enough to satisfy the rising demand of the industrial and urban economy. Therefore, the performance of agriculture, whose production increased at only a third of the speed of that of industry during the eighteenth century, does not deserve the extreme praises that have been heaped upon it.[123]

[121] Forster, pp. 1605–15 (whose views are accepted by Rostow, *How It All Began*, pp. 82–5); also see above, p. 66–7.
[122] Morineau, *Les Faux-semblants*, pp. 73, 76–9, 82, 84; 'History and Tithes', pp. 474–6.
[123] O'Brien, 'Agriculture and the Industrial Revolution', pp. 173–5.

To return now to France, Michel Morineau, who can hardly be said to belong to the 'optimistic' school, has argued that in 1789 French population had not broken through the ceiling of food supplies, the asymptotic limit determined by cereal production (in other words, it had not fallen into a Malthusian trap); he does, admittedly, allow that a process of generalised immiserisation was in progress, that is, a 'contraction' of the people's standard of living, but he declares that the starving point had not been reached. Morineau speaks of 'rural French provinces that were by no means too backward in their technical development and their performance, but were overloaded by a middling peasantry which was not well-off, and by a swelling proletariat of journeymen'. A number of other passages in *Les Faux-semblants* and a subsequent article are, admittedly, more sombre, and suggest that, after an economically satisfactory first half of the century, the second half reverted to catastrophe; France, Morineau asserts, was a country which did not experience genuine growth (that is, a rise in per capita product), and where the increase in food production lagged behind the rise in population.[124]

Regardless of the precise discrepancy between the performances of British and French agriculture in the eighteenth century, a major problem is that of the effects of the indisputable differences between them upon the industrial development of the two countries, as it is often held that there exist close links between agrarian systems and 'the state of preparation for industrialisation'.[125]

Thus, Eric Jones has argued that the progress of agriculture in Southern England from the middle of the seventeenth century, and the resulting increase in cereal production, had affected farmers in regions less suited to grain cultivation, such as the West and North of the country, and even the Midlands; they responded by turning towards stockbreeding and above all by looking for non-agricultural sources of income, that is, by developing domestic industry. Jones admits that proof that such specialisation increased between 1650 and 1750 is hard to come by, but he claims that the advent of cereals surpluses stimulated rural industrialisation. The regions which benefited from these surpluses accumulated capital and business expertise, a semi-

124 Morineau, *Les Faux-Semblants*, pp. 68–9, 321, 325, 329–31, 334; 'Cambrésis et Hainaut', pp. 331–2, note 4, 342.

125 Pollard, p. 47.

proletariat formed, which gradually was 'divorced' from the land and became available for the purposes of factory industry.[126] More recently, Sidney Pollard has applied to England the 'protoindustrialisation' model, which had been meanwhile developed; he has identified a striking negative correlation between those regions which were favoured from the agricultural point of view and those in which industry developed, at first in the traditional form of rural domestic industries, organised according to the putting-out system, then in its modern form; the latter were almost all either areas of mediocre agricultural potential (the Pennines), or zones which had barely been touched by the Agricultural Revolution (the belt of clay soils in the Midlands). The Pennines and the Midlands were both regions in which cultivators obtained a poor return for their labour (the inhabitants had therefore to supplement their incomes with industrial work) and in which wages were low (which tended to attract the putters-out).[127] Another factor was the existence in neighbouring regions of a commercial agriculture, whose surpluses made it possible for protoindustrial workers to be fed. Subsequently, if one agrees with the model developed by Franklin Mendels and Charles Tilly, a dynamics internal to protoindustrialisation brought about the birth of the factory system and the gradual substitution of it for the domestic system, and therefore industrialisation *stricto sensu*.

However, Donald Coleman, who is sceptical of the value of the concept of protoindustrialisation, at least as far as Great Britain is concerned, has denied that the regions in which this phenomenon occurred were all poor, hilly, barren ones. This was certainly not true of East Anglia or of the Black Country; in the Pennines, moreover, industry was established in the valleys and not on the hills. It is possible that in certain areas the population turned towards industrial work because poor soils did not give them a sufficient income, but the rise of English rural industries began at the end of the Middle Ages, during a period when real wages were rising, and Coleman claims that

[126] E. L. Jones, 'Agricultural Origins of Industry', in his *Agriculture and the Industrial Revolution* (Oxford, 1974), pp. 130–1, 137–9: the original version of this article had been published in *Studi Storici* in 1968, and therefore predated Franklin Mendels' famous article on 'protoindustrialisation'.

[127] Pollard, p. 5. For a valuable discussion of the relationship, in the West Riding of Yorkshire, between the agrarian system and the development of industry, see Pat Hudson, *The Genesis of Industrial Capital* (Cambridge, 1986), Chapter 3 (pp. 57 ff.), 128.

the causes for the development of these industries were quite diverse and could not be reduced to the low wages factor only.[128]

Coleman's observations may just as readily be applied to France, where 'protoindustrialisation' occurred both in 'rich' provinces (Normandy, Picardy) and in very poor regions (inland Brittany, Cévennes). Furthermore, since France had a plentiful supply of agricultural journeymen and of plot-holding peasants, conditions there were at least as favourable for protoindustrialisation as in England, as far as the cost of labour was concerned. If it is true that an Industrial Revolution can spontaneously emerge from protoindustrialisation, why was it that this phenomenon did not arise in France?

There is another, older conception of the relationship between agriculture and the Industrial Revolution. To begin with, it is argued that the Agricultural Revolution was a necessary condition of the Industrial Revolution: if there had not been an increase in the productivity of agricultural labour, it would not have been possible for England to feed its industrial and urban population, whose rapid increase industrialisation was bringing about, both in absolute terms and, crucially, as a percentage of total population; indeed, without this increase in agricultural productivity, the Industrial Revolution would have aborted. Secondly, agriculture, owing to the progress it had made, is held to have made a positive contribution to the growth of industry, by providing it with labour, capital and markets. Paul Bairoch has produced the most elaborate formulation of this argument, especially with respect to markets; he has argued that it was the increased demand for textiles and for iron implements, on the part of an expanding agriculture, which had set the Industrial Revolution in motion. Since the Agricultural Revolution began in England much earlier than it did in France, it was only to be expected that the Industrial Revolution should have started up first in Great Britain.[129] Indeed, given that in the eighteenth century (and even for a part of the nineteenth century) it was not *materially* possible for a country the size of England to feed a significant part of its population with imported foodstuffs, it must be admitted that the Agricultural Revolution was a

[128] D. C. Coleman, 'Proto-Industrialization: A Concept Too Many', *The Economic History Review*, XXXVI, 3, August 1983, pp. 440–45. See also L. A. Clarkson, *Proto-Industrialization: The First Phase of Industrialization?* (Basingstoke and London, 1985).

[129] P. Bairoch, *Révolution industrielle et sous-développement* (Paris, 1963), pp. 77 ff.; Bairoch, in C. M. Cipolla, *The Fontana Economic History of Europe*, vol. 3, *The Industrial Revolution* (London, 1973), p. 470.

necessary or, rather, an enabling or 'permissive' condition for large-scale industrialisation. I would not go any further than that, however.[130]

Besides, it is reasonable to suppose that the increase in agricultural productivity helped to boost demand in eighteenth-century England (although I would not attribute so decisive a role to it as Bairoch does), whereas the low and stagnant productivity of French agriculture tended to depress the standard of living of the mainly peasant masses and, consequently, the demand for manufactured goods.[131] On the other hand, there have been lively discussions (although I will not enter into the details here) concerning the impact that the course of British agricultural prices may have had upon the demand for manufactured articles.[132] I will simply observe here, in agreement with Patrick O'Brien, that it is by no means proven that the redistribution of incomes which took place, to the detriment of wage-earners and to the benefit of tenant-farmers and landowners, as a result of the rise in agricultural prices during the second half of the eighteenth century, significantly enlarged the market for industrial products. According to O'Brien, it was in the expanding towns that industry found new outlets; moreover, conspicuous consumption by landowners was in part of imported goods. I have discussed in some detail the transfer of capital from the agricultural to the industrial sector in another chapter, where the reader will see that it was not really very substantial, at any rate directly. As for labour, there can be no doubt that agriculture made some available for the industrial and urban sector, but it should be borne in mind that the surplus rural population which left the land was by no means 'driven off' by enclosures but was mainly the product of natural increase; in industrial areas (where age at first marriage was lower than elsewhere), the latter increase was so large,

130 Morineau, 'History and Tithes', p. 447.

131 Crouzet, 'Agriculture', pp. 78–82; Coleman, *The Economy of England*, p. 200.

132 See Chapter 1, p. 32 (and Crouzet, *De La Supériorité*, p. 216), with respect to the thesis of A. H. John, according to which the fall in the prices of foodstuffs during the first half (and the second quarter, in particular) of the eighteenth century stimulated the consumption of manufactured articles and enlarged their market in an enduring way; there is basically the same view in Jones, 'Agricultural Origins', p. 138, where it is argued that, given the slow demographic growth between 1650 and 1750, demand from industrial regions kept agricultural prices up, and regions of progressive agriculture bought more manufactured articles. Conversely, J. V. Beckett, 'Regional Variations and the Agricultural Depression, 1730–50', *The Economic History Review*, XXXV, 1, February 1982, pp. 35–51, maintains that the depression hit cereal-growing regions not stockbreeding ones, and that there is no proof that it stimulated the demand for industrial products. It is hardly likely that industry would have benefited first from a fall and then from a rise in agricultural prices.

indeed, that industry to a great extent generated its own manpower. We should bear in mind, however, both the view of many English-speaking historians, who hold that in England agrarian 'institutions' did not favour as much as in France the retention of a population surplus in agriculture,[133] and the work of Jean-Pierre Poussou, who has shown how much migratory movement there was, and how mobile the work-force was, in eighteenth-century France.[134]

If it is correct to argue, as I would, that the causal relationship in England between the Agricultural Revolution and the Industrial Revolution was a weak one (for the impetus to industrial change did not come from the agricultural sector), the lack of the former in France cannot have been an important factor in the non-existence of the latter.[135]

When, in 1966, I had considered the problem from several different angles, I came to the conclusion that the most pronounced contrast between England and France consisted in the fact that the former suffered from relative shortages in vegetable fuels and labour, which did not exist in the latter country. Crafts later made the astute observation that these shortages boil down to an energy crisis: a labour shortage is tantamount to a shortage in animal power, and the steam engine was at first a way of making it easier to exploit mines, and therefore facilitated the substitution of mineral fuels for those of vegetable origin.[136]

This hypothesis had been inspired by a paper given by David Landes in 1961; in his *Unbound Prometheus*,[137] he returned to, and developed the argument that the search by the British, in the eighteenth century, for machines and other inventions, had its origins in the more and more pronounced inability of the traditional systems of production to keep pace with increasing demand: 'It was in large measure the pressure of demand on the mode of production that called

[133] O'Brien, 'Agriculture and the Industrial Revolution', pp. 175–80. See also E. L. Jones, 'Agriculture and Economic Growth in England, 1650–1815: Economic Change', in *Agriculture and the Industrial Revolution*, pp. 85–127; also the introduction to the same author's *Agriculture and Economic Growth in England, 1650–1815* (London, 1967).

[134] J.-P. Poussou, *Bordeaux et le Sud-Ouest au XVIIIe siècle. Croissance économique et attraction urbaine* (Paris, 1983).

[135] Morineau, *Les Faux-semblants*, p. 95, appears sceptical about such a connection, even in 'agronomically' progressive countries.

[136] Crafts, in Floud and McCloskey, vol. 1, pp. 11–12.

[137] The first version of which appeared in volume VI of the *Cambridge Economic History of Europe*, which was published in 1965, when my article was at the printers.

forth the new techniques in Britain.'[138] Crafts has observed that there are only two purely economic hypotheses that can account for the innovations of the eighteenth century, namely, the one which invokes relative shortages and the energy crisis (where it is argued that the rise in production and in the stock of capital tended to overtake that of the supply of labour and of natural resources; changes occurring as a consequence in relative prices would then have stimulated the search for innovations) and the one which invokes an increase in demand, across a broad front and very rapidly, which would also have put pressure upon the factors of production and provoked 'a vigorous innovative response'.[139] The mechanism involved would in part have been the same in either case, but I hope that I have demonstrated that demand did not rise much faster in England than in France. There is some broad consensus regarding the explanation of the technical lag between England and France, in terms of 'bottlenecks' and obstacles on the supply side. 'It is a commonplace', writes Peter Mathias, 'to remark that Britain had been fortunate in her scarcities since the sixteenth century' and, in John Harris's view, she had the right bottlenecks.[140]

This hypothesis seems to me to be still valid today, but it needs to be qualified, if it is to match the chronology of inventions and innovations, and also the fluctuations in demand (and we must begin by making sure that these 'bottlenecks' really did exist!).

If one takes the long-term view, it is beyond dispute that technical progress in the British iron industry arose out of inadequate wood resources, while demand was growing, and of the fact that the iron industry could not develop without resorting to coal. If, however, one adopts a short-term view, a number of difficulties arise. There is no reason whatsoever to think that Abraham Darby's invention was connected to any worsening in the wood 'shortage', which had existed for the best part of a century; in addition, there is something perplexing about the half-a-century (or thereabouts) timelag between the invention of the coke-smelting process and its diffusion. The latter phenomenon was, admittedly, conditioned by economic factors, by changes in the relative costs of using charcoal and coke in blastfurnaces, and

138 Landes, *The Unbound Prometheus*, pp. 65, 77, 112, also 43, 84–9, 166, 194.
139 Crafts, 'Industrial Revolution in England and France', p. 435; Crafts in Floud and McCloskey, vol. 1, p. 11.
140 Mathias, 'British Industrialisation', in Léon *et al.*, *L'Industrialisation*, p. 508; Harris, *Industry and Technology*, p. 3.

therefore in the 'availability' or 'shortage' of these two fuels. Nevertheless, the link between bottleneck and invention (or even innovation) only seems to be close in the case of a few large ironworks with coke-fired blastfurnaces, which were set up to supply the Army and Navy during the Seven Years War.[141]

Likewise, no one can dispute that there is some causal connection between the invention of the steam engine and mining of tin and copper in Devon and Cornwall, but of coal also. But no one has managed to demonstrate that Savery and Newcomen built their engines as a direct response to increasing difficulties encountered in pumping out seepage water from pits. As for Watt's machine, it was the work of a genius, and one can never give a strictly economic explanation for the timing of inventions of that order. It is worth noting, nevertheless, that, prior to his association with Boulton, James Watt was backed financially by Dr John Roebuck, who was involved in the development of coal-mines in the vicinity of Edinburgh.

These examples lend credence to Crafts' thesis that economic factors were more important for the spread of inventions than for the very fact of inventing, that it is difficult to establish a correlation between these factors and inventions, and that there was a random element in the dating of specific inventions.[142]

The notion that a relative shortage of labour in England was responsible for the invention of textile machinery has been rejected by several historians. Ralph Davis, for example, had argued that the main region where the cotton industry flourished in France was Normandy, where several industries were competing for labour, which had therefore become rare.[143] It is certainly true that one cannot speak of a shortage of labour at the national level in eighteenth-century England, where many were underemployed, even in years of prosperity. Moreover, Wrigley and Schofield's index of real wages, which had risen from the 1620s up until the early 1740s, displays a fairly flat

[141] See n. 29 and 67 of Chapter 1, for references to C. K. Hyde's works. As for puddling, it was only the last and most satisfactory of several processes for transforming pig-iron into iron.

[142] Crafts, in Floud and McCloskey, p. 12.

[143] Davis, *The Rise of the Atlantic Economies*, p. 312. See also Crafts, 'Industrial Revolution in England and France', p. 435; Rostow, *How It All Began*, pp. 181–3, is sceptical on this point, claiming that nothing suggests that the plentiful supply of labour in France rendered unprofitable the introduction of new textile machines; this is clearly correct, given the huge increases in productivity that these machines brought about, but my argument was concerned with the invention and not the diffusion of these machines. For the same reason, I would contest his argument that mechanisation resulted in a large increase in employment in the cotton industry and that the necessary manpower was none the less found.

tendency during the middle decades of the eighteenth century; then, after a brief rise and a peak in 1781, the index falls for a whole generation. These fluctuations do not lend credence to the idea of a shortage of labour, for such a shortage would have brought about a rise in wages. This index is, admittedly, a national one, heavily weighted in favour of wages in Southern England and therefore perhaps unduly pessimistic.[144]

Nevertheless, it is difficult to deny the existence of local labour shortages, of pockets of 'overemployment', in particular in the North of England, where money wages increased rapidly between 1750 and the late 1760s, and in an activity as labour intensive as cotton-spinning. In this respect, the boom in cotton exports from 1750 to 1763 may have fostered the quest for cotton-spinning machines. Moreover, these same years saw migrations of manpower towards areas which were poor from the agricultural point of view but where industrial activity was increasing, such as Lancashire, the West Riding and the hosiery districts in the East Midlands; still merchant-manufacturers had to put out spinning outside their own counties.[145] As far as cotton printing is concerned, S. D. Chapman and S. Chassagne have found that France and Germany were suffering from the same shortage of highly-skilled workers as Britain, but that the latter was hit by a sharp rise in labour costs, because these workers organised themselves in clandestine unions which imposed 'restrictive practices', and to such effect that their employers lost no time in adopting the newly invented cylinder printing.[146] Finally, it is worth quoting François Caron's observation that 'the whole of French growth in the eighteenth century gives the impression of being based upon an enormous accumulation of human labour, precisely to the extent that it was readily available everywhere. This is probably the main difference between France and England.'[147]

A new variant on the explanation in terms of labour costs has

[144] Wrigley and Schofield, pp. 432–3; these authors regard the end of the eighteenth century as a 'malign' period, when England was threatened by a Malthusian trap (pp. 408–12, 439–42). Their wages index has been criticised by Flinn (see above note 84).

[145] Pollard, *Peaceful Conquest*, p. 6. E. E. Hunt, 'Industrialization and Regional Inequality: Wages in Britain, 1760–1914', *Journal of Economic History*, XLVI, 4, Dec. 1986, pp. 937–8, 941, 947, stresses the wages rise in Northern England from 1760 to 1800, which changed the wages geography of Britain.

[146] S. D. Chapman and S. Chassagne, *European Textile Printers in the Eighteenth Century. A Study of Peel and Oberkampf* (London, 1981), pp. 192–3, 198–9.

[147] Caron, *Histoire économique*, p. 13.

recently been advanced, which reminds us of the crucial but often forgotten fact that the cotton industry originated in the East, that it developed in Europe as a way of providing substitutes for imports, by producing ersatz East India goods, and that, given the skill of Indian spinners and weavers and the much lower wages paid there than in Europe, this development could not have been a spontaneous one.

The traditional textile industries had been threatened at the end of the seventeenth century, both in England and in France, by the flooding of the market with East India goods, which were the height of fashion; the governments of both countries responded in largely the same way, by banning the importation of East Indian goods and also the manufacture of cotton textiles. However W. W. Rostow has observed that this policy was taken much further in France, where not only pure cotton textiles but also mixed ones, such as linen and cotton, were banned. In England, on the other hand, the manufacture and printing of fustians and other mixed articles were tolerated, so that a 'cotton industry' did in fact develop and was of some importance by the middle of the eighteenth century; English yarns were still, however, inferior to Indian ones, and this led to the invention of machines which could produce good quality yarn, especially for warp. In France, the prohibition of cotton goods was lifted in 1759, but it was too late to catch up with the English.[148] K. N. Chaudhuri has also emphasised that the Europeans could not compete with the Indians in both the quality and the price of handspun yarns, and he has suggested that the competition of authentic East Indian goods might have been 'a contributory factor' to the setting in motion of the Industrial Revolution, for, without machines to achieve a cut in costs, it would have been impossible to counter the comparative advantages enjoyed by India.[149]

This suggestion was welcomed and elaborated by Fernand Braudel.[150] Moreover, Chapman and Chassagne have observed that

148 Rostow, *How It all Began*, pp. viii, 63–4, 66, 130–1. His observation on France is open to dispute; anyhow numerous cotton-printing works were established just outside the French borders and were able to set up in business within the kingdom itself after 1759. See note 43 of Chapter 1.

149 K. N. Chaudhuri, *The Trading World of Asia and the English East India Company 1660–1760* (Cambridge, 1978), pp. 237–8, 273; 'Some Reflections on the World Trade of the XVIIth and XVIIIth Century: A Reply', *Journal of European Economic History*, VII, 1, Spring 1978, pp. 227–8.

150 F. Braudel, *Civilisation matérielle, Economie et Capitalisme, XVe–XVIIIe siècle*, vol. 3, *Le Temps du Monde* (Paris, 1979), pp. 490, 495.

the Peels, by contrast with Continental cotton printers, such as Oberkampf, ventured into cotton-spinning quite early on, because they wanted to lower the prime cost of the grey cloth they were using, in order to compete with India. Since the English had the closest relations with India, they were by the same token most vulnerable to its competition, and they therefore had the strongest incentive to seek out alternatives to Indian artisanal techniques (not only in spinning but also in printing, in which field they perfected copper plate, and then cylinder processes).[151]

The merit of the above explanations is that they refuse too insular an approach and interpret the progress and mutation of the English cotton industry in terms of the international context. One must, however, bear in mind that cotton goods coming from India were banned on the English domestic market; some were of course smuggled in, but such luxury articles were not meant for the same market as fustians and other British 'cotton goods', which were in effect well protected. It is true that Indian competition was not held back by such obstacles on third markets, including those of the British colonies, where East India goods could be re-exported. Yet this rivalry had not prevented a rapid rise of exports of English 'cotton goods' during the two decades prior to the invention of spinning machines. Furthermore, textile imports to London by the East India Company were both large and steady, in spite of some cyclical fluctuations, from around 1725 up until the 1740s, then they tailed off as a result of the wars in India and in the West; as to their prices, they rose on trend. These data do not suggest fierce competition between East India and English cotton goods.[152]

There is some risk of historians' discussions going around in circles, always reprocessing the same information, until new research into primary sources brings out fresh raw materials. We therefore owe a great deal to John Harris's studies, based upon archive sources, on the problem of the transfer of technology between England and France in the eighteenth century; he has thus produced many precise data

151 Chapman and Chassagne, pp. 13, 18–19, 198, 214.

152 Chaudhuri, *The Trading World of Asia*, pp. 105–8, figs. 15–20, 272, 284, 286, 293, figs. 21–6, 297–302, 547–8, Table C.24; 'The Economic and Monetary Problems of European Trade with Asia during the Seventeenth and Eighteenth Centuries', *Journal of European Economic History*, IV, 2, Autumn 1975, pp. 352, 354–7.

From 1750 to 1763, there was an increase in the share of British-produced cotton goods which were exported; the growth of exports then slowed down, but production accelerated, thanks to the domestic market; see Chapter 6 below, p. 253.

documenting the former's technological lead over the latter, the modalities of its 'divergence' and, as a consequence, the origins of the Industrial Revolution.[153] John Harris has revived, but refined, elaborated and corrected J. U. Nef's hypothesis regarding the importance of the early development of the coal industry in Britain, and of the gradual spread, beginning in the sixteenth century, of the use of coal in an increasing number of industries.[154] This diffusion had significant, though subtle effects, for it demanded technical innovation: one cannot introduce coal into an industry without at the same time effecting some change or other in the equipment or methods that one is 'importing' from another branch. This is how what Harris calls coal–fuel technology was developed, a technology which comprised a tried-and-tested ensemble of know-how and a whole panoply of plant and tools; this ensemble would be gradually perfected, through a series of 'spin-offs' as some elements of technology were transferred from one industry to another. This was how the art of building furnaces was improved and reverberatory furnaces in particular, which were used in glassworks from 1612 on, adapted to the smelting of non-ferrous metals in the 1680s, and judged by foreign visitors in the eighteenth century to be one of the distinguishing marks of British technology; the same was true of the boiler-maker's art, of the techniques involved in the making of crucibles in refractory clay capable of standing up to high temperatures (a very important feature in a number of industries), and also, at

[153] J. R. Harris, *Industry and Technology*; 'Skills, Coal and British Industry in the Eighteenth Century', *History*, LXI, 202, June 1976, pp. 167–82; 'Saint-Gobain and Ravenhead', in B. M. Ratcliffe (ed.), *Great Britain and her World, 1750–1914. Essays in honour of W. O. Henderson* (Manchester, 1975); 'Attempts to transfer English Steel Techniques to France in the Eighteenth Century', in S. Marriner (ed.), *Business and Businessmen, Studies in Business, Economic and Accounting Policy* (Liverpool, 1978), pp. 199–233; 'Michael Alcock and the Transfer of Birmingham Technology to France before the Revolution', *Journal of European Economic History*, XV, 1, Spring 1976, pp. 7–57; 'The Diffusion of English Metallurgical Methods in Eighteenth-Century France', *French History*, II, 1, March 1988, pp. 22–44; 'The Transfer of Technology between Britain and France and the French Revolution', in C. Crossley and I. Small, eds., *The French Revolution and British Culture* (London, 1989), pp. 156–86. I would also emphasise the interest of Chapman and Chassagne's recent study, *European Textile Printers*; one of its virtues is to bring down comparison from the macro- to the microeconomic level; the authors thus make a clean sweep of some accepted ideas, and while they bring out many resemblances between the two undertakings, they also demonstrate that productivity was far higher at Peel's than at Oberkampf's (pp. 211–14).

[154] Nef had made the mistake of seeing coal as the agent of a so-called 'first industrial revolution' (see Chapter 1, p. 12–13), whereas it was used mainly for domestic consumption, and also of restricting himself to the period 1540–1640, while important innovations in the use of coal were achieved during the half-century after 1660.

a more humble level, the practices involved in the cutting of coal or in the stoking of furnaces.

In all of these different domains and, just as crucially, in metal-working, England made great progress in the course of the seventeenth century; indeed, by 1700–10, she was assured of a lead that could hardly be challenged. Yet the process was to go on and it was in the 1780s that the last significant lacunae in coal–fuel technology were filled, with puddling, rolling of iron and the rotative steam engine.[155]

Harris reckons that the introduction of coal–fuel technology played a key part in setting England on the path to the Industrial Revolution. In particular, it gave a powerful stimulus to the inventiveness of the English, inasmuch as it familiarised employers, managers, foremen and skilled workers with technological change, it showed them the advantages that accrued from observation and experimentation, and it gave them the aim of cutting costs; it also created a 'technological optimism' and brought innovation even to the attention of the ordinary worker. Indeed, the skills needed to operate this technological system were perfected in a purely empirical manner, by practical men rather than by scientists, and at shop floor level; it was a question of craft skills, which were the prerogative of skilled workers and could only be 'learned by doing', through a long apprenticeship on the job.[156]

This technology was of course peculiar to England, for the other countries of Europe, and France in particular, only used coal on a much smaller scale and continued to employ wood or charcoal as the main fuel in industry. For about two centuries, England pursued a lone path. Harris emphasises how long term the process of developing coal–fuel technology was, but he also makes it clear that these refinements were interconnected and achieved on a broad front, with a mutual cross-fertilisation of techniques in various industries and with the transfer from one to the other of a whole set of processes and equipment which were all bound up with the use of coal.

Harris has also shown just why it was that France had such difficulties, right up until the beginning of the nineteenth century, in adopting coal–fuel technology. His argument is that France was trying to compress into a short period of time the extended apprenticeship undergone by the British, that the technology involved formed an interconnected whole, which could not be imported bit by bit, and that

[155] See, in particular, Harris, *Industry and Technology*, pp. 3–9.
[156] Harris, 'Skills, Coal', p. 168.

only men who were used to it were capable of introducing it abroad (they only arrived in sufficient numbers after 1815).[157]

Given these conditions, it is tempting to suppose that the early emergence and development in England of technologies for the use of coal and/or metal-working, and their concentration in this country, at a time when they were ignored in France (for the simple reason that England had a large number of rich, easily workable coalfields, well served by waterways, whereas France was much poorer in such resources), may to a great extent account for England's 'divergence' and for its being 'first'. This would bring us back to a 'geological' explanation, of the kind so scorned by social scientists.[158]

Admittedly, Peter Mathias has levelled an objection at Harris's hypothesis. He is prepared to admit that the 'divergence' established by the progress of coal–fuel technology had its roots in the seventeenth century and that a 'strategic' sector was involved, of great importance for the future; Mathias emphasises, however, that this sector continued to be a small one for a long time, and that it was in no position to have an effect upon growth rates and industrial performance at the national level at any point prior to 1780; after that date, however, this technology made possible a whole range of advances, which gathered momentum as it progressed. Harris responded that a technological gap ought not to be measured in terms of growth rates, and that 75 per cent of the gap separating France from England in 1790 had already been hollowed out by 1750.[159] One can also observe that Harris has neglected the textile industries, and cotton in particular; but he has suggested that the coal and metal-working technologies, which had been set up before the invention of textile machinery, had made it possible for the latter to be developed and perfected. Harris also states that he had not meant to advance a monocausal explanation, that he

[157] Harris, *Industry and Technology*, pp. 9–17; 'Attempts to transfer', p. 277, and also 199, 218, 228. This latter article gives a detailed account of the failures of the numerous attempts made in France in the course of the eighteenth century to produce genuine steel, in spite of support by government and of the employment of first class scientists; one humble, but in fact crucial cause of this failure was an incapacity to master the technology of refractory clays. Harris's article on Saint-Gobain and Ravenhead also shows the difficulties a 'high tech' French firm encountered when it tried to use coal (pp. 384 ff.).

[158] According to Pollard, *Peaceful Conquest*, pp. 4, 21, there has been too much underestimation of the part played by natural resources, and by coal in particular.

[159] Mathias, 'La Révolution industrielle', pp. 37, 43; see also his intervention, and that of J. R. Harris, in Léon and others, *L'Industrialisation*, pp. 505, 511.

was concerned with just *one* factor and that this did not rule out those mentioned by other historians.[160]

Be this as it may, Mathias agrees with Harris that, from the beginning of the eighteenth century, the English enjoyed a clear superiority in several technological domains: mining, the use of coal as a fuel and metal-working in particular. Those who specialised in these craft techniques were relatively few in number – especially for the more sophisticated among them – but groups of them were to be found in large towns and in the various industrial mining districts. Since these specifically British techniques were attuned to England's endowment in natural resources (coal and metallic ores in particular), there were kinds of technological 'pyramids', combining raw materials, equipment, processes and know-how.[161] Thus, one must stress not only geology but also human capital – the workmen and craftsmen skilled in coal and metal-working techniques – and also the pool of entrepreneurs which England could call upon in the middle of the eighteenth century, in relatively greater numbers, it would seem, than was the case in France. There were both big and small businessmen, who were alert to any opportunity for making money, skilled in organising production in response to market signals, and also endowed with confidence, energy, and, indeed, with aggressiveness. The combination of this human capital and of the new technologies was bound to be an explosive one, but these technologies were themselves the creation of the workers' skill and the businessmen's astuteness.

One can argue that the Industrial Revolution arose out of the confluence of two different currents of technical progress. The first current was that of coal–fuel technology, which involved the use of coal by a widening range of different industries, together with associated technical improvements; this current was an old one, continuous but slow, and it did not involve many opportunities for rapid increases in productivity. The second current was that of textile machinery, which emerged quite suddenly in the form of the spinning jenny and the water-frame (although they had been anticipated by the knitting machine, the Dutch loom for ribbon-making, the silk-throwing mills,

160 Harris, 'Skills, Coal', p. 181; *Industry and Technology*, p. 9.

161 Mathias, in Léon and others, pp. 504–6; 'La Révolution industrielle', pp. 41–4; 'Skills and the Diffusion of Innovations from Britain in the Eighteenth Century', an article first published in 1975, and republished in *The Transformation of England*, pp. 33–6; Hartwell, *The Industrial Revolution and Economic Growth*, p. 295, observes that the important thing was not science, where France was ahead of England, but skills, where the latter was superior.

the flying shuttle and the Lewis and Paul machine); these machines brought about huge increases in productivity and that of Arkwright imposed a new form of organisation of work – the factory system. This breakthrough had revolutionary consequences, bringing about the rapid creation of a new and substantial industry, namely, the machine-spinning of cotton, which also served as a model for the other textile industries. The Industrial Revolution was truly on its way when these two currents converged, in the 1780s and 1790s, when, for instance, Watt's engine was used to power textile machines, which were themselves constructed of cast and wrought iron, and not anymore of wood (as they had been before), and when the engineering industry was born.

A full two decades after I had drafted my original article, research into the origins of the Industrial Revolution still seems somehow akin to the quest for the Holy Grail, to borrow a phrase from Peter Mathias. Perhaps Crafts was right to have argued that a comparison of the British and French economies in the eighteenth century gave no chance of arriving at a valid conclusion on this problem of origins. Furthermore, historians are still divided as regards the performances of the two countries, whether they are considered separately or compared one with the other.

There is more dispute over these matters in the case of France, for, where England is concerned, the debate tends only to be between the 'insularists', who insist upon the uniqueness of English economy and society, and those who admit that, in many respects, there was nothing particularly exceptional about them. At the beginning of this chapter we saw that Ralph Davis had judged my views on the economic position of eighteenth-century France to have been too optimistic; Tom Kemp's sombre perspective on these matters was also delineated. As for Michel Morineau, he holds an extremely pessimistic view, reckoning that France at the end of the *Ancien Régime* was moving backward, unlike the rest of Europe, that agricultural production was declining, as were the traditional industries, so severely hit by the treaty of 1786; as population kept on rising, France was caught in a trap which led inexorably to her impoverishment, to a 'generalised immiserisation'. In Morineau's view, France's technical backwardness placed her in a situation of inferiority; she was 'powerless, in the short term, to come to terms with the new needs of new inhabitants ... unprepared for the changes that the future would bring'.[162]

Conversely, some writers have gone much further than I had done in

[162] Morineau, *Les Faux-semblants*, pp. 331, 334; and see above p. 86.

'rehabilitating' the French economy of the eighteenth century. This is the case, as I have already observed, with Crafts, and the same is true of Richard Roehl. The latter has used the criteria of backwardness formulated by Alexander Gerschenkron, and he has claimed that in no sense can they be said to apply to eighteenth-century France. Indeed, it is quite the reverse and Roehl reckons that France set out very early on the path to modern industrialisation, around the middle of the eighteenth century, that is a little before England; at any rate, he believes that the industrialisation processes of the two countries had many aspects in common.[163] This may seem to be an extreme position, and yet J. Marczewski had suggested as early as 1960 that if there had been a take-off in France, it could be placed around 1750, that is, before the British one. T. J. Markovitch has, more recently, written in a similar vein: 'France was, alongside England, the main pioneer of the modern Industrial Revolution.'[164] And Pierre Léon considered that France at the end of the eighteenth century was an 'industrial nation ... in the grip of powerful forces for renewal, undergoing a complete transformation internally'.[165]

Other historians hold a median position. This is the case with Denis Richet who has, on the one hand, identified a Malthusian trap, with population continuing to expand right up until the French Revolution, while agricultural production had no elasticity; but, on the other hand, he has judged that industrial and commercial activities were making a much more substantial contribution than in previous periods (one of the enduring achievements of the Enlightenment).[166] Clive Trebilcock wonders first of all if 'the promising French developments of the eighteenth century' might not have been merely 'surgical improvements to an arthritic frame', which in the end rejected them, and he reminds us that the *Ancien Régime*'s attempts at economic modernisation helped to unleash the Revolution, which was itself catastrophic for the economy; but he concludes that 'clearly, the late French edition of the ancien régime was not in itself incompatible with economic progress'.[167] These testimonies may be rounded off with a quotation

[163] Roehl, pp. 238, 255, 268; Asselain describes Roehl's ideas as paradoxical, in *Histoire économique de la France*, p. 15.

[164] J. Marczewski, in W. W. Rostow (ed.), *The Economics of Take-Off into Sustained Growth* (London and New York, 1963), p. 129; T. J. Markovitch, 'La croissance industrielle sous l'Ancien Régime', *Annales ESC*, XXXI, 3, May–June 1976, p. 647; 'La Révolution industrielle: le cas de la France', *Revue d'histoire économique et sociale*, LII, 1974, no. 1, p. 123.

[165] Léon, in *HESF*, vol. II, p. 422.

[166] Richet, pp. 786–7.

[167] Trebilcock, pp. 115–18, 125.

from a young scholar: 'to believe that traditional capitalism was lethargic, one would need to be a Jacobin or a Colbertist'.[168]

It would clearly be prudent not to push revisionism too far, for it is itself liable to be revised in turn. Charles Kindleberger has reminded us that eye-witness accounts of contemporaries, together with a number of simple tests, do not bear out the notion that the per capita income of the French was higher than that of the British, on the eve of the Revolution. Kindleberger stresses, moreover, that France's financial institutions were a century behind those of Great Britain: the hundred and six years gap between the founding of the Bank of England and that of the Bank of France is not just a symbolic one.[169] If we adopt a different point of view, and consider demography, this too is not in France's favour at the end of the *Ancien Régime*: mortality was stable, while it had been falling in England since 1731 (and life expectancy had been rising there), nuptiality and fertility of marriages were falling, and the age of first marriage rising, whereas it had been falling in England. Wrigley and Schofield record that, at the mid eighteenth century, population growth rates were the same on both sides of the Channel but, after 1770, the English rate increased 'powerfully' and overtook that of France, where demographic growth was clearly weakening when her rival was embarking upon an impressive expansion.[170] Nor should we forget that there was a 'crisis of the French economy at the end of the *Ancien Régime*'.[171]

In the last resort, one can readily understand why it was that the Industrial Revolution did not begin in France, for it is hard to discern any force there which was tending to effect a thoroughgoing transformation of economic structures. Yet it remains somewhat difficult to see what set it going in England; we lack, in particular, an adequate explanation of the timing of its first, tentative steps, that is the causes behind the constellation of inventions and innovations which cluster around 1770 and, more precisely, the causes behind the invention of cotton-spinning machinery, since it is there that, in the last analysis, the Gordian knot lies.

[168] F.-X. Caron, *Quarante négociants anoblis au XVIIIe siècle. Etat, Elites et Question Sociale au siècle des Lumières* (*3e cycle* thesis, University of Paris–Sorbonne, 1984), p. 618.

[169] C. P. Kindleberger, 'Financial Institutions and Economic Development: A Comparison of Great Britain and France in the Eighteenth and Nineteenth Centuries', *Explorations in Economic History*, XXI, 2, April 1984, pp. 104–6, 109–10, 120.

[170] Chaunu, *Lumières*, pp. 138–9, 155–6; Dupâquier, *La Population*, pp. 105, 107; Wrigley and Schofield, *Population History of England*, pp. 215, 236, 262. The mortality rate was lower in England than in France. Weir, pp. 33–5, takes a more optimistic view.

[171] This was the title of Ernest Labrousse's classical book.

It is admittedly true that, by the middle of the eighteenth century, England had an 'advanced' economy, which was making progress on a broad front, though slowly and without any Draconian change. It is likewise true that the social, political and intellectual environment favoured innovation and, besides, according to the hypothesis which I advanced in 1966 and which I would still, in all essentials, wish to defend, progress was checked by bottlenecks, in the cotton industry in particular. Now, Hargreaves and Arkwright's inventions were realised only a short time after a period of very rapid expansion in the export of cotton goods; in more general terms, an acceleration in the growth of England's total exports (which took place between 1744 and 1760) preceded the innovatory 'constellation' observable around 1770. One can perhaps follow Rostow in identifying 'some oblique linkage' of foreign trade and technical progress,[172] but one cannot be certain.[173] The explanation of the changes that took place around 1770 cannot help but be a tentative one; indeed, this is so much the case that it is tempting to subscribe to Crafts' view that innovation had for a part a random character. After all, Hargreaves' and Arkwright's inventions could have been fortuitous.

Conversely, once the first technological breakthrough was achieved, everything is as clear as day. The success of Richard Arkwright and his associates, who had been the pioneers of the factory system in the 1770s, demonstrated the remarkable opportunities for profit-making that the new technology offered,[174] and many followed in their tracks,[175] especially when the depression, political in origin, which had prevailed during the 1770s, had come to an end and when demand recovered dramatically at the end of the American War of Independence, which reopened large markets to British industry (France did not benefit from the peace of 1783 to the same degree and the treaty of 1786 was very shortly to shake a number of industries there). At this

[172] Rostow, *How It All Began*, p. 130. See also below, Chapter 6, pp. 253, 260–2.

[173] See the rather negative conclusions of C. P. Kindleberger, 'Commercial Expansion and the Industrial Revolution', *Journal of European Economic History*, IV, 3, Winter 1975, pp. 623–32. One can also observe that the diffusion of these innovations began during a period in which exports fell, at the end of the 1760s and in the 1770s: it would have been stimulated by such difficulties. But you cannot have your cake and eat it too (as is the case also with the argument regarding the effect of agricultural prices, see note 132, above).

[174] Chapman and Chassagne, p. 215, have also pointed out that the success of a few large calico-printing works was the occasion and the stimulus for the transition from the proto-factory to the factory proper.

[175] We should not forget the annulment, in 1785, of Arkwright's patents, though, admittedly, they had not stopped the pirates.

same moment, the growth in population was accelerating, and the convergence (as noted above) was effected between the new current of technical progress in textiles (which, beginning with cotton, soon had imitative effects in other textile industries) and the traditional current of coal, metal-working and machinery technologies. One therefore moved from a breakthrough in one relatively narrow sector to an advance on a much broader front and, consequently, to an acceleration of global growth;[176] the cumulative and self-sustaining process of innovations, which can be compared to a chain reaction, was unleashed. The Industrial Revolution was both or, rather, in succession, unbalanced and balanced growth. It has been compared to a two stages rocket, but the curious thing is that the first was the smaller one and its firing is hard to explain; on the other hand, the detonator of the 'explosion' of the 1780s was undoubtedly the peace of Versailles. France had not undergone the first stage of change; it was therefore only natural that she should have experienced some difficulty in her attempts to imitate England's second innovatory push, a difficulty that was exacerbated by the troubled condition of her economy in the 1780s.

[176] Even if this acceleration was less pronounced than one had thought, as Harley and Crafts' calculations suggest (see above, p. 51).

The Persistence of Open-Field Farming in Nineteenth-Century France

GEORGE W. GRANTHAM

The enclosure of the open fields is an example of Europeans' willingness to alter long-standing social and economic institutions in the interest of higher living standards. In Scandinavia, England, and Germany the rise in the value of enclosed relative to unenclosed land induced widespread abandonment of open-field forms of agrarian organization by the middle of the nineteenth century. In France, on the other hand, the traditional patterns of landholding maintained themselves until after the First World War. This paper examines some of the ways French farmers responded to the possibilities of agricultural change within the traditional framework of open-field agriculture.

McCLOSKEY'S studies of the enclosure movement in England provide an exhaustive and careful analysis of the factors determining the rate of enclosure.[1] Enclosure raised agricultural productivity by limiting free access to land, by reducing the costs of making decisions, and by lowering the cost of undertaking more intensive and flexible crop rotations. The disadvantage of compact farms was mainly that the risks of variation in crop yields were higher than for divided and scattered holdings. Rising demand for food in the eighteenth century greatly increased the value of technical improvements, while the increased use of flexible rental contracts absorbed some of the tenants' risk. The demand for compact holdings, however, could not be met through normal processes of purchase and sale. Exchange of smaller parcels was impeded by the burden of fixed transactions costs, while the exchange and consolidation of larger ones was often blocked by the costs of strategic bargaining for the fruits of enclosure among a small number of landowners. In these circumstances the land market could not fully accommodate the growing demand for more compact farms.

The simultaneous extinction of old property rights and the consolidation of holdings by general survey and enclosure cut through these private costs of exchange, and it is clear that the recourse to expropriatory

The Journal of Economic History, Vol XL, No. 3 (Sept. 1980). ISSN 0022-0507.

George W. Grantham is Associate Professor in the Department of Economics, McGill University, 855 Sherbrooke Street West, Montreal PQ, H3A 2T7 Canada. Research for his article was made possible by a generous grant from the Canada Council. He has benefitted greatly from the comments of Jon Cohen, Eric Jones, Ian McClean, Donald McCloskey, William Parker, Richard Roehl, and the referees and editors of this JOURNAL.

[1] Donald N. McCloskey, "The Economics of Enclosure: A Market Analysis," in William N. Parker and Eric L. Jones, eds., *European Peasants and Their Markets: Essays in Agrarian Economic History* (Princeton, 1975), pp. 123–60; and McCloskey, "English Open Fields as Behavior towards Risk," *Research in Economic History*, vol. 1 (1976), pp. 124–70.

procedures in the late eighteenth century was a result of their cost-effectiveness as compared with market exchange. The main problems of general enclosure were how to prevent a minority of proprietors from blocking change, and how to secure reasonably adequate compensation for extinguished property rights. Each of these issues required expropriatory solutions, which in turn required delegation of governmental authority to alter property rights.

Table 1 lists some of the major legislation concerning enclosure in Europe since the eighteenth century. Rising demand for food in the eighteenth century stimulated legislation to extinguish common rights, and in Scandinavia and England gave impetus to full enclosure. In Germany and France there was a flurry of activity towards the end of the eighteenth century followed by a period of quiet, which was ushered in partly by the restoration of conservative governments and partly by the relative abundance of food on European markets. This pattern changed after 1850, when most German states adopted general enclosure laws while France retained the unanimity rule for general consolidation of holdings.

To what extent can this difference be attributed to the divergence in the industrial and demographic histories of France and Germany in the late nineteenth century? Ruttan suggests that the slow rate of structural transformation in French agriculture after 1850 can plausibly be linked to the relatively weak pressure of demand on French agricultural resources.[2] Yet the land-man ratios in the two countries and the relative scarcities of land and labor were not far apart in 1880: the average amount of land per agricultural worker in France was 6.96 hectares as compared with 6.34 in Germany, although the price of land relative to the daily wage was 20 percent lower in France than in Germany. More detailed figures from around 1900 suggest that there was no significant difference in rural population densities in open-field Lorraine and the adjacent states in Germany, despite sharp contrasts in the extent of enclosures.[3] Thus, although land was somewhat more scarce in Germany than in France, the differences do not seem great enough to explain the almost total absence of enclosure activity in France before the First World War.

I

Aggregate ratios are crude instruments for detecting specific economic incentives. Any assessment of institutional failure in France must be based on direct estimates of the costs and benefits from enclosure. Such evaluations are extremely difficult to make for France because of the pau-

[2] Vernon Ruttan, "Structural Retardation and the Modernization of French Agriculture: A Skeptical View," this JOURNAL, 38 (Sept. 1978), 714–28.

[3] Hans P. Binswanger, Vernon Ruttan, et al., *Induced Innovation: Technology, Institutions, and Development* (Baltimore, 1978), p. 48; and Folke Dovring, *Land and Labor in Europe, 1900–1950* (The Hague, 1956), pp. 69–70.

TABLE 1
ENCLOSURE LEGISLATION

Denmark	1769	Enclosure by unanimous consent of landowners.
	1781	Individual landowners empowered to demand an enclosure. Agricultural board empowered to take decisions in absence of unanimity.
Sweden	1749	Enclosure by unanimous consent, carried out by government officials.
	1758	Individual proprietors given right to separate their own lands and consolidate them.
	1783	Any landowner can demand a general enclosure.
	1803	All land in Skane required by law to be enclosed.
England	18th century	Enclosure by private act. No standard procedures for initiating enclosure, but ownership of substantial part of land usually required.
	1801	General enclosure act, simplifying procedures. A majority of owners can petition for enclosure.
	1845	Petition needs only owners of one-third of assessed valuation.
Germany	1718	Duchy of Lauenburg: Seigneurs can evict tenants with compensation in order to consolidate holdings.
Prussia	1751	Legislation promoting enclosure on Crown lands.
	1811	Up to one-half of plowland could be removed from common pasture.
	1821	25 percent of landholders holding one-third of land can petition to separate their land from the commons.
Saxony	1834	Owners of three-fourths of land can determine an enclosure.
Saxony	1843	Two-thirds of owners can determine enclosure.
Hanover	1842	Two-thirds of the area and assessed value determine.
Hanover	1856	One-half of the area and assessed value determine.
Prussia	1850	Majority of landowners determine.
Baden	1856	Two-thirds of land and two-thirds of landowners determine.
Hesse	1861	One-half of the area determines.
France	1791	Individuals have right to enclose their own land.
	1865	Unanimous consent of landowners associated in a *syndicat libre* determines enclosure.

Sources:
Denmark: K. Stovgard, "Consolidation of Agricultural Land in Denmark," *International Journal of Agrarian Affairs*, 1 (May 1952), 9–14; Alex Nielsen, *Danische Wirtschaftsgeschichte* (Jena, 1933), pp. 328–29.
Sweden: Eli F. Heckscher, *An Economic History of Sweden* (Cambridge, 1956), pp. 156–58.
England: E. C. Gonner, *Common Land and Enclosure* (London, 1912), pp. 60–69, 90–92.
Germany: Jerome Blum, *The End of the Old Order in Rural Europe* (Princeton, 1978), p. 266; Theodor von der Golz, *Geschichte der Deutschen Landwirtschaft*, Bd. 2 (Stuttgart, 1903), p. 147; Heinz Haushofer, *Die deutsche Landwirtschaft im technischen Zeitalter* (Stuttgart, 1963), pp. 57–58; August Meitzen, *Die Boden und die landwirtschaftlichen Verhaltnisse des Preussischen Staates* (Berlin, 1868–1908), Bd. 5, pp. 201, 250.

city of documented enclosures between 1850 and 1914. One is forced to rely on a variety of estimates of the increase in rental values that contemporaries expected to derive from enclosure or from partial consolidations known as *abornements généraux*.

Contemporary estimates of the rise in land rents due to consolidation ranged between 20 and 30 percent in the nineteenth century, which is somewhat lower than the 40 percent common for English enclosures in the eighteenth century, but within the range for Germany in the nineteenth.[4] These values do not seem to vary much through time and are commonly attributed to improvements in drainage and more efficient utilization of men and animals. Cost estimates of a *remembrement* in Somme in the 1920s indicate that at least 75 percent of the savings from consolidation derived from the reduction in labor and animal costs.[5]

Estimates of the direct costs of enclosure can be gleaned from the records of *abornements généraux*. An abornement was a resurvey of holdings and some rearrangement and consolidation of plots in order to give each farmer direct access to a path. Such procedures required the unanimous consent of the affected proprietors and were usually carried out on all the land in the village in order to secure the scale economies of title search and survey. The costs of this operation are therefore similar to the direct outlays of enclosure. Title search, surveying, and construction of new paths would be roughly the same and at least one expense, the purchase and positioning of boundary markers, would be less than under enclosure. Negotiating a full enclosure undoubtedly took more time and perhaps necessitated larger outlays, but many of these costs could be avoided by legislating formulas to expropriate recalcitrants. Thus the costs of an abornement are at least indicative of the likely direct costs of enclosure.

Average costs in the late nineteenth century were 20 to 30 francs per hectare.[6] These costs were powerfully affected by subdivision of holdings. Detailed figures of a commune of 800 hectares having 3,350 parcels indicate that the expenses proportioned to area were 17.55 francs per hectare and 5 francs per parcel. Given the average plot size in open-field France of a fifth to a quarter of a hectare, costs per hectare directly attributable to the excessive subdivision of plots ran as high as 20 to 25 francs.[7]

[4] The most frequently cited reference is the Count of St.-Vallier's assertion in 1884 that *remembrement* would raise yields by 20 to 30 percent; Claude Chauveau, *Le Remembrement de la propriété rurale* (Paris, 1918), p. 105. At the beginning of the century De Neufchâteau estimated that the losses from excessive dispersion of plots amounted to 25 percent of the crop. Jean Peltre, "Les Remembrements à l'époque moderne (XVIIe–XVIIIe siècles)," *Annales de l'Est*, 5e série, 28 (1976), 238. For similar estimates see Georges Hottenger, *Morcellement et remembrement* (Paris, n.d.), p. 83; Alfred de Foville, *Etudes économiques et statistiques sur la propriété foncière: le morcellement* (Paris, 1885), p. 155; Jean Colonna, *La Petite Propriété et le remembrement* (Alger, 1914), p. 14, and Louis N. Grandeau, *L'Agriculture et les institutions agricoles du monde au commencement du XXe siècle* (Paris, 1905–1911), vol. 3, pp. 128. Evidence for Germany and one commune in France at the end of the 1850s, reported in the *Journal d'agriculture pratique* (1860, p. 78; 1865, pp. 74–75), shows increases in rents from about 20 to 35 percent. I am indebted to Professor McCloskey for allowing me to see his evidence on the gains from enclosure in the eighteenth and early nineteenth century.

[5] Eugène Vandervynct, *Le Remembrement parmi les améliorations foncières rurales* (Paris, 1936), p. 71. Studies cited by Vandervynct show that one hectare divided in six separated parcels and sown in wheat cost 79 percent more to farm than a one hectare plot, and that a three parcel holding cost 32 percent more.

[6] *Journal d'agriculture pratique* (1860), p. 268; Grandeau, *Agriculture et les institutions agricoles*, vol. 3, p. 132; Foville, *Le Morcellement*, p. 181; Hottenger, *Morcellement*, pp. 74–75, 116; Colonna, *La Petite Propriété*, p. 126; Chauveau, *Le Remembrement*, p. 70.

[7] Hottenger, *Morcellement*, pp. 179–80, gives a detailed breakdown of the typical costs of abornement.

When a commune consisted mainly of large properties, costs were lower. In the commune of Pulnoy, surveyed in 1912 and 1913, expenses were as low as 10 francs. On the other hand, Chauveau estimated that a complete consolidation of holdings in most communes would cost upwards of 50 or 60 francs per hectare.[8]

The fixed costs of registering titles and legal fees meant that the expense of private exchange varied more than the expense of general consolidations. The need to use two different documents to identify ownership—the cadastral register, based on an accurate survey but usually out of date, and the register of titles held by the mortgage bureau—greatly increased the cost of private land sales, as the two documents frequently failed to agree.[9] The cost of searching and rectifying titles, of locating the exact boundaries of plots in unmarked open fields, and the various legal fees associated with the transfers often drove the costs of private exchange as high as 25 percent of the value of holdings worth less than 1,000 francs (about half a hectare).[10]

If we accept contemporary estimates that consolidation of divided farms in France raised rents by 20 to 30 percent, it is clear that consolidation by private agreement was at best marginally profitable. On the other hand, simultaneous exchange to consolidate open fields was highly advantageous. Assuming a cost of 20 to 50 francs per hectare, and an average price of land of 2,000 francs, the costs amounted to only five to ten percent of the expected increase in land values. Contemporary estimates of the rate of return to consolidation all point in this direction. A remembrement in Lorraine in the late 1850s is said to have paid for itself within three years. Grandeau estimated the rate of return to an abornement to be about 18 percent. The figures collected on abornements at the turn of the twentieth century suggest rates of return of 20 to 30 percent.[11] Thus there is sketchy but consistent evidence that by the late nineteenth century enclosure in France was socially profitable.

[8] Hottenger, *Morcellement*, p. 118; Chauveau, *Le Remembrement*, p. 60.

[9] Jacques Dumas, "The Present State of the Land System in France," *Economic Journal*, 19 (1909), 42. The high costs of surveying individual parcels meant that plots in the open fields continued to be identified by reference to neighboring plots. The location of precise boundaries was normally given over to arbitrators or to the few men in a village who were believed to have exact knowledge of the precise location of each boundary or the placenames and parcels that identified particular plots. Hottenger, who witnessed many such surveys as a notary in Lorraine, called them "le triomphe de l'à-peu-près"; Hottenger, *Morcellement*, pp. 68, 76–77.

[10] The high cost of surveys was probably the main reason why land was not consolidated more often by exchange. De Neufchâteau argued that it was the main reason why peasants divided property strip by strip, rather than by establishing plots of equivalent value. Nicolas François de Neufchâteau, *Voyages agronomiques* (Dijon, 1806), p. 12. See also M. J. Diogey, *Du Morcellement du sol en France* (Paris, 1857), p. 133; and Foville, *Le Morcellement*, p. 161. A cost of 10 percent of the value of large holdings was usual at the beginning of the twentieth century. Hottenger, *Morcellement*, p. 184.

[11] Hottenger, *Morcellement*, pp. 74–75, 116, 179–180; *Journal d'agriculture pratique*, 1860, p. 268; ibid., 1865, p. 75; Foville, *Le Morcellement*, pp. 181, 172; Grandeau, *Agriculture et les institutions agricoles*, vol. 3, pp. 129–32; Colonna, *La Petite Propriété*, p. 126.

II

High social rates of return to consolidation of holdings in late-nineteenth-century France can be taken as evidence of impediments to a smooth reorganization of the land. The history of open-field agriculture in the nineteenth century can be largely written in terms of the market responses consistent with maintaining the open-field pattern of scattered holdings.

Open field communes were administered under the provisions of the rural code passed in September, 1791, the last major piece of legislation undertaken by the National Assembly before it disbanded. The act is imbued with physiocratic doctrines of economic individualism. The preamble declared the land of France to be as free as its people; the body of the act implemented this profession of faith by abolishing all non-contractual limitations of farmers' choices of crops, stocking levels, and agricultural techniques. It reflected the belief that economic freedom was necessary and sufficient to bring about the full exploitation of existing technical and commercial opportunities.[12] It thus completed the institutional reforms that had commenced with the dismantling of the seigneurial regime. The rights of common belonging to seigneurs were abolished outright, as was customary inter-commoning between villages. Common pasture itself was not outlawed, but the creation of new commons except by contract was expressly forbidden. The act gave each proprietor the right to enclose his holding, while permitting open-field communities to retain customary practices and regulations on unenclosed lands.[13] The legal framework, then, gave maximum latitude for local and individual choice as to the form of agrarian organization. The history of this evolution therefore approximates a counterfactual history of how the agrarian structure in Europe might have evolved in the absence of an enclosure movement.

The surviving published material describing open-field agriculture in

[12] See *Procès-verbaux de l'Assemblée Nationale*, vol. 63 (Paris, n.d.), "Décret de l'Assemblée Nationale sur les biens et usages ruraux, et sur la police rurale, 28 septembre 1791." The legislative history of this measure is reviewed in Octave Festy, *L'Agriculture pendant la Révolution française: Les conditions de récoltes des céréales. Etudes d'histoire économique, 1789–1795* (Paris, 1947); and Jacques Godechot, *Les Institutions de la France sous la Révolution et l'Empire* (Paris, 1951), pp. 158–70. Earlier administrative attempts to relax the constraints imposed by collective regulation of pasture are described and analyzed in Marc Bloch, "La Lutte pour l'individualisme agraire dans la France du XVIIIe siècle," *Annales d'histoire économique et sociale*, vol. 2 (1930), pp. 329–81, rpt, in Marc Bloch, *Mélanges historiques*, vol. 2 (Paris, 1963). The responsiveness of local custom to new patterns of cropping is surveyed by Jean Meuvret, "La Vaine Pâture et le progrès agronomique avant la Révolution," in Jean Meuvret, *Etudes d'histoire économique* (Paris, 1971), pp. 193–96.

[13] Nineteenth-century legal texts treat common pasture as a relationship obtaining between properties rather than between persons, along the lines of a public easement. Jean-François Fournel, *Les Lois rurales de la France* (Paris, 1819), pp. 262–63; Joseph-Laurent Jay, *Traité de la vaine pâture* (Paris, 1863), p. 36; Ferdinand Jacques, "De la propriété et des servitudes rurales d'après la loi de 1791," *Revue critique de législation et de jurisprudence*, n.s. 22 (1890–1891), p. 557; and Edgard Bourgueil, *La vaine pâture, commentaire théorique et pratique* (Charleville, 1893), p. 22.

France after 1791 is very limited. Studies based on close scrutiny of the minutes of village councils are as yet too few to provide any basis for generalization. There was an attempt to collect statistics on the extent of common pasture in 1859, but most of the records of this enterprise have disappeared. The account below is mainly based on information contained in manuals of local custom (*usages locaux*) compiled for justices of the peace. These by-laws are not as sharply focussed as one might wish, since judges had considerable latitude in interpreting them and many customs persisted despite their legal abolition. Nevertheless, they are numerous enough to survey differences in time and space and to generalize about the broad lines of evolution, especially with respect to the dates when customary practices were abandoned.[14]

Open-field agriculture was less like a puzzle of tightly interlocking practices than an onion that kept its form as its surface institutions peeled away. The key institutional elements were the common crop courses, communal rights of pasture on fallows, meadows, and wastes, the rights of the poor to glean, and the collective regulation of certain key agricultural activities such as the harvest. These regulations were administered by the villagers themselves, and enforced in the seigneurial court. When the Revolution abolished seigneurial justice, administration and policing of local customs were transferred to the village council. There is evidence that in the early years after the Revolution the old customs were rigorously enforced.[15]

At the beginning of the century the perceived dangers of local famine were still strong enough to make village councils require all able-bodied residents to help bring in the crop. Regulations growing out of the externalities of open-field farms, such as the need to control harvest dates as a means of controlling the spread of fungal disease or to limit the damage of men and animals to the standing crop, were also issued as necessary.[16] The spread of migratory labor and the development of new methods of harvesting and hay-making around the middle of the century led to the

[14] For a bibliography of these works, consult [Henri?] Stein, *Bibliographie des usages locaux* (Besançon, 1907). (This work is in the Musée des Arts et Traditions Populaires, Paris.) Some of the problems of interpretation of these lists of customs are discussed in Alexandre Bouthors, *Les Usages locaux du département de la Somme* (Amiens, 1861), p. xxxiv; Arthur Daguin, *Usages locaux; étude historique, juridique, critique et bibliographique* (Paris, 1905), pp. 9, 15; and Henri Clément, *Essai sur les usages locaux du département du Pas-de-Calais* (Arras, 1866), pp. 242–44.

[15] André Meynier, "La Commune rurale française," *Annales de géographie*, 58 (1945), 166–70; J.-P. Moreau, "Une Cause de transformation du paysage agraire en Bourgogne: La dislocation de la communauté paysanne," *Revue géographique de Lyon*, 35 (1960), 178–80.

[16] The right to issue *bans* stemmed from problems associated with "voisinage" or geographical externalities; Fournel, *Les Lois rurales*, pp. 345–49. Some examples are cited in the compilation of Joseph de Verneilh-Puyraseau, *Observations des commissions consultatives sur le projet de Code Rurale* (Paris, 1810–1814), vol. I, pp. 199–200, 333. See also Jean Meuvret, *Le Problème des subsistances à l'époque de Louis XIV. La production des céréales dans la France du XVIIe et XVIIIe siècle*, vol. I (Paris, 1977), pp. 165–66.

abandonment of the traditional harvest *ban*.[17] Only the date of the grape harvest continued to be proclaimed in order to maintain the quality and reputation of local wines. Gleaning rights disappeared for the same reasons: migratory workers left little to glean, and incentives to glean were reduced by the rise in rural incomes after 1850 combined with the disappearance of the small flour mills where the poor had taken their grain. The practice was rare by the 1890s and had all but disappeared by 1930. The value of the right to gather the standing straw and rake left-over hay also declined after 1850, when the fall in transport costs made tile cheaper than straw for roofing and coal became the most economical fuel.[18]

Common pasture on fallows and common crop rotations survived into the twentieth century. In contrast to the regulations governing harvest dates and rights of gleaning, these practices stemmed directly from the scattering of individual holdings in the open fields. Although legally binding crop rotations were abolished in 1791, the advantages of synchronizing crop courses under conditions of open-field farming were so great that the common crop rotation survived to the Second World War.[19] When farmers followed the same crop rotations, the rights of access to isolated plots were unrestricted. Otherwise a farmer had the right to reach his plot only from the nearest path, which imposed substantial costs if his plots were widely separated. The need for room to turn the plows at the

[17] See Jacqueline Moquelet, "Les Pratiques communautaires dans la plaine vendéene au XIXe siècle," *Annales: économies, sociétés, civilisations*, 18 (1963), 609; Antoine Granveau, *Usages locaux ayant force de loi dans les quatre arrondissements du département de la Marne* (rpt. of the usages of 1855; Bar-le-Duc, 1922), p. 11; Aimé Chérest, *Usages locaux . . . de l'Yonne*, p. 270; Aimé Dumay, *Usages locaux de département de la Côte d'Or; procès verbal de la commission centrale* (Dijon, 1859), pp. 66–67; Pierre Perrenet, *Les Usages locaux de la Côte d'Or, suivis de la nomenclature des communes et hameaux du département, avec l'indication de l'ancienne coutume observée chacun d'eux* (Dijon, 1901), p. 75. In Languedoc the harvest *bans* fell into disuse in the eighteenth century. Victor Fons, *Usages locaux ayant force de loi dans le département de la Haute-Garonne* (Toulouse, 1875), p. 57. This may reflect the early recourse to migratory labor in this district; Meuvret, *Le Problème des subsistances*, vol. 1, pp. 179–82.

[18] Département de Seine-et-Marne, *Recueil des usages locaux publiés sous les auspices du Conseil Général et de l'administration départemental* (Melun, 1910), p. 281; *Usages locaux du département de Seine-et-Oise* (1938), p. 228; Paul de Grully, *Le Droit de glanage* (Montpellier, 1912), pp. 138–39; *Recueil des usages locaux dans le département de l'Aisne* (Laon, 1863); Edgard Bourgueil, *Usages locaux du département des Ardennes* (Charleville, 1898), p. 183; *Usages locaux reconnus dans les cantons de Chartres Nord et Sud* (Chartres, 1874), pp. 26, 43; Département de la Seine et Oise, *Usages locaux* (1939), p. 56; Egmont Berthelin, *Usages locaux ayant force de loi dans le département de l'Aube* (Troyes, 1856), pp. 162–65; Fréderic Le Roy, *Receuil des usages locaux, particulier au canton de Calais en matière civile et commerciale* (St.-Pierre-les-Calais, 1880), pp. 10–11; Maurice Tailliandrer, *Les usages locaux en matière agricole dans le département du Pas-de-Calais* (Arras, 1939), pp. 56–57; Chambre d'Agriculture du Calvados, *Usages locaux à caractère agricole* (Caen, 1936). General adoption of the scythe occurred between 1840 and 1870 as a result of tighter rural labor markets. Even before these dates, however, landlords were inserting clauses into leases that required their tenants to scythe the standing grain; Jean-Baptiste Hénault, *Usages locaux du département d'Eure-et-Loir* (Chartres, 1861), pp. 149, 172, 178; Hénault, *Recueil des usages locaux suivis dans le canton de Bréjolles* (Chartres, 1845), p. 24; *Usages locaux de canton de Brulon* (Le Mans, 1856); Clément, *Usages du Pas-de-Calais*, p. 220.

[19] Jean Gottman, ed., *Documents pour servir à l'étude de la structure agraire dans la moitié occidentale de la France* (Paris, 1964).

end of furrows also encouraged farmers to synchronize farming operations.[20] Thus the advantages of coordination were significant, especially in communities where there was little economic incentive for neighbors to differ in their choice of crops or agricultural techniques. One village in Lorraine in the 1850s imposed a common rotation for the first time on one of its fields in order to reduce losses caused by uncoordinated movement of men and animals.[21] In practice these rules were flexible enough to allow for considerable innovation in cropping, as only small portions of the open fields had to be coordinated to secure the benefits of easy access.

The common flock was the last open-field institution to be abandoned. In the eighteenth century the minimum economical size of a flock of sheep was 100 head and since few farms were large enough to support this number (at a maximum of four sheep per hectare), farmers grouped their animals in flocks of 200 to 700 to exploit the limited pastoral resources of the fallow and meadows.[22] The law of 1791 guaranteed each open-field resident the right to keep 6 sheep and one cow on the commons. The value of these rights to poor people in the early 1860s may have amounted to 20 percent of their income, but of this more than three-fourths is attributable to the family cow, which as a rule did not have access to the open fields.[23] Keeping sheep was mainly a matter of large-scale intensive hus-

[20] As long as farmers followed the same rotations, access to isolated strips raised few difficulties. In the words of a saying in Picardy, "Terre en labour, terre emblavée,/Doivent passage à terre enclavée"; Alexandre Bouthors, *Les Proverbes, dictons, et maximes du droit rural traditionnel* (Paris, 1858), p. 108. See Louis Pyot, *Usages locaux de la Côte-d'Or* (Dijon, 1934), p. 36; Egmont Berthelin, *Usages locaux . . . Aube*, pp. 90–91, 114; Préfecture de Meurthe-et-Moselle, *Usages locaux à caractère agricole* (Nancy, 1933), pp. 19–20; Préfecture des Vosges, *Usages locaux à caractère agricole* (St.-Dié, 1934), p. 295. In some districts where long narrow furrows did not dictate plowing methods, unrestricted access to isolated plots depended on farmers maintaining the traditional furrows; André Didoux de la Madrière, *Recueil des usages locaux ayant force de loi dans . . . Pontarlier* (Pontarlier, 1937), p. 17.

[21] G. E. Colin and A. Blanc, "Extension et déclin de l'assolement triennal au XIXe siècle: L'évolution d'une commune Lorraine," *Revue géographique de l'Est*, 6 (1966), 91–93.

[22] The minimum efficient size of a flock of sheep in the nineteenth century was determined by two indivisibilities. The first was the fixed cost of the shepherd, which was a relatively skilled occupation in areas that practiced sheepfolding or where the flock had to be moved frequently; the second was the size of flock that was needed to manure a piece of land adequately. At a salary of 300 francs and sheep producing wool and meat worth 6 to 10 francs per year, the minimum flock size was above 50 sheep. The same minimum also seems to hold for sheepfolding, which before the late nineteenth century was the most efficient means of moving manure onto the fields. To ensure even distribution of manure a fold had to be densely packed, which implied minimum flocks of 50 to 75 head. At the end of the eighteenth century it required flocks of 120 to 150 sheep to keep 15 hectares of land well manured. See Meuvret, *Le Problème des subsistances*, vol. I, pp. 133–34, and vol. II, pp. 113–14. In the 1830s and 1840s flocks of 150 to 500 sheep were normal in Meuse. At the then common stints to 4 head of sheep per hectare of arable, this implies a minimum farm size of from 75 to 100 hectares to support a single flock; Mariel Delamarre, *Le Berger dans la France des villages* (Paris, 1970), pp. 200, 215–17, 231. These were giant-size enterprises by French standards.

[23] This estimate is based on calculations of the income obtainable from 6 sheep and a cow taken from data in the census manuscripts of the 1862 agricultural census. Six sheep would yield wool and meat (assuming sheep were killed at 4 years) worth 40 francs a year. A cow in milk and bearing one calf would yield about 150 to 160 francs a year. The income of a rural landless family at this time would have ranged from 800 to 1000 francs. See Ministère de l'Agriculture, *Statistique de la France . . . résultats généraux de l'enquête décennale de 1862* (Strasbourg, 1868). Detailed calculations

bandry. Thus, contrary to expectation, it was large landowners who tended to exercise the rights of *vaine pâture* (common fallow pasture). During the first four decades of the nineteenth century the high price of wool helped to maintain common pasture in the open fields. By the late 1830s many communities had to impose stinting.[24] The saturation of the wool market after 1850 relaxed this pressure and by the 1860s farmers were withdrawing land from common pasture by means of such simple signs as a shallow furrow or a stake tied with a wisp of straw, despite a legal requirement that an enclosure be sufficient to stop roving animals.[25] Never quantitatively important as a source of fodder, the *vaine pâture* could have supported at most five or six percent of the sheep in northern France in 1882.[26] Even so, repeated proposals to eliminate the practice stirred up enough resistance to dissuade governments from acting. In the early 1890s over 8,000 of some 12,000 open-field communes petitioned to retain common pasture. The common flock survived in some communes down to the mid-1960s when a combination of herbicides, increased traffic, and new kinds of harvest machinery finally eliminated all its advantages.[27]

---

for the departments of Eure-et-Loir and Yonne were made from the manuscripts, *Enquête agricole décennale, 1862: questionnaires cantonnaux*, Archives Nationales, Paris, F[11] 2693–2697.

[24] See Henri Sée, "La Vaine Pâture en France sous la Monarchie de Juillet d'après l'enquête de 1836–1838," *Revue d'histoire moderne*, 1 (1926), 202, 205. There are frequent references in the usages to the abuse of common grazing by large farmers. See *Recueil des usages locaux du département du Nord* (Lille, 1856), p. 65: Chérest, *Usages locaux de l'Yonne*, p. 239; Hénault, *Usages locaux d'Eure-et-Loir*, p. 204; André Paris, "Les Conditions du progrès agricole dans le centre du bassin parisien: Droits seigneuriaux, jachère, et vaine pâture dans la région de Montford-l'Amaury (XVIIIe à XIXe siècles)," in *Ethnologie et histoire* (Paris, 1975), pp. 236–38.

[25] Egmont Berthelin, *Usages locaux encore en vigueur dans le département de la Marne* (Chalons, 1857), p. 84; Antoine Grandveau, *Usages locaux du département de la Marne*, 2nd ed. (rpt. of 1855 usages; Bar-le-Duc, 1922), p. 38; Alexandre Bouthors, *Les Usages locaux du département de la Somme, précédés d'un essai d'application des usages ruraux du nord de la France au projet de code rural* (Amiens, 1861), pp. 70–73; Département du Loiret, *Recueil des usages locaux* (Orléans, 1905), p. 18. In others a light picket fence or a furrow marked off the area to be withdrawn from common pasture. See, for example, Hénault, *Usages locaux d'Eure-et-Loir* (1861), p. 204; Bourgueil, *La vaine pâture* (rpt. of 1855 usages of the department of Ardennes), pp. 101, 107; and Berthelin, *Usages locaux de l'Aube*, p. 65.

[26] This estimate is based on the assumption that all of the fodder potentially obtainable from fallow grazing was consumed by sheep. Estimates of production are derived from figures on value and output per hectare on fallows published in *Statistique de la France*, 1er série, *Agriculture* (Paris, 1840–1841), and *Statistique de la France*, 2eme série, *Statistique agricole décennale de 1852* (Paris, 1858, 1860). These estimated yields were then multiplied by the areas reported to lie fallow in 1882. Estimated consumption of all sheep in northeastern France was computed on the basis of mid-century reports that sheep consumed roughly 5 percent of their live weight daily in hay-equivalent fodder; Louis Moll and Eugène Gayot, *La Connaissance général du boeuf* (Paris, 1860), p. xxxix. Numbers of sheep are taken from the 1852 and 1882 agricultural censuses. Live weights of sheep are a weighted average of figures reported in the 1882 agricultural census. In 1852 fallows supported at most 25 percent of the sheep in northeastern France.

[27] Pierre Barral, *Les Agrariens français de Méline à Pisani* (Paris, 1968), pp. 28–29; J. Coudert, "La Vaine Pâture dans les pays de la Meurthe au XIXe siècle," in *Mélanges Paul Voirin* (Nancy, 1967), pp. 152–54. Delamarre, *Le Berger*, pp. 159, 171–73; and Tina Jolas and Françoise Zonabend, "Gens du finage, gens du bois," *Annales: E.S.C.*, 28 (1973), 298.

Common pasture is the best documented communal practice, and is associated with dispersion of private plots and small holdings. The presence of common pasture can thus serve as an index of the extent of communal farming in its attenuated nineteenth-century forms. A regression testing the relation between crop mix and common pasture in 30 cantons in the department of Yonne in 1859 failed to show any relation between common pasture and the fodder productivity of the land. The dependent variable is output of fodder per hectare in hay-equivalent units and the test variable is the percentage of communes in the canton recorded as having common pasture. The control variables are rural population density, distance from the nearest railway, the supply of natural meadows, and the presence of a *comice agricole* (an institution for diffusing technical knowledge). Examination of a smaller sample of the 1859 returns from the open-field cantons of the department of Vosges shows a similar lack of pattern. In view of the documented flexibility of open-field customs, these statistical results support the opinion that the surviving open-field customs did not significantly impede technical change.

The broader geographical patterns of common pasture, as reconstructed from the accounts in the *usages locaux*, show more clearly the impact of market opportunities on the evolution of this institutional form. Figure 1 shows the extent of common pasture between 1800 and 1900 from the lists of cantons where the *vaine pâture* continued to exist. Although boundaries are necessarily imprecise, the outline is clear enough. This institution of common farming eroded quickly in urban and industrial regions where intensive mixed husbandry had become highly profitable. It hung on most tenaciously in the outlying regions of small peasant farms.[28] The early abandonment of *vaine pâture* near Paris also reflects the predominance of large farms having their origins in the grange farms of ecclesiastical estates.

The absence of general measures for enclosure and high transactions costs in the land market could be circumvented in part by cheap verbal transactions in the rental market for land. In districts where leasehold predominated, farmers often concluded verbal agreements exchanging parcels without the knowledge or consent of the ultimate landowners.[29] This

[28] See George W. Grantham, "The Diffusion of the New Husbandry in Northern France, 1815–1840," this JOURNAL, 38 (June 1978), 311–37.

[29] This procedure, which was apparently very common in districts of large-scale farms near Paris, was extensively described in the report of the French land system at the Exposition Universelle of 1900. "A kind of exchange which occurs sometimes on highly subdivided property cultivated mainly by leasehold farmers is exchange by sub-lease. Two farmers exchange leaseholds of parcels in order to round out their farms. No lease is signed, the agreement is verbal, and its duration is determined by the length of the original leases. Most often the actual landlords of the respective plots are unaware of these exchanges; but even if they were it would not be in their interest to oppose them. Their rights bear on the harvest of their own land, cultivated by another farmer, rather than on the harvest of their own farmer on other lands. . . . The most frequent cases . . . are verbal and because of this escape attention; for there exists no trace of these agreements either in the municipal archives or in the land registry"; Louis N. Grandeau, *L'Agriculture et les institutions agricole*, vol. 3, p. 118. For a discussion

FIGURE 1
ABANDONMENT OF COMMON PASTURE, 1800–1900

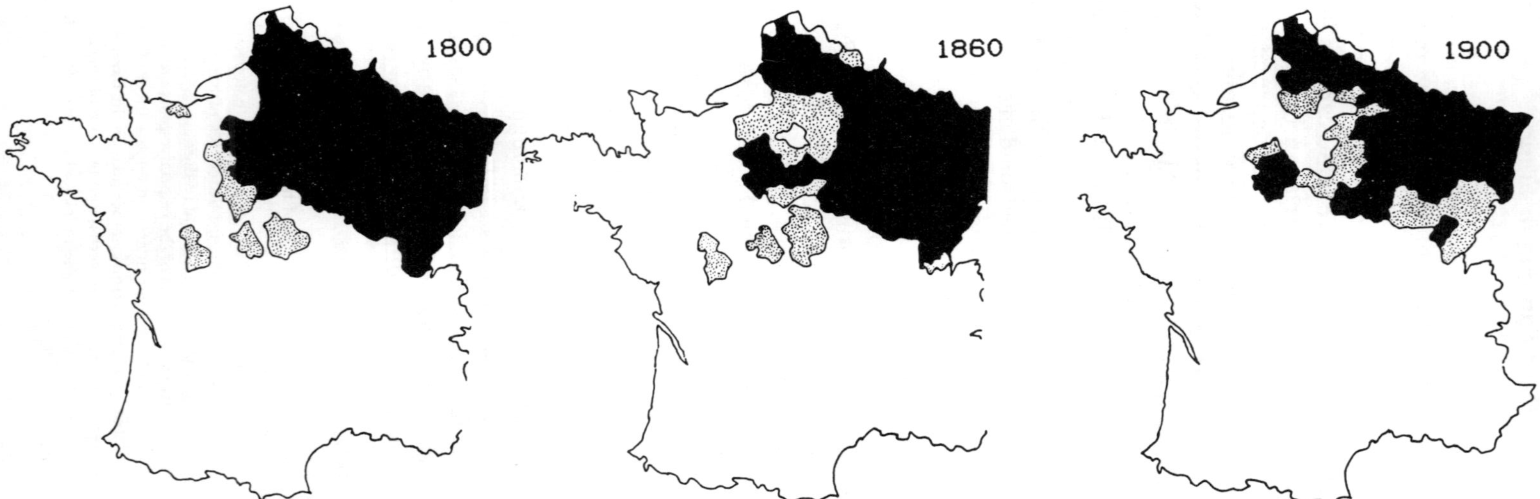

Sources: Roger Dion, *Essai sur la formation du paysage rural français* (Tours, 1934); Joseph de Verneilh-Puyraseau, *Observations des commissions consultatives sur le projet de Code Rurale* (Paris, 1810–1814); Henri Sée, "La Vaine Pâture en France sous la Monarchie de Juillet d'après l'enquête de 1836–1838," *Revue d'histoire moderne*, 1 (1926), 198–213. In addition to the works on *usages locaux* cited in the footnotes to this text, I have examined the *usages* deposited in the Bibliothèque Nationale, the Archives Nationales, and the library of the Musée des Arts et Traditions Populaires in Paris. In constructing these maps I have inferred earlier presence of the custom when it was mentioned in a later *usage*.

■ Over half the cantons in arrondissement reporting common pasture

▨ Less than half

□ None or negligible

pattern seems to have been very pronounced in the plains around Paris, where a nucleus of large farmers already existed and where many small farmers who had emigrated to Paris leased their property. There is a weak correlation between the extent of leaseholding and farm size in the aggregate statistics of landholding in France's northern departments.[30] Given the relatively low risks of farming around Paris and strong incentives to adopt intensive husbandry in this region, the more frequent appearance of large farms (over 40 hectares) and leasehold tenure is comprehensible. Since common flocks were unnecessary when farms were united in large parcels, the *vaine pâture* was abandoned in these regions relatively early. Thus in France's agricultural heartland the institutions of 1791 seem to have provided sufficient flexibility for a smooth evolution to larger and more efficient farm sizes. It was otherwise in outlying districts, where market incentives were weaker, the land market narrower, and where farm sizes were on the average significantly smaller. Here reorganization required more drastic methods.

III

Had landownership in France been more concentrated it is certain that individual landowners would have worked harder to secure consolidation of their holdings. Most students of France's agrarian institutions point to the wide distribution of property and the relatively small size of estates as the main reason why fundamental tenurial changes were delayed.[31] To the extent they required state intervention, enclosure and consolidation were intrinsically political acts. Wide distribution of property made it inadvisable for governments forcibly to rearrange customary property rights. Had landowning been more concentrated, the size of the gains from consolidation might have induced more political support for enclosure legislation.

There is evidence for this hypothesis in the pattern of eighteenth-century enclosure in France. Of the half dozen or so that are known, one was carried out by the Duc de Bourgogne in the village not far from Dijon where he had his chateau. In the 1760s and 1770s the *intendants* of Lorraine, the Marquis de la Galazière and his son, enclosed two villages near Nancy in the face of strong local resistance. At the beginning of the nineteenth century the mayor and legislative deputy of a village near Chatillon-sur-Seine carried out an enclosure.[32] Such concentrations of local

of the development of large tenant farms in this area, see Pierre Brunet, *Structure agraire et économie agricole entre la Seine et l'Oise* (Caen, 1960).

[30] The correlation between the percentage of farms over 40 hectares in size and the percentage of farms held on leasehold tenure as reported in the 1862 agricultural census is .47.

[31] For example, Patrick O'Brien and Caglar Keyder, *Economic Growth in Britain and France, 1780–1914: Two Paths to the Twentieth Century* (London, 1978), pp. 127–37.

[32] See Hottenger, *Morcellement*, p. 90. The enclosure at Neuville created a new park, apparently at the expense of land previously farmed by the peasants, since they reclaimed it during the Revolution.

authority and landed wealth were rare in France, and few landowners had the power to overcome the resistance of their tenants or the dense network of overlapping legal and political jurisdictions which protected traditional sources of income and customary rights. Even when the seigneurs had a *carte blanche* to reorganize the landscape, as they did on lands abandoned during years of warfare in Lorraine at the beginning of the eighteenth century, concern for their traditional rights led them to reimpose the old field system.[33]

Authoritarian regimes were perhaps best suited to changing rural law. Under Napoleon a law to unite dispersed parcels was under discussion in the Corps Legislatif when the Empire foundered.[34] During the July Monarchy an attempt to revise the law on common pasture stirred up so much resistance that it was quietly abandoned.[35] The re-emergence of dictatorship in 1851 led to new proposals to secure consolidation.[36] In 1863 the Conseil d'Etat proposed to include consolidation (*remembrement*) in the list of public works for which proprietors could associate in *syndicats autorisées* to exercise powers of expropriation and local taxation. Upon the vote of two thirds of the landowners representing at least half of the land in a commune, a group could petition the prefect for permission to carry out public works. Legislative debate revealed great resistance to using the public power to force "proprietors to do things against their will in the general interest," and the proposal to include *remembrement* as a work of public utility was dropped. The compulsory *syndicat* was limited to works protecting the value of property.[37] For other improvements proprietors had to join in a *syndicat libre*, which required unanimous consent.

France's post-Revolutionary constitutions, then, provided only limited institutional means for local provision of public goods. Local officials were chosen or nominated by the central government, which closely scrutinized local finances. Under the Second Empire and Third Republic the municipal budget had to be approved by the prefect, who also approved all municipal borrowing and taxation. The result was that local authorities depended on the central government for subsidies and grants to finance extraordinary expenses and capital improvements. The reasons for

In 1831, as part of the Restoration settlement, the heirs of the enclosures had to cede back 40 hectares to the village in compensation for land taken during the enclosure. Peltre, "Les Remembrements," p. 245; and Chauveau, *Le Remembrement*, p. 363.

33 Peltre, "Les Remembrements."

34 Paul Marmottan, "Un Projet de Code Rural sous de 1er Empire," *Revue des études napoleoniennes*, 6 (1913), 321–46.

35 Sée, "La Vaine Pature sous la Monarchie de Juillet," pp. 198–213.

36 Napoleon III appointed a commission to study a reformulation of the Code Rural in 1855. Articles on the German states' legislation facilitating consolidation (*flurbereinigung*) began to appear at the end of the 1850s.

37 This is well brought out in the debates conducted in the Corps Législatif over the authority of the *syndicat autorisée*. See *Le Moniteur universelle*, 1865, pp. 628–29. A full account is to be found in Vandervynct, *Le Remembrement*, pp. 134–36; and Colonna, *La Petite Propriété*, p. 128.

such tight control over local finance are well known. Financial autonomy and political autonomy went hand in hand. Political security required control of local finance to prevent local notables from using the villages and small towns as sources of patronage and political power. This distrust of local authority extended to matters of agrarian reform, since the village government was a possible agency for carrying out consolidation.[38]

Distrust ran both ways. If the law and a series of judicial decisions placed final authority over open-field routines in the hands of the prefects, the same law erected almost insuperable barriers to governmental tampering with property rights. By a series of self-denying ordinances, post-Jacobin governments limited their rights of expropriation to works of unquestioned public interest, such as the national roads, common paths, dikes, canals, and other projects needed to protect property and persons. Projects raising the economic value of property were expressly excluded as valid reasons for compulsory shifting of property rights.[39]

The reason for this timidity in the face of obvious opportunity and the example of most other industrializing societies is to be found in the record of the Revolution and its aftermath. The confiscations, dictatorships, and the unsettling experience of price controls and forced food deliveries made peasants and the middle classes wary of any extension of state authority that might create real power to alter the distribution of income. The peasants also feared the emergence of power that might be used to reverse the revolutionary land settlement. Therefore, there was in France no constituency for new means of expropriation to consolidate farm holdings.[40] Most political Frenchmen, faced with the example of German legislation on enclosure, regarded German procedures as unacceptable interference with the right of every man to keep his own. The economist Alfred de Foville characterized German legislation as a means of lining up land the way a corporal lined up his men. At the beginning of this century France's most eminent agronomist questioned whether it was necessary to:

[38] Theodore Zeldin, *France 1848-1945*, Vol. I, *Ambition, Love and Politics* (Oxford, 1973), pp. 522–23, 530–31. In 1890 per capita debt incurred by local authorities in France was 37 francs, excluding the large towns. In England local authorities indebted their citizens to the amount of 180 francs per head; ibid., p. 524.

[39] In contrast, American state governments were almost indecent in their readiness to expropriate private property for purposes of economic development. See Harry N. Scheiber, "Property Law, Expropriation, and Resource Allocation by Government: The United States, 1789–1910," this JOURNAL, 33 (March 1973), 232–51.

[40] In this and the preceding paragraph I follow David Thomson's interpretation of post-Revolutionary French politics as an expression of the tension between the Revolutionary ideal of popular sovereignty in an omnicompetent state, with overtones of social as well as political egalitarianism, and the embodiment of this ideal in ministerial government. The question to whom the executive was to be responsible was never adequately resolved, with the result that in 1791, 1830, and 1848 governments were changed by popular uprisings. See his *Democracy in France since 1870*, 4th ed. (London, 1964), pp. 13–14. For a similar verdict, see Stanley Hoffmann's brilliant analysis of the "stalemate society" in "Paradoxes of the French Political Community," in Hoffmann et al., *In Search of France* (Cambridge, Mass., 1963), pp. 12–18.

introduce into our jurisprudence the obligation of a citizen to abandon his paternal fields for another, perhaps more valuable, holding because the majority of the residents of a commune will have decided it. Is it necessary to consecrate in law the right of a majority to dispossess a proprietor against his will?[41]

The answer from the countryside was emphatically no. Surveys in the late nineteenth century showed that farmers recognized the value of consolidating their holdings, but the same polls showed they opposed the creation of mechanisms for accomplishing this change against the will of individual landowners.[42]

Despite evidence of high social returns to consolidation, the structure of government before the First World War did not favor decisive action. In contrast to education, agrarian reform was not required to secure the republican regime. Failure to act had few political consequences, while the possibility of reforming France's land law did. The difficulties of legislating in the general interest against concentrated local resistance were exacerbated by the weakness of the executive. Deputies did not fall with governments but rather owed their seats to control of local patronage. As a result ministers had little leverage on deputies of their own party, so that governments could fall on minor issues raised during question period, the dreaded *interpellation* which undid many governments before the final defeat in 1940. Governments, then, were weak in relation to particular vested interests and had to husband their resources for major political trials. This, too, was a product of the early history of the Third Republic, which steered its way between the designs of the dukes and Gambetta's rhetoric to create a government that offended few by doing little.[43] Enclosure was a minor issue in late-nineteenth-century France. Its benefits were substantial but diffuse, while its costs were concentrated. Some general legislation to facilitate enclosure would undoubtedly have yielded benefits, but there was no coalition to pass it. The most successful practical reforms were due to a surveyor named Gorce who induced over one hundred communes to undertake an *abornement général*. In all, 200 villages at most out of about 12,000 in open-field districts undertook some formal rearrangement of land holdings before the First World War.[44]

The wars of the twentieth century had a decisive influence on France's open fields. After the first, the devastation in eastern France created op-

[41] Grandeau, *Agriculture et les institutions agricole*, vol. 3, p. 133.

[42] Vandervynct, *Le Remembrement*, p. 175.

[43] Zeldin, *France*, vol. I, pp. 582–84. Deputies in the Third Republic were professional agents of their districts in Paris, and owed their election more to their ability to provide patronage and small-scale public works than to their stand on "national" issues. General elections were infrequent and the turnover of deputies was very low. At any time in early twentieth century, over a quarter of them had served more than 20 years. Thus the frequent turnover of governments was matched by the infrequency of changes in the personnel of the Assemblée Nationale. The deputies could thus block strong executive actions, while their hold on patronage prevented the emergence of a "ministerial party" that might have provided some room for government to maneuver.

[44] Chauveau, *Remembrement*, pp. 40, 102–03; Grandeau, *Les Institutions agricole*, vol. 3, p. 128; Vandervynct, *Le Remembrement*, p. 168. The numbers of consolidations actually carried out are impossible to determine, owing to the apparently common device of exchange in leasehold, as discussed above. There is no question, however, that consolidation of land by sale and exchange was very rare.

portunities for more rational farm layouts. The Chauveau Law of 1919 made it possible for the majority of landowners in these districts to impose consolidation. But most peasants refused this mechanism and maintained traditional field patterns down to the 1940s.[45] During the Occupation the Germans encouraged consolidation, but it was not until the late 1950s that *remembrement* began to be carried out on a large scale. Under a law of 1960 consolidation became a duty of the state. It could be instituted by the commune, prefect, individual farmers, or by the corps of agricultural technicians (*génie rurale*). No vote was needed to begin enclosure, though there was consultation, and the state had the power to force landowners to exchange holdings to increase the size of individual plots.[46] Since then the rate of disappearance of open-field farms has been swift.

It thus took France nearly two hundred years to follow the example of Great Britain and Scandinavia. It is unlikely that this lag in institutional innovation can be attributed solely to an unfavorable economic environment. A more compelling set of reasons is to be found in political institutions and political history. For enclosure was always a public act. The high cost of private transactions predetermined a collective solution to the problem of dispersed fields. In France the links between economic opportunity and political decisions were often too weak to generate utilitarian legislation. In this case other motives determined the outcome of an essentially technical problem.

The history of enclosure in France has a bearing on general models of institutional innovation in Western societies which treat political institutions as surrogate firms that aggregate and act on the preferences of their customers. One might argue that the wide dispersion of property made the costs of aggregation excessively high in France. But this cannot be the whole story. People in institutions behave in part because of what has happened before. The purely political history of an imperfectly centralized society which by revolution destroyed some of the wealth and much of the authority of its ruling elite created institutions and patterns of behavior that were inimical to changing the structure of property rights. The political histories and institutions of the different societies of Western Europe diverged more widely than their technical and economic histories. In the industrial area these variations seem to have had little impact on the historical pattern of resource allocation. In agriculture, on the other hand, variations in the willingness of governments to intervene in land law had lasting effects.

On the eve of the First World War Chauveau estimated that only one commune (probably Teutonville) had taken advantage of an 1865 law making it possible for landowners to form themselves into a syndicate for the purpose of carrying out consolidation of their land. The number of resurveys of open-field villages combined with a reduction in the number of parcels and a reorientation of paths seems to have been under one hundred. See Grandeau, *L'Agriculture et les institutions agricole*, vol. 3, p. 128. Hottenger lists the communes having carried out full or partial *abornements généraux* in *Morcellement*, Appendix.

[45] "The Consolidation of Farms in Six Countries of Western Europe," *International Journal of Agrarian Affairs*, 1 (1952), 25–28; Barral, *Les Agrariens français*, pp. 199, 258.

[46] Lucien Valentin, *L'Action administrative dans la vie rurale* (Paris, 1961).

The Regional Economy of Flanders and Industrial Modernization in the Eighteenth Century: a Discussion

Christian Vandenbroeke
University of Ghent

Flanders has recently attracted particular interest for historians of the early modern period and the nineteenth century *. The concept of proto-industry, put forward by F. Mendels and others, was first developed in the context of this region [1].

Nowhere were higher employment levels in domestic industry to be found, and by the end of the Ancien Régime nearly half of the population was engaged in the rural linen industry, at least on a part-time basis (see map). The most important concentration of linen workers was particularly in the area of Ghent,

* We should like to thank F. Mendels and in particular J. Riley who made comments on an earlier version of this article. However, responsibility for errors and extrapolations is ours alone.

1 F. Mendels, Proto-industrialization: the first phase of the industrialization process, in *Journal of Economic History*, 1972; Agriculture and peasant industry in eighteenth century Flanders, in E. Jones-W. Parker (eds.), *European peasants and their markets*, Princeton, 1975; Proto-industrialization: theory and reality. General Report, in *Eighth International Economic History Congress*, Budapest, 1982; *Industrialization and population pressure in eighteenth-century Flanders*, New York, 1981. See also P. Deyon. L'enjeu des discussions autour du concept de proto-industrialisation, in *Revue du Nord*, 1979, pp. 9-18; H. Kellenbenz, Les industries dans l'Europe moderne, 1500-1750, in P. Leon-F. Crouzet-R. Gascon (eds.), *L'industrialisation en Europe au XIXe siècle*, Paris, 1972, pp. 75-114.

2 G. Jacquemyns, *Histoire de la crise économique des Flandres, 1845-1850*, Brussels, 1929; C. Vandenbroeke, Mutations économiques et sociales en Flandre au cours de la phase proto-industrielle, 1650-1850, *Revue du Nord*, 1981; Analyse critique de la phase proto-industrielle en Flandre. L'evolution sociale et le comportement démograpi-

Tielt, Kortrijk and Oudenaarde [2]. Every week woven linen was carried to the urban markets and from there shipped, via Spain and France, to America [3]. In the second half of the eighteenth century about 100,000 pieces of linen, i.e. 5 million metre, were brought every year to the market of Ghent [4].

The situation deteriorated at the close of the Ancien Régime, when proto-industrialists had to compete with mechanization in the textile sector. Flanders also provides an example of an area experiencing de-industrialization and rapid impoverishment in the course of the nineteenth century[5]. Between 1840-50 the decline of rural industries became disastrous. Many sources describe the whole area in northwestern Belgium at that time as "Poor Flanders" or as "The Ireland of the Continent" [6]. An interesting summary of the causes and effects of this process of impoverishment was given by L. Varlez: "The peasants had agreed to a brutal cuts in pay. But there are limits to everything, even to the resignation of Flemish peasants. There is a moment when the wage ceases to be human and when the worker prefers to abandon his trade" [7].

Yet this impoverishment was not the result of a linear trend in the evolution of rural industry. Only in the first and final phases of proto-industrialization was social decline clearly evident [8]. Such decline was especially evident in the late sixteenth and the seventeenth centuries on the one hand and during the first half of the nineteenth century! A brief sketch of the rural wages of agricultural

que, 17e siécle - début du 19e siécle, in *Proceedings of the Eighth International Economic History Congress*, Budapest, 1982; Sociale en konjunkturele facetten van de linnenijverheid in Vlaanderen (late 14e - midden 19e eeuw), in *Handelingen Maatschappij Geschiedenis en Oudheidkunde Gent*, 1979, pp. 117-174.

3 J. Bastin, De 'Gentse lijnwaadmarkt en linnenhandel in de XVIIe eeu, in *Handelingen Maatschappij Geschiedenis en Oudheidkunde Gent*, 1967; F. Mendels, Agriculture and peasant industry... *op. cit.*; E. Dubois, *L'industrie du tissage du lin dans les Flandres*, Brussels, 1900; J. Mokyr, *Industrialization in the Low Countries, 1795-1850*, London 1976; E. Sabbe, *De Belgische vlasnijverheid*, Kortrijk, 1975.

4 W. Haagen, Uitbuiting door handel als verklaringsfaktor voor de vertraagde industrialisering van de linnenijverheid in Vlaanderen, in *Handelingen Maatschappij Geschiedenis en Oudheidkunde Gent*, 1983, p. 218.

5 G. Jacquemyns, *op. cit.* The most important information can be found in the famous "*Enquête sur l'industrie linière*", Brussels, 1941.

6 For a more general view about the process of pauperization, cf. C. Lis-H. Soly, *Poverty and capitalism in pre-industrial Europe*, Hassocks, 1979.

7 L. Varlez, *Les salaires dans l'industrie gantoise. II. Industrie de la filature du lin*, Brussels, 1904, p. LX.

8 In this context, see the critical articles of P. Jeannin, La protoindustrialisation: développement ou impasse?, in *Annales. Economies-Sociétés-Civilisations*, 1980 and D. C. Coleman, Proto-industrialisation: a concept too many, in *Economic History Review*, 1983. Other criticisms were put forward in *Geschichte und Gesellschaft* in review of P. Kriedte-H. Medick-J. Schlumboihm, *Industrialisierung vor der Formationsperiode des Kapitalismus*, Göttingen, 1977.

workers and linen weavers over the century (see graph) reveals the situation [9]. From the fourteenth century to the beginning of the twentieth century there are only two periods of relative prosperity and high purchasing power of the people in the countryside. These occurred in the middle of the fifteenth century and the second and third quarter of the eighteenth century. The latter period was indeed described as an "*âge d'or*" by some contemporaries [10]. A comparative analysis of physical product also confirms the relative prosperity of the Flemish people around 1800. These values can be compared with reconstructions for neighbouring countries such as France, England and the United Provinces [11].

We must look more closely at the reasons for industrial development during the eighteenth century. It is frequently argued that domestic purchasing power had a decisive influence in economic growth. According to H. Perkin "the most immediate and necessary of all economic factors (is) the factor of demand... the Industrial Revolution was a social revolution with social causes and social effects" [12]. The same conclusion was reached by W.A. Cole: "growth of demand is now widely regarded as one of the essential elements in the transformation of the economy" [13].

9 C. VANDENBROEKE, *Mutation economiques..., op. cit., pp. 85-88; De leefbaarheid van het platteland in Vlaanderen, 15e-19e eeuw, in Bijdragen tot de Geschiedenis*, 1981, pp. 131-166. Analogous remarks were made by M. P. GUTMANN, *War and rural life in the early modern Low Countries*, Assen, 1980, p. 109; D. TERRIER, Mulquiniers et gaziers: les deux phases de la proto-industrie textile dans la région de Saint-Quentin, 1730-1850, in *Revue du Nord*, 1983, p. 540; P. SERVAIS, Les structures agraires de continuité, in *XLVe Congrès Féderation des Cercles Archéologiques et Historiques de Belgiques, Comines*, 1980, p. 148; Les structures agraires du Limbourg et des pays d'Outre-Meuse du XVIIe au XIXe siécle, in *Annales. Economies-Sociétés-Civilisations*, 1982, p. 304.

10 N. BRIAVOINNE, *Mémoire sur l'état de la population, des fabriques, des manufactures et du commerce dans les provinces des Pays-Bas, depuis Albert et Isabelle jusqu'à la fin du siécle dernier*, Brussels, 1840, p. 85.

11 Cf. *the publications of the I.S.E.A. in France and more in particular the studies of* J. C. TOUTAIN, J. MARCZEWSKI and T. J. MARKOVITCH. For England, see P. MATHIAS, *The Transformation of England. Essays in the economic and social history of England in the eighteenth century*, London, 1979; P. DEANE-W. A. COLE, *British economic growth, 1688-1959*, Cambridge, 1962; P. O'BRIEN-C. KEYDER, *Economic growth in Britain and France, 1870-1914*, London, 1978; F. CROUZET, *L'économie de la Grande-Bretagne Victorienne*, Paris, 1978; R. FLOUD-D. McCLOSKEY (eds.), *The economic history of Britain since 1700*, Cambridge, 1981. For the Netherlands; J. DE VRIES, *The decline and rise of the Dutch economy, 1675-1900*, paper prepared for the Conference on the Economic History of the Netherlands, UFSAL Brussels, 1983; J. H. VAN STUYVENBERG, De Economie in de Noordelijke Nederlanden, 1770-1970, in *Nieuwe Algemene Geschedenis der Nederlanden*, 10, Haarlem. 1981.

12 H. PERKIN, *The origins of modern English society, 1780-1880*, London, 1969, pp. 73 and 100.

13 W. A. COLE, *Factors in demand, 1700-80*, in R. FLOUD-D. McCLOSKEY, *op. cit.*, p. 36.

Christian Vandenbroeke

However, in addition to demand, many other variables were at work and each author emphasizes different specific characteristics existing in England on the eve of the Industrial Revolution. A few years ago, R. M. Hartwell listed the variables cited most often, and found that they were more than thirty [14]. There is not any lack of debate then about the factors that gave size to industrialization at the end of the Ancien Régime, and many of the causes previously cited are now in doubt or seem less important than was once supposed [15]. A good illustration of this is the sub-heading N.F.R. Crafts gave in 1977 to one of his review-articles: "Some thoughts on the question: Why was England first?" [16].

As the number of comparative regional studies grows, it becomes increasingly important to bring Flanders into the discussion. For too long research has focused on France and England alone. What is needed is a broader perspective and a regional analysis of economic development. D.C. Coleman pointed this out in 1969: "We need comparative studies of economy and society in the varied political entities which made up Europe" [17].

Recently this approach has again been emphasized by S. Pollard when he confirmed that the industrial revolution in the West was a process of the industrialization of regions [18].

F. Mendels' conclusions pointed in the same direction when he developed the concept of proto-industrialization [19].

A reconstruction of physical product in East-Flanders, a region which was remarkable because of the diffusion of the linen industry, can provide the starting point for further research into industrial development at the end of the Ancien Régime. There is no lack of statistical information. Detailed censuses of agricultural production, cattle, dairy industry and industrial production were conducted during this period, when Belgium was temporarily integrated in the French empire [20].

[14] R. M. Hartwell (ed.), *The causes of the Industrial Revolution in England*, London, 1967.

[15] P. O'Brien-C. Keyder, *op. cit.*; R. Roehl, L'industrialisation française. Une remise en cause, in *Revue d'Histoire Economique et Sociale*, 1976; Britain and European Industrialization: pathfinder pursued, in *Review*, 1983; N. F. R. Crafts, British economic growth, 1700-1831. A review of the evidence, in *Economic History Review*, 1983.

[16] N. F. R. Crafts, Industrial Revolution in England and France: some thoughts on the question "Why was England first?" in *Economic History Review*, 1977.

[17] D. C. Coleman, *The economy of England, 1450-1750*, Oxford-London, 1977, p. 200.

[18] S. Pollard (ed.), *Region und Industrialisierung*, Göttingen, 1980, p. 21.

[19] F. Mendels, *Industrialization and population pressure...*, *op. cit.*, pp. 1-5.

[20] J. Craeybeckx, Les débuts de la Révolution Industrielle en Belgique et les statistiques de la fin de l'Empire, in *Mélanges G. Jacquemyns*, Brussels, 1968; B. Gille, *Les sources statistiques de l'histoire de France: des enquêtes du XVIIIe siécle à 1870*, Paris, 1964; H. Coppejans-Desmedt, Bijdrage tot een kritische studie over de nijverheidsstatistieken uit de jaren 1795-1846, in *Handelingen van de Koninklijke Commissie voor*

As a result we have excellent information for the period 1801-1813 and the quality of the census data is not in doubt: "the Napoleonic Statistics without any doubt provide the best instrument for studying economic development in the early years if the 19th century" [21].

An exhaustive analysis of all the sources, in the archives at Paris or in the provincial dépots at Ghent and partly published by Faipoult in his "*Mémoire Statistique du département de l'Escaut*" is the best way to study economic development shortly after 1800 [22]. For East-Flanders the composition and the value of the physical product can be summarized as follows [23]:

Agriculture:				*Industry:*		
Tillage	=	48,537,711	(33.9%)			
Cattle-breeding	=	25,137,652	(17.2%)			
Total	=	73,675,363	(50.2%)	Total	=	73,190,397 (49.8%)
Main Total	=	146.865.760	francs			
Average Income per head	=	235	francs			

When we first examine the components of productions, we notice that the share of cattle and the dairy industry was exactly half that of tillage. It is striking indeed that F. Quesnay, the most famous spokesman of the physiocrats, described such a propostion as optimal for a proper agrarian economy: "livestock equals at least one half of the annual product of the harvest" [24]. Yet the equal contribution of the primary and secondary sector is also striking and demonstrates the advanced character of the Flemish area at the end of the Ancien Régime [25].

Geschiedenis, 1960; W. VANDERPIJPEN, *De landbouw en de landbouwpolitiek in het Leie-en het Scheldedepartement (1794-1814)*, unpublished Ph., Brussels, 1983.

[21] *La statistique en France à l'époque Napoléonienne*, Brussels, 1981, pp. 42-43.

[22] FAIPOULT, *Mémoire statistique du département de l'Escaut, 1801 (edited by* P. DEPREZ, *Ghent, 1960).* An exhaustive analysis of the statistics for the agricultural sector can be found in W. VANDERPIJEN, *op. cit.*

[23] We presented an elaborated survey of physical product in Flanders at the Conference in Brussels (Musée Royal de l'Armée, october 1983) of the Révolution Brabançonne. This contribution is to be published shortly: *De sociaal-economische context van de Brabantse Omwenteling in de Vlaamse regio's. Het fysisch product in Vlaanderen op het einde van de 18e - begin 19e e eeuw.*

[24] J. C. TOUTAIN, *Le produit de l'agriculture française de 1700 à 1958. I. Estimation du produit au XVIIIe siécle*, Paris, 1961, p. 161.

[25] For a comparison with France and England, *Cf.*, R. FLOUD-D. MCCLOSKEY, *op. cit.*; J. C. TOUTAIN, *Le produit de l'agriculture française de 1700 à 1958. II. La croissance*, Paris, 1961, pp. 200-201.

It is even more interesting to put these approximate values of physical product in a wider context by adding a few comparisons in time and space [26]:

Approximate values of the physical product (in francs)

	Flanders	*Belgium*	*France*	*England*	*Netherlands*
1760-80	(245)	—	—	—	—
1780-90	—	—	192	—	—
1801-10	235	—	247	293	280(x)
1811	—	—	—	303	—
1815-24	—	—	252	266	—
1825-34	—	—	286	—	—
1840-50	—	266(x)	—	495(x)	—
1845-54	—	262(x)	—	—	—

The evolution of the physical product in Flanders is difficult to follow for the nineteenth century, but a marked decline seems clear. By about the middle of that century something between a quarter and a half of the population became indigent and needed assistance from public or private funds in one way or an other [27]. The direct cause of this spectacular impoverishment lay in the process of de-industrialization mentioned earlier and in the disastrous failure of grain and potatoes in the years 1845-47. But there was also a longer-term decline after the end of the eighteenth century, and this crisis has been described by several historians [28]. Flanders during the first half of the nineteenth century very quickly became one of the most impoverished regions of the Continent [29].

The contrast with the second and third quarters of the eighteenth century is remarkable. At that time physical product per head must have been higher than in 1800-1810. It is, however, difficult to provide an exact value for the middle of the century because of the lack of detailed statistical material before the French occupation of the Austrian or Southern Netherlands. But in an approximate way we can follow the evolution of the most important economic sectors in the

[26] The value with a sign between brackets, are reduced by about 20% from the GNP. For this there are many publications. For Belgium, see P. Leon, *op. cit.*; P. Lebrun *et. al.*, *Essai sur la Révolution Industrielle en Belgique, 1770-1847*, Brussels, 1979; J. Gadisseur, Contribution à l'etude de la production agricole en Belgique de 1846 à 1913, in *Revue Belge d'Histoire Contemporaine*, 1973. Most of the indications for the neighbouring countries can be found in the publications cited in note 11.

[27] P. C. Van Der Meersch, De l'état de la mendicité et de la bienfaisance dans la province de Flandre Orientale depuis le règne de Marie-Thérèse jusqu'à nos jours, in *Bulletin de la Commission Centrale de Statistique*, V. Brussels, 1852; C. De Keverberg, *Essai sur l'indigence dans la Flandre Orientale*, Ghent, 1819.

[28] G. Jacquemyns, *op. cit.;* B. Verhaegen, *Contribution à l'histoire économique des Flanders*, Louvain-Paris, 1961.

[29] C. Vandebroeke, *Sociale Geschiedenis van het Vlaamse Volk*, Beveren, 1981, pp. 241-282.

eighteenth century, and it is possible to draw some rough impressions about physical product and per capita income, since partial information is available for cereal production, stock of cattle and linen output, which were the leading sectors of economy before the end of the Ancien Régime.

At the end of the century and certainly after the 1820's, grain supplies could not any longer keep up with population growth [30].

We also know that the linen industry had to contend with growing competition from mechanized textiles abroad[31], and the response was to reduce prices for finished products. In addition the surplus value extracted from the linen sector increased less quickly than the population [32]. There was also a parallel stagnation and even a relative decline in cattle-breeding and the dairy industry, since population expansion meant that more and more households had to manage without cattle. In the seventeenth century almost every household had cattle in the area of Aalst (the so-called "Land van Aalst"), as the probate inventories in this area show; at the end of the Ancien Régime there were cattle in fewer than 60% of the households [33].

Again and again the evidence demonstrate that physical product per head must have reached its maximum in the middle of the eighteenth century, and an approximate income of about 240 to 250 francs seems a fair estimate for the years 1750-60, which were described as a golden age by contemporaries [34].

Putting all these data together, the homogeneous character of the nominal values, expressed per head of the population, is clear. By 1800 the difference between Flanders, England, France or the United Provinces was certainly not more than 20%. P. Bairoch has also argued that the countries of Western Europe were very similar at the end of the Ancien Régime [35]. But the nominal values,

[30] G. Bublot, *La production agricole belge. Etude économique séculaire 1846-1955*, Louvain, 1957; J. Vandervaeren, *Les faits principaux de l'histoire de l'agriculture belge durant un siécle, 1830-1930*, Brussels, 1930; C. Vandenbroeke-W. Vanderpijpen, The agricultural revolution in Flanders and in Belgium: Myth or reality?, in H. E. Van Cauwenberghe (eds.), *Productivity of land and agricultural innovation in the Low Countries (1250-1800)*, Louvain, 1978.

[31] M. Levy-Leboyer, *Les banques européennes et l'industrialisation internationale dans la première moitié du XIXe siécle*, Paris, 1964, p. 188.

[32] C. Vandenbroeke, *Sociale en konjunkturele facetten... op. cit.*, p. 155.

[33] A. Verhulst-G. Bublot (eds.), *L'agriculture en Belgique, hier et aujourd'hui*, Brussels, 1980, p. 36.

[34] C. Vandebroeke, *Sociale Geschiedenis..., op. cit.*, pp. 185-188.

[35] P. Bairoch, Ecarts internationaux des niveaux de vie avant la Révolution Industrielle, in *Annales. Economies-Sociétés-Civilisations*, 1979; Estimation du revenu national dans les sociétés occidentales préindustrielles et au XIXe siécle, in *Revue Economique*, 1977; Europe's gross national product, 1800-1975, in *Journal of European Economic History*, 1976.

used in this comparison give a rather distorted picture of reality[36]. If we want to understand social conditions we have to transpose these values into real terms. In general, an initial correspondence can be derived from grain equivalents, but such a simplification does not hold very well for the end of the eighteenth or for the first years of the nineteenth century, since extremely high grain prices were quoted in England at that time as a consequence of the Continental Blockade [37]. Therefore we have used a different comparison, based upon the average meat prices at the turn of the century [38]:

Approximate values of physical product in real terms around 1800

	Flanders	*France*	*England*
Wheat, litres	1,282	1,140	857
Meat, kilograms	255	230	254

This indicates that whatever basis we use for the transpositions of the nominal values into real terms, the final impression remains the same. In England, contrary to what is so often stated, there was no particularly favoured position concerning material supplies of foodstuffs at the end of the eighteenth or the beginning of the nineteenth century. On the other hand, there is no doubt that Flanders was one of the most prosperous areas in the world in this period.

What was true for the period 1800-1810 must certainly have been true for the middle of the eighteenth century. In his description of a journey through the Austrian Netherlands in 1782-3, Dérival testified: "All the inhabitants of the towns of the Low Countries live if not in opulence at least in great abundance: but this is even more to be remarked in the countryside than in the towns; in towns and in the countryside one cannot but be struck by the good fortune enjoyed by the inhabitants [39].

This picture contrasts with the usual images of social life in France recorded by A. Young [40]. It is also worth remembering that many English contemporaries spoke negatively about social conditions in their own country.

In that sense an observer in 1717 recorded: "they of Great-Britain who must work, be, steal or starve are five times as many as those who have Estates or Professions, whereupon to live without working" [41]. This description of mate-

[36] In this context, see also the critical remarks presented by M. Morineau, Richesse et pauvreté des régions. Une nouvelle approche, in *Congreso de Historia Rural, Siglos XV al XIX*, 1984, p. 386.

[37] B. R. Mitchell, *Abstract of British Historical Statistics*, Cambridge, 1962; H. Phelps Brown - S. V. Hopkins, *A perspective of wages and prices*, London, *op. cit.*

[38] P. O'Brien - C. Keyder, *op. cit.*, pp. 44-47.

[39] Derival, *Le Voyageur dans les Pays-Bas Autrichiens*, I. Amsterdam, 1782-1783., pp. 9-10.

[40] A. YOUNG, *Voyages en France en 1787, 1788 et 1789*, Paris 1931 (edited by H. See).

[41] R. W. Malcolnson, *Life and labour in England, 1700-1780*, London, 1981, p. 19.

rial conditions in England hardly differs from what G. King some decades earlier was saying about the social structures in his country [42].

These impressions together with the approximate values of physical product create serious doubts about the importance of the demand factor, and suggest that purchasing power cannot be seen as the deciding variable in explaining industrial take-off in Great-Britain. Nonetheless a temporary rise in the standard of living remained an important factor stimulating economic expansion during the eighteenth century in most of the Western countries and areas. We may follow F. Braudel, in a paraphrase of J. M. Keynes, in saying: "It was an increase in demand which, so to speak, started the motor. The need to increase production set the machines in motion..."

But what about the other explanations and factors, which are frequently said to explain the beginning of the Industrial Revolution?

What is the result of a comparative analysis using the Flemish region as a base? Anticipating the results of our research, it can be repeated that the so-called advantages of Great Britain were more apparent than real. In most respects, oportunities for industrial growth were at least as good in Flanders as in England and in other neighbouring countries.

The "human capital" factor may have been essential to economic development after the middle of the nineteenth century. However, in the context of the process of modernization in the eighteenth century, the influence of this variable was rather weak [44]. This is clearly expressed by G. N. Von Tunzelmann, who has noted that: "science was in vastly better shape on Continental Europe... than in Britain" [45]. If we take literacy as a criterion, we find more similarities than differences among the countries in North Western Europe. By about 1785 some 50% of men and 25% of women signed their marriage acts in Flanders [46], a similar proportion to that in France or England [47]. Yet this average disguises regional differences: in Flanders the contrast between the areas with commercial agriculture in the west and the north on one hand, and the industrial area in the south and the interior part of the province on the other hand (see map) was

[42] See also the estimates made by Colquhoun for the year 1803 (H. PERKIN, *op. cit.*, table 21).

[43] F. BRAUDEL, *Les Jeux de l'échange*, Paris, 1979, p. 150.

[44] P. BAIROCH, *Le Tiers-Monde dans l'impasse. Le démarrage économique du XVIIIe siécle*, Paris, p. 25.

[45] G. N. VON TUNZELMANN, *Technical progress during the industrial revolution*, in R. FLOUD - D. MCCLOSKEY, *op. cit.*, p. 149.

[46] J. RUWET - Y. WELLEMANS, *L'analphabétisme en Belgique, XVIIIe - XIXe siécles*, Leiden-Louvain, 1978.

[47] J. HOUDAILLE, Les signatures au mariage de 1740 à 1829, in *Population*, 1977; L. STONE, Literacy and education in England, 1640-1900, in *Past and Present*, 1969; F. FURET-W. SACHS, La croissance de l'alphabétisation en France, XVIIIe - XIXe siécle, in *Annales. Economies-Sociétés-Civilisation*, 1974.

sharp. In districts where rural industry was common and where child labour was widespread, literacy was lower [48].

More important in this cross-country comparison are the contrasts in population growth and structure. From recent demographic research, we can conclude that higher growth rates were seldom found ouside Flanders. This is particularly true for southern and inland Flanders, precisely the area with the most important extension of proto-industry [49]. Even when looking at regional population figures elsewhere, for example in England, and making a division between so-called agrarian and industrial areas, we still come to the same conclusions. Maximum growth occurred in Flanders during the eighteenth century. This was to change after the turn of the century and from that moment the most rapid population growth occurred in Great Britain [50].

Population growth during the XVIIIth century and the first half of the XIXth century (indices).
(1700 = 100)

Period	Flanders: *Area of Tielt* [51]	Flanders: *Area of Aalst* [52]	*Belgium* [53]	*France* [54]	*England: Total* [55]
1700	100	100	100	100	100
1750	180	159	—	119	114
1800	256	202	189	142	171
1850	345				331

Period	England: *Agricultural Areas* [56]	England: *Mixed Areas* [56]	England: *Industrial Areas* [56]
1700	100	100	100
1750	100	100	115
1800	134	145	193

48 C. VANDENBROEKE, *Mutations économiques... op. cit.*, p. 93.

49 P. DEPREZ, The demographic development of Flanders in the eighteenth century, in D. V. GLASS-D. E. C. EVERSLEY (eds.), *Population in History*, London, 1965.

50 N. L. TRANTER, *Population since the Industrial Revolution. The case of England and Wales*, London, 1973.

51 Unpublished study, presented at the University of Ghent in 1981 by C. STEVENS.

52 J. DE BROUWER, *Demografische evolutie van het Land van Aalst, 1570-1800*, Brussels, 1968.

53 L. PRENEEL, *De demografische ontwikkeling in België sedert 1800. Aspecten van een transitieproces*, Louvain, 1981; P. DEPREZ, The Low Countries, in W. R. LEE, *European demography and economic growth*, London, 1979, pp. 236-283.

54 J. DUPAQUIER, Les caractères originaux de l'histoire démographique française au XVIIIe siécle, in *Revue d'Histoire Moderne et Contemporaine*, 1976, pp. 182-202.

55 N. L. TRANTER, *op. cit.;* M. W. FLINN, *The European demographic system, 1500-1820*, Brighton, 1981.

56 P. DEANE - W. A. COLE, *op. cit.*, p. 103.

Marriage behaviour also shows a fundamental difference. While a decrease in marriage ages occurred in England toward the end of the Ancien Régime, Flanders — like France — was distinguished by a restrictive marriage pattern [57]. The average age at first marriage was about 29 years for men and 27 years for women around 1800 [58]. A further indication of the restrictive pattern of nuptiality in Flanders can be seen in the proportion of men and women who never married. This proportion increased from 15% at the beginning of the eighteenth century to more than 25% shortly after the middle of the nineteenth century [59]. In most areas of Flanders the index of nuptiality (Im) was even lower than 0.45 [60].

All this resulted in a gradual ageing of the population. In that sense it is striking that the proportion of "communicants" in the total population was only 60% around 1700 as against 70% in 1800 [61]. Direct proof of gradual ageing in Flanders is found when we split up the age groups recorded in the censuses for the end of the eighteenth and the first decades of the nineteenth century. The differences between Flanders, France and England become very clear when we bring together the results of these first censuses:

Age structures in Flanders, France and England 1791-1851 [62]

	England 1791	*England 1821*	*Flanders 1796*	*Flanders 1846*
0-19	42.4	48.3	42.1	40.2
20-29	14.4	16.4	15.1	16.8
30-69	38.4	32.4	38.3	39.7
70+	4.8	2.9	3.4	3.3

	France 1816	*France 1851*
0-19	41.2	36.1
20-59	49.4	53.7
60 +	9.4	8.7

57 C. VANDENBROEKE, Caractéristique de la nuptialité et de la fécondité en Flandre et en Brabant aux XVIIe - XIXe siécles, in *Annales de démographie historique*, 1977, pp. 7-20; Karaksteristieken van het huwelijks - en voortplantingspatroon in Vlaanderen en Brabant, 17de-19de eeuw, in *Tijdschrift voor Sociale Geschiedenis*, 1976, pp. 111-115.

58 In England an opposite evolution can be seen at the end of the eighteenth century. Cfr. R. D. LEE - R. S. SCHOFIELD, *British population...*, in R. FLOUD - D. MCCLOSKEY, *op. cit.*, p. 27; E. A. WRIGLEY, *The growth of population in eighteenth-century England: a conundrum resolved, in Past and Present*, 1983, pp. 121-150.

59 C. VANDENBROEKE, *Karakteristieken..., op. cit.*, pp. 116-117.

60 R. LESTHAEGHE, *The decline of Belgian Fertility, 1800-1970*, Princeton, 1977; E. VAN DE WALLE, La nuptialité en Belgique de 1846 à 1930 et sa relation avec le déclin de la fécondité, in *Population et famille*, 1965.

61 During the period 1630-1845 the proportion of "communicants" increased by more than 20%:

1630-39 = 59.7%	1700-19 = 63.6%	1770-89 = 69,0%
1660-79 = 59.7%	1710-29 = 63.3%	1780-99 = 71,8%
1670-89 = 64.7%	1740-59 = 67.3%	1810-19 = 71,0%
1780-99 = 64.5%	1750-69 = 66.0%	1840-45 = 72,2%
1690-09 = 65.7%	1760-79 = 66.2%	

Christian Vandenbroeke

There was a stronger representation of adults in Flanders and France than in England, and during the first half of the nineteenth century this difference became more pronounced. Thus the dependency ratio, with an average of 0.6 was notably lower and better adapted to economic development. If we also suppose that the participation of women and children was at its maximum in the rural and linen industry of Flanders, there will be less doubt for the higher rate of activity [63]. Nowhere was there more labour power available in proportion to the population than in the countryside of Flanders [64].

On the eve of the Industrial Revolution, when output was still largely provided by muscle power, this was an undeniable advantage.

In many ways it helps explain the prosperity of Flanders' regional economy during the eighteenth century. As J.D.Chambers put in 1972: "Increased output was not due entirely to new technology, but increasing effort by those working in the traditional industries played its part" [65]. However, the saving and spending possibility of the family budget was strongly influenced by the age structure and the low dependency ratio [66]. The smaller the weight of the passive population, the larger the opportunities to save [67].

Concerning the agricultural economy there is no doubt at all that Flanders was one of the most highly developed areas and enjoyed an intensive and progressive agricultural system without fallow and providing two crops a year. Average grain yields of 1500 to 2000 lires per hectare had been common since the late Middle Ages [68]. In spite of the high population density of 250 to 300 inhabitants per km^2, Flanders was able until the end of the Ancien Régime to export a large part of grain production. During the second half of the eighteenth century about a quarter of the so-called bread-grain (wheat, meslin or rye) was exported. Moreover, about 20% of the harvest was used for beer and gin production[69]. The best proof of the reputation that the Flemish agricultural

[62] J. D. CHAMBERS, *Population, economy and society in pre-industrial England*, Oxford, 1972, p. 121; N. L. TRANTER, *The labour supply, 1780-1860*, in R. FLOUD - D. McCLOSKEY, *op. cit.*, p. 210. The age structure in France, worked out by J. BOURGEOIS-PICHAT, is cited in P. LEON *et. al.*, *Histoire économique et sociale de la France. III: l'avènement de l'ère industrielle (1789-années 1880)*, Paris, 1976, p. 185. An exhaustive study of the census of 1796 in East-Flanders is published by J. DE BELDER-L. JASPERS-C. STEVENS-C. VANDENBROEKE, *Arbeid en tewerkstelling in Oost-Vlaanderen. Een demografische en socio-professionele analyse*, Ghent, 1984.

[63] B. VERHAEGEN, *op. cit.*, passim.

[64] See also the notion of "productivity of labour", worked out by F. MENDELS.

[65] J. D. CHAMBERS, *op. cit.*, p. 149.

[66] Cf. R. D. LEE - R. S. SCHOFIELD, *op. cit.* p. 30: "changes in the age structure of the population, due primarily to its higher fertility late in the (XVIII) century, did increase the consumption presure on the typical worker".

[67] M. W. FLINN, *The origins of the Industrial Revolution*, London, 1966, pp. 30-34.

[68] B. H. SLICHER VAN BATH, The yields of different crops (mainly cereals) in relation to the seed c. 810-1820. In *Acta Historiae Neerlandica*, 1967, pp. 26-106.

[69] F. MENDELS, *Agriculture and peasant industry...*, *op. cit.*, C. VANDENBROEKE,

economy enjoyed is found in the fact that so many foreign agronomists (R. Weston, J.Sinclair, T.Radeliff, I.Thys, D.De Gomicourt (Dérival), T.A.Mann, A.Thaer, J. Cordier, J.N.Schwerz and S.Von Grouner) went to this country to learn about the production results and the rotation-system [70].

The situation was totally different in England. In the course of the seventeenth century great progress was made in agriculture[71]; this could not prevent England from becoming a grain-importing nation after 1760-1770 [72]. The repercussions of this evolution are found in the higher prices of grain at the end of the eighteenth century; as we know already from the calculation of physical product, the absolute maxima were reached during the years of the Continental Blockade. It is one of the reasons why the real value of the physical product, expressed in litres of wheat, was so low in England compared to Flanders or France.

In terms of economic policy a promising strategy was applied to the Southern Netherlands during the reign of Maria Theresia (1740-1780). The tariffs on international trade were no longer seen as part of a fiscal system, but also as a protective support for the domestic economy.

For this reason at the initiative of Coblenz, trade policy was revised completely [73], and simultaneously initiatives were taken to stimulate home production as well as foreign trade [74].

In order to apply economic policy as effectively as possible, different ad hoc committees and special "bureaux de régie" or "jointes" were established [75].

The most spectacular results were obtained in the infrastructure [76]. Even the most critical contemporaries agreed that the best road system in Europe was to be found in Flanders. Voltaire's praise of its quality is famous: "amongst the modern nations, it is France and the little country of Belgium which have roads

Agriculture et Alimentation dans les Pays-Bas Autrichiens, Ghent, 1975, pp. 85-90; W. Vanderpijpen, *op. cit.*

[70] A. Verhulst-G. Bublot, *op. cit.*, p. 29.

[71] E. L. Jones, *Agriculture and economic growth in England, 1650-1815*, Suffolk, 1967; *Agriculture, 1700-80*, in R. Floud-D. McCloskey, *op. cit.* pp. 66-85.

[72] P. Deane-W. A. Cole, *op. cit.*, p. 65.

[73] H. Van Houtte, L'essor économique de la Belgique sous le règne de Marie-Thérèse, 1740-1780, in *Revue Générale*, 910; *Histoire économique de la Belgique à la fin de l'Ancien Régime*, Ghent, 1920.

[74] P. Moureaux, *La statistique industrielle dans les Pays-Bas Autrishiens à l'epoque de Marie-Thérèse, Brussels*, 1974; C. Vandenbroeke, *Agriculture et Alimentation..., op. cit., passim.* In France, such a system of "enquétes" was also in use during the second half of the eighteenth century. Cf. C. E. Labrousse, *La crise de l'économie française à la fin de l'ancien régime et au début de la révolution*, Paris, 1944.

[75] P. Moureaux, Un organe peu connu du Government des Pays-Bas Autrichiens: le Bureau de Régie des Droits d'entrée et de sortie, in *Revue Belge de Philologie et d'Histoire*, 1986.

[76] L. Genicot, *Histoire des routes belges depuis 1704*, Brussels, 1948.

worthy of Antiquity"[77]. Analogous indications are given in the statistical reports from the beginning of the nineteenth century. At that time Belgium possessed 1 km (navigable) waterway per 14 km^2 against 1 km per 33 km^2 in England[78]. The contrast of the country roads is illustrated by a comment from X. Heuschling: «Belgian roads can be considered the finest in Europe: they are better than those of England and better by far than those of France[79]. In 1850 it was still accepted that there were three times more roads and canals in Belgium than in England[80].

Another variable concerns foreign trade. It cannot be denied that England had a good lead in this, largely because of the strong expansion of colonial trade since the seventeenth century[81]. About the year 1800, foreign trade represented ± 30% of the physical product of England[82], considerably higher than in France or the Southern Netherlands. However, the contribution of foreign trade to industrial growth cannot be seen as a leading force before the end of the Ancien Régime[83], and it was some time later, as R. P. Thomas and D. N. Mc Closkey argue, before foreign trade became really significant: «the strongest effect between commerce abroad and industry at home was from industrialisation to commerce, not the reverse. Trade was the child of industry»[84]. Furthermore the rhythm of foreign trade was not at all linear. Sharp interruptions characterized the years 1770-80's in France as well as in England or the United Provinces[85]. Against this there was an extraordinarily favourable evolution in a few smaller countries and regions. This is also the case with the Southern Netherlands, where after the middle of the eighteenth century traditional trade gaps disappeared and even changed into a positive balance with some neighbouring countries[86].

The increase in foreign trade, expressed in constant prices, would have been

[77] L. DECHESNE, *Histoire économique et sociale de la Belgique*, Liège-Paris, 1932, p. 275.

[78] I. T. BEREND - G. RANKI, *The European Periphery and Industrialization, 1780-1914*, Cambridge, 1982, p. 92.

[79] P. LEBRUN, *op. cit.*, p. 605.

[80] L. DECHESNE, *op. cit.*, p. 377.

[81] R. DAVIS, *The Industrial Revolution and British Overseas Trade*, Leicester, 1979; P. BAIROCH, *Commerce extérieur et dàeveloppement économique de l'Europe au XIXe siécle*, Paris, 1976; N. F. R. CRAFTS, *British economic growth..., op. cit.*, F. CROUZET, Angleterre et France au XVIIe siécle. Essai d'analyse comparée de deux croissances économiques, in *Annales. Economies-Sociétés-Civilisations*, 1966; R. P. THOMAS-D. MCCLOSKEY, *Overseas trade an empire, 1700-1860*, in R. FLOUD-D. MCCLOSKEY, *op. cit.*, pp. 87-102.

[82] P. DEANE. *The Industrial Revolution and economic growth: the evidence of early British National Income Estimated, in* R. M. HARTWELL (ed.), *op. cit.*, p. 86.

[83] R. ROEHL, *Britain and European Industrialization..., op. cit.*

[84] R. N. THOMAS-D. MCCLOSKEY, *op. cit.*, p. 102.

[85] F. BRAUDEL, *op. cit.*, p. 175; F. CROUZET, *Angleterre et France... op. cit.*, p. 262.

[86] V. JANSSENS, *Het geldwezen der Oostenrijkse Nederlands*, Brussels, 1957.

about equal in France and England. Between 1716-20 and 1784-88, F. Crouzet estimates an increase of respectively 200% and 140 to 175% [87]. Less complete indications are available for the Southern Netherlands. We know only the total amount of foreign trade that took place with France and England between 1716-20 and 1776-88 [88]. With France there was an increase from 7.2 million to 33.7 million Livres Tournois (= x 4.7); with England foreign trade rose from an average of 347,435 to 1.82 million Pounds Sterling (= x 5.2). For both countries together, which make up more than half the trade relations of the Southern Netherlands, this represents a fivefold increase. In other words foreign trade, expressed in constant prices, increased at least as quickly as in the most important neighbouring countries up to and including the third quarter of the eighteenth century. A maximum growth arose in the course of the 1770s, when the Southern Netherlands profited by neutrality during the American war. On the other hand, the years 1783-89 indicate again a decline in the trade activities with England and France. The peak of economic expansion was over. The Flemish economy and the foreign trade in the Southern Netherlands peaked just in the second and third quarter of the eighteenth century [89].

Considering the different explanations and variables that are generally used by economic historians, we are left comforted by more problems than solutions. Many explanations and generalisations prove weak and doubtful when they are related to the economic perspectives and performances of smaller countries and regions. The so-called «privileged» starting point in England, which is always emphasized by economic historians, does not hold any longer. Indeed, in most cases, we find that the Flemish regional economy was at least equally productive and in many respects enjoyed a more favourable position. This is illustrated by the following summary of the most frequently cited variables:

Variables cited to explain industrial development at the and of the Ancien Régime

	In favour of Flanders	*Equal*	*In favour of England*
"Human Capital"		X	
Population growth	X		
Age structure	X		
Activity rate	X		
Agriculture	X		
Home industry	X		
Economic policy		X	
Infrastructure	X		
Importance of foreign trade			X
Growth of foreign trade		X	
Domestic purchasing power		X	

87 F. Crouzet, *Angleterre et France..., op. cit.*, p. 261.

88 R. Davis, English foreign trade, 1700-1774, in *Economic History Review*, 1962-63); R. Romano, Documenti e prime considerawioni intorno alla 'Balance du commerce' della Francia dal 1716 al 1780, in *Studi in onore di Armando Sapori*, II, Milano, 1957, pp. 1265-1300.

89 C. Vandenbroeke, *Sociale Geschiedenis..., op. cit.*, p. 185.

Christian Vandenbroeke

These eleven variables suggest that the balance was six times in favour of Flanders, that the situation was comparable in four cases and that only foreign trade revealed a better outlook for England.

The favourable starting position of the Flemish regional economy is revealed even more strongly when we take into account its competitive position and the wage costs. Such comparisons are too frequently neglected [90]. Still, the cost of wages forms one of the most essential components in the final level of prices. In a society where labour-saving devices were few and where therefore few differences in productivity were possible, it is logical that the relative importance of the costs of wages will be great.

Because good wage surveys for industrial labourers are lacking, this issue cannot be resolved in a comprehensive form. But an approximate picture can be derived from the situation of unskilled or less qualified workers in the building sector [91]. The wage disparity found for this group is *a fortiori* true for industrial labourers and textile workers:

Wages of bricklayers, expressed in grams of silver.

Period	*Flanders*	*France*	*England*	*United Provinces*
1651-75	6.20	6.93	5.60	7.78
1676-00	6.05	6.19	5.62	7.75
1701-25	5.33	—	6.77	7.69
1726-50	5.29	5.73	7.18	7.69
1751-75	4.94	5.88	7.37	8.34
1776-00	5.10	7.40	9.19	8.65
1801-25	6.17	8.77	13.47	8.65
1826-50	6.14	10.26	13.91	8.58

Following the wage data from 1651 to 1850, we find some striking differences, especially between wage levels in England and Flanders [92]. At the end of

[90] A comparative analysis of the (nominal) price levels was set up by F. BRAUDEL-F. SPOONER, Prices in Europe from 1450 to 1750, in *The Cambridge Economic History of Europe*, IV, Cambridge, 1967. Less research was done to bring together the wage costs in the different countries. However, see the studies of J. DE VRIES, *An inquiry into the behaviour of wages in the Dutch Republic and the Southern Netherlands, 1580-1800;* J. MOKYR, *Industrialization in the Low Countries, 1795-1850*, New Haven, 1976, and J. M. M. DE MEERE, *Economische outwikkeling en levenstandard in Nederland gedurende de eerste helft van de negentiende eeuw*, The Hague, 1982.

[91] C. VERLINDEN - E. SCHOLLIERS (eds.), "*Documents pour l'histoire des prix et des salaires en Flandre et en Brabant*", Bruges, 1959-1973; C. VANDENBROEKE, "Prijzen en lonen als sociaal-economische verklaringsvariabelen (14e-20e eeuw)", in "*Handelingen Maatschappij Geschiedenis en Oudheidkunde Gent*", 1982.

[92] For the decrease of the standard of living in England during the XVIIth century, see C. HILL, "*Reformation to Industrial Revolution*", Harmondsworth, 1976, p. 84; "1620-1650 were probably among the most terrible years through which the country has ever passed".

the eighteenth century the wages of unskilled or semi-skilled labourers in Flanders were 60 to 70% lower than in England. At that moment, when productivity differences were exceptional, factor costs were essentially fixed by the wages. This holds particularly for labour-intensive sectors such as textiles, which is also the basic sector of industrial activity in the Ancien Régime [93]. In this way, it is no longer a surprise to find that the export trade in general and the export of linen from Flanders in particular were characterized by an exponential growth in the course of the eighteenth century [94]. Even the sharpest protectionist measures abroad could not outweigh the lower cost of labour. No wonder that Flemish linen played a leading part in international trade at the end of the Ancien Régime [95]. On the other hand, England, the United Provinces and France were becoming less competitive. In many ways, the interruption of foreign trade during the third quarter of the eighteenth century was a direct result of the high level of prices and wages in these countries [96].

But lower nominal wages should not be identified with a lower standard of living in Flanders [97] They provide only the means for comparing the economic competitiveness among different countries and nations. Comparisons of purchasing power can be made only by turning the basic data into real prices and real wages, and all kinds of fiscal and social deductions must be taken into account [98]. We indicated above that there were not any essential differences concerning purchasing power among the Western nations before the end of the

[93] F. Crouzet, Essai de construction d'un indice annuel de la production industrielle française au XIXe siécle, in *Annales. Economies-Sociétés-Civilisation, 1970.* Concerning the problem of factor costs, cfr. P. Leon et. al., *Histoire économique et sociale de la France.* III, Paris, 1976, pp. 538-539.

[94] An approach of the export figures for linen pieces can be given in this way:

1670 = 120.000	1800 = 195.000
1700 = 120.000	1815 = 250.000
1720 = 130.000	1825 = 275.000
1750 = 145.000	1833 = 210.000
1780 = 195.000	1845 = 185.000

[95] In general about 80% of the production was exported to Spain and France and from these countries to the New World!.

[96] This is especially true for the United Provinces. For France and England the problem is not so clear, because the interruptions in foreign trade were influenced by the American war.

[97] "Il faust substituer à la monnaie défaillante un étalon de valeur nouveau, assis sur la notion de richiesse réelle" (A. Sauly, *Histoire économique de la France entre les deux guerres,* Paris, 1965, p. 314).

[98] F. Braudel, *Le temps du monde,* Paris, p. 329: "le contribuable français est soumis à un lourd prélèvement social, au bénéfice des segneurs et de l'Eglise. Et c'est cet impôt social qui limite à l'avance l'appétit du Trésor Royal".

Ancien Régime. The table below, expressing the nominal wages of bricklayers in wheat-equivalents, brings an additional illustration of this conclusion:

Real wages of bricklayers, expressed in litres of wheat

Period	*Flanders*	*France*	*England*	*United Provinces*
1651-75	8.6	8.5	6.9	13.1
1676-00	8.9	8.1	8.4	14.8
1701-25	9.7	—	10.4	15.6
1726-50	10.7	10.6	11.9	16.4
1751-75	9.3	9.5	10.2	15.0
1776-00	7.8	9.9	8.9	11.9
1801-25	7.3	12.2	9.2	9.8
1826-50	7.3	15.1	13.5	10.0

The differences in the middle of the eighteenth century fade away almost completely when tax burdens are taken into account (see below). In this sense, we can repeat that purchasing power was about equal in all these countries and areas at the end of the Ancien Régime. In other words: the improving standard of living after 1750 certainly was a strong stimulus to economic expansion. However, this cannot be seen as a final factor explaining the industrial take-off in England. Everywhere in Nort-West-Europe a similar improvement of the social climate was felt between 1725-1775 [99]. On the other hand, competitive power differed very strongly from country to country. And here Flanders had better opportunities for economic growth during the eighteenth century.

The deeper reason for Flanders' favourable position may be explained by four factors: lower food prices, lower wage costs, less fiscal duties and less inflation.

In the first place we have to take the price of foods into account. While about half of family expenses went to the purchase of grains and bread until well into the nineteenth century, it is logical that the level of the grain prices must have had an enormous influence on wage levels [100].

A summary of wheat prices shows very sharp contrasts among the different countries in North-West-Europe [101]:

[99] Cf. E. Le Roy Ladurie, "*Le territoire de l'historien*", II, Paris, 1978.

[100] See for instance abbé Mann: "le prix de la journée de travail fût fixé en tout temps à la valeur d'un tiers d'un boisseau de froment; par-là ses gains seroient constamment proportionnés à ses besoins" ("Mémoire sur la question: dans un pays fertile et bien peuplé, les grandes fermes sont-elles utiles ou nuisibles à l'Etat en général?", in "*Memoires de l'Académie Impériale et Royale de Bruxelles*", IV, 1783, p. 219).

[101] C. Verlinden - E. Scholliers (eds.), *op. cit.;* C. Vandenbroeke, "*Prijzen en lonen...*", *op. cit.*

Wheat prices per hectolitre, expressed in grams of silver.

Period	*Flanders*	*France*	*England*	*United Provinces*
1651-75	73.4	87.3	81.3	61.2
1676-00	72.2	79.7	69.9	55.7
1701-25	59.0	67.2	66.7	50.3
1726-50	50.8	54.4	61.9	47.5
1751-75	53.4	64.5	76.5	56.4
1776-00	67.2	75.0	106.7	75.7
1801-25	88.0	95.8	151.1	97.5
1826-50	84.9	88.2	103.8	86.8

In the course of the eighteenth century there was a price advantage in Flanders of 10 to 15% with respect to France and of 25 to 30% with respect to England. Only at the end of the eighteenth century were wheat prices also lower in the United Provinces. Contemporaries also noted in their travel books the lower prices prevailing in Flanders,[102] and good example can be found in the travel book of Dérival: «at Brussels it is possible to find all sorts of food-stuffs; in general they are rather cheap; life is less expensive in Brussels than in Paris and probably about half that in London or Amsterdam...».[103] The lower grain prices in Flanders are more noticeable still when we remeber that the largest population density and the quickest population growth were to be found there. This is another proof of the extraordinary evolution of its agricultural economy and food production. However, this evolution was also sustained by a radical change in the production and consumption pattern and above all in the early spread of potatoes in Flanders. In the course of the eighteenth century the consumption if grain was halved. One person consumed only 0.5 or 0.6 litres of grain per day instead of one litre at about 1700. The loss of calories was compensated eating one kg of potatoes. [104]

The lower wage costs can also be explained by differences in the tax burden. At the end of the Ancien Régime the tax burden was about 20 to 25% of physical product in England and the United Provinces. In France this proportion can be

[102] Cf. J. Gardner, *Voyage pittoresque par Manheim, Mayence, Aix-la-Chapelle, Bruxelles.* Translated by J. Coudroy, London, 1972, p. 150; *Lettres and journals of Mrs. Calderwood of Polton from England, Holland and the Low Countries in 1756,* edited by A. Fergusson, Edinburgh, 1884, pp. 267-268.

[103] Derival, *op. cit.,* I. p. 26.

[104] C. Vandenbroeke, "Cultivation and consumption of the potato in the XVIIth and XVIIIth century", in "*Acta Historiae Neerlandica*", V, 1971, pp. 15-39.

put at 10 to 15%.[105] In the Southern Netherlands it was no more than 5 to 10%, a level that would persist up to the beginning of the twentieth century. L. Franck, of Minister of State and governor of the National Bank, noticed: «Before the war of 1914, Belgium was a fortunate country in which taxation placed only a light burden on the economy.[106] The economic historian B.S. Chlepner shares a similar view and claims that «the Belgian tax-payer was of the least heavily taxed in the world».[107]

We must also look at the contrasts in employment. It is clear that in England, as a direct result of enclosures, the problem of expropriation was more acute for the lower classes than in France or the Southern Netherlands. Full-time wage-labourers were not at all unusual in eighteenth century England. In Flanders, on the contrary, the majority of the people still owned a piece of land.[108] This meant that people sought first of all an income from agriculture, while rural industry was considered a part-time activity. The prefect of the Département de l'Escaut, Faipoult, affirmed this situation in his «*Mémoire Statistique*»: «the first commitment is to work on the land, which is done with diligence, but all the time that cannot be devoted to agriculture is taken up with the spinning and weaving of linen and wool».[109] Full-time weavers and spinners were to be found in Flanders only after the second quarter of the nineteenth century. The effects on wages are obvious: part-time workers, who have an income from other activities (for instance, agriculture), can live with smaller earnings.[110]

Finally, monetary relations could also influence the evolution of prices and wages, but J. Riley and J.J. McCusker have demonstrated recently this was not a strong factor during the eighteenth century.[111] There are also considerable difficulties involving the accuracy of the money supply estimates in the different

105 P. Mathias, *op. cit.*; P. Mathias-P. O'Brien, "Taxation in Britain and France, 1715-1810. A comparison of the social and economic incidence of taxes collected for the central governments", in "*The Journal of European Economic History*", 1976; M. Allais, "Classes sociales et civilisations", in "*Economies et Sociétés*", 1974; F. Braudel, "*Le temps du monde*", *op. cit.*, p. 329; F. Hincker, "*Les Français devant l'impôt sous l'ancien régime*", paris, 1971; M. Morineau, "Budgets de L'Etat et gestion des finances royales en France au dix-huitième siécle", in "*Revue Historique*", 1980.

106 L. Franck, *La stabilisation monétaire en Belgique,* Paris, 1927, p. 18.

107 B. S. Chlepner, *Cent ans d'histoire sociale en Belgique,* Brussels, 1972, p. 196.

108 Just as in Flanders, the process of expropriation was also in the Walloon area less pronounced before the beginning of the nienteenth century. See the publications, cited in note 9.

109 Faipoult, *op. cit.*, p. 103.

110 F. Mendels, *Seasons and regions in agriculture and industry during the process of industrialization,* in S. Pollard, *op. cit.*, p. 189.

111 J. Riley- J. J. McCusker, Money supply, economic growth and the quantity theory of money: France, 1650-1788, in *Exploration in Economic History,* 1983, pp. 274-293.

countries, but an approximate idea can be indicated for the end of the Ancien Régime. Once again the values are related to physical product:

	Southern Netherlands [112]	*France* [113]	*England* [114]
Amount of money per head	82.5 francs	116.4 francs	150.0 francs
Physical product per head	235 francs	230.3 francs	287 francs
Velocity of money	2.8	1.9	1.9

Especially in England, the abundant creation of money and bank-notes caused a higher inflation, and this again is confirmed in many contemporary accounts. Dérival, or example, wrote in the 1780s: «It was following the creation of the *Banque de Londres* that labour in England became 15% to 16% dearer than in France which now as a result holds a 15 to 16% advantage over the English in every market».[115] The comparison holds well for Flanders and the Southern Netherlands, where the amount of money per head of the population was 30 to 40% lower than in France!

Using Fisher's equation M x V = P x T, we can infer that the smaller stock of money species in the Southern Netherlands was compensated by a greater velocity. And knowing that the average price level was lower than abroad, we can further also conclude that the factor T was greater than in France or England. Equilibrium in Fisher's equation is only made possible by a greater turnover of goods in relative terms. This is doubtless the best proof of the economic expansion in the Southern Netherlands and Flanders at the end of the Ancien Régime. Some years ago E. Héling came to the same conclusions: «Flanders and Brabant experienced a prosperity which was quite exceptional in Europe at that time».[116]

For all these reasons, it is remarkable that economic historians, with the exception of the new interest in the process of proto-industrialization (which is all too often seen in relation to impoverishment) have paid so little attention to the Flemish regional economy. All manner of economic relations have been put forward to explain industrial modernization and the causes of the Industrial Revolution, and it has always been accepted that England enjoyed some advan-

[112] V. JANSSENS, *op. cit.*, pp. 164-165.

[113] F. BRAUDEL, *op. cit.*, p. 212; see also P. LEON, *op. cit.*, p. 409.

[114] S. JONES, The first currency revolution, in *The Journal of European Economic History*, 1981, pp. 589-592; J. RILEY - J.J. MCCUSKER, *op. cit.*, pp. 277 and 289.

[115] DERIVAL, *op. cit.*, IV, pp. 311-312.

[116] Presentation by E. HELIN of article of J. POLASKY, La Révolution Brabançonne, in *Cahiers de Clio*, 1979, p. 54.

tage vis-à-vis other countries. The favourite reasoning is well known: *post hoc, ergo propter hoc.*[117] Because England was the pioneer of the Industrial Revolution, it was taken for granted in the older literature that its evolution opportunities must have been optimal!

However, this one-sided vision has become more and more open to doubt in recent years. A recent comparative analysis of England and France proved that the supposed contrasts between the two countries were more apparent than real. A further study of regional cases, including the economic evolution of Flanders, leads to further revision of the older conclusions and interpretations. Industrial growth of the eighteenth century in fact seems to have owed less to a favoured starting position in England: indeed, the reserve is true. The best prospects were to be found in smaller countries like Flanders, where lower wage costs were the decisive force in economic expansion in the course of the eighteenth century. This meant also that Flanders had no reason at all to change its economic structures and strategy, since lower nominal wages were at that moment the best weapon in the struggle for more competitiveness. However, with the technical innovations at the end of the century, this advantage of lower nominal wages produced a phase of recession and de-industrialization. On the other hand, the situation in England was quite the reverse. Lack of competitiveness threatened to throw the country in a hopeless crisis, just as was to be the case in the United Provinces. The only way — and the best way — to cope with the prosperous development of proto-industry in Flanders, and in some other regions like the Rhineland, Westphalen or Silesia was to increase productivity through mechanization.[118] In other words, the real reason for the industrial take-off has to be found in a threatening crisis! For England, the Industrial Revolution was much more a *MUST* than a historical coincidence or a logical result of economic leadership through the eighteenth century.[119] It was an attempt to overcome the competitiveness of rivals on the Continent.

[117] N. F. R. Crafts, *Industrial Revolution...*, *op. cit.*, p. 433.

[118] By the development of the linen industry in the area of Bielefeld, see W. Mager, Protoindustrialisierung und agrarisch-heimgewerbliche Verflechtung in Ravensberg während der Frühen Neuzeit, in *Geschichte und Gesellschaft*, 1982.

[119] N. F. R. Crafts, *op. cit.*, pp. 429-433.

J. MOKYR

The Industrial Revolution in the Low Countries in the First Half of the Nineteenth Century: A Comparative Case Study

The comparative method will accomplish great things.... The historical specialist asks for a method which is a technical instrument, generally used, easily manageable and capable of giving positive results. The comparative method is precisely such an instrument.... I believe that this method can and must penetrate monographic studies. Marc Bloch

NORTHWESTERN Europe has always enjoyed a prominent and perhaps slightly disproportionate role in the discipline of economic history. It is therefore surprising that the Low Countries in the nineteenth century have not as yet received from English-speaking economic historians the attention they deserve by virtue of their location, their size, and the uniqueness of their case.[1]

This neglect is the more surprising because Belgium and the Netherlands offer a unique opportunity for comparative economic history: they are countries of comparable size, located in the same corner of Europe, surrounded by giants, with considerable overlapping in linguistic and cultural backgrounds. One of them, Belgium, underwent a process of rapid industrialization in the first half of the

This research is made possible by a grant from the Concilium on International and Area Studies. I should like to express my thanks to my dissertation advisers John C. H. Fei and William N. Parker for their help and encouragement. In addition, I am indebted to many of my friends and colleagues, especially Stephen DeCanio, Jan deVries, Nachum Gross, Yoav Kislev, and Rick Levin. The usual warning about the author's sole responsibility is in order.

[1] Belgium is covered to some extent in D. Landes, *The Unbound Prometheus* (Cambridge: University Press, 1969), Ch. 3 and in W. O. Henderson, *Britain and Industrial Europe, 1750-1870,* 2nd ed. (Leicester: Leicester University Press, 1965), pp. 102-38. (The latter deals only with the interaction of Britain and Belgium.) The basic facts with respect to Belgium can also be found in the somewhat unbalanced contribution of Jan Dhondt and Marinette Bruwier to the *Fontana Economic History of Europe,* Vol. 4, Sec. 1. A more useful, though brief, summary of Belgian development in this period can be found in A. S. Milward and S. B. Saul, *The Economic Development of Continental Europe, 1780-1870* (London: Allen and Unwin, 1973), pp. 292-296, 432-453. The Netherlands has been virtually untouched by English-speaking economic historians, and nonspecialists have to rely on translated work, sometimes in abridged form. A recent article of a very general nature is J. A. Van Houtte, "Economic Development of Belgium and the Netherlands from the Beginning of the Modern Era," *The Journal of European Economic History,* I (Spring 1972), 100-120.

nineteenth century, emerging in the 1840's as the most industrialized country on the Continent. On the other hand, very little industrialization can be observed in the Netherlands during the same period. It may be important to inquire into the underlying causes of this gap not only in order to understand the particular economies of the Low Countries, but also to shed some light on the mechanisms of economic growth and industrialization in certain types of economies.

The first section of this article will be devoted to a summary description of the industrial sectors of the two countries between 1798 and 1850. Then some of the traditional explanations that have been put forward to explain their divergent patterns will be presented. This will be followed by a simple model of economic growth that will help in focusing on one crucial element in the differential development of the two countries. Next, the basic relevance of this model will be demonstrated, and finally, some evidence will be presented and some suggestions for future research made.

I

For the purpose of this article it is useful to define the discontinuous element of the industrial revolution as the adoption of a new and more capital-intensive technology to produce the same (or very similar) commodities that were formerly produced. The new technology is exogenously given to the economy (in this case, imported from England). According to this definition, the beginning of the industrial revolution in Belgium can be dated at the close of the eighteenth century. From the beginning, three main centers of industrialization can be distinguished: (1) the cotton center in and around Ghent in East Flanders; (2) the wool center in the département de l'Ourthe (now the Province of Liège) along the Vesdre, with Verviers and Eupen its main focal points; (3) the heavy industry in the city and surroundings of Liège and in the département de Jemappes (now the Province of Hainault).

These three industrial centers underwent remarkable expansion between the late 1790's and 1812. This development can be illustrated by a few statistics that are, of course, a poor substitute for an exhaustive study on Belgian industry in the French period that still remains to be written. The mechanized cotton-spinning industry in Ghent grew from zero in 1798 (the year in which Lievin Bauwens smuggled his first mule from England to the Continent) to 115,000 spindles employing about 10,000 workers in 1810.[2] There is reason to believe that Ghent accounted for about two thirds of the total cotton industry in Flanders.[3] The dimensions attained by this industry in a short period can thus easily be seen.

In 1798, the same year that mechanized cotton-spinning was introduced on the Continent, the wool-producing firm of Biolley and Simonis in Verviers hired William Cockerill to construct the first spinning mills and carding machines at their plant. As in the cotton industry, the discontinuous element in this development is not the emergence of a new industry *ex nihilo,* but rather the application of a new technique to the production of old goods and the associated quantitative expansion. The annual rate of growth of the output of cloth between 1800 and 1810 is estimated at 6 percent, compared to .46 percent between 1752 and 1784.[4]

The data on heavy industry are less easily interpreted. Two of the major breakthroughs in the English iron industry, namely, the use of coke in the smelting process and Cort's rolling and puddling process, were introduced only after the French period.[5] Nevertheless, the number of blast furnaces, estimated at 63 in the period before 1795, rose to 83 in 1814. The output per furnace grew from less than 1,000 kg. a day to as much as 3,000 kg. a day.[6] Total output of cast iron in the département Sambre et Meuse (today the Province of Namur) grew, according to one source, from 10,671 tons in 1789 to 15,240 tons in 1811.[7]

2 *Municipal Archives,* Ghent, File K2-1 (G).

3 N. Briavoinne, *De l'Industrie en Belgique, Causes de Décadence et de Prosperité, sa Situation Actuelle* (Brussels: A. Wahlen, 1839), Vol. II, p. 374.

4 P. Lebrun, "Croissance et Industrialisation, l'Expérience de l'Industrie Drapière Verviétoise 1750-1850," in: *First International Conference of Economic History (Stockholm 1960), Contributions and Communications* (Paris: Mouton et Cie, 1960), p. 566.

5 Briavoinne, *De l'Industrie,* Vol. I, pp. 252-262. On the relative backwardness of the Belgian iron industry around 1820 see G.M. Roentgen, "Berigt van den Toestand van Ijzerwerken in de Waalsche Provinciën," [Report on the Situation of the Iron Works in the Walloonic Provinces] (1823). Edited with an introduction by M. G. De Boer, *Economisch-Historisch Jaarboek,* Vol. IX (1921), pp. 103-149.

6 N. Briavoinne, "Sur les Inventions et Perfectionnements dans l'Industrie," *Academie Royale de Belgique, Mémoires Couronnés,* XIII (1838), pp. 121, 124.

7 J. P. B. de Montalivet, *Exposé de la Situation de l'Empire* (Paris: Imprimerie Impériale, 1813), pp. 48-49. Montalivet's figures are, however, somewhat inconsistent and his data for the *départment* of Jemappes (Hainault) for example seem too low. For an analysis of Montalivet's data see J. Craeybeckx, "Les Débuts de la Revolution Industrielle en Belgique et les Statistiques de la Fin de l'Empire," in: *Mélanges offerts à G. Jacquemyns* (Brussels: Université Libre de Bruxelles, Éditions de l'Institut de Sociologie, 1968), pp. 115-144.

The collapse of the Napoleonic Empire and the incorporation of Belgium in the Dutch monarchy caused the young Belgian industry considerable strain. By the end of the second decade, however, it seems that recovery was by and large complete. In the mid-1820's many important technological changes were introduced, including widespread use of steam power and mechanized weaving in cotton. In the woolen industry mechanized weaving came much later, but other important advances were made in the twenties and thirties, especially in fulling, teaseling, shearing, and pattern-design.[8] The total number of spindles in the Flemish cotton industry doubled in the fifteen years of Dutch rule.[9] The total amount of fixed capital in the cotton-spinning industry in East Flanders was estimated by a contemporary writer to have grown from about 1.5 million guilders in 1817 to 2.7 million in 1826.[10] Similarly the woolen industry recovered after suffering a severe setback in the last years of the Empire: the *arrondissement* of Verviers produced about 88,000 *pièces* in 1811, 65,000 in 1815, and between 100,000 and 120,000 in 1830.[11] Likewise, the metallurgical industry grew considerably: 1817 marks the founding of John Cockerill's famous machine factory at Seraing, soon to become the most prominent plant of its kind on the Continent. Total output of cast iron was estimated in 1816 to be 65 million pounds (29,000 metric tons).[12] In the mid-1830's estimates vary be-

8 H. Coppejans-Desmedt, "Incidenten rond de Constructie van de Eerste Mechanische Weefgetouwen te Gent," [Some Details Concerning the Construction of the First Mechanical Looms in Ghent], *Handelingen der Maatschappij voor Geschiedenis en Oudheidkunde te Gent,* XIII (1959), pp. 163-177. J. S. Rénier, "Histoire de l'Industrie Drapière aux Pays de Liège," *Mémoires de la Societé Libre d'Emulation de Liège,* n.s. Vol. VI (1881), pp. 181ff, 267, 273, 278. Briavoinne, "Sur les Inventions," pp. 91-93.

9 Briavoinne (1839), *De l'Industrie,* II, p. 374 puts the number of spindles in 1830 at 440,000. In 1835 the number is estimated at 383,320 of which more than 300,000 are "active." Cf. X. Heuschling, *Essai sur la Statistique Générale de la Belgique,* 2nd ed., (Brussels: Établissement Géographique Faubourg de Flandre, 1841), p. 9.

10 H. Coppejans-Desmedt (ed.), "De Statistieken van E. C. Van Der Meersch over de Katoenindustrie in Oost Vlaanderen," [The Statistics of E. C. Van De Meersch on the Cotton Industry in East Flanders], *Bulletin de la Commission Royale d'Histoire,* CXXVIII (1962), 121-181.

11 R. Demoulin, *Guillaume Ier et la Transformation Économique des Provinces Belges* (Liège: Bibliothèque de la Faculté de Philosophie et Lettres de l'Université de Liège, 1938), p. 322.

12 "Mémoires sur les Houillères des Provinces de Hainaut, Namur, Liège et Limbourge," Mons, 1816, cited in A. Warzeé, "Exposé Historique et Statistique de l'Industrie Métallurgique dans le Hainaut," *Mémoires et Publications de la Societé des Sciences, des Arts et des Lettres du Hainaut,* 2nd Series, Vol. 8 (1860-1862), p. 43.

tween 135,000 and 150,000 metric tons.[13] Input of coal in iron production increased from 16,000 tons in 1815 to 180,000 in 1830.[14]

After the revolution of 1830, industrialization seemed to lose some momentum. There is some evidence that the rate of growth of the cotton industry was slowing down substantially, due to a fall in the price of the final goods as well as to a rise in raw material prices.[15] Some growth, however, still went on, despite difficulties on the demand side. In 1846 the number of spindles in the cotton industry was about the same as in 1836,[16] but raw cotton imports increased[17] so that it is possible that excess capacity was reduced and that old machines were replaced by new and better ones. The woolen industry fared much better and was able to quadruple its exports between 1832 and 1844 (though the starting point of this series is unusually low).[18] The greatest expansion was experienced by heavy industry, doubtlessly due in part to the construction of a railway network. Thus, in 1850, Belgium emerged as the most industrialized country on the Continent.

It should be emphasized that the new industry did not supplant the old industry right away, but rather *coexisted* with it for most of the period under discussion. The industry in Belgium before the industrial revolution was a typical "protoindustry"—a rural-domestic industry of peasants who had to supplement their incomes, which had become insufficient because of population pressure on the land.[19] The three main centers of the protoindustrial sector coincided to a large extent with the three centers of modern industry: the linen industry in Flanders, the woolen industry around Verviers, and the metallurgical putting-out industry (nail-making, cutlery, and so on) in Liège, Hainault, and Namur. That this is no coincidence will soon become clear.

13 Briavoinne (1839), *De l'Industrie,* II, p. 290. Heuschling, *Essai,* p. 100.

14 Demoulin, *Guillaume 1er*, p. 315.

15 P. Schöller, "La Transformation Économique de la Belgique de 1832 à 1844," *Bulletin de l'Institut de Recherches Économiques* (Louvain), XIV (1948), 585.

16 *Statistique de la Belgique, Récensement Générale (1846), Industrie* (Brussels: Ministère de l'Intérieur, 1851), pp. x, xi. The figure cited is more than 380,000. Cf. footnote 9.

17 Imports average about 5,000 tons in the first years of the 1830's, rising to double that figure in 1850. *Statistique de la Belgique, Tableau Générale du Commerce avec des Pays Étrangères* (Brussels: Ministère de l'Intérieur, 1838 and subsequent years).

18 Schöller, "La Transformation," p. 594.

19 This is pointed out by F. Mendels, "Industrialization and Population Pressure in Eighteenth-Century Flanders," unpublished dissertation, University of Wisconsin, 1969.

The availability of data for the Netherlands is even more restrictive than for Belgium. But it is beyond doubt that industrialization was considerably slower. It is important to keep in mind that at the beginning of the period the Netherlands was far from being a non-industrial country. Much industry existed, especially in the maritime provinces (Holland). Most of these industries were "traffics," in other words, activities associated directly or indirectly with Dutch commerce and shipping (which were still substantial at the end of the eighteenth century). In addition to ship-building, sail-making, and food-packing, the most important industries were paper, glass and earthenware, sugar refineries, distilleries and breweries, tobacco, and luxury textiles (calico printing). Some of these industries had already declined considerably in the eighteenth century, others much less so.[20] It is clear, however, that these industries suffered very severely during the French period, especially during the years of the continental blockade, and declined to a fraction of their initial size.[21] The impoverished Dutch cities experienced a marked decline in population, and many of the remaining urban dwellers were reduced to charity.[22]

The post-Napoleonic period was one of slow recovery for the traffic industries. The lack of data does not allow the determination of the extent to which recovery in most traffics was completed by 1830. The paper industry, perhaps the most prestigious traffic industry, all but disappeared, since it was unable to adopt the new techniques that came into use among its competitors between 1815 and 1850.[23] The type of heavy industry that was rapidly growing in Belgium was virtually nonexistent in the Netherlands. In 1830 the largest and best-known Dutch machine factory employed only 80 to 100 workers.[24]

The secession of Belgium in 1830 provided the initial impetus for

[20] J. G. Van Dillen: *Van Rijkdom en Regenten,* [On Wealth and Regents, Handbook of the Economic and Social History of the Netherlands During the Republic] (The Hague: Martinus Nijhoff, 1970), pp. 540-557.

[21] F. J. B. D'Alphonse, *Aperçu sur la Hollande* (1811), published by the Dutch Central Bureau of Statistics (The Hague: Belinfante Bros., 1900), pp. 223-309.

[22] Déliberation du Conseil du Département de Zuiderzee, 2 Sept. 1811. *R.G.P.* [National Historical Publications],Vol. 16 (1912), p. 1164.

[23] B. W. De Vries, *De Nederlandse Papiernijverheid in de 19e Eeuw,* [The Dutch Paper Industry in the Nineteenth Century] (The Hague: Martinus Nijhoff, 1957), pp. 182-190.

[24] I. J. Brugmans, *De Arbeidende Klasse in Nederland in de 19e Eeuw (1813-1870),* [The Working Class in the Netherlands in the Nineteenth Century], 8th ed. (Utrecht-Antwerp: Het Spectrum, 1971, orig. pub. in 1925), p. 33.

the adoption of new and more efficient techniques in the Netherlands. A few signs of modernization can be observed in the industries in the maritime provinces. More important were the developments in the areas in which there was some protoindustry. These areas were located in the east (the Twente area) and in the south (especially around the towns of Tilburg and Helmond). It is striking to what extent these areas resembled the centers of industrial development in Belgium: diluvial soils, population pressure, the existence of a rural industry, and so forth.[25] The relative size of these areas was, however, insufficient and growth was slow. In the period before 1830, the cotton and fustian industries in Twente stagnated, while the woolen industry in the south grew very slowly and only with considerable government help. The total capacity of the steam engines used by the cotton industry in the Twente area rose between 1830 and 1850 from a negligible 18 HP to a slightly less negligible 220 HP.[26] Exports of this industry increased from about 30,000 pieces in 1834 to 730,000 pieces in 1840.[27] The influx of a few Belgian Orangist entrepreneurs, as well as the demand of the Dutch Indies for cotton goods no longer supplied by the Flemish industry, explains this growth. In the Tilburg and Helmond areas in the south, where these beneficial effects were by and large absent, growth was much slower.

Thus one could expect a large gap between the two countries around 1850, although divergent rates of growth do not prove this a priori. Unfortunately the lack of comparable national income accounts makes such a comparison impossible. Some very crude indicators can be used here, however, to hint at the orders of magnitude involved. The total number of steam engines in the Dutch economy in 1837 is estimated at 73 with a capacity of 1,120 HP, climbing to

[25] On the protoindustry in the Twente area see, e.g., Z. W. Sneller, "La Naissance de L'Industrie Rurale dans les Pays Bas aux XVIIe et XVIIIe Siècles," *Annales d'Histoire Économique et Sociale*, I (1929), pp. 193-202. Also: J. A. P. G. Boot, *De Twentsche Katoennijverheid 1830-1873,* [The Cotton Industry in the Twente Area] (Amsterdam: H. J. Paris, 1935), pp. 1-105. On the North Brabant industry see, e.g., A. W. M. Keune, "De Industriële Ontwikkeling Gedurende de 19e Eeuw," [Industrial Development in the Nineteenth Century] in: H. F. J. M. Van Den Eerenbeemt and H. J. A. M. Schurink (eds.), *De Opkomst van Tilburg als een Industriestad,* [The Rise of Tilburg as an Industrial Town] (Nijmegen: N. V. Centrale Drukkerij, 1959), pp. 11-60. On Helmond see: W. A. J. M. Harkx, *De Helmonds Textielnijverheid in de Loop der Eeuwen,* [The Textile Industries in Helmond during the Centuries] (Tilburg: Stichting Zuidelijk Historisch Contact, 1967).

[26] Boot, *Twentsche Katoennijverheid,* Appendix 2.

[27] Ibid., p. 71.

392 machines with 7,193.25 HP in 1853. The corresponding figures for Belgium for 1846 are 1,514 machines, with a capacity of 37,007 HP.[28] The total value of Belgian textile exports in 1844 was about 60 million francs in 1833 prices, or a few percent less in current prices.[29] Dutch textile exports in current prices (average 1846-1850) was 9.4 million guilders, or some 20 million francs.[30] A very different but equally suggestive illustration is the participation in the famed Crystal Palace exhibition in 1851. Belgium was represented by 512 entries and won 2 gold medals and 97 distinctions. The Netherlands sent 114 exhibits and received 1 gold medal and 16 distinctions.[31] The Belgian population at the time was approximately 50 percent larger than the Dutch population.

II

Some explanations of this quite remarkable example of uneven development have been suggested in the literature, though the problem itself has not frequently been posed in explicit form.[32] The most obvious difference between the two countries is their respective physical endowments: Belgium has rich deposits of coal and iron, whereas the Netherlands has no iron and virtually no coal, but relatively more fertile agricultural land. This argument can hardly explain the whole phenomenon, since two of the pivotal sectors in Belgian industry, wool and cotton, used imported raw materials. Steam power began to be used widely in Belgium only after 1825. On the other hand, the Dutch did have rich endowments of peat, which was widely used as a fuel in Dutch industry.[33] In addition, the

28 For the Netherlands: *Staatkundig en Staathuishoudkundig Jaarboekje,* [Yearbook of Politics and Political Economy], VIII (1855), p. 69; for Belgium: *Statistique de la Belgique, Industrie,* pp. x-xi.

29 Schöller, "La Transformation," pp. 592-594.

30 *Statistisch Jaarboekje,* [Statistical Yearbook of the Netherlands], I (1851), p. 218 (2).

31 Figures cited by H. Baudet, "De Dadels van Hassan en de Start der Nederlandse Industrialiteit," [The Start of Dutch Industrialism], in: *Bedrijf en Samenleving,* [Enterprise and Society], *Festschrift* presented to Prof. I. J. Brugmans(Alphen on the Rhine: N. Samson, 1967), p. 3.

32 Some of the arguments presented here appear in Van Houtte, "Economic Development," and in *Economische en Sociale Geschiedenis van de Lage Landen,* [Economic and Social History of the Low Countries] (Antwerp: Standaard Boekhandel, 1964). Other explanations have been drawn from various contexts.

33 On the importance of peat see P. Van Schaik, "De Economische Betekenis van de Turwinning in Nederland," [The Economic Significance of Peat Mining in the Netherlands], *Economisch-Historisch Jaarboek,* XXXII (1968-69), pp. 141-205 and XXXIII (1970), pp. 186-235. The first steam engines in the Twente area seem to

importance of wind as an energy source cannot be discounted. The paper, oil, and sawmill industries in North Holland relied heavily on this cheap source of power.[34]

The difference in historical backgrounds is of somewhat greater importance. Wars and political settlements had destroyed all Belgian shipping and commerce and most of its industry in the last decades of the sixteenth century, whereas these activities prospered in the Netherlands. The result was that a politically dominant commercial class à la Amsterdam—probably a hindrance to modern industry—was absent in Belgium. The Dutch *haute bourgeoisie* consistently resisted the introduction of protective tariffs and viewed their country as primarily a commercial nation with no future in industry. These attitudes hampered Dutch industrial development. Of equal significance was the absence in Belgium of a heritage of technological traditionalism and entrepreneurial conservatism. These were two of the factors that made it so difficult for the Dutch traffics to transform themselves into a modern industry. Losses sustained by the traffic industries between the 1770's and 1815 may have discouraged the already very risk-averse Dutch investors from establishing industrial projects.

Linked to this argument is the emphasis put on the demand side. The Belgian provinces were annexed to France in 1795 and enjoyed a large market of 50 to 60 million consumers (including military demand, an important component) until the collapse of the First Empire. This demand more than compensated for the loss of the overseas market. On the other hand, the Netherlands were incorporated into the Empire as late as 1810, and the tariff barrier between the Dutch provinces and the rest of the Empire was not lifted until 1812.[35] Obviously, advantages on the demand side provided the Belgians with a considerable edge over the Dutch. However, this cannot fully explain the widening of the gap during the Dutch pe-

have been operating on peat. Cf. N. W. Posthumus (ed.), "Bijdragen tot de Geschiedenis der Nederlandsche Grootindustrie," [Contributions to the History of Dutch Industry], *Economisch-Historisch Jaarboek*, XI (1925), p. 185, for an eyewitness report. See also J. Boessenkool: "De Eerste Stoommachine in de Twentsche Textiel-Industries," [The First Steam-engine in the Textile Industry in Twente], *Textiel-Historische Bijdragen*, IV (1962), pp. 67-72.

[34] A. M. Van Der Woude, "Het Noorderkwartier," [The Noorderkwartier, A Study of the Demographic and Economic History of Western Parts of the Netherlands] *A.A.G. Bijdragen* 16, II (1972), pp. 476, 488, 496.

[35] I. J. Brugmans, *Paardenkracht en Mensenmacht, Sociaal-Economische Geschiedenis van Nederland 1795-1940*, [Horsepower and Human Might, A Social-Economic History of the Netherlands] (The Hague: Martinus Nijhoff, 1961), p. 39.

riod and the post-revolutionary period, and it is certain that the industries that grew up in Belgium during the Empire were not "hothouse industries."[36]

Similar to this vein is the line that emphasizes the various roles the respective governments of the two countries played in industry. The Austrian, and later the French governments awarded Belgian industry significant help and encouragement in the form of tariff protection, prizes for inventions, technological contests and exhibitions, the opening of technical schools, and subsidies. More unexpected is the attitude of King William I of Orange, who tried to encourage Belgian industry as much as he could and committed part of his private wealth to this purpose. The founding of the famous *Societé Générale* is only one example of this policy.[37] But this in a sense underlines rather than resolves the problem. Why would a Dutch (and Protestant) king literally bet all his money on Belgian industry rather than try to emulate it by establishing similar enterprises in the northern provinces?

An argument frequently encountered in the writings of Dutch economic historians as an explanation for their country's relative backwardness has been the lackadaisical mentality of Dutch entrepreneurs in the eighteenth and nineteenth centuries. It is interesting to note that a kind of "Landes-Gerschenkron debate" on the importance of social attitudes and entrepreneurship can be observed among Dutch specialists.[38] One authority writes:

> The industrialist in the early capitalist period [i.e. 1813-1870] . . . is content with a normal profit and does not even consider changing his methods as long as he is not compelled to by his consumers . . . . He does not engage in organization or calculations, leaving this matter to his foreman . . . . The industrialists of this period cannot complain about lack of leisure and could devote themselves to what was then considered as the noblest pastime: poetry . . . .[39]

Prototypes and caricatures of narrow-minded, cautious, and conservative entrepreneurs can be found in nineteenth-century Dutch literature, for example in the writings of Hildebrand and Potgieter.

36 This is emphasized by F. Crouzet, "Wars, Blockade and Economic Change in Europe, 1792-1815," JOURNAL OF ECONOMIC HISTORY, XXIV (Dec. 1964), 579-580. Cf. Landes, *Unbound Prometheus*, p. 144.

37 Demoulin, *Guillaume 1er*, pp. 157-181.

38 For a summary see J. H. Van Stuijvenberg, "Economische Groei in Nederland in de 19e Eeuw: een Terreinverkenning," [Economic Growth in the Netherlands in the Nineteenth Century: A Reconnaissance] in: *Bedrijf en Samenleving*, pp. 195-225 and works quoted there.

39 Brugmans, *Arbeidende Klasse*, p. 67.

Implicit in these caricatures are the technological backwardness, high risk aversion, and high leisure preference of entrepreneurs. If one extends "entrepreneurial attitudes" to include saving and investment behavior, a satisfactory explanation for the tardiness of Dutch industrialization could be obtained. However, the differences in this respect between the two countries are a matter of degree rather than of essence. Moreover, entrepreneurship was to some extent an internationally mobile factor—many of the pioneers of Belgian industry were English or of English descent—so that scarcity of entrepreneurial talents cannot fully account for long-run trends.

Finally, the existence of a large protoindustrial sector in Belgium and the relative density of rural population have recently been put forward as an explanation of industrialization in Belgium.[40] This view is correct, but merits some elaboration, since it is by no means evident which mechanism is operating here. Capital accumulation in the protoindustry, cited by Mendels, seemed relatively minor, since there were severe difficulties in transforming accumulated circulating capital into fixed capital in the absence of good capital markets. Nor is there much reason to believe that the protoindustry created a technological and entrepreneurial infrastructure that facilitated the growth of modern industry. Inventions were by and large all imported from England, and it is far from clear how the protoindustry facilitated their adoption. Most entrepreneurs, except in the woolen industry, were *homines novi* or of commercial backgrounds and seldom directly connected to protoindustrial production.[41] It will thus be useful to develop a more formal framework in which the various factors in this process can be rigorously analyzed.

III

The interrelationship between income distribution and industrialization, long a subject of controversy among theorists, has been taken up in the last decade by historians as well.[42] The issue may

[40] Van Houtte (1972), "Economic Development," p. 108. F. Mendels, "Protoindustrialization: The First Phase of the Industrialization Process," JOURNAL OF ECONOMIC HISTORY, XXXII (March 1972), 241-261.

[41] P. Lebrun, "La Rivoluzione Industriale in Belgio," *Studi Storici*, II, (Dec. 1961), p. 589. J. Dhondt, "The Cotton Industry of Ghent During the French Regime," in F. Crouzet, et al. (eds.), *Essays in European Economic History* (London: Edward Arnold, 1969), pp. 16-18.

[42] H. J. Habakkuk, *American and British Technology in the Nineteenth Century* (Cambridge: University Press, 1962); S. B. Saul (ed.), *Technological Change. The United States and Britain in the Nineteenth Century* (London: Methuen, 1970).

be formulated as follows: Assume two preindustrial revolution economies to which a new technology becomes available. Suppose the initial structures of the two traditional societies are such that one entered the new era as a high-wage economy, the other as a low-wage economy. Is it possible to argue on a priori grounds which of the two will adopt the new technology at a faster rate and which of the two will complete the "traverse" first? The answer, obviously, depends on the particular model used.

One type of model might be the simplest microeconomic setup. A high-wage level means that labor is expensive relative to capital. It is thus expected that the high-wage economy will adopt a more capital-intensive technique, since the incentive to substitute machines for dear labor is higher. A more sophisticated version of the same idea is that high wages stimulate the development of labor-saving inventions.[43] In other words, technology itself is made a function of factor prices and is not exogenously given. In addition to many other problems that this approach raises,[44] it seems that such models are not applicable to the present issue. The question at this stage is not so much the *choice* of technique, which is a static model, as *how fast* an exogenously given technology was adopted, which is a dynamic problem. It does not make much difference, moreover, whether the new technology consisted of one or of many techniques.

In order to demonstrate this, consider two economies, both of which suddenly face a new technology that is superior to the old one at any set of factor prices (so that the factor-price-frontier associated with the new technique lies everywhere above and to the right of the frontier associated with the old technique). In order to shift to the new and better technique, it is necessary to accumulate a large quantity of new capital goods in which it is embodied.

Assume now that one of the two economies has, for some reason, a high wage level while the other is a low-wage economy. The duration of the "traverse" will depend, among other things, on the relationship between capital formation and the wage level. Habakkuk concedes that high wages were in all cases a severe restraint on investment simply because higher wages implied lower profits in a

[43] Habakkuk, *American and British Technology*, pp. 51-53.

[44] P. Temin: "Labor Scarcity and the Problem of American Industrial Efficiency in the 1850's," JOURNAL OF ECONOMIC HISTORY, XXVI (Sept. 1966), 277-295; W. E. G. Salter, *Productivity and Technological Change* (Cambridge: University Press, 2nd ed., 1969), pp. 43-44.

one-technique model as well as in a multitechnique one.[45] In other words, since there was a ubiquitous trade-off between profits and wages and since the rate of accumulation is assumed to depend on the rate of profit, high wages implied a slower process of "traverse." Moreover, let us assume that the new technology consisted of two techniques, one relatively labor intensive and the other relatively capital intensive but both fully superior to the traditional technology. The high-wage economy will choose a higher capital-labor ratio than the low-wage economy, so that it will have to accumulate relatively more capital per worker. But since its rate of accumulation is in fact lower, it will complete the transition to the new technique in a much longer time span than the low-wage economy. Thus the speed of transition from a traditional, nonmechanized industrial sector to a modern, capital-intensive industry depended on the wage rate regardless of whether factor-substitution within the new technology was possible at all. The availability of a choice of techniques reinforced this dependency.

What determined the rate of accumulation? The dependency of the rate of investment on the rate of profit is more than an assumption. It is wholly independent of the assumption of perfect capital markets, perfect information and foresight, and notions that capital was somehow allocated in an optimal fashion so as to equalize its rates of return. Capital markets in nineteenth-century Europe were so imperfect that industrialists, for the most part, were unable to obtain long-term credit from savings generated in other sectors. Hence modern industry was dependent on its own savings, in other words, retained profits.[46] Under these conditions the interest rate loses its relevance as an input price. In fact, it is not easy to say what "the price of capital" is at all.[47] On the other hand, the wage level becomes a crucial variable. If countries with different wage levels

[45] Habakkuk, *American and British Technology,* pp. 43-45.

[46] This is emphasized for Belgium in R. Cameron: *Banking in the Early Stages of Industrialization* (New York: Oxford University Press, 1967), pp. 130-131. For the Twente area in the Netherlands this is argued by Boot, *Twentsche Katoennijverheid,* pp. 265-268, 276. Similar arguments have been made for France by Landes, *The Unbound Prometheus,* pp. 130, 131n, and for England, among others, by F. Crouzet, "Capital Formation in Great Britain During the Industrial Revolution," in *Capital Formation in the Industrial Revolution* (London: Methuen, 1972), p. 188.

[47] Habakkuk, *American and British Technology,* p. 28 argues that capitalists behaved as if capital had cost them nothing. In these situations it may be preferable to abstract altogether from the concept of a "return to capital" and to view profits as a (quasi) rent. Cf. Salter, *Productivity and Technical Change,* p. 61.

were facing similar technologies and selling in a competitive world market, low-wage countries could be expected to have higher profit rates and thus accumulate capital faster. The rate of interest does not enter here in a direct way at all—capital markets in this period traded almost exclusively in government bonds and short-term credit.

The assumption of similar technologies and a competitive world market may not seem very realistic for a comprehensive analysis of Western Europe in the nineteenth century. Yet it reflects to a large extent the fact that the Netherlands and Belgium were in comparable situations, especially after 1813. They faced the same markets and were trying to apply the same technology. Therefore, if it can be shown that wages in Belgium were considerably lower than in the Netherlands, an alternative explanation for Belgium's spectacular industrialization would be found. Since Belgium's success was unique on the European continent, it may prove useful to analyze the structure of its economy in some detail. A more formal analysis appears in the Appendix.

In the eighteenth century the population of what was later to become Belgium increased very rapidly. This growth was confined to rural areas and was of such dimensions that it created considerable population pressures on the land, especially in Flanders, Liège, and Hainault. The failure of Malthusian checks to prevent this rapid growth is explained by two phenomena. The first was the introduction of the potato, which increased the output of food in terms of nutritional units per acre. The second and more important factor was the rapid expansion of the rural-domestic industry or protoindustry.

The protoindustry is characterized by the fact that it produces its output, the Z-good,[48] with minimal capital inputs.[49] The first implication of a one-input industry is that, with exogenously given prices, every peasant has the alternative to engage in Z-good production when and where his agricultural productivity falls beneath the value of his productivity in Z-good production. In a simple model of time allocation, every peasant has a unique equilibrium in which he devotes part of his time to agriculture and the rest to the Z-good. With an increasing population, the aggregate number of hours spent on

[48] This term is borrowed from S. Hymer and S. Resnick, "A Model of an Agrarian Economy with Non-agricultural Activities," *American Economic Review*, LIX (Sept. 1969), 493-506. In the present context the Z-good plays a somewhat different role, since it is sold in the market rather than consumed on the spot.

[49] Mendels, "Protoindustrialization," p. 243.

the Z-good will increase at a faster pace than the total devoted to agriculture. A second implication of the special features of the Z-good is that, as long as the price of the Z-good does not fall, population can continue to grow indefinitely because the check on diminishing returns has been removed. If the price of Z-goods falls with increasing output, the income of the peasants may eventually be reduced to the level of no population growth. This stationary state will have, however, a significantly larger population than one in which there is no Z-good.

It can be concluded that in areas with a large Z-good sector it is reasonable to expect that labor was less scarce than in other areas. Thus, not only were wages in general lower, but also the supply of labor was more elastic. On this level of abstraction the supply of labor must have been perfectly elastic, the supply price of labor being equal to the productivity of labor in producing Z-goods times the price of Z-goods. It cannot be asserted that wages must have been constant, since the price of Z-goods on the world market fluctuated considerably. But it turns out that the rate of accumulation is independent of these fluctuations.[50] The model is thus similar to W. A. Lewis's labor surplus models.[51] The present model differs from Lewis's in one aspect: there is no "surplus" labor in this economy—all labor is employed and none is superabundant in any sense. The wage level is therefore fully determinate in an economic sense, and it is not necessary to resort to an institutionally determined wage.[52] Apart from that, the growth process is very much like the one described by Lewis. Capitalists hire workers at a wage equal to their productivity in Z-good production. Because their productivity in the modern sector is higher, a surplus is created that accrues largely to the capitalist. The capitalist takes a part of this surplus for his own needs and plows the rest back into his business, buying new machines, hiring additional workers, and so on.

For the purpose of industrialization and accumulation it is not strictly necessary to have a large reservoir of labor in the Z-good sector. But it helps. Consider a country without a Z-good sector. One

50 For a proof, see Appendix.

51 W. A. Lewis, "Economic Development with Unlimited Supplies of Labor" *The Manchester School*, XXII (May 1954), repr. in A. N. Agarwala and S. P. Singh (eds.), *The Economics of Underdevelopment* (New York: Oxford University Press, 1958), pp. 400-449. See also, "Unlimited Labor: Further Notes," *The Manchester School*, XXVI (January 1958), 1-32.

52 For a definition see John C. H. Fei and Gustav Ranis, *Development of the Labor-Surplus Economy* (Homewood, Ill.: Richard D. Irwin, 1964), pp. 21-22.

possibility is that the country does not need the Z-good because its productivity in agriculture is sufficiently high. In that case the wage level would be higher than that of a country with a Z-good sector. On the other hand, it is also possible that the objective conditions that made Z-good production possible (such as the availability of flax for linen, water resources for wool manufacturing, the tradition of technical skills among peasants) were absent. In these economies the wage level was equal to or lower than the wages in Z-good countries, but the supply of labor was more inelastic. Equally important for countries with Z-good sectors was the fact that their workers already possessed some technical skills and were not entirely alien to industrial activity.

The process of reinvestment of quasi-rents over time does not go on forever. Accumulation will be subject to two counteracting forces: on the one hand wages will start to rise sooner or later as the reservoir of cheap labor approaches depletion. On the other hand technology is proceeding, raising the productivity of labor on machines of later vintages. The industrialization process can thus be viewed as a race between rising wages and rising productivity of labor. If rising wages had prevailed in this race, the economy would have ended up in a classical stationary state with no net capital formation and zero profits. It was thus of critical importance for industrializing countries that wages be not only low but also slow to rise. In this fashion Z-good countries can be seen to have an advantage over other countries.

Finally, there is the role of demand. It might be argued that high wages are associated with a high level of demand for standardized products, which is conducive to mass production. This argument cannot withstand scrutiny in the present context. As indicated before, the two countries faced essentially the same markets during most of the period under discussion, and tariffs as well as transport costs were largely the same for both. As for local demand, poor consumers had a preference for cheap, low-quality goods whether they were mass-produced or not.[53] Nor is there much evidence that the respective levels of aggregate demand were strongly affected by income distribution.[54]

[53] Landes, *The Unbound Prometheus*, p. 127.

[54] This point has been raised by P. W. Klein in *Kapitaal en Stagnatie tijdens het Hollandse Vroegkapitalisme*, [Capital and Stagnation during the Period of Early Capitalism in Holland] (Rotterdam: Universitaire Pers, 1967), pp. 8-9. Curiously enough, Klein makes the argument in reference to Holland, ignoring the comparatively high wages there.

IV

Why should one expect a substantial wage differential between Belgium and the Netherlands? There was a large protoindustrial sector in Belgium but not in the Netherlands, and although there is no strong a priori reason why the wage rate derived from the Z-good sector should be particularly low, there is little doubt that in fact it was. Agricultural productivity in Belgium was lower than in the Netherlands, in part because the labor/land ratio was much higher. In 1824 total population of the northern provinces of the United Kingdom of the Netherlands (not including Dutch Limburg) was 2,262,712 persons on a land area of 30,386 km^2, which yields an overall density of 74.4 persons per km^2. The corresponding figures for Belgium (including sparsely populated Luxembourg) were a population of 3,750,766 on a land area of 34,217 km^2, giving a density of 109.6 per km^2. The northern provinces were, however, far more urbanized, and if one subtracts urban population, the ratio of rural population per square kilometer becomes 46.7 in the northern provinces and 83.6 in the southern provinces.[55] The gap in agricultural productivity would be smaller than these figures imply because the Belgians did have the Z-good; therefore agricultural productivity would not fall below the productivity in Z-good production. But the fact that there was little Z-good production in the Netherlands except in two areas might be taken as an indication that agricultural activity was, in reality, more productive than in Belgium.[56] The noted Dutch statesman and political economist Van Hogendorp was once asked by a Flemish gentleman what the Dutch peasants did if they neither spun nor wove. The reply was that they made butter and cheese and that this provided them with sufficient work.[57] He should have said "sufficient income."

The extraordinarily low wages paid in the protoindustry impressed contemporaries of the time as well as historians. In the Flemish linen industry, for example, a memorandum of 1765 estimates the daily

55 All figures from E. Smits, *Statistique Nationale, Développement des 31 Tableaux Publiés par la Commission de Statistique* (Brussels: H. Tarlier, 1827), Annex Table I.

56 This refers to alluvial clay soils only. Cf. Z. W. Sneller, "Anderhalve Eeuw in Vogelvlucht," [A Century and a Half in Overview], part A, in *Geschiedenis van de Nederlandsche Landbouw 1795-1940,* [History of Dutch Agriculture] (Groningen: J. B. Wolters, 1943), pp. 26-32. There seems to be no absence of the objective conditions for protoindustry in the Dutch alluvial provinces.

57 K. G. Van Hogendorp, *Bijdragen tot de Huishouding van Staat in het Koninkrijk der Nederlanden,* [Contributions to the Political Economy of the Kingdom of the Netherlands] (Amsterdam: K. H. Schadd, 1825), Vol. V, p. 167.

income of a weaver to be 7 or 8 sous (63 to 72 centimes), which enabled the weaver to subsist on a diet of rye bread, potatoes, buttermilk, a little bacon on Sundays, and water. Even lower figures are quoted in a letter dating from 1789.[58] In the nineteenth century, after a short boom during the Napoleonic Empire, the wages of linen weavers and spinners fell to new lows, reaching catastrophic depths in the 1830's and 1840's.[59] It should be noted that during most of the period under discussion the Flemish protoindustry was for its greater part a self-employed rather than a putting-out industry, so that the terms of trade between final output, raw materials, and food wholly determined the income of protoindustrial workers, given a constant productivity coefficient in Z-good production.

The woolen and metallurgical rural industries in the Verviers-Liège area and in Hainault were, on the other hand, mostly putting-out industries. Putting-out in the Z-good complicates the picture somewhat, especially since the truck system (payments in kind) and embezzlement of materials by the workers tend to distort what little information on wages we have.[60] It is certain nevertheless that these wages were very low. A source from 1741 estimated the ratio of wages in Leyden, Tilburg, and Verviers to be 3 : 1.5 : 1 .[61] Wages in the woolen industry remained more or less stable throughout the eighteenth century, rising somewhat in the first decade of the nineteenth century, but less than proportionally to the rise in output prices.[62] The situation in the woolen industry was succinctly described by Van Hogendorp in 1817:

[58] Cited by L. Varlez, *Les Salaires dans l'Industrie Gantoise* (Brussels: J. Lebègue, 1901-04), Vol. II, p. xix.

[59] E. Ducpétiaux, *Mémoire sur le Pauperism dans les Flanders* (Brussels: M. Hayez, 1850), pp. 66-67; P. C. Van Der Meersch, "De l'État de la Mendicité et de la Bienfaisance dans la Province de la Flandre Orientale," *Bulletin de la Commission Centrale de Statistique*, V (1853), 237; G. Jacquemyns, "Histoire de la Crise Économique des Flanders (1845-1850)," *Académie Royale de Belgique, Mémoires*, XXVI (1929), 136-137, 205-228.

[60] On the use of the truck system see C. Génart, *L'Industrie Cloutière en Pays Wallon* (Brussels: J. Lebègue, 1900), p. 28, and P. Lebrun, *L'Industrie de la Laine à Verviers* (Liège: Bibliothéque de la Faculté de Philosophie et Lettres de l'Université de Liège, 1948), p. 264. The quantitative importance of embezzlement is, by its very nature, impossible to appraise, but it may have been important as a supplement to wages. Cf. sources quoted by Mendels (1972), "Protoindustry," p. 244.

[61] Cited in N. W. Posthumus, "De Industrieele Concurrentie tussen Noord- en Zuid-Nederlandsche Nijverheidscentra in de XVIIe en XVIIIe Eeuw," [The Competition between the Industrial Centers in the Northern and Southern Netherlands in the Seventeenth and Eighteenth Centuries], in *Mélanges d'Histoire offerts à H. Pirenne* (Brussels: Vromant et Cie., 1926), p. 376. As noted above, Tilburg was one of the few areas in the Netherlands in which protoindustry was important.

In Verviers I noted the wealth of the industrial entrepreneurs as well as the extreme poverty of the workers. When business booms and new machines are being introduced, so that goods can be produced at a cost lower than the market price, the workers continue to receive the same wages, while the industrialists gain thousands.[63]

In the nail industry in the Charleroi area, a typical domestic industry, the best workers earned an off-season wage of 6 to 7 sous, whereas others made no more than 3 to 4 sous. (The sou, or patard de Liège, was about 70 percent of the Flemish sou.) During peak seasons wages were much higher, but this situation obtained for only six to eight weeks of the year.[64] The conspicuously low wages in the Walloonic provinces are confirmed by a source from 1820.[65]

The situation in the Netherlands was more complicated. It is possible, of course, to explain the wage differential between the Netherlands and Belgium entirely in terms of the absence of a protoindustry in the Netherlands and the implied higher agricultural productivity. But it seems that wages in the Netherlands were high compared not only to Belgium but to other countries as well.[66] "The wages of labour are said to be higher in Holland than in England, and the Dutch, it is well known, trade upon lower profits than any people in Europe," writes Adam Smith.[67]

Some additional conjectures in regard to the causes of the high wage level in the Netherlands are thus in order. The fact that urbanization was so intense may alone be a factor, since mortality in the cities was high and tended to underline the need to bid away workers from the agrarian sector at relatively high wages. Also, the existence of a large structure of welfare and charity organizations, especially in the cities, may have had considerable effect on the level of wages in general and its downward stickiness in particular. In addition, it could be that high wages were established during the zenith of Dutch prosperity and had since outlived it, maintained by inertia and institutions preventing them from falling during less favorable periods. It is also possible that, since employment in the

62 Lebrun, *De L'Industrie de la Laine*, p. 325.
63 Van Hogendorp, *Bijdragen*, II, p. 269.
64 Génart, *Industrie Cloutière*, p. 26.
65 Roentgen, "Berigt," p. 149.
66 Van Dillen, *Van Rijkdom*, p. 553. Joh. De Vries, *De Economische Achteruitgang der Republiek in de Achttiende Eeuw*, [The Economic Decline of the Dutch Republic in the Eighteenth Century] (Amsterdam: J. Van Campen, 1959), p. 107.
67 Adam Smith, *The Wealth of Nations* (Andrew Skinner, ed., Harmondsworth: Penguin Books, 1970), p. 194.

Dutch cities was either directly or indirectly connected to the commercial-maritime sector, employment was subject to rather severe fluctuations and that wages therefore included a "risk premium."

An interesting note is struck by Charles Wilson. The large government debt in the Netherlands, Wilson argues, caused an exceptionally high level of taxation. Most taxation, because it was indirect and levied on necessary consumption goods, tended to drive the wage level up and profits and production of exports down.[68] The process fits in well with the framework described above, especially since taxes remained high in the nineteenth century. On the other hand, the emphasis on taxation raises two problems. First, an indirect tax will in general be borne only partially by the consumers in the form of higher prices. The supply curve of labor (as a function of nominal wages) shifts to the left, but equilibrium wages will rise less than proportionally to the rise in prices. In addition, the employers, facing higher costs, may be able to raise output prices unless they face perfectly elastic demand curves. In short, the actual influence of indirect taxes on the wage level and on profits depends on the elasticities of supply and demand of the goods consumed by workers, of labor, and of the export product. A second problem arises from the fact that the Dutch national debt was largely domestically held. It is necessary to assume that the *rentiers* who owned the debt were a different subset of the population than the industrialists whose profits were curbed by the high wages, and that they had no interest in investing in industry. Otherwise, the high-wage-low-profit result of the Wilson effect was insignificant, since industrialists are assumed to reinvest part of their profit.

Finally, one could return to the entrepreneurship argument put forward earlier. Microeconomic theory postulates wages to be equal to the value of the marginal product of labor if and only if the firm maximizes profits. But the essence of the entrepreneurship argument seems to be that these "incompetent" entrepreneurs were in fact not maximizing profits. It does not matter whether they were in fact maximizing some utility function in which profits only appear as one argument among many, or whether they were not maximizing anything at all; in either case wages must have been somewhere between

[68] Charles Wilson, "Taxation and the Decline of Empires, an Unfashionable Theme," in *Economic History and the Historians: Collected Essays* (London: Weidenfeld and Nicolson, 1969), pp. 114-127.

the average and the marginal product, and thus higher than in a country in which entrepreneurs were more aggressive.

V

In summary, there is reason to believe that lower wages may have been important in determining rapid industrialization in Belgium and high wages in determining Dutch stagnation. It was demonstrated that there were structural differences between the two countries that could widen such a gap. It is necessary to show, however, that wages *were in fact* significantly lower in Belgium. To look at nominal wages would be sufficient in this case since for our purpose the interesting problem is not the standard of living of the workers, but rather the slice of the revenue pie that is left over for the producer after wages have been paid out.

However, aggregate income distribution data for this period are virtually unavailable. For example, the data for England before 1880 are too scattered and incomplete to make aggregate figures possible.[69] As to the Low Countries, for most of the nineteenth century the data are of such a nature as to cast heavy doubt on the validity of interspatial comparisons. It is thus fortunate that there is one source that allows us to perform the comparison without the usual caveats pertaining to the comparison of data assembled by different statistical services. This is the Dutch industrial survey of 1819.[70] The survey, actually undertaken in 1820, contains valuable information about the number of industrial firms, the number of adults and children employed, and the daily wages earned. In addition, some qualitative information as to the "state of business" compared with previous periods is supplied and some general remarks added. Since the returns are organized by province, it was possible to calculate aggregates that could be used to contrast and compare the north and south.

Some of the main defects of this survey should be mentioned in order to stress the fact that these data are crude approximations and unsuitable for more refined econometric tests. For one thing, the data shown in Tables 1 and 2 are provincial averages, weighted by

69 P. Deane and W. A. Cole, *British Economic Growth, 1688-1959*, 2nd ed. (Cambridge: University Press, 1967), p. 148.

70 I. J. Brugmans (ed.), *Statistieken van de Nederlandse Nijverheid uit de Eerste Helft de 19e Eeuw*, [Statistics of Dutch Industry from the First Half of the Nineteenth Century], *R.G.P.* Vols. 98, 99 (1956).

the number of workers in each industry in that province. However, the original wage entries for each industry in each province as they appear in the returns are themselves means of all the firms in that industry. There is strong reason to believe that *unweighted* averages were employed by the officials in charge of aggregating the original returns. This procedure tends to bias the figures downward, since it seems that wages were positively correlated with the size of firms (though the correlation is weak). Other sources of possible bias and inaccuracy are payment of wages in kind, the seasonal nature of some forms of employment, the fact that many wages were paid at a piece rate rather than at a time rate, and the particular timing of the survey at the end of a prolonged depression. Some other shortcomings of the data, stemming from clerical or administrative error, have been corrected for as far as possible.[71]

Nonetheless, the 1819 survey constitutes a unique source of information for this period. It should be mentioned that by checking provincial returns (as far as they were available) against aggregate returns, it could be verified that most municipalities in charge of the actual collecting of the data conducted the survey in a responsible and efficient way, and that the compilation and editing were carried out with care. In spite of its weaknesses, the 1819 survey thus provides us with a unique opportunity to test the hypotheses advanced in this paper. The main aggregates computed from the returns are presented in Table 1.

It can readily be seen that adults' wages, taken as a whole, are almost 55 percent higher in the northern provinces. Moreover, the two northern provinces in which wages are relatively low, Overijssel and North Brabant, are the same provinces in which protoindustry existed and where, indeed, the nuclei of modern industry started in the 1830's. In the Belgian provinces wages are exceptionally low in Flanders and Antwerp, somewhat higher (but still well below the northern average) in Liège. The only exceptions, Namur and Hainault, can be explained by the importance of coal mines in these provinces, which traditionally paid higher wages.

It is also possible to ascertain from these data whether the wage differential differed from industry to industry. This is significant, since if the gap in wages was confined to the consumption good, the only source of the Belgian advantage was the difference in profits. If

[71] J. Mokyr, "An Early Nineteenth Century Industrial Survey," unpublished manuscript, Yale University, 1972, pp. 9-15.

TABLE 1
TOTAL NUMBER OF FIRMS, ADULTS AND CHILDREN EMPLOYED, AND DAILY WAGES IN CENTS

(1) *Province*	(2) *Firms*	(3) *Adults*	(4) *Children*	(5) *Adults' Wages*	(6) *Children's Wages*
N. Brabant	8,659	12,716	2,400	55.2	16.5
Gelderland	5,130	6,692	1,267	62.7	18.2
S. Holland	6,764	20,446	1,739	86.0	25.8
N. Holland	8,493	25,674	2,184	94.2	20.2
Zeeland	3,094	3,653	621	82.5	19.0
Utrecht	3,126	8,169	2,438	82.0	21.2
Friesland	4,991	8,785	1,384	76.5	13.5
Overijssel	4,636	12,209	3,261	58.2	32.6
Groningen	4,606	6,606	852	76.0	22.3
Drenthe	1,234	1,967	n.d.	73.4	n.d.
Total North	50,739	106,917	16,146	74.8	22.4
S. Brabant	6,732	13,608	538	59.8	9.5
Limburg	6,770	6,371	236	52.4	19.6
Liège	5,482	27,911	4,018	65.6	23.9
E. Flanders	37,288	101,601	10,780	40.3	17.5
W. Flanders	59,336	74,675	3,806	32.3	12.8
Hainault	8,415	32,893	2,591	77.1	29.0
Namur	2,712	6,915	163	70.6	30.6
Antwerp	7,192	23,167	1,391	48.6	24.1
Luxembourg	8,395	12,295	273	54.5	23.6
Total South	142,322	299,436	23,796	47.8	19.5
Total	193,061	406,353	39,942	54.1	20.6

Source: See footnote 70.

the Belgian wages in the machine-producing sector were lower as well, the Belgians would have enjoyed the additional advantage of cheaper machines. Note, however, that this advantage would only be effective if the Dutch did not buy Belgian machines. To the extent that they did, the only disadvantage of higher wages in this respect was higher transportation costs of machines.

The wage data in four industries are presented in Table 2. The industries are textiles (cotton, linen, and wool only), capital goods (for which construction, metallurgy, and instrument-making were chosen), foodstuffs (a small-scale artisan industry), and paper (a relatively capital-intensive industry, which was in the process of being mechanized in Belgium while in a process of stagnation in the Netherlands).[72] The difference in wages is more pronounced in tex-

[72] B. W. DeVries, *Nederlandse Papiernijverheid*, pp. 191-192.

TABLE 2

DAILY WAGES OF ADULTS IN SELECTED INDUSTRIES, IN CENTS

(1) *Province*	(2) *Textiles*	(3) *Foodstuffs*	(4) *Capital Goods*	(5) *Paper*
N. Brabant	52.5	68.6	50.7	57.9
Gelderland	58.3	71.1	60.4	56.9
S. Holland	72.8	96.5	69.3	84.3
N. Holland	76.8	116.6	121.7	118.0
Zeeland	55.8	90.6	78.7	98.9
Utrecht	52.5	77.7	76.7	73.6
Friesland	60.9	75.1	63.3	66.4
Overijssel	45.2	64.8	57.3	59.9
Groningen	62.0	88.8	64.8	61.0
Drenthe	50.1	59.9	47.5	60.0
Total North	55.0	83.7	72.7	91.8
S. Brabant	53.6	71.7	48.4	59.0
Limburg	47.6	54.0	51.0	74.2
Liège	70.1	64.0	63.3	78.5
E. Flanders	37.4	67.3	54.8	68.8
W. Flanders	27.0	70.4	50.2	82.3
Hainault	35.6	87.6	86.2	71.2
Namur	59.1	76.5	68.9	48.4
Antwerp	34.8	67.3	76.0	82.0
Luxembourg	45.2	66.1	54.9	61.2
Total South	36.0	70.9	62.3	64.7
Total	37.9	74.9	66.9	78.6
ratio North/South	1.528	1.182	1.151	1.419

Source: See footnote 70.

tiles and paper than in capital goods and foodstuffs, but the overall trend of lower wages in Belgium is consistent.

The rationalization may be that paper and textiles could more readily draw upon the reserves of labor in the Z-good sector, whereas in the capital goods and foodstuffs industries skills or other barriers to free entry of labor kept the wage level relatively higher. Indeed it is not entirely clear that the capital-goods industry needed more skills than the textile industry, but it may be conjectured that it needed *different* skills—skills that were not readily available in the traditional sector. It is also possible that the interregional trade in capital goods was of more consequence than the trade in textiles, so that the trends toward factor-price-equalization are stronger in the capital-goods industry.

The implication of the interindustry differences in the wage-gaps between the Netherlands and Belgium is that, insofar as the relative success of Belgium in industrialization can be attributed to wage-level differences, they were more consequential in the consumption-goods industries than in the capital-goods industry. Therefore, the emphasis is placed on the larger profits generated in the consumption-goods industry rather than on cheaper capital goods. The latter was, however, not altogether insignificant, since in the capital goods too, wages in Belgium were lower than in the Netherlands.

Other hypotheses, testable in principle, can be derived from the simple model presented in section III. For example, one could test the hypothesis that as long as the protoindustry exists, wages in the modern sector do not rise significantly. In the Netherlands we should see a stable or declining wage so long as population growth is unaccompanied by industrialization. It should also be possible to correlate the relative importance of the protoindustry with the wage level, but the availability of data is a major obstacle here, compounding difficulties defining the relevant variables.

Yale University

Appendix

Assume an open economy which initially produces two commodities—agricultural goods, A, and industrial goods produced in the protoindustry, Z. The respective production functions are:

$$(1)\quad A = A(L_A, \overline{T}), \frac{dA}{dL_A} > 0, \frac{d^2A}{dL^2} < 0$$

$$(2)\quad Z = b \cdot L_Z$$

where $\overline{T}$ is the (fixed) amount of land, L_A and L_Z the aggregate number of labor-units devoted to A and Z respectively, and b the productivity (average and marginal) of labor in the production of Z. Equilibrium requires:

$$(3)\quad bP_Z = P_A \cdot \frac{dA}{dL}$$

that is, the value of the marginal product of labor is the same in both sectors. Choosing A as our numéraire and assuming that P_Z is exogenously determined, it is clear that bP_Z is the "going wage" in this economy. For any given P_Z the labor supply curve will be horizontal in part at a wage w, where

$$(4)\quad w = bP_Z.$$

Now suppose a new technique becomes available. The consumption good Q, assumed to be identical to Z, is produced by labor and capital:

$$(5)\ Q = c \cdot \min\,(L_Q, nK)$$

which is a simple, fixed coefficients production function with c the productivity of labor and n the labor-capital ratio.[73] For simplicity, assume that capital goods are produced by labor alone:

$$(6)\ \dot{K} = mL_k.$$

Since Q and Z are identical, $P_Q = P_Z$. Hence the total amount of (gross) profits accruing to the capitalist can be written as:

$$(7)\ R = P_Z Q - wL_Q$$

which, by using (4) and (5), can be written as:

$$(8)\ R = P_Z cL_Q - P_Z bL_Q = P_Z nK(c\text{-}b)$$

since efficient production implies $L_Q = nK$.

Assume now that capital markets are wholly absent. All purchases of new machines are financed by profits plowed back into the firm. Let s_π denote the plowing-back coefficient. The value of new machines purchased each period is equal to total profits earned in that period times s_π. Thus:

$$(9)\ \dot{K}P_k = s_\pi R = s_\pi P_Z nK(c\text{-}b)$$

where P_k is the price of machines. Assuming homogeneous labor, the price of a machine can be seen to be (using [6] and [4]):

$$(10)\ P_k = \frac{w}{m} = \frac{bP_z}{m}.$$

Substituting (10) in (9) and solving for $\frac{\dot{K}}{K}$, the rate of capital accumulation, we observe:

$$(11)\ \frac{\dot{K}}{K} = s_\pi nm(\frac{c}{b} - 1).$$

From (11) we can draw the following conclusions. First, the rate of accumulation is independent of the terms of trade between industry and agriculture. Second, the rate of accumulation is inversely related to b, that is, the lower the productivity in the Z-good industry (or any other determinant of the wage level for that matter), the faster the growth rate. Third, the question whether accumulation will take place at all is wholly determined by whether $c > b$ or not, that is, whether the productivity of labor is higher in the modern industry than in the Z-good

[73] The formal setup has been borrowed from R. M. Solow, *Capital Theory and the Rate of Return* (Amsterdam: North Holland, 1963), although the context is of course different.

industry. The productivity of labor in machine production, m, as well as the labor-capital ratio, n, do not affect this basic decision, though they do affect the growth rate. Finally, if the reinvestment coefficient, s_π, is a positive function of the profit rate, the negative relationship between wages and capital accumulation is reinforced.

THE DECLINE AND RISE OF THE DUTCH ECONOMY, 1675–1900

J. de Vries

I. CONNECTING TWO HISTORIES

In the seventeenth century the economy of the Dutch Republic reached a position of international influence and material prosperity that elicited the envy of its neighbors and, since then, the curiosity of generations of historians. That "Golden Age" ended, as they all do. Opinions vary about the dating of the Golden Age's expiration, but it is generally agreed to end sometime in the last third of the seventeenth century.

The Kingdom of the Netherlands began to experience modern economic growth, growth closely linked to industrialization and associated technological transformation, no earlier than the last quarter of the nineteenth century. Indeed one of the few aspects of its nineteenth century economic advance to attract international attention is its lateness.[1]

In this paper I am concerned with the intervening two centuries of Dutch economic history. My objective is to assess what can now be said about the

Technique, Spirit, and Form in the Making of the Modern Economies: Essays in Honor of William N. Parker
Research in Economic History, Suppl. 3, pages 149–189

ISBN: 0-89232-414-7

course of national income and related quantitative indicators of economic activity between the end of the Golden Age in the late seventeenth century and the onset of modern industrial growth at the end of the nineteenth century.

The eighteenth and nineteenth centuries are not often studied together. An obvious reason for this is the general perception that the French and Industrial Revolutions constitute a great watershed separating two utterly dissimilar epochs. This dogma is now occasionally challenged, and rightfully so,[2] but a second obstacle remains to make difficult the unified study of this period. The various "schools" of economic history are often regarded as in competition with each other, but in fact each has staked out a "temporal turf" that seems most congenial to the methodology and ideology of the school. Thus, the French historical school, or Annales school, regards the eighteenth century as the end of an era of "quasi-immobile history," and its practitioners tend to grow weary of their monitoring of the respiration of the social order well before the French Revolution, which is the terrain of another sort of historian.[3] Likewise, the New Economic History might be thought to enjoy general applicability. In practice it is closely associated with a specific era, in this case the nineteenth century. It is the exceptional cliometrician who has pushed back beyond the British Industrial Revolution; more surprisingly, few have proceeded beyond World War I into the twentieth century.[4]

It is perhaps no accident that I should embark on the task of examining Dutch economic performance over a span of years divided between such dissimilar historical schools. As an economic historian of preindustrial Europe I have been much influenced by the French historical school, although I must admit that it inspires more readily than it instructs. On the other hand, my training in economic history emphasized the work of the then still new New Economic History. Economic history lives at the intersection of two proud disciplines, and my guidance as a graduate student came from someone prepared to work in industry as well as agriculture, in Europe as well as in America. Now it must be seen whether the rigor of the New Economic History can be crossbred with the insight of the French historical school.

II. LOOKING BACK FROM THE TWENTIETH CENTURY

The first step in plotting the likely course of Dutch national income from the end of the Golden Age to the onset of modern economic growth must be an assessment of the literature, to establish what is now thought to be true and to identify areas of controversy and ignorance.

In a reconnaissance of this kind a sensible strategy is to begin from a known point and move from there to the unknown. In this case that requires looking back into the past from the twentieth century, for the Dutch series of national income accounts extends back no further than 1900. The Dutch economists Jan

Tinbergen and J. B. D. Derksen produced this series of net national income on the eve of World War II, and it has been incorporated into the official Central Statistical Bureau series of national accounts.[5] All estimates of pre-1900 national income are back projections from an early twentieth century base; none is founded upon direct estimations of either output or income. Derksen, a participant in the project that yielded the official twentieth-century series, tried his hand at estimating national income at 10-year intervals back to 1860. Regrettably, he has never published his study. The results of his efforts have been published, and I will allude to them later in this survey, but since it is not possible to evaluate his estimates, it is also not possible to lend them much credence.[6]

The back projections of Dutch national income most worthy of consideration are the product of J. Teijl, who in 1971 offered two separately calculated series of estimates extending back at five-year intervals of 1850.[7] One series is based on Dutch government tax revenues. Teijl observed that total tax revenues strayed but little from a fixed relationship to national income in the period 1900–1915. He then satisfied himself that such changes as were introduced to the structure of taxation between 1850 and 1900 did not affect the relative tax burden. Armed with such assumptions, the estimation of national income became a straightforward matter: it was a constant multiple of total tax revenue.

Teijl's second estimate was based on the estimation of a Cobb–Douglas production function for the Dutch economy. With parameters estimated on the basis of 1900–1940 data, he "postdicted" net domestic product by introducing nineteenth-century data on the size of the labor force and, as a proxy for the capital stock, the annual consumption of energy. Net domestic income became net national income with the addition of independent estimates of foreign income.

The two series follow very different courses on a decade-by-decade basis, but they cross each other and end up in close agreement in 1850. This has led some observers to suppose that the two series confirm each other; indeed, J. H. van Stuijvenberg offers averages of the two series as the best available estimates of national income.[8]

It is hard to suppress the thought that the rough conformity of the two series is sheer coincidence. The tax-based estimate is an ad hoc procedure that depends entirely on the assumption that the tax burden remained constant. In discussing the reasonableness of this assumption, Teijl rests his case on the fact that the economy was relatively static and the role of government changed but little in the nineteenth century.

Even if Teijl's justification is accepted, a second problem remains: is the estimate of national income measured in constant or current guilders? If the tax system were based primarily on a proportional income tax, the results of Teijl's exercise would be national income in current guilders. Before 1893 there was no income tax. If most tax revenues were raised by indirect levies such as fixed excise taxes and license fees (as opposed to ad valorem taxes), the result might approximate national income in constant guilders. Teijl *appears* to interpret his results in this latter way, and the users of his estimates assume that they are expressed in constant 1900–1910 guilders.[9] If tax revenues actually grew at the rate of real national income, they would have become a steadily larger percentage of money income during the sharp deflation of the period 1873–1896. Since Teijl did not discuss this problem, the meaning he imputes to the concept of "constant tax burden" must remain uncertain.

The production function–based estimates depend on time series of labor inputs and energy consumption. Since the input of labor is simply assumed to grow at the rate of population growth (unemployment is assumed to be constant), the estimates of per capita national income depend entirely on the energy series, a product of Teijl's own research. Richard Griffiths has correctly remarked of this approach that it is likely to underestimate the growth of productivity in an economy, such as the Netherlands', in which agriculture, services, and transportation (where technological change was energy-saving) are of great importance relative to industry.[10]

Teijl's estimates describe the Dutch economy as growing at 0.7 percent per annum over the period 1850–1910. Our reservations suggest the possibility of more rapid growth, implying a lower 1850 income than Teijl estimates. Derksen, whose never-published estimates go back to 1860, set national income at a level 15 percent above those of Teijl.

Since the estimates of 1850 national income are anything but firm, projections further back would appear foolhardy; nevertheless, two Dutch scholars have ventured to make them. J. H. van Stuijvenberg, in a recent essay on the Dutch economy during the past two centuries, concluded that between 1770 and 1850 productivity advances were so minor that their effect on national income were fully negated by a long-term rise in the unemployment rate. He guesses that real 1770 net national income was roughly equal to Teijl's estimate for 1850.[11] Th. P. M. de Jong, whose research has concentrated on the late-eighteenth-century economy, made rough estimates of the capital stock in 1780 that he interpreted as support for the view that real per capita income was stagnant between then and 1860 (he preferred to rely on Derksen's estimates).[12] Both Van Stuijvenberg and De Jong restate a view widely shared by historians and deriving from nineteenth-century liberal polemicists that the Dutch economy broke out of a long-term stagnation only after 1850 (that is, after promulgation of the liberal constitution of 1848).

A revisionist position has recently been staked out by Richard Griffiths. In his 1980 inaugural address at the Free University of Amsterdam he extended the analysis of his book *Industrial Retardation in the Netherlands 1830–1850* to argue that the Dutch economy experienced balanced growth—he called it *stiekeme groei* (surreptitious growth)—that had not been appreciated fully because of the relative absence of dynamic industrial expansion.[13] Griffiths offers no national income estimates, but he makes two assertions relevant to our subject: (1) Teijl underestimates 1850–1910 growth, and (2) the growth path of

1850–1910 should be extended back to 1830, since there is no basis for supposing the midcentury to be an economic turning point.

Since Teijl's estimate of the long-run rate of per capita growth is 0.7 percent per annum, it seems reasonable to interpret Griffiths's statements to mean that the Dutch economy grew at a rate of approximately 1.0 percent per annum over the period 1830–1910, i.e., that per capita income roughly doubled during the interval.

The literature on the nineteenth century now offers something for everyone. At one extreme, real income stood at *f* 250 (i.e., 1900 guilders) in 1770, stood no higher until after 1850, and rose to *f* 353 in 1900; at the other extreme, it stood at about *f* 175 in 1830, on the order of *f* 215 in 1850, and *f* 353 in 1900.

III. LOOKING FORWARD FROM THE GOLDEN AGE

The view backward from 1900 is still shrouded in mist. If we now position ourselves at the apogee of the Golden Age and gaze forward into the eighteenth and nineteenth centuries, is it possible that some orienting landmarks will be revealed?

The literature on the aggregate economy in these centuries is primarily concerned with establishing trends, turning points, and crises. These efforts are based primarily on time series of demographic, price, and trade statistics and are not readily usable as indicators of national income. Indeed, no modern historian has ever attempted the estimation of the level, that is, the valuation, of national income or product. Only contemporaries of that era had the temerity to venture onto that treacherous terrain.

Gregory King, the celebrated English political arithmetician, made estimates of Dutch and French national income for 1688 and 1695 in conjunction with his more familiar efforts concerning England.[14] Economic historians have an obvious interest in defending King's reputation for sagacity. It is convenient to imagine that he had access to voluminous and accurate information that is regrettably no longer extant. Indeed, his writings on the quantitative dimensions of England's population and economy remain well regarded after a generation of scholarly scrutiny.[15]

The same cannot be said for his French estimates, which have been savaged by E. Le Roy Ladurie.[16] What of King's national income estimates for the Dutch Republic, where he sets per capita income in 1688 at £8.5, or 7 percent above his estimate for England? King's own published works reveal next to nothing about the basis for his figures, but his manuscript notebook, happily preserved and published, permits an evaluation of his method.[17]

King's approach to national income accounting in this case was to estimate the returns to the factors of production, albeit in a desperately abbreviated way. The total rent receipts, including dwellings and urban property, he set at £4 million, the result of multiplying the 8 million acres surface area of the Republic (an accurate figure) by an assumed average rent of 10 shillings per acre. He had reckoned English rents to average 5 shillings per acre and reasoned that the higher population density of the Republic warranted a doubling of that figure. In this regard it is worth noting that in 1835–1840, just after the completion of the first Dutch cadastral survey, the Netherlands had 2.1 million hectares (5.25 million acres) of cultivated land, which had an average rental value of *f* 31.2 per hectare.[18] Average rents before the decline in agricultural prices that sets in by the 1670s may not have stood far below the 1835 level.[19] Even if they were set at two-thirds of the 1835 level, which is 20 percent below the average 1650 rental price found in a survey of 177 parcels in Holland, the agricultural rents of the Netherlands would have totaled some £4 million. To that figure one would need to add the rental value of woodlands and rural dwellings, not to mention all urban property, which accommodated some 40 percent of the population in 1675. An indication of the probable importance of such property is available in 1808, when the nation's 400,947 residential dwellings were taxed on the basis of their rental value. The total assessed value was then *f* 21 million, or nearly £2 million. This figure may well have been substantially higher in the seventeenth century.[20]

The returns to labor and capital King collapsed into a single estimate of "trade and business," elsewhere labeled "improvements by labor." There is nothing in his notes to hint at the source of this estimate except perhaps some marginalia about the importance of the fishing industry and a tenuously reasoned estimate of the total capital stock (£71 million, implying a capital–output ratio of 5.5:1).

The incomes to trade and business he estimates at £13 million for 1688 and £14 million (and also apparently £15 million) in 1695, making his total estimate for national income range from £17 million to £19 million. King divided these national income figures by the population of the Republic, which he set at 2 million in 1688 and thought had risen to 2.2 million by 1695. Modern estimates are very close to the 2 million figure. However, there is no evidence that the population was growing; the reverse is more likely. King's per capita income figures for 1688 through 1695 range between £8.14 and £8.64, equivalent to 90–94 guilders at the ruling exchange rates of those years.

King's next step was to split the total into its component parts of national expenditure: consumption, government, and savings. Here we come to the motivation for King's excursion into comparative national income accounting. He wished to advance an argument about how best to finance the wars then being waged against France. For present purposes it should be noted that his figure for Dutch tax revenues is apparently based on contemporary evidence and that his estimate of savings was likely dictated by the needs of his political argument. Private consumption expenditure is almost certainly a residual figure. Although King estimated Dutch per capita income to be 7 percent above the English level, he was left with a figure for Dutch consumption that was only three-quarters of the English figure. This, despite his remark in the notebook that "Ye necessaries in Holland in many things, especially Food and Rayment and other matters of Importance being generally a 5th or 4th part dearer than in England, Excepting Fish. . . ."[21]

It suited King to put this image of Dutch frugality before his English readers, but his estimates cannot stand as an adequate characterization of Dutch national income. The fact is that King's notebook reveals little more knowledge of the Republic than its size, population, and government revenues.

The only other estimates of Dutch national income date from the early nineteenth century.[22] The more informative of the two is W. M. Keuchenius' *Nationaale Balans der Bataafsche Republiek,* which refers to an unspecified year around 1800. Keuchenius furnishes his study with a table itemizing the national income in one column and national expenditure in another. Unfortunately, Keuchenius shows himself to be an inconsistent physiocrat. His assessment of national income and "expenditure" is concerned primarily with the value of domestic agricultural and fishing output, to which he adds exports. All domestic trade, transport, and manufacturing he dismisses, "since the profit passes from hand to hand without enriching the country or increasing the [quantity of] money." Ironically, he does include under national income the classic transfer payment, interest on the public debt.

Little can be done with Keuchenius's study to estimate national income, but, assuming that the component estimates can be trusted, his work can serve to reconstruct the chief components of the Dutch balance of payments. Table 1 shows that Keuchenius—either directly or by implication—believed foreign earnings to total 96 million guilders, only 14 million of which consisted of domestically produced exports. On the other hand, foreign payments totalled 79 million guilders, all of which went for commodity imports. Were one to accept Keuchenius' estimate of national income, 221 million guilders, or *f* 126 per capita, this would imply that 43 percent of national income was earned abroad. The conclusion cannot be avoided that either his individual estimates are very far from the mark, or national income stood far above *f* 126 per capita, or both.

A second source, written at about the same time but with knowledge of Keuchenius's estimates, is Rutger Metelerkamp's *De toestand van Nederland in vergelijking gebragt met die van enige andere landen van Europa.* He begins by estimating the average annual payments to the owners of real property (*f* 45 million—compare to Gregory King's £4 million ≅ *f* 44 million in 1688), to the owners of domestic and foreign government bonds (*f* 70 million, *f* 30 million of which is from domestic debt) and then plunges into the commercial sector. Foreign trade and shipping, by Metelerkamp's reckoning, yielded *f* 36 million on a capital investment of *f* 350 million. Further calculations were apparently beyond him, for he then moves on to offer an overall estimate of national income of *f* 250 million, or *f* 125 per capita, and an estimate of the "national wealth"—the capital stock—of *f* 3500 million! These estimates he placed in the context of a

Table 1. The Balance of Payments of the Batavian Republic, circa 1800, According to the Estimates of W. M. Keuchenius (in millions of guilders)

Receipts from Abroad			*Payments to Foreigners*		
Exports			*Imports*		
Agricultural products	10		Agricultural products	33	
Fish	1		Raw materials	11.5	
Manufactures	3		Manufactures	26.5	
Colonial commodities	12[a]		Colonial commodities	8[a]	
Subtotal		26	*Subtotal*		79
Invisibles					
Shipping, banking, insurance		30[b]			
Investments					
Income from foreign government debt		40			
Total		96			79

[a]Keuchenius estimates that the Dutch East Indies and West Indies possessions yielded an income from the sale of delivered commodities of *f*20 million. He then reckons that the domestic consumption of these goods accounted for *f*8 million. I am treating the *f*20 million as net proceeds from all colonial activities. I suspect that this results in a substantial overestimate of colonial earnings.

[b]Keuchenius' estimate of these earnings from "invisibles" is for "years of peace."

Source: W. M. Keuchenius, *De inkomsten en uitgaven der Bataafsche Republiek voorgesteld in eene nationale balans* (Amsterdam: 1803), p. 139.

large table entitled "Comparison of the situation of the Netherlands and several other countries in various periods." He compared eight European nations in a variety of quantitative dimensions (among others, value of foreign trade, size of armies) in the course of which he displays evidence of familiarity with the work of Gregory King and Patrick Colquhoun. Unfortunately, his Dutch estimates give even less evidence than those of Keuchenius of an understanding of the elements of national income.

Neither King nor Keuchenius and Metelerkamp can be of much use in pegging the level of Dutch national income in the seventeenth and eighteenth centuries. The first reveals too little knowledge; the two others, too much ignorance.

We must turn then to the modern scholars who have sought to establish the trends in the volume of seventeenth- and eighteenth-century economic activity. Still the most comprehensive study of the eighteenth-century Dutch economy is Johan de Vries's *De economische achteruitgang der Republiek in de achttiende eeuw.*[23]

In a sector-by-sector assessment, De Vries emphasized the structural changes taking place in the Dutch economy and concluded that the economy was overshadowed by others, that is, had suffered relative decline. A loss of position seemed unquestionable; a loss of real income was more doubtful. De Vries ventured to guess that real national income remained constant into the 1780s but that per capita income may have fallen slightly.[24]

Since Johan de Vries wrote, several monographs and articles have added a great deal to our knowledge of trends in the domestic economy. A. M. van der Woude demonstrated that the region of North Holland experienced a severe demographic and economic contraction between the 1670s and the 1740s.[25] J. A. Faber noted a somewhat milder contraction in the primarily agricultural economy of Friesland in the same period. In both areas the second half of the eighteenth century brought some respite, but only in Friesland was a marked recovery observable. These regional studies portray an economy of more discontinuous change than did Johan de Vries, and they regard as likely an absolute decline of the economy up to the mid-eighteenth century. But in a complex, urbanized economy, studies of several rural regions cannot alone reveal the overall trends.

I have claimed that the acceptance of Gregory King's estimate of Dutch national income for 1688–1695 cannot be justified. But, in order to get a clear picture of the present status of our knowledge, I now ask the reader to suspend disbelief for the moment and accept King's figures as roughly correct. They suggest a per capita income of 90–95 guilders. No reliable price deflator is available to express this amount in 1900 guilders, since no comprehensive Dutch price index has ever been constructed for the nineteenth century. Van Stuijvenberg and De Vrijer have put together a series that links the seventeenth- and eighteenth-century price history of N. W. Posthumus with twentieth-century data by splicing them to English and German wholesale price indices.[26] The open character of the nineteenth-century Dutch economy gives a certain justification for this expedient, but the fact remains that this spliced series is subject to substantial error, particularly in the calculation of long-term price trends. Compounding this problem is the fact that 1688–1695 was itself a period of rapidly increasing prices. The composite wholesale price series, indexed to 1900–1910 = 100, sets prices in 1685–1689 at 66 and in 1690–1694 at 85. It is indisputable that prices in 1688 were substantially below those of the Napoleonic era, and it is probable that the 1900–1909 level, while lower than a century earlier, had not descended to the level of the 1680s. In King's day 90–95 guilders certainly possessed greater purchasing power than it did in 1900; it may have been the equivalent of 105–115 guilders of the 1900–1910 period.

If we attach the King estimate of income *level* to the Johan de Vries assessment of income *trend*, the conclusion must be that Dutch per capita income at the end of the eighteenth century could not have stood far above *f* 100 (in 1900 guilders) but was certainly above the level set in Keuchenius's incomplete analysis.

IV. AN AWKWARD JOIN

A large gap yawns between the late-seventeenth-century estimate of Gregory King (105–115 guilders) and the several mid-nineteenth-century estimates

(215–250 guilders). One need not place a great deal of faith in the price adjustments to be led to the observation that per capita income appears roughly to have doubled. How is that gap to be bridged?

In fact, few scholars have addressed that problem; while many have expressed themselves about income trends in one period or another, they have rarely considered the implications of their views for later or earlier periods. It is obvious that the orthodox belief of the early modernists in eighteenth-century stagnation and that of the specialists in the nineteenth and twentieth centuries in stagnation before 1850 is not compatible with the available estimates of income level. What is not obvious is which element—the estimates of trend, level, or both—is most in need of revision.

Recently two scholars have ventured to trace a path linking the Dutch Golden Age to the twentieth century. One is Angus Maddison, an economist whose chief interests are in the industrial era; the other is James Riley, an historian of the early modern period.

Angus Maddison, a veteran analyst of national income data, addresses the question of Dutch growth in his study *Phases of Capitalist Development.*[27] He focuses on the successive leaders in productivity in order to develop an argument about gaps between performance and potential, or actual and best practice in the various capitalist economies. He proposes that an envelope curve of per capita output—ideally gross domestic product per man-hour—in the successive leaders in economic performance defines the long-run pace of technological progress in the global economy. In order to establish the contours of his envelope of growth rates from the seventeenth century onward, Maddison had to identify the leading economies in each period and construct for them time series of national income. He asserts that in the twentieth century the United States growth curve traced the uppermost national growth trajectory, having surpassed that of the United Kingdom around 1890. Britain, in turn, held the top honors for approximately a century, having surpassed the Dutch around 1780. The Dutch lead extends back into the seventeenth century, the starting point of Maddison's analysis.

Maddison, now a professor at the University of Groningen, inaugurates his study of long-term economic growth with an analysis of the Dutch Republic, regarding it to have been the first economy to advance beyond the economic constraints of an agrarian society on a national scale. This seventeenth century achievement endowed the Republic with productivity levels that he estimates as 50 percent above the level established by Gregory King for England in 1688. Maddison justifies this steep upward revision of King's implied ratio of Dutch to English incomes by observing that (1) upwards to two-thirds of the Dutch labor

---

Figure 1. Locus of Productivity Leadership, 1700–1979. Gross domestic product (GDP) per man-hour in 1970 U.S. dollars. [From Angus Maddison, *Phases of Capitalist Development* (Oxford: 1982), pp. 30–31.]

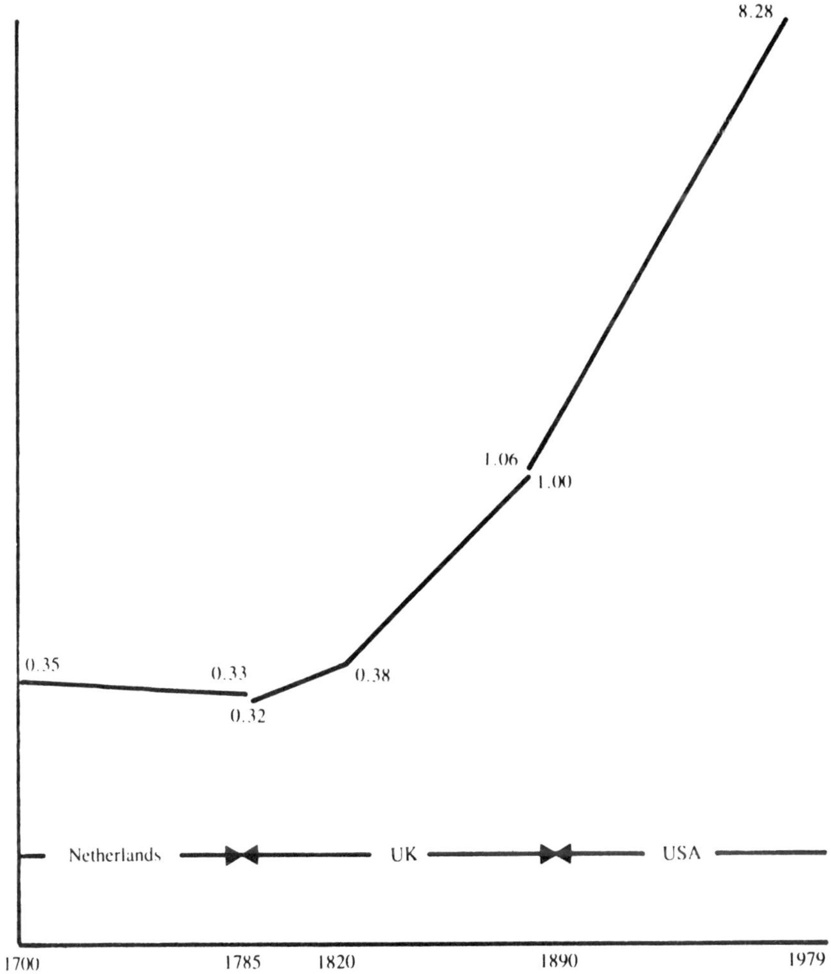

Lead Country	*Date*	*Annual Average Compound Growth Rate of:* *GDP per Man-Hour*	*Gross Fixed Nonresidential Capital Stock per Man-Hour*
Netherlands	1700–1785	−0.07	N.A.
UK	1785–1820	0.5	0.0
UK	1820–1890	1.4	0.9
USA	1890–1979	2.3	2.4

force found employment outside of agriculture, where incomes are usually higher; (2) the high-income foreign trade and services sectors were equal in size to England's, and on a per capita basis much larger; (3) agriculture was highly specialized and presumably more productive; (4) the industrial structure was diversified and benefited from a low-cost energy source in the form of peat; and (5) the investment rate ran higher in the Republic than in England, and this must have provided for some measure of technical progress.[28]

Having set Dutch income 50 percent above the English level in 1700, Maddison proceeds by accepting Johan de Vries's assessment of the eighteenth century economy: national income remained about the same, but per capita gross domestic product, which excludes income from foreign investments, fell by about 10 percent between 1700 and 1760, remaining constant from then until 1820. After 1820 Maddison makes no direct estimates of Dutch growth. Since Dutch national income by then stood well within the envelope curve of the lead countries, he understandably turns his attention elsewhere. But an unavoidable consequence of connecting Maddison's eighteenth-century scenario to the available estimates for the late nineteenth century is an amazingly high rate of growth, one of the highest in all of Europe.[29] Much the same comment can be made about the GNP estimates for the Netherlands presented by Paul Bairoch.[30] Bairoch does not describe the basis for his Dutch estimates, which is all the more puzzling when one notes that they reflect a growth rate in the period 1820 to 1870 of 1.5 percent per annum, higher than any other European national economies except Switzerland and Belgium. Even Richard Griffiths, the most vigorous exponent of the nineteenth-century growth hypothesis, writes only of "surreptitious growth" and "some measure of economic growth."

James C. Riley suggests a very different growth path for the Dutch economy between the seventeenth and twentieth centuries. He is suspicious of the orthodox image of a lethargic, outmaneuvered eighteenth-century economy. His own work on the Amsterdam capital market showed a rapid growth of Dutch foreign investment in the second half of the century. In addition recent work on Dutch shipping suggests that the decline of Holland's entrepôt function found compensation in the rise of direct Dutch shipping services that bypassed the Republic's ports. On these twin pillars he bases his case for eighteenth-century growth.[31]

Riley, just as Maddison, starts with Gregory King, and just as Maddison he suspects that the gap between England and the Republic was greater than King allowed. Riley suggests a maximum figure for Dutch per capita income about 25 percent higher than King's figure for England. Riley then turns to the mid–nineteenth century, calling on Derksen rather than Teijl, and proposes that the apparent doubling of real income between 1688 and 1860 was chiefly an achievement of the eighteenth century. He agrees that stagnation ruled in the first half of the nineteenth century and suggests that undramatic but persistent growth characterized the eighteenth century, as continued commercial vitality and its multiplier effects plus the addition of a flourishing financial sector sustained employment, increased incomes from abroad, and secured lower prices for Dutch consumers.

Figure 2 brings together the various more or less explicit estimates and guesses embodied in the literature. No graceful way has yet been found to join the Golden Age to the Industrial Age.

V. AN AWKWARD COMPARISON

The estimates of Dutch national income made by Gregory King, Angus Maddison, and James Riley depend on their respective views about English national income. That is, their judgments about Dutch income are essentially statements about the relationship of Dutch to English national income. It follows that the level at which Dutch income is set in 1688 and later is not independent of our understanding of England's national income at the same time. Unavoidably, this

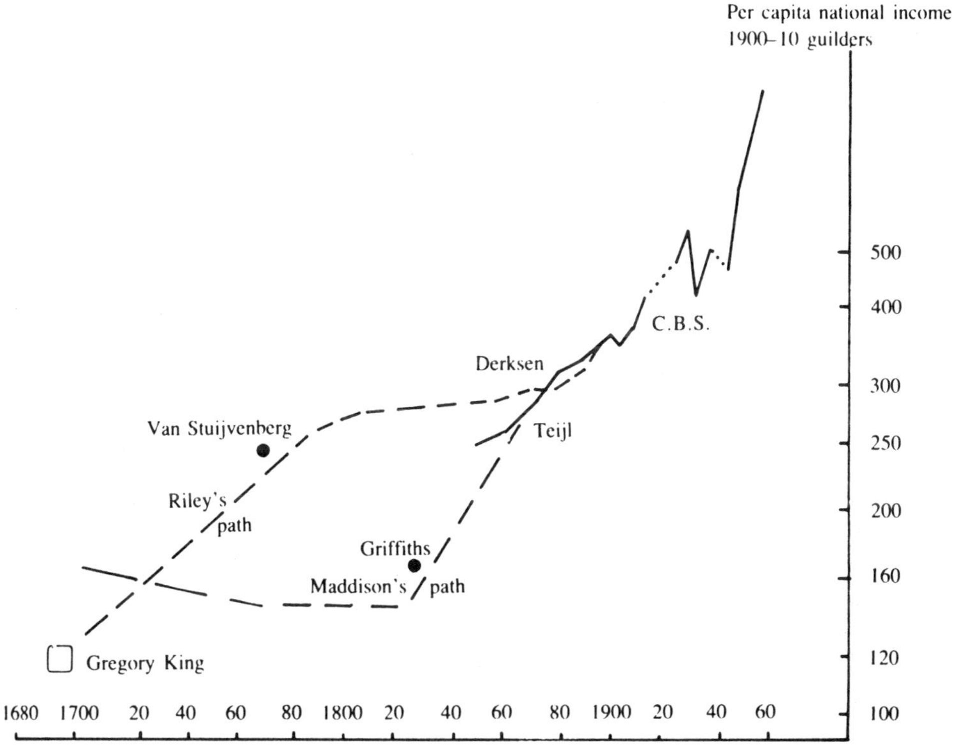

Figure 2. Alternative Proposed Growth Paths of Dutch National Income, 1688–1900.

investigation of the Dutch economy must turn its attention to the other side of the North Sea. What is now known about British national income in the eighteenth and nineteenth centuries, that is to say, during the Industrial Revolution?

The standard source on this subject is Phyllis Deane and W. A. Cole's *British Economic Growth 1688–1959*.[32] There the authors produced an index of real output for the eighteenth century comprising decennial estimates of activity in agriculture, export industries, domestic industries, government, and rents. They concluded that total output had grown 2.5-fold over the century, and on a per capita basis by 60 percent. Among the critics of this output series, Nicholas Crafts has argued cogently that they underestimated the growth of agricultural output in the first half of the eighteenth century.[33] The publication of the *Economic History of Britain Since 1700*, edited by Roderick Floud and Donald McCloskey, provided an opportunity for Cole to offer a revised total output series incorporating Crafts's suggestions.[34] Total output is now thought to have grown by 2.7-fold. But since new calculations of English population show more rapid growth in the early eighteenth century than Deane and Cole had originally assumed, the per capita output estimates remain essentially unchanged, although it is not clear that this was the intention of either Cole or Crafts.

Deane and Cole made no effort to cast their output index in monetary terms, but when the 1800 estimate is linked to their nineteenth-century national income series, the implied 1700 income is roughly consistent with Gregory King's 1688 estimates, which they discuss approvingly. Cole's revisions are also capable of incorporating King's vision of the English economy and Cole ventures to estimate total and per capita national product. Table 2 displays Cole's series and converts it into guilders at prevailing exchange rates.[35]

From the early nineteenth century onward the Deane and Cole estimates of national product of *Great Britain*, and from 1855 their series for the net national product of the *United Kingdom* (i.e., including Ireland), can serve our purposes, once they are expressed in 1900–1909 prices and converted to guilders at current exchange rates.[36] Cole's eighteenth-century estimates yield a 57 percent growth of *English* per capita real income. Deane and Cole's nineteenth-century estimates yield, for *Great Britain*, a 408 percent increase. The entire two-century period therefore experienced per capita income growth, in real terms, of 640 percent. When expressed in guilders the rise is just under sevenfold, reflecting a diminution in the exchange value of the guilder after the period of French domination ending in 1814.

On the basis of this evidence Dutch per capita national income in the first decade of the twentieth century stood at approximately 63 percent of Great Britain's (somewhat higher relative to U.K. national product). If we accept the accuracy of the twentieth-century relationship, then the issues involving Dutch growth and stagnation and their timing will be affected sensitively by what is thought to be true of the Dutch–English income ratios in the late seventeenth century, and this, in turn, will hinge on one's perceptions of the eighteenth-century English economy.

Table 2. W. A. Cole's Estimates of English Per Capita National Product in 1700 Pounds and Guilders

Year	*In Pounds*	*In Guilders*
1700	£9.45	*f* 103
1720	10.16	111
1740	10.80	118
1760	12.40	135
1780	12.20	133
1800	14.80	161

	In Current Pounds and Guilders		*In 1700 Guilders*	*Source*
1688	£7.91	*f* 84	*f* 98	Gregory King

Sources: W. A. Cole, "Factors in Demand, 1700–80," in Roderick Floud and Donald McCloskey, eds., *The Economic History of Britain Since 1700*, Vol. I (Cambridge: 1981), p. 64; Gregory King, "Natural and Physical Observations Upon the State and Condition of England," in Peter Laslett (ed.), *The Earliest Classics . . .* (London: 1973).

If one accepts Gregory King's figures and holds to the view that Dutch income at the end of the eighteenth century was no higher than in 1688, then Dutch income circa 1800 would be no more than 67 percent of the British level. Since Dutch income in 1900–1909 stood in about the same relationship to British income, an implication of this view is that the nineteenth-century Dutch economy grew at the same overall rate as Britain's, or 1.4 percent per annum throughout the century. No student of the nineteenth-century Dutch economy has ever advanced such a view.

Implications of Deane and Cole, and King, Regarding Eighteenth-Century Stagnation

	1700	*1800*	*1900*
Britain	100	157	698
Netherlands	107	107	440
Neth./Britain	107	67	63

If one accepts Maddison's position that Dutch per capita income was half again the English level in 1700 and declined by 10 percent thereafter, England surpasses the Dutch in 1780, as Maddison claimed. In 1800 Dutch income is at 81 percent of the British level. This assumption, too, requires very rapid nineteenth-century growth. The major problem here is to find a period after 1800 when Dutch per capita income could have grown 3.6-fold. Accordingly to Teijl the second half of the nineteenth century booked only a 43 percent increase. That is, in the period when industrial growth and structural change began to take hold, Dutch growth was less than half that of Britain. If one follows Griffiths in rejecting Teijl's estimates as too low, and assumes 1.0 percent per annum growth from 1830 to 1900, a gap still looms in the period 1800–1830 requiring a rate of growth exceeding Britain's in those years, and far exceeding later Dutch growth.

Implications of Deane and Cole, and Maddison, Regarding the Eighteenth-Century Decline and 1% Per Annum Growth after 1830

	1700	*1780*	*1800*	*1830*	*1900*
Britain	100	130	157	254	698
Netherlands	150	135	135	220	440
Neth./Britain	150	104	86	87	63

If we now turn our attention to the nineteenth century, the comparison of Dutch and English data uncovers several different but equally unlikely growth scenarios.

Teijl's back projections imply that British per capita income exceeded the Dutch level only after 1860—after a century of vigorous growth that is witness to the Industrial Revolution. A comparison with U.K. income, would push the crossing point forward even further. Could it be that Britain's time at Maddison's productivity frontier (Figure 1) lasted only 25 years, years that provoked so much wringing of British hands about industrial retardation?

An equally iconoclastic conclusion must be drawn if one embraces the orthodox position that the Dutch economy achieved no, or next to no, per capita growth in the 70 or 80 years before 1850. Van Stuijvenberg's analysis requires us to believe that Dutch per capita income in the 1770s stood at two and a half times the English level.

Finally, we must consider Riley's argument that eighteenth-century growth bridged the gap between the Golden Age and the mid-nineteenth century. By his own estimate of the size of the gap (he uses Derksen's nineteenth-century estimates, inflates King's 1695 Dutch income figure by some 25 percent, and believes there was little growth after 1800) real per capita income needed to grow at 0.6 to 0.8 percent per annum from 1695 to 1800.[37] Teijl and the official twentieth-century national income series yield no more than 0.7 percent per annum for the period 1850–1939, and the highest estimates of English growth in the eighteenth century yield no more than 0.5 percent overall. It is difficult to accept that such an impressive "preindustrial" growth rate could have been sustained for over a century without anyone noticing.

My purpose here can be compared to clearing out the attic of an elderly and absent-minded relative. Many of the accumulated possessions obviously should have been disposed of years ago. Others reputed to be valuable heirlooms are found to be as worthless as the rest. In such a situation it is desirable, albeit painful, to be candid about just how little of value there is. It is best to face one's tasks without illusion. Gregory King's estimates of Dutch income are without merit; Teijl's back projections rest on dubious interpretations and data, while his results are difficult to explain. The assumptions made by most economic historians about Dutch income trends are mutually incompatible and at certain points implausible in the extreme.

We now see more clearly what could not have happened; there remains the task of establishing what did. In the course of dusting off and scrutinizing the artifacts in the attic of Dutch economic history I came more and more to feel that the greatest obstacle to achieving a better appreciation of their true value was, ironically enough, one of the treasured relics in a neighbor's attic: the long-reigning interpretation of the English Industrial Revolution as a momentous discontinuity bringing about rapid economic growth in a brief period of time. No plausible account of the Dutch economy's history can be made to rhyme with that interpretation.

I came to the (perverse) conclusion that English quantitative economic history would have to be changed in the interest of achieving a better understanding of the Dutch experience. No sooner had this heretical thought asserted itself than there appeared within one year, in a remarkable example of intellectual convergence, three articles produced by four economic historians working in three countries, all arguing variously and vigorously that the English economy grew much less rapidly in the eighteenth century than has hitherto been believed. Directly or by implication they all regard England before the Industrial Revolution to have been more industrial and enjoying a larger income than the literature reviewed above could admit to.

Peter Lindert and Jeffrey Williamson, reexamining the work of the political arithmeticians Gregory King, Joseph Massie, and Patrick Colquhoun (who all estimate flows of income rather than Deane and Cole's volumes of output), came to the conclusion that King underestimated English national income by 25 percent, that Massie did so by 16 percent, and that Colquhoun's totals were approximately correct.[38] The result, obviously, is to reduce greatly the estimated rates of eighteenth-century income growth.

By their calculations the adjustments to Deane and Cole's series must come almost entirely in the period before 1760. By inflating King's figures by 25

Table 3. Revisionist Estimates of English Rates of National Income Growth, 1700–1841

A. Average annual rates of growth:

Lindert and Williamson	*1700–1760* .39		*1760–1800* 1.01	
Harley	*1700–1770* .56		*1770–1815* 1.31	*1815–1841* 2.23
Crafts (1982)	*1700–1760* .66	*1760–1780* .83	*1780–1801* 1.27	*1801–1831* 1.93
Dean and Cole	.66	.65	2.06	2.86
Cole (revision)	.84	.65	2.06	

B. Average annual rates of per capita growth:

Lindert and Williamson	*1700–1760* .19		*1760–1800* .14	
Harley	*1700–1770* .27		*1770–1815* .33	*1815–1841* .86
Crafts (1982)	*1700–1760* .28	*1760–1780* .14	*1780–1801* .30	*1801–1831* .48
Dean and Cole	.45	−.04	1.08	1.41
Cole (revision)	.46	−.09	1.00	

C. Chronology of per capita income growth (1688–1700 = 100):

Year	*Lindert and Williamson*	*Harley*	*Crafts*	*Deane and Cole*
1688	100			
1700		100	100	100
1760	115		118	130
1770		119		122
1780			122	129
1800	122		130	160
1815		140		
1831			150	237
1841		175		265

Sources: Peter H. Lindert and Jeffrey G. Williamson, "Revising England's Social Tables, 1688–1867," *Explorations in Economic History* 19 (1982) 385–408; C. Knick Harley, "British Industrialization before 1841: Evidence of Slower Growth During the Industrial Revolution," *Journal of Economic History* 42 (1982): 267–90; N. F. R. Crafts, "British Economic Growth, 1700–1831," *Economic History Review* 36 (1983): 177–99; Phyllis Deane and W. A. Cole, *British Economic Growth, 1688–1959* (Cambridge: 1967).

percent, Lindert and Williamson set the rate of growth in 1688–1760 at 0.39 percent per annum, whereas Deane and Cole had it 0.64 percent. Per capita growth Lindert and Williamson estimate at under 0.2 percent per annum while Deane, Cole, and Crafts (1981) all put it at more than double that level.

C. Knick Harley set about recomputing industrial output indices as well as consumption estimates. He concluded that "the growth of industrial production was much slower between 1770 and 1815 than either most narrative accounts of the industrial revolution or the quantitative research of Walter Hoffman and of Phyllis Deane and W. A. Cole have suggested."[39] His computation of English national income reveals a per annum growth rate of 0.56 percent from 1700 to 1770 and 2.23 percent from 1815 to 1841. Both of these are within striking distance of the Deane and Cole estimates. He departs from this orthodoxy, however, in the classic period of English industrialization, 1770–1815. Here he sets per annum growth at 1.31 percent, little more than half the rate advanced by Deane and Cole. In Harley's words: "The indices imply that the industrial sector in the eighteenth century was nearly twice as large as previous estimates indicated."[40]

Finally, Nicholas Crafts reviewed afresh the evidence of British economic growth. While his earlier critique of Deane and Cole (discussed above) had the effect of increasing their estimates of early-eighteenth-century growth, his most recent utterances have the opposite effect, particularly on the estimates for the post–1760 period.[41] Table 3 displays a summary of the estimates put forward by these exponents of the "slow-growth" school.

It is apparent that the three revisionist works are not of one mind on certain critical issues, but they share the position, of importance for present purposes, that the English economy before 1760 was more industrial and generated a higher national income than has hitherto been accepted. Instead of an economy with a per capita income level in 1700 that was less than 40 percent of the 1840 level (as Deane and Cole had it), we must now consider the well-supported view that England's 1700 per capita income stood at nearly 60 percent of the 1840 level.

VI. A NEW ASSESSMENT

This hydra-headed reassessment of the quantitative dimensions of the English economy opens the way to the construction of a plausible account of the long eighteenth- and nineteenth-century readjustment process of the already "modern" Dutch economy. Figure 3 presents a sketch of my hypotheses concerning the trends and levels of Dutch national income, comparing this to Harley's reassessment of pre-1840 English growth. I will argue that (1) the Dutch economy declined absolutely during the period 1675–1813, with most, if not all, of the decline occurring before 1750; (2) from 1813 to 1850 the economy recovered

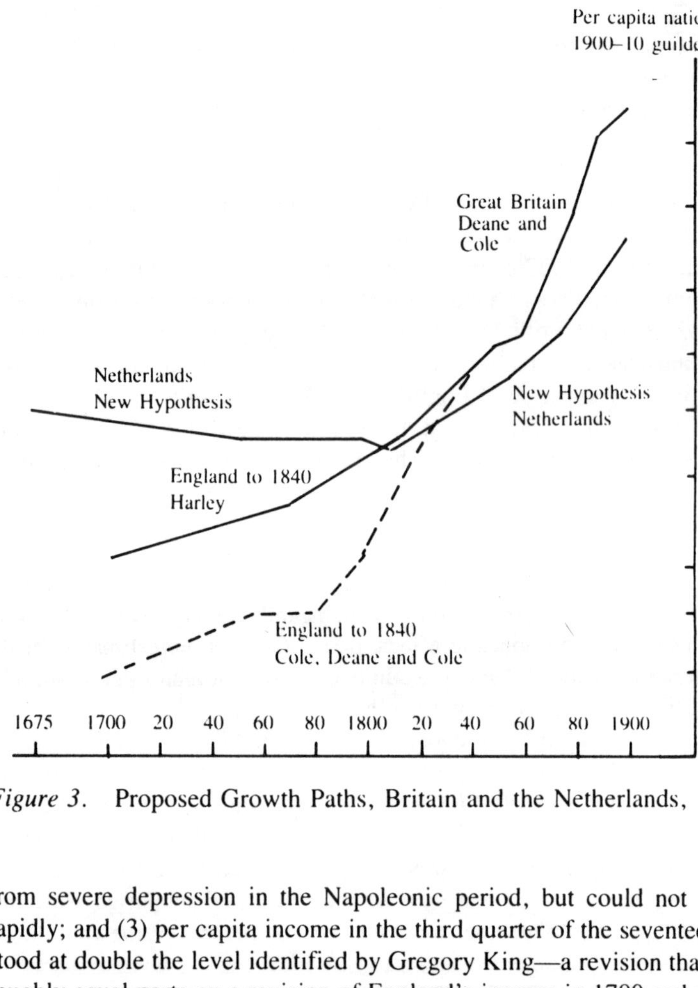

Figure 3. Proposed Growth Paths, Britain and the Netherlands, 1675–1900.

from severe depression in the Napoleonic period, but could not have grown rapidly; and (3) per capita income in the third quarter of the seventeenth century stood at double the level identified by Gregory King—a revision that is based in roughly equal parts on a revision of England's income in 1700 and an argument that the Republic's income could have exceeded England's by two-thirds. The reader must not expect a comprehensive set of Dutch national income estimates; the foundations for such an enterprise simply have not yet been built. What one can expect is a marshaling of evidence, some new, to assess the plausibility of the hypotheses.

A. Trends

One approach to the question of Dutch economic trends is to survey the economy sector by sector, industry by industry, noting the fortunes of each. This is not the place to review the available evidence; several works do an adequate job of that.[42]

Industry and regional studies have established beyond a reasonable doubt that commodity production declined in the course of the eighteenth century, as such major industries as Leiden woolen textiles, Haarlem linen bleaching, Gouda and Delft ceramics, and Zaandam shipbuilding fell into irremediable decay, while sailmaking, fishing, and the activities of industrial windmills declined substantially. Certain industries resisted these trends—papermaking, tobacco processing, and gin distilling are perhaps the most important—and many smaller crafts may have shown greater stability than the export industries mentioned here, but the basic direction of change is clear.[43]

Eighteenth-century commentators often wrote of the flood of imports in the Republic.[44] Indeed, Keuchenius, in his incomplete assessment of Dutch national income circa 1800, set final imports at 36 percent of national income and estimated that the value of imported manufactures exceeded that of exported manufactures by a factor of nine.[45]

At present the course of agricultural output can only be inferred from evidence about the timing of investment, land abandonment, volumes of marketed dairy products, and incidental information. In the western half of the Republic it seems probable that output declined substantially between the 1660s and 1740s. The growing pressure of taxation forced land abandonment; reclamation activities sank to zero; the epizootics of 1713–1719 and 1744–1765 severely depressed dairy output, and the pressure of costs encouraged more land-extensive production techniques. Certain compensatory activities in the eastern provinces, particularly tobacco production, relieve this litany of gloom.[46] In the second half of the eighteenth century the evidence all points to expansion, a trend that is not broken by the political disturbances of the Napoleonic era and that continues into the nineteenth century.[47]

Commercial, banking, and shipping services form the most debated sector of the eighteenth-century economy. There is no doubt that these activities suffered severely during the Napoleonic period, never again to enjoy their earlier importance. But the course of affairs before 1795 is not at all clear. The financial role of Amsterdam grew throughout the second half of the eighteenth century, although it too suffered greatly during the Napoleonic period.[48] It is arguable that Amsterdam's commercial activities did not decline until 1780 (although the emphasis shifted from commodity trade to financial services), but it is difficult to be as charitable about the numerous lesser ports of the Republic. There are many indicators that eighteenth-century Amsterdam attracted to itself the remnants of a previously more broadly distributed commercial life.

Our perceptions about the various sectors of economic life, even if they were unerring, cannot form the basis of national income estimates when the relative size of these sectors is unknown and changing. With so many imponderables standing in the way of direct estimation of national income, it seems advisable to turn our attention to less direct but aggregative economic indicators.

One such indicator is the time series of intercity passenger travel by barge on the canal network connecting almost all the major cities of the Republic's maritime provinces. From the mid–seventeenth century until the mid–nineteenth century, when railways replaced the barge services, hundreds of thousands of passengers per year made use of this backbone of the Republic's transportation system. The long-term pattern of demand for these services is displayed in Figure 4. In *Barges and Capitalism* I sought to identify the variables that most affected the demand for intercity passenger transportation and to measure their impact.[49] A reduction of per capita urban incomes emerged as the most important variable, although in the absence of reliable knowledge of the income elasticity of demand for passenger transportation no firm estimate of the decline in income could be made. This decline was concentrated in the period 1675–1745. Thereafter the demand for passenger travel revived slightly, suggesting that in the second half of the eighteenth century Dutch cities did not experience further income decline.

This indicator of the direction of change illuminates only the urban sector of the Republic's maritime provinces, and must certainly be used with caution. Layer upon layer of assumption and inference envelop the model that yielded these insights, and it would certainly be comforting to be able to reinforce these findings with independent evidence.

In the modern economy economic growth is generally associated with changes in structure that find their reflection in urbanization and a reduction of the relative importance of agriculture. It has also been associated strongly with population growth.[50] From this perspective the case for substantial economic growth in the Netherlands during the two centuries after 1675 must appear weak indeed. Table 4 summarizes our present understanding of Dutch population growth and urbanization.[51] By preindustrial standards the total population grew rapidly from about 1520 until the mid–seventeenth century, and the cities absorbed most of the increase. The rural sector grew at only half of the total rate, while cities grew dramatically, accounting for over 40 percent of the total population by 1675. From then until the first decade of the nineteenth century overall population growth was all but imperceptible. After 1750, as the rate of population growth rose throughout Europe, the Dutch population remained all but stationary, such growth as there was being located primarily in the poorest, most rural regions.

Nineteenth-century population growth was substantial, but the 1800 level of urbanization was not exceeded until after 1870, implying that the rural sector had to accommodate ever larger numbers at just the time that industrial mechanization, chiefly beyond the Dutch borders, reduced rural nonagricultural employment possibilities.

In 1849 the first occupational census showed 44 percent of the male labor force to be engaged in agriculture.[52] It is possible to doubt that agriculture

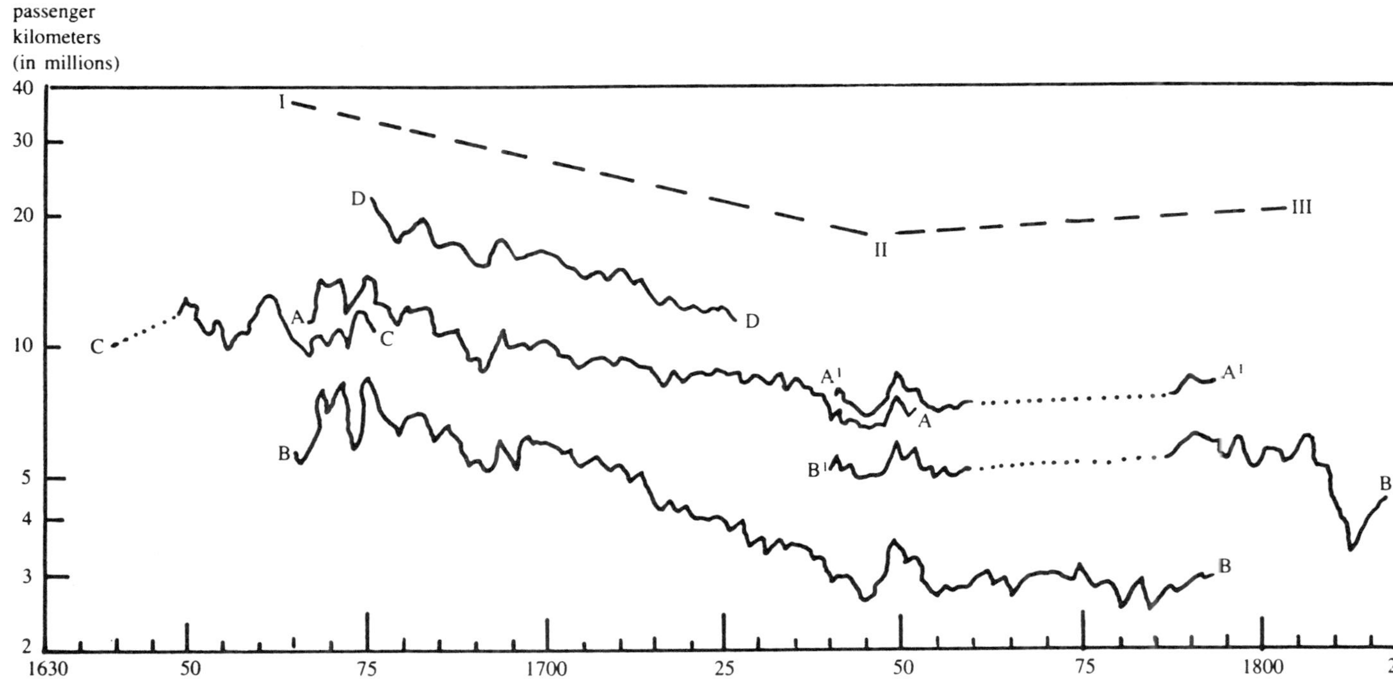

Figure 4. Annual Volume of Passenger Travel of Dutch Intercity Barge Network. [From J. de Vries, *Barges and Capitalism*. (Utrecht: 1981).] p. 223. A, B, C, and D refer to various combinations of routes; I, II, and III represent estimates of the total volume of traffic on the entire network.

Table 4. Dutch Population and Urbanization, 1500–1900 (in thousands)

Year	Total Population	Population of Holland	Rural Population	*Percentage Urban Population* >2500	>10,000	>20,000
1500	900–1000					
1525		275[a]	800	27	15	5.5
1550	1200–1300				15	
1600	1400–1600	672[b]			24	
1650	1850–1900				32	
1675		883	1085	42		24.5
1700	1850–1950				34	
1750	1900–1950	783	1150	39	30	21.8
1800	2100	783	1300	37	29	20.8
1815	2202	764		35	26	17.5
1830	2613	894		39	26	
1840	2860	969		39	28	21.3
1850	3049	1039	1830	39	29	
1860	3309	1143				24.7
1870	3580	1266	2140			28.9
1880	4013	1484				30.5
1900	5104	2113				35.6

[a] In 1514.
[b] In 1622.
Sources: A. M. van der Woude, "Demografische ontwikkeling van de Noordelijke Nederlanden, 1500–1800," in *Algemene Geschiedenis der Nederlanden*, Vol. 5, pp. 102–168; E. W. Hofstee, *De demografische ontwikkeling van Nederland in de eerste helft van de negentiende eeuw*, Table IA; M. C. Deurloo and G. A. Hoekveld, "The Population Growth of the Urban Municipalities in the Netherlands Between 1849 and 1970," in H. Schmal (ed.), *Patterns of European Urbanisation since 1500*, pp. 247–283.

absorbed so large a percentage of the labor force in the third quarter of the seventeenth century. This is so in part because of the intervening minor deurbanization, in part because post–1675 population growth was concentrated in the more agrarian eastern provinces, and in part because seventeenth- and eighteenth-century regional occupational censuses and estimates show the agricultural population to have been proportionally smaller than in the nineteenth century. Agriculture may well have occupied under 40 percent of the mid–seventeenth-century labor force.[53]

Given the evidence calling attention to the relative prosperity of the agricultural sector in the century after 1750, such an occupational shift should not be altogether unexpected. Moreover, the gap typically observed between urban and rural incomes was undoubtedly relatively small, certainly in the maritime provinces. The mid–ninteenth-century level of urbanization, even after 150 years of stagnation, was still high by European standards. These features help explain why Dutch income *levels* may have been high even after a long stagnation, but they are difficult to reconcile with an argument for rapid *growth* of incomes.

The final approach to be considered here is to examine a principal component of national income, payments to labor. In most preindustrial economies true wage earners were rather uncommon and the wage bill could not have formed a large part of national income. In the Dutch Republic, where the urban population approached 40 percent of the total and agriculture in many regions was commercialized, wage and salaried labor could easily have approached half the adult labor force.[54] What can now be said about the course of wages and salaries and their purchasing power after the mid–seventeenth century?

The most abundant data refer to craftsmen and laborers in the building trades and related outdoor employments such as road, canal, and dike maintenance. A valuable source of wage data is the archives of the regional drainage authority of Rijnland (serving central Holland), which employed numerous carpenters, masons, and common laborers in the maintenance of sluices and dikes at several rural locations.[55] At Spaarndam, a village near Haarlem, the daily pay of such workers reached a plateau in the 1640s of 16 stuivers (st.) for common labor, 24 st. for craftsmen, and 28 st. for master craftsmen. In 1679 each of these wage rates rose by 2 st. per day to compensate the workers for the discontinuation of the beer ration. The wages of common labor fell back by 2 st. in 1739. Otherwise only temporary adjustments interrupted the maintenance of this wage scale until after 1860.

This constancy is typical of the wages of craftsmen and common laborers throughout the Netherlands. Small adjustments were sometimes made, but these did not form a consistent pattern. Table 5 presents a summary of the unweighted mean daily wage paid to common labor and craftsmen in the places for which I have assembled data. The number of observations per entry varies from 7 to 14.

Few sources stand ready to illuminate the course of wages in the industrial sector, or, indeed, in nongovernmental employment generally. However, one might hope that the industrial enterprises of government agencies would pay wages that bear some relation to those paid by their private counterparts. The ropeworks and ship wharves of the Admiralties employed large numbers of workers, and the records of their wages and salary payments are available from the eighteenth century onward. The ropeworks of the Amsterdam Admiralty employed scores of spinners and ropemakers whose standard daily pay stood at 16 st. in the summer and 14 st. in the winter from 1719, when the sources begin, to 1795.[56] In the following year (the first of the Batavian Republic) the wages rose by 25 percent to 20 st., at which level they remained until 1870; then the nominal wages were raised at frequent intervals, reaching a new plateau of 32 st. (double the eighteenth-century level) by 1887.

The naval shipyards of Amsterdam and its counterpart at Rotterdam employed many hundreds of workmen.[57] Besides numerous shipwrights and unskilled

Table 5. Unweighted Mean Daily Summer Wages Paid to Common Labor and Skilled Labor in the Building Trades in the Western and Eastern Netherlands, 1550–1854, in Current Stuivers (20st = 1 guilder)

	Common Labor		*Skilled Labor*	
Period	*Western*	*Eastern*	*Western*	*Eastern*
1550–1558	4.5	3.0	6.5	5.0
1583–1592	9.0	6.9	11.5	10.5
1650–1679	18.7	13.6	25.1	18.6
1745–1754	18.1	13.6	25.2	18.9
1790–1799	17.9	13.1	24.8	19.5
1820–1829	17.8	—	25.2	—
1840–1854	18.0	—	26.2	18.0

Source: Dutch labor market study, in progress.

laborers, the Admiralty payrolls recorded the pay of a wide variety of building crafts. Here we will content ourselves with averages of annual earnings (assuming full employment) of all skilled and unskilled employees. The eighteenth-century pay scales were established well before 1700, but the initiation date is not known. From then until after 1847 nominal wages held constant for skilled workers and rose slightly after 1815 for unskilled workers. From 1847 to 1902 they all rose by two-thirds. This post–1850 experience is consistent with a survey of wages in the building trades which found the 1900–1910 level of nominal pay to exceed that of 1869–1873 by two-thirds (see Table 6).

The final series to be introduced concerns salaried employees. Assembling data on salaried members of the labor force is a daunting task, since the number of separate occupations was enormous and salary variations among individuals within an occupation was common. All that is attempted here is an overall salary index that represents the unweighted index of 9 separate series, each for a group of differently salaried individuals ranging in number from 7 to 27. The index is composed of the salaries of city officials, teachers, pastors, orphanage employees, and rural windmill operators. It differs from the wage indices in that the salaries crept upward throughout the late seventeenth and eighteenth centuries. By the beginning of the nineteenth century the composite index stood 32.6 percent higher than in the mid–seventeenth century (see Table 7).

For the purpose at hand it is not sufficient to establish the course of nominal wages; annual earnings are, of course, what one needs to estimate the size of the wage bill in national income. In the case of salary earners we can risk making the convenient assumption that annual salaries equal annual earnings; analogous assumptions cannot be made with regard to wage earners. Fortunately, the detailed wage records of the drainage authority and the Amsterdam ropeworks

Table 6. Nominal Pay of Workers at the Admiralty Shipyards of Amsterdam and Rotterdam
(in current guilders; pay is expressed as annual earnings at full employment)

Period	*Average for Skilled Workers*	*Index*	*Average for Unskilled Workers*	*Index*
Before 1695 to after 1780	*f*391.44	100	*f*238.42	100
1815	377.66	96	248.64	104
1847	401.52	103	297.64	125
1902	660.15	169	498.05	209

Source: Dutch labor market study, in progress.

make possible the specification of how, in these two cases, nominal wages were related to annual earnings.

At the sluices at Spaarndam the permanently employed craftsmen and laborers worked 260 to 270 days per year throughout the eighteenth century until 1784. Then the drainage authority made the first of a series of adjustments that had the effect of transforming a day of work from a real to a notional "day." By 1798 workers were being paid for at least 325 "days" per year, even though they did not physically labor more than before. The long summer days then counted for more than one day for pay purposes. In short, the wage rate had risen by 22 percent in a series of steps between 1784 and 1798.

The Amsterdam ropeworks introduced a similar but smaller hidden wage increase in 1774. The observed wage increase of 1796 abolished the fictive days, with the effect that full-employment earnings remained constant. But after 1816 piece rates were introduced for certain activities that had the effect of raising total earnings once again. The next wage increase in 1872 was, just as the last one in 1796, coupled with cancellation of the piecework practices, negating much of the effect of the increase.

Two organizations do not form a sufficient basis for generalization about labor compensation practices in the Netherlands, but these findings should certainly warn us that the stated daily wages may understate actual pay and that the timing of effective nominal wage increases could differ substantially from the apparent pattern.

We have seen that the pattern of maximum annual earnings of ropeworks employees differed significantly from the stated daily wages. The actual earnings—taking unemployment into account—alters the picture once again. The early-eighteenth-century work force typically labored 250–270 days, and in the 1740s and 1750s regular employees worked even less, often under 200 days per year. After 1770 the real days of employment rose sharply, and in the first half of the nineteenth century most workers logged upward of 300 days per year. The relative importance of changes in the demand for labor and changes in the supply

Table 7. Trends in Salary Earnings, 1582–1894

Year	*Middelburg, Seven Occupations, Municipal Government*	*Middelburg, Teachers at Latin School*	*Goes, Ten Occupations, Municipal Government*	*Amsterdam, Eight Employees, Burger's Orphanage*	*Amsterdam, All Teachers in Municipal Schools*	*Amsterdam, Rectors and Correctors of Municipal Schools*	*Utrecht, Six Employees, Burger's Orphanage*	*Amsterdam, Reformed Clergymen*	*Smaller Cities of the Netherlands, Reformed Clergymen*	*Rural Areas and Western Provinces, Reformed Clergymen*	*Central Holland, Seven Windmill Operators for Hoogheemraadschap van Rijnland*	*Unweighted Average of All Series*
1582		25										
1584		38							39	36	41[b]	
1600		52			40	32						
1605		60										
1620		60	59		59	38		75			72	
1630		75		58	65							
1650	94		85	81	73	53	105				106	86
1664					77		94	91			107	
1680		86		103			83				98	
1700	105	100	86			94	96	91			99	96
1720–1744	100	100	100	100	100	100	100	100	100	100	100	100
1750	100	100		105	100	111	100				101	
1778		111									120	
1790	132	118	103		100		108	100			126	114
1815				103	120[a]	121[a]					134	
1849			128									
1894				220				124	208	218		

[a]1800.
[b]1590.
Source: Dutch labor market study, in progress.

cannot now be determined. For whatever reason, the ropeworks employees experienced large swings in their actual annual earnings, which often rose most in periods when the purchasing power of wages was under pressure.

This brings us to the final step in the analysis of the earnings of Dutch labor, the calculating of purchasing power. N. W. Posthumus laid the foundation for the study of Dutch prices in two great studies: a two-volume compendium of prices and a three-volume history of the Leiden textile industry that included a wealth of price data and a consumer price index. That index and the several wholesale and other commodity price indices in his price series span the seventeenth and eighteenth centuries. They either end in 1800 or peter out in the course of the first half of the nineteenth century. We have already noted that no broadly based price index exists for the nineteenth century. The single effort to link Posthumus's work with the twentieth century series relied on German and British wholesale price series.[58]

The various series display substantial differences during the periods of rapid price increases in the 1690s, 1860s, and, most seriously, during the French Revolution and Napoleonic Wars.

On the basis of Posthumus's compilations, I have constructed a cost-of-living index based on a basket of goods designed after the manner of the Phelps-Brown and Hopkins index of consumables prices.[59] Unfortunately, the full index ends in 1800, after which date industrial prices become scarce. After 1800 the foodstuffs portion of the index continues to 1850, when it too must end. A separate index of food and drink prices constructed by J. A. de Jonge is available for the period 1852–1911.[60] Regrettably, it cannot yet objectively be linked to the earlier series.

Figures 5, 6a,b, and 7 present the real wage and real earnings trends for the wage data introduced above, adjusted by the new cost-of-living index and, after 1850, the index of food and beverage prices prepared by De Jonge. Given the constancy of nominal wages, the real wage is a faithful mirror of price trends, peaking in the 1680s and again in the 1720–1740 period, when prices are low, and plunging during the high price periods associated with the Napoleonic Wars, the First and Second Anglo–Dutch Wars and the "hungry forties" of the nineteenth century. When the surreptitious wage increases described above are taken into account in Figures 5 and 6, the late-eighteenth-century earnings crisis is somewhat moderated. If the consumer price index were to be adjusted after 1780 so that potatoes substituted for half the previous level of rye bread consumption, the crisis would be further moderated, particularly in the worst years.[61] Finally, when actual real annual earnings are plotted, which can be done only for the employees of the Amsterdam ropeworks, a pattern arises that deviates markedly from the real wage trend.

The basic trends in the real wage can be summarized as follows:

1. Real wages rose erratically from the mid–sixteenth century to the 1680s,

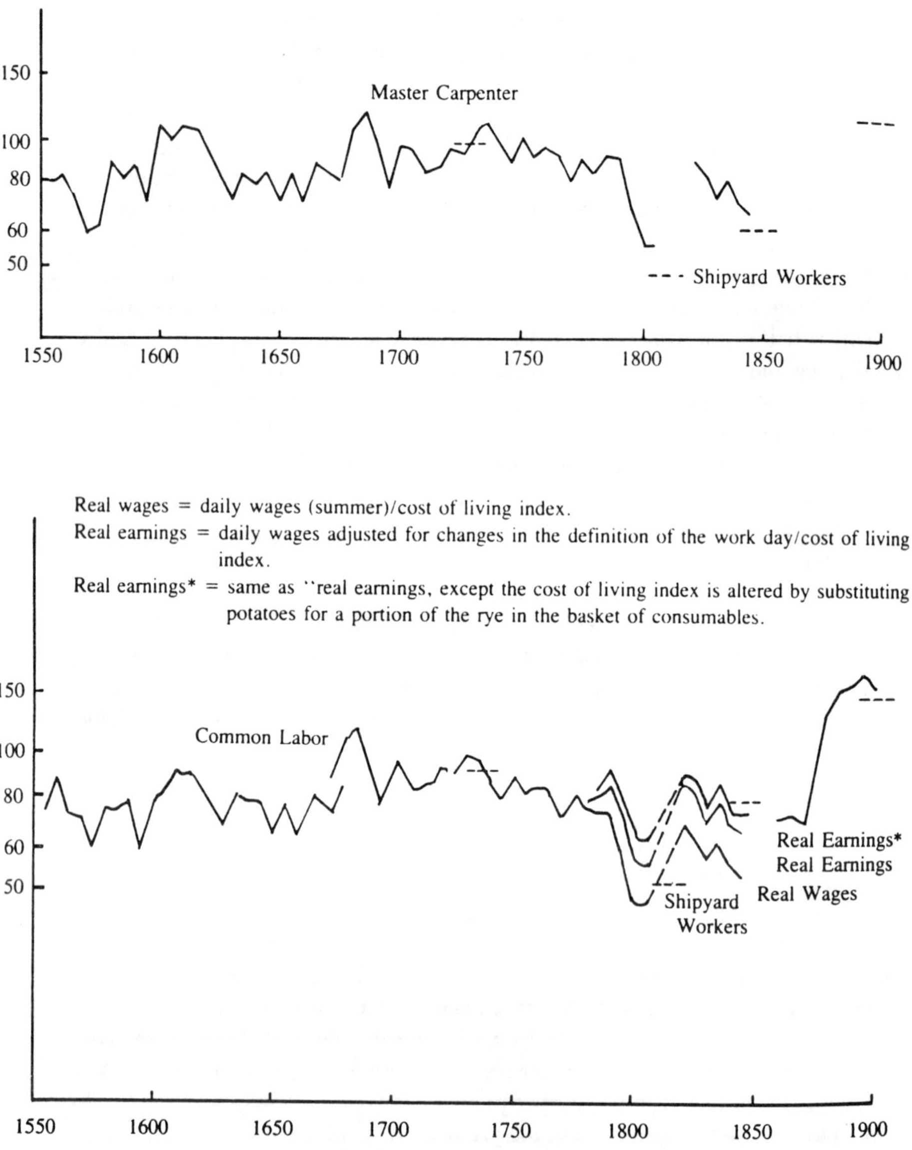

Figure 5. (a) Real Wages of Master Carpenters Employed by the Hoogheemraadschap van Rijnland at Spaarndam, 1550–1849. Real wages of all skilled workmen employed at the Amsterdam shipwharves of the Dutch Navy, 1720–1902 (1720–1744 = 100). (b) Real Wages of Common Laborers Employed by the Hoogheemraadschap van Rijnland at Spaarndam and Katwijk, 1550–1900. Real annual earnings of all unskilled workmen employed at the Amsterdam shipwharves of the Dutch Navy, 1720–1902 (1720–1744 = 100).

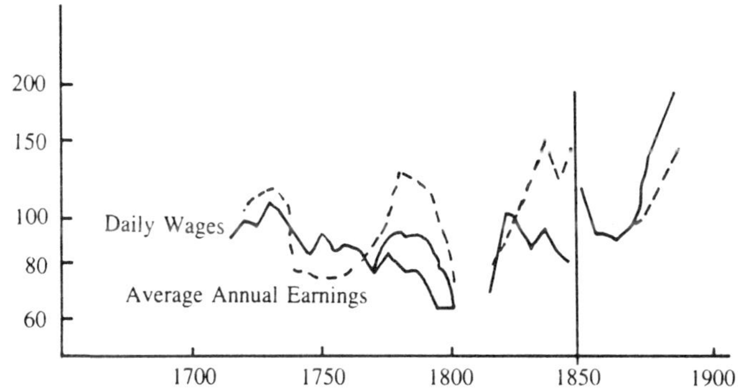

Figure 6. Real Wages and Earnings of Laborers at the Ropeworks of the Amsterdam Admiralty, 1715–1890 (1720–1744 = 100; 1850–1859 = 100).

the rise being interrupted most notably by the many years of high prices during the wars of the 1650s and 1660s. The wage increases dominated the pattern to the 1640s; thereafter price changes generated the changes.

2. From the real wage peaks extending from the 1680s through the 1730s, a new trend of declining real wages set in that reached its nadir shortly after 1800. Price changes dominated this trend, but from the 1770s onward various stratagems to increase earnings reduced the erosion of purchasing power for at least some workers.
3. The recovery of real wages after the Napoleonic era was only partial and was determined by price changes. Only after the 1860s does a combination of rising wages and falling prices push up the real wage to unprecedented levels. This breakthrough was concentrated in the 1870s and 1880s.

The attentive reader will have noted that real wages stand at their highest levels in precisely the period, 1680–1740, for which I hypothesize declining national income. What is more, they tend inexorably downward in the second half of the eighteenth century, when I suspect that the decline of national income had already run most of its course.

The contemplation of the ironic character of these trends in real wages, and similar trends at other times and places, has more than once led historians into debates where one scholar's depression is another's Golden Age.[62] Indeed, the apparent contradictions of these and similar findings are at the root of the skepticism expressed by "Annales school" historians concerning the concept of national income when applied to the preindustrial economy.[63]

A proper understanding of these real wage trends require that we address the causes of the price movements that play the dominant role in driving real wages between 1640 and 1870. When prices fall as a response to growth in the volume

180 J. DE VRIES

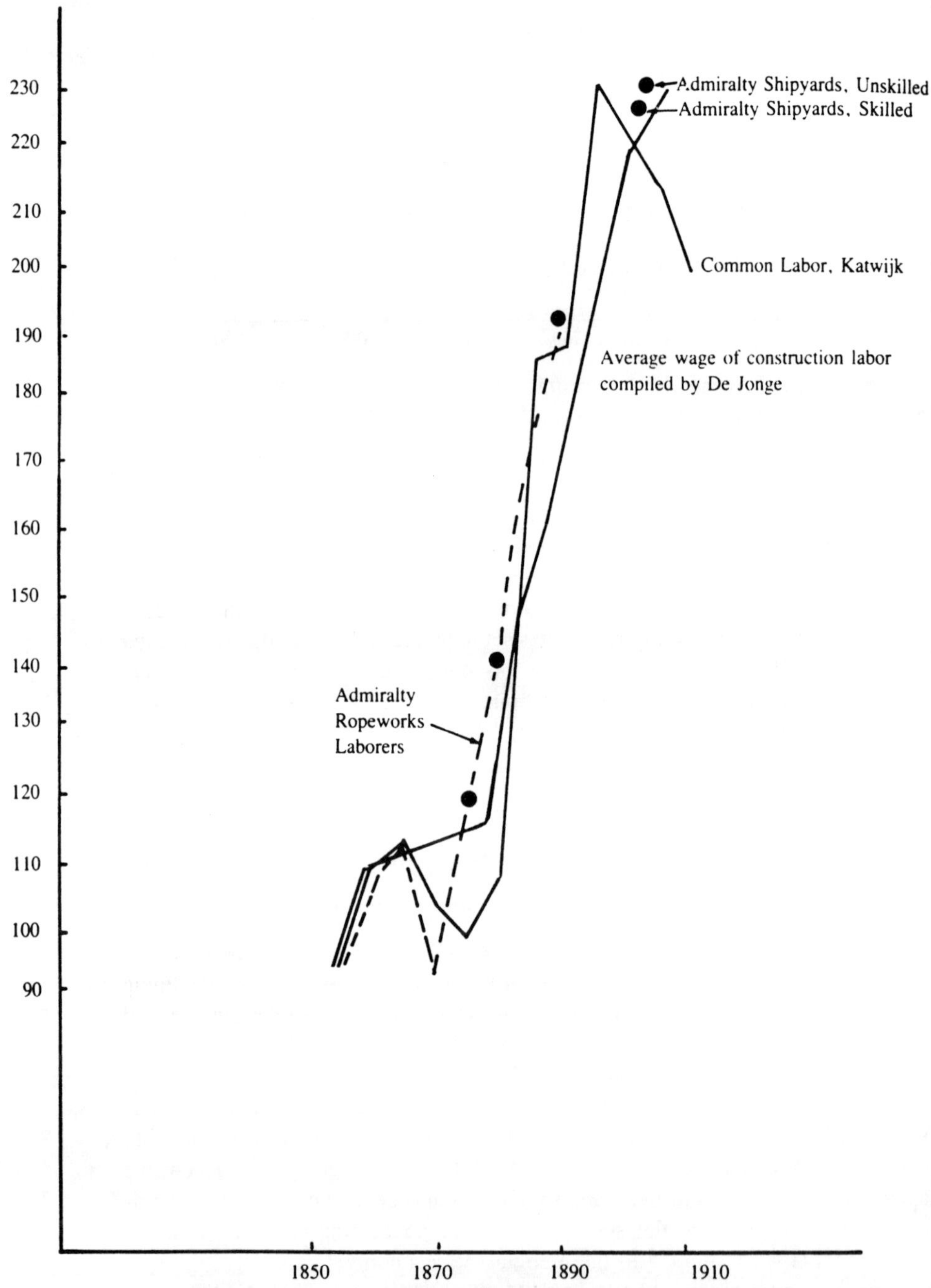

Figure 7. Real wages, 1850–1910 (1852–1861 = 100).

of output or because of cost-reducing investment or both, there is reason to expect rising real wages will be associated with rising per capita national income. The argument that the decline in grain prices in early-eighteenth-century England was paired with the diffusion of the "new husbandry," reducing the costs of production as it increased output per acre, would be consistent with this view.[64] When a fall in prices is not associated with output growth and cost-reducing investment, it takes the form of an exogenous shock to the economy. In the absence of adjustments in nominal wages, such exogenous price changes have the effect of redistributing income, in this case from capital and rents to the wage bill. The high real wage is compensated for in depressed land values and low returns to capital and, eventually, unemployment.

This scenario seems consistent with what is known about the Dutch economy from 1680 to 1740. In these decades commodity production declined as capital was withdrawn particularly from labor-intensive industries. Agriculture, too, was depressed, as high costs pressed on the farmers, forcing the abandonment of land, the adoption of less labor-intensive production techniques, and bringing about the collapse of rental values.

In the following period of falling real wages, exogenous factors again dominated the movement of prices, shifting income away from labor. The population growth that might have pressed down on agricultural productivity in most preindustrial societies was largely absent in the Netherlands at this time. The increased number of real days of work and the effective increases in nominal wages observed at the ropeworks in Amsterdam and at the workplaces of the Rijnland drainage authority are consistent with one's expectations in a situation where the real cost of labor of constant productivity was falling.

What still needs explanation, of course, is the labor market structure that held nominal wages all but fixed through these long swings in price level. For present purposes it is enough to observe its probable effect: labor's share of national income rose to a high level during the period 1680–1740, undermining the demand for labor in the agricultural and industrial sectors. Thereafter labor's share declined, reviving the possibilities for agricultural prosperity and investment, but the latter came to be directed almost entirely abroad, limiting the extent to which the domestic demand for labor could rise. Only after 1870 is there a growth of real income that seems to have sources in domestic productivity improvements and increased production.

B. Levels

The third hypothesis of this study concerns the *level* of Dutch national income. Indeed, it was present by implication in the preceding discussion of the relationship of real wage trends to national income.

In the absence of sufficient data for direct estimation of national income, I will focus attention here on the level of Dutch wages in relation to neighboring economies. Table 8 represents crude estimates of the nominal wage for skilled and unskilled labor in the building trades. The wages are unweighted averages drawn from such time series as I have found for Germany, the Southern Netherlands, and England. The English wages for 1583–1592 are based exclusively on the Phelps-Brown and Hopkins series for southern England; all the others are based on at least four independent series. The wages are expressed in Dutch currency, the basis for conversion being current exchange rates in all years except the 1580s, when the intrinsic silver content of the currencies was relied upon. One might think of the resulting wage estimates as expressing the relative cost of labor to a prospective employer producing for an international market.

Table 8. Unweighted Mean Daily Summer Wage Paid to Labor in the Building Trades in Four Countries, Expressed in Current Stuivers

Place	*1582–1592*	*1650–1679*	*1745–1754*	*1790–1799*
		Unskilled Labor		
Holland	9.0	18.7	18.3	17.9
E. Netherlands	6.9	13.6	13.6	13.1
Belgium	9.5	13.8	12.0	11.9
Germany	5.3	9.0	7.9	7.6
England	6.0	12.1	14.2	19.9
		Journeymen		
Holland	11.5	25.1	25.2	24.8
E. Netherlands	10.5	18.6	18.9	19.5
Belgium	16.1	25.0	22.5	22.3
Germany	7.5	15.0	12.0	11.4
England	9.0	18.7	20.4	27.9
		Wage Rate Index (100 = highest average wage in each time period)		
W. Netherlands	95	100	100	90
E. Netherlands	73	73	74	66
Belgium	100	74	66	60
Germany	56	48	43	38
England	63	65	78	100
		Journeymen		
W. Netherlands	71	100	100	89
E. Netherlands	65	74	75	70
Belgium	100	99	89	80
Germany	47	60	48	41
England	56	75	81	100

Source: Dutch labor market study, in progress.

The estimates must not be pressed very far, but they are probably sufficiently accurate to permit the observation that wages in the maritime provinces of the Netherlands surpassed the Belgian level after 1600 and reigned as the highest in northern Europe until the end of the eighteenth century, when English wages assume that position.

Since the comparison with England has functioned as an orientation point in this study of Dutch national income, we should now look more closely at the course of wages in the two countries. Figure 8 shows the course of real wages in England and the Netherlands. It includes a 25-year moving average of southern English craftsmen's real wages as calculated by Phelps-Brown and Hopkins and adjusted by Wrigley and Schofield,[65] a 25-year moving average of real wages of common labor at Spaarndam, and the more broadly based Dutch real wage levels at selected intervals. Since each national series is expressed in its own currency and adjusted by its own cost of living index, comparisons of the level (as opposed to the trends) of the purchasing power of English and Dutch laborers cannot avoid a degree of arbitrariness. But two facts can help reduce the likelihood of misinterpretation. First, the silver values of wages were roughly equal in the first half of the sixteenth century and again, more briefly, around 1800. Second, the close economic relations of the two countries enforced a high degree of correlation in their commodity prices.

These considerations have emboldened me to join the English and Dutch series as shown in the figure. It must be emphasized that their position on the graph is purely a matter of judgment. While this exercise is not without its dangers, it has the redeeming academic virtue of calling attention to the important differences in the histories of Dutch and English real wages.

The central feature of the graph is the dramatic divergence between Dutch and English wages between 1580 and 1620. English real wages fall by half while Dutch wages rise. England was not alone in suffering falling real wages in this period; students of the sixteenth-century price revolution have established that plunging real wages characterized most of Europe. In this perspective a great achievement of the Dutch Golden Age was to prevent the fall of real wages and, indeed, allow for a measurable increase in the purchasing power of many categories of labor.

The effect was to open an enormous gap between real wages in the two countries that was gradually reduced to zero over the course of the eighteenth century. The gap diminished as English nominal wages rose step by step to reach the level attained by the Dutch in the mid–seventeenth century. Seen in this light, the English economy of the eighteenth century appears to be proceeding through an extended catch-up or recovery phase. It would be surprising if in the 1680s Dutch per capita income stood only 7 percent above the depressed English levels, as Gregory King had it.

In the "good" years (i.e., nonwar years) of the seventeenth century, Dutch real incomes stood far above the English level and European levels more gener-

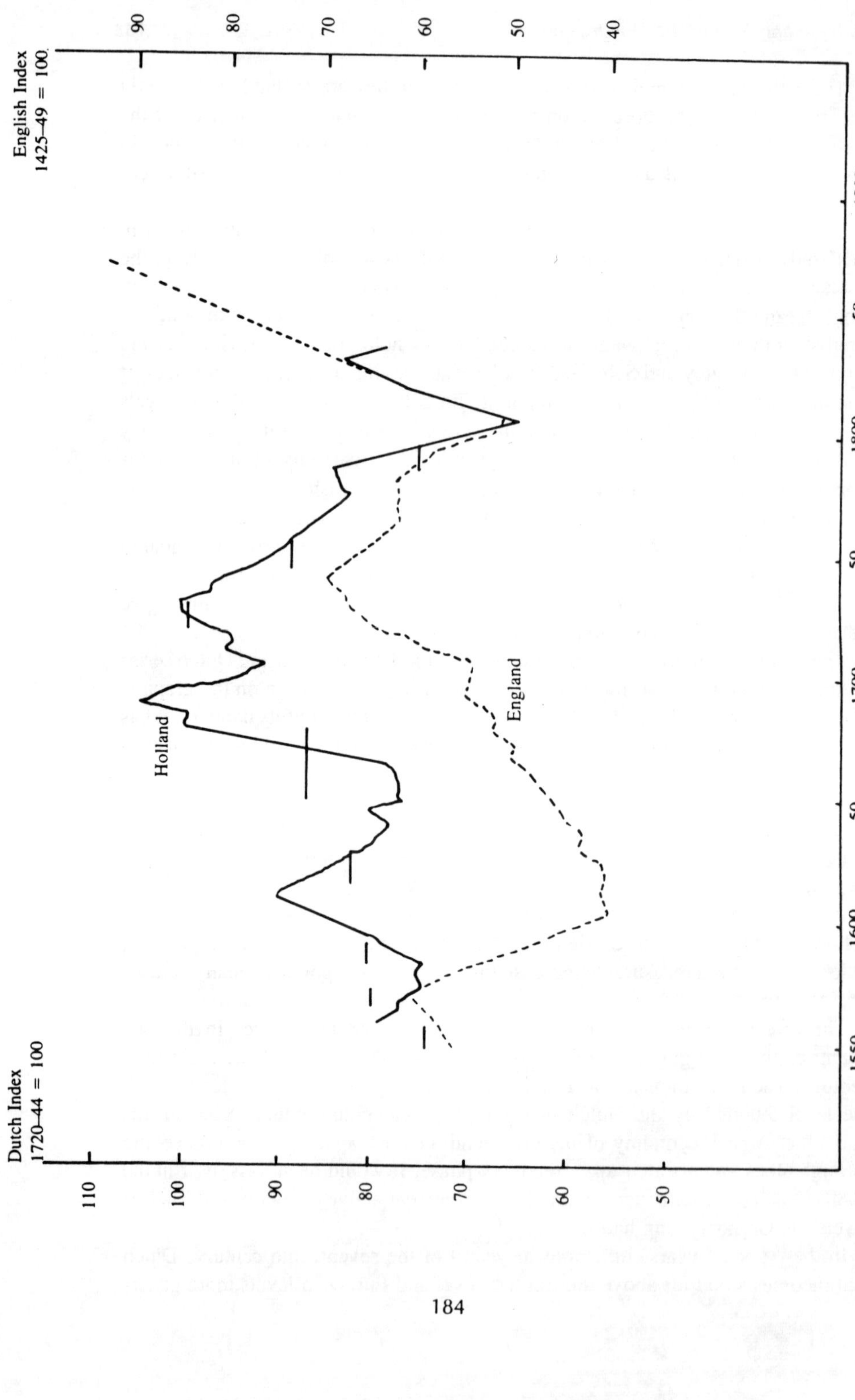
English Index
1425–49 = 100
90
80
70
60
50
40
Dutch Index
1720–44 = 100
110
100
90
80
70
60
50
1550
1600
50
1700
50
1800
50
1900
Holland
England

ally because the Republic's economic expansion had sheltered it from the sharp decline in purchasing power experienced by neighboring countries. The high wage position of the Republic proved to be a problem during the long-term deflation culminating in the period 1680–1740. Just as real wages reached their apogee, employment-generating activities became unprofitable in the Republic while the economies that had not been able to raise wages in earlier decades were well placed to expand. If this scenario has merit, then a central issue of Dutch economic history must be why the restoration of equilibrium to the Dutch labor markets took well over a century to achieve.

VII. CONCLUSION

Of the tasks I set for myself in this study, establishing the *level* of Dutch natural income is the most difficult. Much more evidence is needed to confirm my hypothesis that Dutch per capita income in the third quarter of the seventeenth century stood at double the level estimated by Gregory King for 1688. But its plausibility now rests on two legs. The first is a revision in our understanding of the growth of the English economy before and during its Industrial Revolution. The chasm separating economic life before and after the initial decades of classic industrialization appear to have been overdrawn, resulting in an underestimation of the size of the early-eighteenth-century economy. The second element rendering this hypothesis plausible is an improved understanding of the substantial real wage differences then prevailing between the Republic and England, and of the scope of the structural differences between the economies.

The issues relating to the *trends* of Dutch national income between 1675 and 1900 have elicited sharply contrasting viewpoints, but I believe the general contours of decline and growth to be evident.

The Dutch faced, by the late seventeenth century, problems requiring structural change in their economy, as the deterioration of their ability to export commodities idled labor and the rising wage bill squeezed profits. Capital then left labor-intensive sectors to be reinvested, more often than not, abroad. Labor reacted in ways more difficult to fathom, but the response included a diminution in the size of the urban labor force and long-term structural unemployment. This process was interrupted by intervals of stabilization and regional prosperity, but the economic dislocations brought on during the Revolutionary and Napoleonic periods certainly depressed incomes further across a broad front.

Figure 8. Real Wage Indices for Southern England and Holland in 25-Year Moving Averages. (Also, average real wage of construction labor in the western Netherlands in selected periods.)

The high initial Dutch level was such that it took a long period of vigorous British growth for it to be surpassed. However, it would be surprising if the British surpassed Dutch income levels only after 1860, as Teijl's GNP estimates imply. By then British nominal wages were easily double those of the Dutch, which had continued to stand where they had been set more than two centuries earlier. Likewise, British income from foreign investments had become a large multiple of Dutch foreign investment income, which had shrunk to a fraction of its late-eighteenth–century level.

In assessing this long period of Dutch economic history, two distinct emphases are possible, and legitimate. On the one hand, there is the loss of leadership and innovativeness across a broad range of activities, leading to idle resources and creating over time an economy that was obsolete and subordinate. On the other hand, there is the survival of the basic attributes of a diversified economy and the preservation of its capital in a situation where there was both much to defend and few means to defend it. Which of these two realities one wishes to emphasize is largely a question of taste about which observers will continue to differ.

ACKNOWLEDGMENTS

The author wishes to express his thanks to Richard Griffiths, Peter Lindert, Angus Maddison, and James Riley, whose corrections and suggestions have improved this article at many points. Indeed, recent publications by each of these scholars (referred to below) created the intellectual environment in which this article could be written. I also benefitted from the comments of many seminar participants at Indiana, Harvard, Berkeley, Amsterdam, Groningen, Leuven, and Brussels. The research on the Dutch labor market presented here is being supported by a National Science Foundation grant.

NOTES

1. The standard authority on nineteenth-century industrialization is J. de Jonge, *De Industrialisatie in Nederland tussen 1850 en 1914* (Amsterdam: 1968).

2. See, for example, Roger Price, *The Economic Modernization of France, 1730–1880* (London: 1975).

3. See, for example, A. M. van der Woude, *Het Noorderkwartier* 3 vols. (Wageningen: 1972); J. A. Faber, *Drie eeuwen Friesland,* 2 vols. (Wageningen: 1972); B. H. Slicher van Bath, *Een samenleving onder spanning* (Assen: 1957).

4. Works of this genre on the Dutch economy are not numerous. See Joel Mokyr, *Industrialization in the Low Countries* (New Haven, Conn.: 1976).

5. Centraal Bureau voor de Statistiek, *Berekeningen over het nationale inkomen van Nederland voor de periode 1900–1920* (The Hague: 1941); C.B.S., *De Nederlandsche Conjunctuur, Speciale onderzoekingen No. 4;* C.B.S., *Het nationale inkomen van Nederland 1921–1939* (Utrecht: 1948).

6. Derksen's estimates, but not his methods, are published in J. H. van Stuijvenberg, "Economische groei in Nederland in de negentiende eeuw: een terreinverkenning," in J. H. van Stuijvenberg (ed.), *Bedrijf en samenleving* (Alphen aan den Rijn: 1967), p. 73. There Van Stuijvenberg introduces an error in presenting Derksen's estimates for 1860. The correct figure is about equal to Teijl's

estimate for that year. I have not corrected the error in Figure 2 or in the text since the subsequent literature always referred to the erroneous transmission of Derksen's views, and not the unpublished original source. See: R. T. Griffiths and J. M. M. de Meere, "The Growth of the Dutch Economy in the Nineteenth Century: Back to Basics?" *Tijdschrift voor geschiedenis* (forthcoming).

7. J. Teijl, "Nationaal inkomen van Nederland in de periode 1850–1900. Tasten en testen," *Economisch- en Sociaal-Historisch Jaarboek* 34(1971), 232–263.

8. J. H. van Stuijvenberg, "De economie in de noordelijke Nederlanden, 1770–1970," *Algemene geschiedenis der Nederlanden* Vol. 10 (Haarlem: 1981), pp. 106–07; J. H. van Stuijvenberg and J. E. J. de Vrijer, "Prices, Population and National Income in the Netherlands, 1620–1978," *Journal of European Economic History* 11(1982): 699–711.

9. This is assumed by Van Stuijvenberg, "economie," pp. 106–107; J. M. M. de Meere and L. Blok, "Welstand, ongelijkheid in welstand en censuskiesrecht in Nederland omstreeks het midden van de 19 de eeuw," *Economisch- en Sociaal-Historisch Jaarboek* 41(1978): 175–293; Richard T. Griffiths, "Achterlijk, Achter of Anders," Inaugural lecture, Free University of Amsterdam, December 4, 1980; and others.

10. Griffiths, "Achterlijk," p. 11. Griffiths also notes that Teijl published elsewhere estimates of energy consumption in the period 1834–1864 that do not agree with those used in his national income estimations. The more recent estimates show a more rapid growth of consumption around the middle of the nineteenth century. J. Teijl, "Brandstoffaccijns en nijverheid in Nederland gedurende de periode 1834–1864," in J. van Herwaarden (ed.), *Lof der historie. Opstellen over geschiedenis en maatschappij* (Rotterdam: 1973), p. 164.

11. Van Stuijvenberg, "Economie," p. 107.

12. Th. P. M. de Jong, "Sociale verandering in de neergaande republiek," *Economisch- en Sociaal Historisch Jaarboek* 35(1972), 3.

13. Griffiths, "Achterlijk," p. 10.

14. Gregory King, "Natural and Political Observations and Conclusions Upon the State and Condition of England," in Peter Laslett (ed.), *The Earliest Classics* . . . (London: 1973).

15. See Phyllis Deane, "The Implications of Early National Income Estimates," and D. V. Glass, "Two Papers on Gregory King," in D. V. Glass and D. E. C. Eversley (eds.), *Population in History* (London: 1965), pp. 159–220.

16. E. Le Roy Ladurie, "Les comptes fantastiques de Gregory King," *Annales E.S.C.* 5(1968): 1086–1102.

17. "'The LCC Burns Journal,' A Manuscript Notebook Containing Workings for Several Projected Works," in Laslett, *Earliest Classics*, pp. 227–229.

18. Blok and de Meere, "Welstand," pp. 187–188.

19. This impression is gleaned from scattered observations cited in Van der Woude, *Noorderkwartier*, Vol. III, Figures 7.2–7.5; C. Baars, *De geschiedenis van delandbouw in de Beijerlanden* (Wageningen: 1973), pp. 114–116; Jan de Vries, "Landbouw in de noordelijke Nederlanden, 1490–1650," *Algemene geschiedenis der Nederlanden* (Haarlem: 1980), Vol. 7, p. 41; J. Kuys and J. T. Schoenmakers, *Landpachten in Holland 1500–1650* (Amsterdam: 1981).

20. I. J. A. Gogel, *Memoriën en Correspondentiën betreffende den Staat van 's Rijks Geldmiddelen in den Jare 1820* (Amsterdam, 1844), Bijlage B, p. 31.

21. King, "Manuscript Notebook," p. 228.

22. W. M. Keuchenius, *De inkomsten en uitgaven der Bataafsche Republiek voorgesteld in eene nationale balans* (Amsterdam, 1803); R. Metelerkamp, *De toestand van Nederland in vergelijking gebragt met die van enige andere landen van Europa* (n.p., 1804).

23. Johan de Vries, *De economische achteruitgang der Republiek in de achttiende eeuw* (1959; 2nd ed., Leiden: 1968).

24. Ibid., p. 167.

25. Van der Woude, *Noorderkwartier*, Vol. I, pp. 256–257; Vol. II, pp. 609–610.

26. Van Stuijvenberg and De Vrijer, pp. 700–701.

27. Angus Maddison, *Phases of Capitalist Development* (Oxford: 1982).
28. *Ibid.*, pp. 31–32; 165–66.
29. *Ibid.*, p. 44.
30. Paul Bairoch, "Europe's Gross National Product: 1800–1975," *Journal of European Economic History* 5(1976): 286. For another view, see: N. F. R. Crafts, "Gross National Product in Europe 1870–1910: some new estimates," *Explorations in Economic History* 20(1983): 387–401.
31. James C. Riley, "The Dutch Economy after 1650: Decline or Growth?" *Journal of European Economic History* (forthcoming). Riley bases his revisionist view on his own work, *International Government Finance and the Amsterdam Capital Market, 1740–1815* (Cambridge, 1980); and also Marten G. Buist, *At Spes Non Fracta: Hope & Co., 1770–1815* (The Hague: 1974); Jake V. T. Knoppers, *Dutch Trade with Russia from the Time of Peter I to Alexander I: A Quantitative Study in Eighteenth Century Shipping*, 3 Vols. (Montreal: 1976); Jake V. T. Knoppers and F. Snapper, "De Nederlandse scheepvaart op de Oostzee vanaf het eind van de 17de eeuw tot het begin van de 19de eeuw," *Economisch- en Sociaal-Historish Jaarboek* 41(1978): 115–153; M. Morineau, "La balance du commerce franco-néerlandais et le reserrement économique des Provinces-Unies au XVIIIeme siècle," *Economisch-Historisch Jaarboek* 30(1963/64): 170–235.
33. Nicholas F. R. Crafts, "English Economic Growth in the Eighteenth Century: A Re-examination of Deane and Cole's Estimates," *Economic History Review*, 2nd ser. 29(1976): 226–235.
34. W. A. Cole, "Factors in Demand, 1700–1780," in R. Floud and D. N. McCloskey (eds.), *The Economic History of Britain since 1700*, Vol. I (Cambridge: 1981), p. 64.
35. Quotations of exchange rates between London and Amsterdam are provided in N. W. Posthumus, *Inquiry into the History of Prices in Holland*, Vol. I (Leiden: 1946), pp. 590–595. The use of foreign exchange market rates for the comparison of national incomes is not without its problems. For a discussion of the "exchange rate problem" see Patrick O'Brien and Caglar Keyder, *Economic Growth in Britain and France, 1780–1914* (London: 1978), pp. 34–47. I have not pursued their suggestions in this study, fearing that any revision of official rates would be as likely to diminish as to improve the accuracy of the conversions. However, I take comfort in calling attention to the fact that the two economies were as unified by trade and by common tastes and patterns of consumption as any two in Europe.
36. Deane and Cole, *British Economic Growth*, pp. 282, 329–331.
37. Riley, "Decline or Growth?", Tables 4 and 5.
38. Peter H. Lindert and Jeffrey G. Williamson, "Revising England's Social Tables, 1688–1812," *Explorations in Economic History* 19(1982): 385–408; *Ibid.*, "Reinterpreting Britain's Social Tables, 1688–1913," *Explorations in Economic History* 20(1983): 94–109.
39. C. Knick Harley, "British Industrialization Before 1841: Evidence of Slower Growth During the Industrial Revolution," *Journal of Economic History* 42(1982): 285.
40. Idem.
41. N. F. R. Crafts "British Economic Growth, 1700–1831," *Economic History Review* 36(1983): 177–199.
42. For surveys of Dutch industry see P. C. Jansen, "Nijverheid in de noordelijke Nederlanden, 1650–1780," in *Algemene geschiedenis der Nederlanden*, Vol. 8 (Haarlem: 1979), pp. 102–123, and Richard Griffiths, "Ambachten nijverheid in de noordelijke Nederlanden, 1770–1844," *Algemene geschiedenis der Nederlanden*, Vol. 10 (Haarlem: 1981), pp. 219–52. Both are supplied with annotated bibliographies.
43. N. W. Posthumus, *De geschiedenis van de Leidsche lakenindustrie*, 3 vols. (The Hague, 1908–1939); S. C. Regtdoorzee Greup-Roldanus, *Geschiedenis der Haarlemmer bleekerijen* (The Hague: 1936); Van der Woude, *Noorderkwartier*, Vol. 2, pp. 457–472.
44. Johan de Vries, *Achteruitgang*, Appendix V, pp. 198–199, provides a summary of contemporary opinion.
45. Keuchenius, pp. 110–111.
46. For summaries of agrarian activity see: Jan de Vries, "Landbouw," pp. 12–43, and H. K.

Roessingh, "Landbouw in de noordelijke Nederlanden 1650–1815," *Algemene geschiedenis der Nederlanden*, Vol. 8 (Haarlem: 1979), pp. 16–72.

47. Griffiths, "Achterlijk," pp. 16–20, for data on early-nineteenth-century expansion.

48. Riley, *International Government Finance;* Buist, *At Spes Non Fracta.*

49. Jan de Vries, *Barges and Capitalism: Passenger Transportation in the Dutch Economy, 1632–1839* (Utrecht: 1981); also published in *A.A.G. Bijdragen* 21(1978), pp. 231–303.

50. These overly familiar generalizations are most comprehensively described and defended in Simon Kuznets, *Modern Economic Growth* (New Haven, Conn.: 1966).

51. For sources, see Table 4.

52. J. A. de Jonge, *De industrialisatie in Nederland tussen 1850 en 1914* (Amsterdam: 1968), p. 296.

53. Jan de Vries, "Landbouw," pp. 36–37.

54. Deane and Cole, *British Economic Growth*, p. 251, estimate the British wage bill in 1801 to be 45 percent of national income.

55. Hoogheemraadschap van Rijnland, oud archief, Nos. 9510–10263, Bijlagen tot de rekeningen; Nos. 10917–10932, werklijsten.

56. Algemene Rijksarchief, The Hague, Archief van de Lijnbaan te Amsterdam, Nos. 8–20, Betalingsboek van spinders en draijers; Nos. 121–124, 186, 188, Betaallijsten.

57. Gemeentearchief van Amsterdam, Part. archief 310, Rijksmarinewerf Nos. 1514, 1516, 2278, 4351–4352; Algemene Rijksarchief, The Hague, Admiraliteitscolleges, No. 786; Verzameling Van der Heim, Nos. 432, 447; Verzameling Van Alphen, No. 4.

58. N. W. Posthumus, *Inquiry into the History of Prices in Holland*, 2 vols. (Leiden, 1946–1964); Posthumus, *Lakenindustrie*, Vols. 2 and 3. The spliced series based on Posthumus and foreign series is in Van Stuijvenberg and De Vrijer, "Prices, population and national income in the Netherlands, 1620–1978," pp. 707–710.

59. The basket of consumables consists of the following items: Rye (1050 kg), yellow peas (143.5 kg), stockfish (25 kg), meat (100 kg), cheese (50 kg), butter (50 kg), beer (621 liters) [for which coffee, tea, and gin are substituted in the second half of the eighteenth century], peat (100 turf tons), and a composite of industrial product prices weighted to equal 25 percent of the food subtotal in the period 1575–1599.

60. De Jonge, *Industrialisatie*, p. 435.

61. In the basket of goods described in note 59, the cost of 1050 kg of rye is replaced with the cost of 525 kg of rye and 1400 kg of potatoes (the caloric equivalent of the missing rye). Potato prices varied locally since this commodity was not widely traded; however, the available price series confirm that the consumption of potatoes did not result in an enormous cost reduction in years of relatively low rye prices in the 1780s but made a great deal of difference in the crisis years of the early nineteenth century.

62. Both the "late feudal crisis" of the fifteenth century and the "great depression of 1873–1896" generated contradictory economic phenomena that have fueled scholarly debate.

63. Van der Woude, *Het Noorderkwartier*, Vol. II, pp. 607–608.

64. N. F. R. Crafts, "English Economic Growth in the Eighteenth Century: A Reexamination of Deane and Cole's Estimates," *Economic History Review* 29(1976): 226–235.

65. E. A. Wrigley and R. S. Schofield, *Population History of England: A Reconstruction* (London: 1982), Appendix 9, pp. 638–644. The real wage series constructed by Phelps-Brown and Hopkins contains gaps, and their nominal wage series often changes abruptly. Wrigley and Schofield introduced linear interpolations to fill the gaps and dampen abrupt changes. As a consequence, the English trends in Figure 8 appear much smoother than the Dutch series, for which no such adjustments have been made.

JAIME REIS

Industrial Development in a Late and Slow Developer: Portugal, 1870-1913*

There is a general recognition that the last few decades leading up to the First World War witnessed an unprecedented measure of industrial growth and structural change in the more backward economies of the European periphery. This included the spread of factory organization and of mechanised production using an increasing amount of steam power, in conjunction with the emergence of concentrations of manufacturing activity in urban locations. Although industrial history has failed to attract the endeavors of Portuguese historians, it would appear that Portugal was no exception to these trends. Despite the lack of published monographic studies or of systematic attempts to quantify levels of output, there is a consensus that between 1870 and 1913, industrial expansion took place along the lines described above.[1] This was part of a wider process of economic change, often referred to as "the penetration of capitalism," which involved the widening of the market, an increasing amount of productive specialisation, the development of a modern transport network, the emergence of banks and the absorption of modern technology by some sectors of the economy. Factory production developed mainly in such industries as textiles, tobacco, flour milling and chemical fertilisers, all of which exhibited unusual rates of growth and called for considerable amounts of fixed capital.

Modern authors agree that on the eve of World War I, even if it had been growing already for several decades, manufacturing was still far less important within the economy than agriculture, and Portugal could hardly be considered a developed economy.

* The author is grateful to Paula Santos for her help with gathering the data on Portuguese industrial production, as well as to the Instituto de Ciências Sociais where the research was undertaken. Thanks are also due to the staff of the Centro de Calculo, Faculdade de Economia, Universidade Nova de Lisboa.

To explain this "failure" different combinations of the following retarding influences have been brought forward: 1) the loss of the colonies, the devastation caused by the Napoleonic invasions and the turmoil of the civil wars, for the earlier part of the century; 2) foreign competition made possible by low tariff barriers; 3) a small domestic market, based on a backward agrarian economy; 4) a weak and divided bourgeoisie, which was therefore incapable of promoting its interests at the level of the State; 5) a shortage of capital, in general, and for industrial purposes, in particular; 6) a lack of elementary and technical education.[2]

Disagreement regarding Portugal's industrial performance during this period revolves around two issues. One of them is the pace of this growth. Drawing on data for raw material and machinery imports, Manuel Villaverde Cabral and Armando de Castro have argued that the expansion of modern industry was fairly vigorous and prolonged and imply that, by 1913, despite its handicaps, this sector had gained relative to the rest of the economy. In her published work to date, Miriam Halpern Pereira appears to deny this view, particularly for the period up to 1891, during which she considers that agriculture was the more "dynamic" sector, and even speaks of "de-industrialization."[3]

The second area of debate has to do with how Portugal's late industrialization is to be interpreted.[4] Although historians have been more concerned with why the manufacturing sector did not develop to a greater extent, two authors have expounded views to account for such industrial growth as there was. Leaning on Celso Furtado's analysis of Brazilian development, Villaverde Cabral has identified two industrial spurts. The first, from 1873 to 1886, occurred in response to an internal demand which was growing and diversifying as a result of the income generated by expanding exports of primary products. The second spurt, from 1891 to the First World War, built on the capacity thus created and was fuelled by an import-substitution process rendered possible by a prolonged situation of balance of payments difficulties which entailed higher tariffs and exchange depreciation.[5] For Joel Serrão, the importance of the agro-export sector for industrial development is also paramount. Using a model devised by Vitorino Magalhães Godinho for an earlier period, he has claimed that during the second half of the nineteenth century, manufacturing could only grow whenever external constraints forced the State to take steps to remedy the loss of export income. During the export boom which lasted until the 1870s and under a regime of free trade, there was no reason for resources to go into this sector. Once this was over, however, most noticeably from 1885 on, conditions were created, particularly in the shape of the opening up of the African colonial market, which "in good measure was the origin of the relative industrial increment at the end of the last century and beginning of this one."[6] Thus, in both interpretations, the competition from foreign manufactures and attractive opportunities in agro-export activities were the chief obstacles to industrial growth and the main disagreement lies in the timing of their removal and therefore of the onset of modern industrialization in Portugal.

Present knowledge of Portuguese industrial development during these crucial decades is hardly more than superficial. In what follows, I shall try to clear up some of these disagreements and go beyond this incipient stage of analysis by presenting some new quantitative evidence regarding the period 1870-1913. I shall review the facts concerning the timing and cycles of industrialization and render more explicit some of the causes of the lacklustre performance achieved. In a concluding section, I shall discuss the comparative advantages and disadvantages of Portuguese industry vis-à-vis that of some advanced economies.

1. *The growth of industrial production, 1870-1913*

Portugal is possibly the only country in Europe where until now there has been no attempt at constructing an index of industrial output for any period prior to World War II. The vague and impressionistic nature of the discussion regarding the country's industrial past is a clear reflection of this. The results of a first effort in this direction are presented in Table 1, with all the reservations which are called for in this type of exercise, in addition to those which arise with all first estimates.[7]

The index covers the following branches: cotton, woollens, linen, tobacco, food, paper, cork, fish canning, metal working and mining. Given the complete lack of direct information on the output of any of these sectors over any period of years, it has been necessary to rely entirely upon indirect indicators. In some cases like cork, fish canning and mining, export data was used, after some corrections. For the most part, however, output has been estimated from data for the values of imported raw materials, since the industries considered depended largely if not altogether on such imports. It was assumed that overall the technology used did not change appreciably in the course of these years with respect to the ratio raw material/finished product values and output values were consequently obtained as a multiple of these known raw material values.

To calculate the overall index, which is a weighted average of the output values for each sector, two sets of weights were used, hence the two series in the table. They represent different estimates of each sector's contribution to net industrial output and are calculated by multiplying value added per worker in 1890 by the sectoral share in the total industrial labor force.[8] The two series of manufacturing output thus obtained at current prices were deflated, using as price series for the various relevant industrial products the unit values in each year, either of exported manufactures (cork, canned fish) or of selected imported goods (textiles, paper, ceramics, light tools, soap and beer) which could be assumed to compete with similar domestic products.

Table 1

Portuguese Industrial Output at Constant Prices, 1870-1914 (1870 = 100)

	I	II		I	II
1870	100	100	1893	178	193
1871	86	88	1894	166	178
1872	104	113	1895	181	194
1873	119	119	1896	172	186
1874	117	124	1897	191	205
1875	121	129	1898	198	216
1876	106	109	1899	205	220
1877	122	122	1900	225	248
1878	118	119	1901	210	236
1879	122	125	1902	219	245
1880	106	111	1903	232	263
1881	118	123	1904	240	269
1882	113	149	1905	211	244
1883	n.a.	n.a.	1906	218	258
1884	n.a.	n.a.	1907	247	283
1885	126	133	1908	235	272
1886	153	163	1909	236	267
1887	157	171	1910	258	296
1888	160	175	1911	299	342
1889	157	181	1912	285	346
1890	179	200	1913	291	347
1891	165	187	1914	247	282
1892	164	183			

Source: J. Reis, "A Produção Industrial Portuguesa, 1870-1914. Primeira Estimativa de um Indíce" (Faculdade de Economia, Universidade Nova de Lisboa, Lisbon, working paper no. 35, 1985), p. 36.

This procedure was followed owing to the absence of published data for the prices of manufactures in the Portuguese market, during the period in question.

As with all partial indicators, the first question to arise is that of coverage. Figures for either gross or net industrial production in any year are hard to come by and are highly unreliable. Armando de Castro has suggested that at the turn of the century, gross industrial product might have reached 50% of the agricultural product, which was then thought to be 200,000 contos. This would imply a not unacceptable coverage for my index of about 60%, since the aggregate value of gross output of the sectors included in it can be estimated at roughly 60,400 contos in 1900. This is perhaps not so surprising considering that it comprises the majority of what were then presumed to be the most significant branches of manufacturing.[5]

In the second place, the methods employed in obtaining these two series involve distortions to which attention must be drawn, even if they cannot be corrected. Probably, two of the most serious of these inaccuracies will have caused the rate of growth to be overestimated. The first arises from the fact that the highly dynamic cotton and metal working sectors are over-represented by the systems of weights, while sectors with a presumably slower growth, such as soap, ceramics and hats are left out altogether. The second is the result of using a single set of weights for the entire period and one which corresponds to the end of the period considered. This means that the sectors which expanded most rapidly over this half century will enjoy an exaggerated influence in determining the aggregate index of industrial production.

Other features of this exercise will have had the opposite effect, however. With woollens and paper, the use of a fairly sizeable amount of home produced raw materials, the value of which had to be assumed constant in the face of a growing consumption of imported inputs, means that the rate of growth for the 1870s and 1880s will appear smaller than it really was. The failure to include any of the "new," rapid growth industries like cement, fertilisers and electricity will have had a similar effect during the early twentieth century, when their presence began to make itself felt. Another source of underestimation may have been present as a result of using a fixed ratio of raw material/finished product values. In the long run, in at least some sectors, e.g. textiles, the value added per unit of raw material increased as a result of the growing emphasis on finely finished goods, such as occurs normally as the import-substitution process of industrialization proceeds. Finally, the fluctuations in manufacturing output induced by the ups and downs of the construction cycle are probably not shown up in these indices owing to the absence of the sectors which were most closely linked to building activity, i.e. cement, glass, bricks and tiles, quarrying and lime. Altogether, it seems likely that the actual performance of Portuguese industry may have fallen slightly below the figures presented in the table, though it is obviously impossible to know the size of the gap.

2. *The path of growth*

The situation portrayed in Table 1 has several features worth noting. The first is the confirmation that steady industrial growth, far from being a twentieth century phenomenon goes back at least several decades into the nineteenth century. Moreover, during this first stage of industrialization, the rate at which this growth took place was hardly a modest one. For the years up to 1913, it was somewhere between 2.5 and 2.8% a year on trend. This is rather less than the various rates which have been suggested by Manuel Villaverde Cabral–his lowest is 4% during 1890-1913, but in other passages of his work he hints at much higher rates–though it supports his view of a fairly vigorous and prolonged industrial expansion during these decades.[10] It also becomes clear that industry must have been growing a good deal faster than agriculture. While we have no figures for the latter, this conclusion seems plausible, not only judging by the few partial indicators available, such as the production of wine, grain and meat, but also because of the preponderant weight of agriculture in gross national product, which itself was certainly growing at much less than 2.5% a year.[11]

Thus, the notion that agriculture was the "dynamic" sector of the economy has no foundation for the period after 1870. It would not be fair to say, however, that the overall economic performance was in any sense "led" by industry. Rather, the point should be made that although manufacturing expanded at a reasonable pace for several decades, this was still too little to either secure a significant improvement in the economy's rate of growth, given the drag represented by the other sectors, or to change the structure of the economy to any serious degree.[12] Industry's role was positive but not positive enough.

Despite the shakiness of such comparisons, it is interesting to note that by international standards, Portugal's industrial performance could hardly be labelled as poor. It did not do as well as Russia, Sweden or Germany but compared favourably with Britain, France, Spain and Italy and was close to Hungary and Austria.[13] This may come as a surprise to some, in particular as it shows that despite its dependence on Britain and the alleged obstacle this was to economic diversification, Portugal's industry appears to have done consistently better than that of the dominant country. On the other hand, the problem here is again that although reasonable, this growth was too slow for such a "latecomer" and that at such rates it was impossible, even in fifty years, for Portugal to "catch-up," either with the more advanced or even with the middle level economies of late nineteenth century Europe.[14]

One of the reasons for this problem is that in a sense the manufacturing sector grew too smoothly. In contrast to the Gerschenkronian notion of industrial spurts for late industrialisers, and although obviously experiencing some ups and downs, Portugal's experience lacked truly intense phases of expansion, with growth rates somewhere between 8% and 12%, such as are to be found in the cases of Sweden (1888-96), Japan (1907-1913) or Russia (1890s).[15] Why this should have been so is not yet clear although two points can be made in this connection. One, which has already been mentioned, is that our series understates fluctuations and we may therefore be facing a false problem. The other is that this lack of industrial spurts was not peculiar to Portugal at this time–Austria and Bulgaria, at least, were similar in this respect.[16] Perhaps we should start considering the usefulness of distinguishing, within the "latecomer" category, a subgroup of countries in which for a variety of reasons, certain branches of manufacturing such as heavy industry or "new" industries, which were likely to be responsible for these phases of explosive growth, either played a minor part or were non-existent.

If we break down the evidence into shorter periods, several interesting findings emerge. The first is that yielded by the classic dichotomy between the 1870s to 1880s and the 1891 to 1913 subperiods. This shows that contrary to Joel Serrão's supposition, not only was industry growing prior to the crisis of 1891 but it was doing so more vigorously than it did later, during the supposed heyday of Portuguese industrial development, when it enjoyed far greater protection, as well as the opening of the colonial market in Africa. The rates are respectively, 2.4% and 2.1% a year, taken from peak to peak of the curve which, given the roughness of the estimates, cannot be considered significantly different. One reason for the weaker performance of this later period is that during most of the 1890s industrial output did not rise at all and when it finally did, in 1897, the rate of growth was not much higher than it had been in the 1870s and 1880s. This stagnation of the 1890s is at variance with the belief entertained by several authors that there was a spurt during these years and suggests that the new, more protective tariff of 1892 was not the great stimulus to industrialization that it has been held out to have been.

Even more striking is the difficulty in reconciling our industrial output series with the current views on the role played by primary product exports in inducing industrialization via the expansion of internal demand. A comparison with the foreign trade data presented by Pedro Lains shows that during most of the first industrial spurt (1873-86), exports by value did not exhibit the slightest upward trend. A second bout of export stagnation, from 1897 to 1909, was also accompanied by an unmistakeable upward trend in industrial activity, which rose at an annual rate of 2%. On the other hand, the only period prior to 1913 during which manufacturing output did not increase (1890-96) was precisely when Portuguese exports were going through one of their most dynamic phases.[17]

The fact that the observed relation between exports and industrial growth is the reverse of what has been claimed in the past, calls for some explanation. One way of solving the problem is to continue to claim that export-induced increments of internal demand were a major determinant but that they functioned with a lag of several years. In this light, the industrial growth of the 1870s and 1880s could be viewed as the consequence of the export boom of the 1850s and 1860s, whereas the industrial expansion from 1897 to 1913 could be seen as having been propelled by the rapid rise in exports of the 1890s.

While not entirely implausible, this approach requires that one then goes on to account for the existence of this lag. One possible account is that Portuguese industry had a very inelastic supply curve, which meant that increases in demand did not lead to immediate significant increases in output when measured at constant prices. Over longer periods, however, this supply curve shifted out to the right, as a result not only of persistently high profits but also of infrastructural development, the reaping of internal and external economies of scale, manpower improvement and so on. It is not even hard to accept that some of these effects might have made themselves felt contracyclically, given that they would often be the result of slowly maturing investments decided upon during periods of optimism induced by export growth. In this way, industrial output at constant prices could be seen to increase by a considerable measure during a period of stagnant export incomes, owing to the slow but sure supply response to an earlier set of circumstances. While there is as yet not much empirical evidence for this hypothesis, it derives some support from the frequently acknowledged lack of dynamism of Portuguese entrepreneurs and their preference for market manipulation over productive modernisation as a source of immediate profit.[18]

An alternative to this way of interpreting the data is to deny the generally acclaimed significance of export incomes in shaping domestic demand for industrial goods, and attribute the latter's evolution to other causes. One surely important one is suggested by the closeness of fit between the fluctuations of Portugal's manufacturing output and those of the local money markets. Thus, while the most important spurts in industrial growth came to an end in 1876, 1883, 1891, 1901, 1906 and 1908, the following were years of depression in the money markets, gold drains, credit restriction and even banking crises: 1876, 1882-84, 1891, 1900 and 1906-08.[19] This sensitivity to local financial conditions is hardly remarkable, in view of the industrialists' frequent complaints about their difficulties in obtaining short and long term credit. It is rendered more plausible still by the discovery now being made that Portuguese banks were active throughout the second half of the nineteenth century in providing funds to a wide variety of manufacturing concerns.[20]

It also emerges from all this that, during these years, a fairly close link did exist between the ups and downs of the manufacturing sector and those of the international business cycle.[21] This did not, however, operate predominantly through trade movements, as many have supposed. The Portuguese economy seems to have been tied into the world economy much more significantly via the short run capital movements which connected the Portuguese money markets with those of the rest of the world (London, in particular) and it was chiefly thanks to this that the manufacturing cycle was synchronised with the world business cycle. In some instances, however, domestic rather than foreign conditions were the dominant influence in bringing about monetary crises with repercussions for industry. Although they had external components as well, the banking collapses of 1876 and 1891 were very much the consequence of local speculative manias which led, in the first case, to an industrial contraction of 20% in one year and, in the second case to a bout of stagnation which lasted for five years.[22]

Only further research can elucidate these matters properly but a further point should nevertheless be made. The two interpretations presented above are not mutually exclusive and, indeed, it is not difficult to accept that in conjunction they have more explanatory power than either of them separately. Taken together, they imply an analysis of industrial performance in which a long term upward trend was the result of gradual and perhaps alternating outward shifts in the supply and demand schedules, as a result of, among other things, surges in export growth, infrastructural development and so on. Overlaying this, there was a series of shorter run spurts and decelerations, principally brought about by changes in monetary and financial conditions and which were mostly though not exclusively induced from abroad.

3. *The limits to industrial growth*

The fairly steady and, for a "late starter," moderately paced industrialization which Portugal experienced between the 1870s and the First World War poses another kind of question: why was this growth not much faster, given the low starting base and the relative backwardness of the country? Why did it slow down over time, instead of gathering speed? Why was there no "take-off" of any sort? Although the problem has been tackled before, it does not seem that all of the "classic" reasons for the so-called failure of Portugal's industrialization are particularly useful. The negative impact of the war, civil war and destruction of the early decades of the last century are a case in point, since by 1870 one could expect an adequate recovery from them to have taken place. The problem of the excessive openness of the economy to competition from abroad is equally hard to accept in view of the long history of tariff protection from at least the 1830s and becomes completely unrealistic from the 1890s on.[23] The argument that the national bourgeoisie was too weak and divided to attempt to implement a development model for the economy is too loose and raises more difficulties than it succeeds in clearing up. On the other hand, it is not hard to accept that the small size of internal demand for Portuguese manufactures, despite a strongly protected market, a deficient stock of skills and knowledge and the high cost of capital inputs were significant retarding factors.

If we compare Portugal with other European countries in similar circumstances at the time, it might also be argued that non-market, institutional influences in favour of development contributed very little. The role of the State in actively encouraging industrialization was weak throughout this period–there were no policies of tax exemptions, subsidies, cheap loans or government contracts such as materialised in some East European countries–although one may wish to question how important this could have been anyway.[24] On the other hand, although the commercial banks were involved in financing industrial activity, Portugal totally lacked anything like the industrial bank sector which in some countries, it has been claimed, played a decisive role in industrial promotion. Even less did it benefit from the influx of foreign capital directed at industry that was recorded in countries, such as Hungary, with better industrial records.[25]

Some important obstacles to a more dynamic advance, however, lay in the very way in which the industrial sector evolved. One such feature was the relatively weak multiplier effect of the most rapidly growing branches of manufacturing, owing to their high content of imported raw materials. This deprived the primary sector of a much needed stimulus, not only to growth but particularly to modernisation and productivity improvement, without which a rapid and sustained industrialization was unlikely over the long run. In the cotton and metal working industries, which together had a substantial share of total manufacturing output and were among the best performers over the entire period, this input corresponded to between 45% and 55% of the value of gross production.[26] Of the other sectors covered by the index, woollens and paper also had significant import contents, while the food industry which consumed mostly domestic inputs was one of the slowest growers. Cork and fish canning, on the other hand, had a rapid growth and were based to a large degree on home produced materials but constituted a small share of the aggregate. A helpful perspective is given to these data by Pereira de Moura's pioneering study of the post Second World War economy. Despite a high rate of industrial development after 1945, structural change leading to further industrialization failed to occur to a sufficient extent largely because of the heavy "external dependence" of Portuguese industry, i.e. a 40% share of gross output went to pay for imported inputs.[27]

Related to this is another negative aspect which has also been identified in the post 1945 era: the weakness of intersectoral links. One particular aspect of this is that, as it grew and equipped itself in order to expand capacity and absorb modern technology, factory production generated a relatively weak demand for the products of the local metal working industry. Thus, a crucial sector like machinery production had little chance of getting off the ground and not surprisingly, the technically more sophisticated industries like cotton, woollens and flour milling acquired most of their machinery abroad. Not only were import duties on capital goods low but foreign suppliers had tremendous competitive advantages in terms of price, quality and promptness of delivery.[28] On the other hand, in industries like cork and fish canning, which were not only very dynamic but whose equipment needs could be easily met by Portuguese firms, the spill-over effect for the engineering sector was slight because of their low capital intensity. This can be judged from the fact that early in this century in the Algarve, one of the principal centers for these activities, the capital-output ratio was 0.07 in cork and 0.25 in fish canning. In the latter industry, capital per worker was less than the equivalent of £25.[29]

Yet another structural problem arose with the fact that the impetus for industrial growth came to a large extent from import-substitution. Output was overwhelmingly composed of consumer goods directed at the home market and was therefore ultimately limited by the level of internal demand which rose only slowly, as befits a poor agricultural base and a not very dynamic demographic situation. Indeed, from the evidence, it would seem that some of the best opportunities for manufacturing growth of this kind had been grasped already during the rapid expansion of the 1870s and 1880s. In the case of textiles, by 1890, their consumption was already overwhelmingly satisfied by local production. Imports represented only 24% of total sales in woollens, 23% in cotton and 10% in linen, it being recognized by contemporaries that a sizeable fraction of these imports was virtually impossible to produce in Portugal, for technical and economic reasons.[30] For many other sectors, this opportunity for substitution was even smaller e.g. garments, food, construction materials, hats, shoes.

In the light of this, it becomes easier to understand why, although conditions should have been more favorable to industrial growth after 1896–a lower exchange rate, higher tariffs, an improving transport network, higher income per capita, the African market–the rate of industrial growth was lower than in the 1870s and 1880s. In particular, this highlights the danger of too much stress being placed on the growth-inducing effect of currency depreciation during this period, as some authors have done. In fact, given the high import content of many Portuguese manufactures, the mere consideration of the rise in the price of foreign manufactures this brought about is misleading, because it overstates the degree of effective protection gained for domestic industry.

This brings us back to the problem posed by the weak aggregate demand which Portuguese industry had to face during these decades. Apart from home consumption, the inadequacy of which has already been noted, there was the market in Angola and Mozambique, which for a while raised great hopes of brilliant new opportunities, both for industry and agriculture. In the event, even after they were strongly protected from foreign competitors, the colonies were never able to absorb more than a small proportion of Portuguese industrial output. Only the cotton industry was helped at all by the emergence of this new outlet and even then, at the peak, the African market took a mere 10% of its total production.[31]

Because they sold principally to the more advanced economies and because the country had a comparative advantage in these fields, the only industries which were not subject to these constraints were cork and fish canning. No matter how vigorous their growth, however, their impact on the overall industrial picture could never be anything but modest. On the one hand, these were relatively small industries, even in local terms, and on the other hand, they faced long run limitations on both the supply and the demand sides, which made any change in this situation unlikely.[32]

The conclusion to be drawn is that unless a sizeable fraction of industrial capacity were able to transcend the narrow confines of the domestic market by exporting on a large scale, manufacturing was bound to remain a sluggish performer. During the years of industrial boom which followed the Second World War, this is precisely what happened. Portugal was able to make efficient use of modern technology which, combined with a considerable advantage in the form of cheap labor, enabled it to compete very successfully in some branches with producers from the traditionally industrial nations. The question this leads to and which I shall proceed to discuss is why this did not occur earlier, particularly during the period 1870-1913, when some of the conditions favorable to this kind of industrial development were already present in the Portuguese economy.

4. *Portuguese competitive disadvantage*

Notwithstanding the already noted sluggish growth, during the formative decades prior to the First World War, the Portuguese industrial structure was quite well adjusted to its social and economic environment. The prevalence of labor intensive processes and the concentration on low quality consumer goods were probably the best response to a situation, the hallmarks of which were low income consumers, scarce capital, cheap labor, a shortage of skills and a lack of industrial tradition. This positive adaptation to circumstances, however, does not appear to have been sufficient to overcome some major difficulties nor, in particular, to have enabled the economy to reap all the benefits of transferring modern technology into a backward setting.

In this early stage of its development, Portuguese industry suffered from three important disadvantages vis-à-vis its foreign competitors, which were abundantly recognized at the time. Too large a share of its raw materials had to be imported and were therefore expensive; credit was difficult to come by and was costly; as might be expected of infant industry, it was less efficient than its counterparts in the advanced nations of Europe. In its favor, labor was cheaper, although it is surprising how little was made of this by contemporaries, given that for most sectors the wages bill represented a large percentage of total costs.

The first of these disadvantages may not have weighed very heavily, not only because by this time international transport costs had become very low (especially for the light raw materials that most concerned Portugal) but also because many of the rival foreign sectors were also importing their raw materials, as happened with cotton, woollens and paper, for instance. With coal, transport made more of a difference but since Portuguese industry was generally not energy-intensive this meant that its share of total costs was small.[33] It is impossible to say at the moment how the cost of capital in Portugal compared with that in France, Britain or Germany, which were the countries that presented the greatest competitive threat, both at home and in the export markets to which Portuguese industry might one day aspire. Not only were interest rates higher in Portugal but there are many suggestions in contemporary accounts that the capital stock was poorly used and was expensive and slow to repair.[34] Since these adverse effects have to be combined with the low capital intensity of industrial production, it seems plausible that the overall impact, although negative may not have been enormous.

A thorough comparison of data on industrial wages across countries has yet to be carried out but there are signs that in Portugal their level was a good deal lower than in the advanced economies. Around 1910, for example, average daily wages, in

Portuguese *milreis*, were respectively 0.705 in France, 0.890 in Germany, and 1.080 in Britain.[35] In the Lisbon district, the largest industrial concentration of the time, wages were typically in the 0.55-0.65 range with extremes of 0.400 in oil pressing and 1.200 in brewing.[36] The ratio of Portuguese to "core" country wage levels, in manufacturing, seems to have been a bit less than 1 : 2, though some systematic research is needed to confirm this.

If, in adopting modern productive methods and techniques from abroad, Portuguese industry had been capable of using its labor input with an efficiency which approached that of the advanced economies, then its competitiveness would have been assured and most of its problems would have been solved, the lack of resources and capital notwithstanding. Unfortunately, by international standards the productivity of labor was extremely low and, as a result, despite the lower wages, unit labor costs ended up being higher.

The figures in Table 2 for value added per worker in a number of sectors of British, French and Portuguese industry represent

Table 2
International Productivity Comparisons: Value Added per Worker (£)

	Britain	France	Portugal
Food (total)	196	158	89
Sugar	367	325	275
Leather Articles	71	120	43
Porcelain and Tiles	68	108	52
Canned Food	81	98	34
Cork[a]	90	43	37
Leather Preparation	118	85	71
Metal Working and Engineering	97	96	86
Paper	167	156	60
Chemicals	86	93	50
Soap	155	210	84
Clothing	63	64	40
Glass	94	28	52
Cotton	79	174	39
Woollens	70	94	56

[a] For Britain and France, we took the values for the wood industry.

Sources: Britain and France: P. K. O'Brien e C. Keyder, *Economic Growth of Britain and France, 1780-1914: Two Paths to the Twentieth Century* (London, 1978), pp. 152-53; Portugal: the figures were calculated from the data in the 1881 and 1890 industrial enquiries and converted at the rate of £1 = 4.5 milreis. Whenever discrepancies between the two enquiries arose in any sector, we chose the highest productivity figure.

an attempt at quantifying this. Using such an approach raises enormous methodological problems, not least of which has to do with the lack of information on inputs and outputs contained in the Portuguese industrial inquiries of 1881 and 1890. The time lag of more than one decade between the Portuguese and the other data is less of a problem given that it does not seem as if there can have been much productivity increase in Portugal at the end of the last century.

It would appear from this table, that the labor productivity of Portuguese manufacturing was generally half or less of that of its more efficient foreign rivals. In fact, the situation was even worse than these figures imply. The high tariffs which sheltered most sectors meant that the figures for Portugal in Table 2, which are based on internal market prices, are inflated compared to what they would have been without such protection. Since it is value added that is the basis of these comparisons, the relevant factor then becomes "effective" rather than "nominal protection" and although no study has yet been carried out on this for Portugal, it seems likely that for most manufactures the former should have been still greater than the latter.[37] If this productivity gap is considered in conjunction with the wage differential relative to France and Britain noted above, it becomes clear that under these conditions, Portuguese products were uncompetitive in foreign markets and industry could never aspire to a take-off stimulated by its export performance.

The causes of this comparative inefficiency constitute a vast and complex topic which cannot be handled here adequately. On a single industry basis, however, it is easier to deal with the main ones. Since it is an important facet of the problem of industrialization in a context of economic backwardness, it is to this that I shall turn our attention, by way of concluding this article.

Although I shall also look at other sectors, the focus will be particularly on cotton textiles, partly because it is one of the sectors about which we know the most, and partly because its characteristic uniformity of product, raw materials and machinery used render international comparisons easier. Being one of the most dynamic sectors in Portugal during the period, it may also be assumed that its defects will have been present even more strongly in other, less successful, industries.

Among the reasons for the low productivity of labor in manufacturing, two that stand out are the small amount of capital per worker and the poor quality of the labor itself. As regards the former, the contrast with Britain is stark as we can judge by comparing one of the largest concentrations of mills–the Porto cotton factories described extensively in the 1881 industrial enquiry–with their British counterparts in 1878.[38] As regards firms which concentrated exclusively on spinning, the British average was 184 spindles per employee, whereas in Porto, in one establishment alone this figure reached 133 spindles and in the remainder it was well below 100. The largest and best equiped cotton mill in the country at this time, the Real Fábrica de Fiação de Tomar, had only 88 spindles per employee. With weaving units the picture was no different, the British average being 2.1 looms per employee and the Porto average 1.8.

Over the next three decades, the expansion of cotton manufacturing in Portugal and the founding of several new establishments did little to change the situation. Whether in spinning, weaving or integrated spinning and weaving factories, an official report compiled in 1911 shows that using the British standards of 1878, the Portuguese mills continued to be grossly overstaffed relative to their processing equipment.[39] It is hardly surprising therefore that physical output per worker should have been so different. In the early 1880s, none of the above mentioned Porto factories was able to achieve as much as half of the 2,509 kilos of cotton spun per worker in British factories at the same time, the best result being that of the Montebello mill, where the figure was no more than 963 kilos per worker.[40]

Although the exact nature of the connection between human capital endowment and labor productivity has baffled many, it is generally recognised that the levels of both literacy and technical education were important in determining the course of development in the nineteenth century western economies.[41] This is the second area in which Portugal and, by reflection, Portuguese industry were seriously disadvantaged, a fact which has not been sufficiently acknowledged in modern studies of the past century but which was of great concern to industrialists and commentators at the time. According to the managers of the Tomar cotton mill, for instance, "one of the greatest difficulties we have to struggle against is the shortage of capable personnel. The lack of general education and the near absence of technical training mean that it is hard not only to find good workmen but also competent foremen."[42]

In general terms, the educational picture was indeed dismal. Even in 1911, only one Portuguese in four was able to read and write and among the industrial labor force the situation was sometimes worse.[43] In the same year, the already mentioned report on the textile sector surveyed the educational attainments of the work force, both in Lisbon and in the provinces, and found, with one exception only, that the literacy rate was never more than 20% and in many cases reached only 10%. In the Porto cotton mills studied in 1881, matters were possibly even worse since in five of them, the entire work force was illiterate and in the other three, the literacy rate fluctuated between 10% and 25%. As for technical education, the very substantial presence of foreign labor as foremen and specialised workers in Portuguese factories bears eloquent witness to the inadequacy of the system in this respect. Meanwhile, in Britain and Germany not only had simple literacy become virtually universal but on average workers in British industry by this time had had schooling of 20 hours a week for seven years and their German counterparts could boast of 32 hours a week for nine years.[44]

One simple but important consequence of this state of affairs was the restriction it placed on the capacity of Portuguese workers to handle industrial equipment on the same scale as elsewhere in the world. According to the Tomar factory management, the trouble was that although their machinery was the best in the world–it was imported–"it is very seldom that we can get our workers to operate more than two looms, while in factories abroad, it is common to see them operating three and even four looms at the same time."[45]

In addition to the relative scarcity of equipment and its poor use by ill-trained workers, a third difficulty which the more modern and on the whole more capital-intensive sectors especially faced, was that posed by the small size of the domestic market, the industry's chief outlet. It was in this period in metal working, and machine making in particular, that "specialisation became the mark of the modern efficient enterprise."[46] Yet in Portugal the option was either to mechanise production and have machinery lying idle for long periods, or to have the work done by hand, which was costly and could entail a lower quality of product. Neither of these options enhanced competitiveness, as entrepreneurs themselves were aware. The owner of one of the largest engineering plants in Porto claimed that a high protective tariff was necessary because "our labor costs tend to be higher, given the smaller output to which we are reduced."[47] Since his firm produced an average of only two steam engines a year between 1865 and 1881, it may be supposed that it was subject, at least during this period, to yet another disadvantage relative to foreign competitors in Britain or Belgium, namely the "learning factor" to which S. B. Saul has referred. "Where bulk orders for steam engines, machine tools, agricultural machines, locomotives, electric motors were available, [in the nineteenth century] the economies from learning were very large."[48]

Cotton textiles was another sector which was hampered by this inability to reap the economies of scale which were available in some other countries. If we take the example of spinning again, the Anglo-Portuguese comparison is revealing. In the 1880s, the average English spinning mill had 28,500 spindles whereas in Portugal, the average was 8,300, with the largest factory reaching only 17,000 spindles. Yet, as the Tomar factory managers pointed out, "if the cotton industry is to provide sufficiently attractive profits, it is essential that it be established on a sufficiently large scale."[49] Besides being on the small side, there was the fact that mills were forced to produce several different counts of thread because the market was too small to absorb easily a single count. As a result, there were costly interruptions of the work schedule every so often, in order to adjust the machinery for new runs. Nevertheless, an enormous variety of products seems to have prevailed, judging from the entries in the exhibition catalogues by the largest mills in the country. In 1888, the Tomar factory offered 28 different counts of cotton thread, for example, and in 1891, the Lisbonense had over 72 different types of cloth and at least 67 kinds of thread.[50]

Portuguese industry was probably affected by additional handicaps which, in the present state of knowledge, we can do no more than outline. Compared with the industrial nations of Europe, there were obvious external economies from which it did not benefit, such as were forthcoming from fully developed and fully utilized transport networks or from a plentiful supply of ancillary services, like repair shops and spare part producers.[51] Being massively dependent on imported technology, one may suppose that the equipment employed was not always the most suited to the prevailing conditions, particularly to the skills and industrial traditions of the work force. To some extent these and other disadvantages could have been surmounted, it has been claimed, had the country been endowed with a more dynamic and inventive entrepreneurial class. As was said at the time and has since been repeated in a variety of ways, industrial management was in the hands of "men who are without knowledge, without intelligence and without good will."[52]

The entrepreneurial failure argument is not only an old favourite in Portuguese economic history but also one of the hardest to handle, not least of all in a comparative perspective, given that it can be found applied just as liberally to other countries with very different industrial performances.[53] This does not mean, however, that it should or can be neglected. Clearly, in most of the manifestations of inefficiency noted above, as well as in some of the causes given for them, one has to recognize that part of the blame must have lain in this quarter. The high cost of capital in Portugal could have been one of the reasons for the low capital/labor ratio but at least in cotton, the latter was so very low that it seems that there was an underinvestment in plant which could be ascribed to wrong decisions. Similarly with the educational issue: one of the general reports of the 1881 industrial enquiry pointed out that although employers clamoured against the ignorance of their employees, they refused to give work to the graduates of the only two technical schools then in existance. The explanation they gave was that they wanted to earn higher wages and were arrogant and stood up to their masters.[54]

The question which remains, however, is, by international standards, how much worse, if at all, were Portuguese entrepreneurs and on this much research will have to be done. In the meantime, the best approach is to consider the entrepreneurial factor as a residual cause of the lack of industrial competitiveness and try to draw out and evaluate other causes which lend themselves better to a rigorous treatment. In some ways, this is what has been attempted here.

Faculdade de Economia, Universidade Nova de Lisboa
and Instituto de Ciências Sociais.

[1] I have based this section mainly on the following: M. Halpern Pereira, *Política e Economia (Portugal nos Séculos XIX e XX)* (Lisbon, 1979); A. de Castro, *A Revolução Industrial em Portugal no Século XIX* (Oporto, 1971); M. Villaverde Cabral, *O Desenvolvimento do Capitalismo em Portugal no Século XIX* (Oporto, 1976) and *Portugal na Alvorada do Século XX: Forças Sociais, Poder Político e Crescimento Económico de 1890 a 1914* (Lisbon, 1979); J. Serrão e G. Martins, eds., *Da Indústria Portuguesa: Do Antigo Regime ao Capitalismo. Antologia* (Lisbon, 1978). The scant list of monographic studies consists of little else beyond the following: M. de Fátima Bonifácio, "Industrialização Oitocentista e Concorrência Externa: A Indústria Chapeleira, 1814-1914," *Cadernos de Documentação GIS*, no. 2 (Lisbon, 1980); J. M. Amado Mendes, *A Área Económica de Coimbra. Estrutura e Desenvolvimento Industrial, 1867-1927* (Coimbra, 1984); M. da Silva Guimarães, *História de uma Fábrica: A Real Fábrica de Fiação de Tomar* (Santarém, 1976); M. das Dores Jorge de Goes, *A Real Fábrica de Lanifícios de Cascais* (Cascais, 1964); J. Ferreira, "Subsídios para a História da Moagem Portuguesa: A Companhia de Moinhos a Vapor Ceres, de 1854 a 1860 no Porto," *Revista de História Económica e Social*, no. 12 (1983), pp. 127-54. The classic work on the eighteenth century is J. Borges de Macedo, *Problemas de História da Indústria Portuguesa no Século XVIII* (2nd ed., Lisbon, 1982) which can be complemented by J. Custódio's introduction to the modern re-edition of

J. Acursio das Neves, *Memória sobre os Meios de Melhorar a Indústria Portuguesa Considerada nos seus Diferentes Ramos* (Lisbon, 1983).

There is a more considerable literature on working class movements which affords some useful information regarding the productive side of industry. See, for example, M. F. Mónica and M. de Fátima Patriarca, eds., *O Movimento Operário em Portugal*, special issue of *Análise Social*, nos. 67, 68, 69 (1981) and M. F. Mónica, *Artesãos e Operários. Indústria, Capitalismo e Classe Operária em Portugal (1870-1934)* (forthcoming).

2 The arguments adduced by modern authors are by no means new but were being rehearsed early in this century and even sooner. See, for example, J. de Oliveira Simões, in *Notas sobre Portugal* (Lisbon, 1908), vol. 1, pp. 366-69.

3 Halpern Pereira, *Política e Economia*, p. 72, though it is not clear from the text exactly when this de-industrialization occurred nor what shape it took.

4 It should be pointed out that in spite of this article's concentration on the years after 1870, earlier though lesser spurts of industrial growth can and have been identified, as for instance the late eighteenth century and the 1840s. See Borges de Macedo, *Problemas de História*, passim, and de Castro, *A Revolução Industrial*, p. 40.

5 See Villaverde Cabral, *O Desenvolvimento do Capitalismo*, pp. 45 and 270 and *Portugal na Alvorada do Século XX*, chaps. 6 and 16.

6 See Serrão and Martins, *Da Indústria Portuguesa*, p. 29.

7 For a detailed account of how these results were obtained, see J. Reis, "A Produção Industrial Portuguesa, 1870-1914: Primeira Estimativa de um Indíce" (Faculdade de Economia, Universidade Nova de Lisboa, Lisbon, working paper no. 35, 1985).

8 F. Pereira de Moura in *Por Onde Vai a Economia Portuguesa?* (Lisbon, 1974) considered the 1881 industrial enquiry "a model, never to be surpassed in Portugal" (p. 105). I have reasons to believe that in some aspects the 1890 enquiry used for this exercise was better, as well as having more quantitative information. Figures on the composition of the labor force were taken respectively from J. de Oliveira Simões, *Escorço dalguns Aspectos da Indústria Fabril Portuguesa* (Lisbon, 1913) and J. H. de Azeredo Perdigão, *A Indústria em Portugal (Notas para um Inquérito)* (Lisbon, 1916).

9 See de Castro, *Revolução Industrial*, pp. 115-18. For a list of the most important sectors at the turn of the century, see R. Ulrich, *Crises Económicas Portuguesas* (Coimbra, 1902), p. 164. Of these the following are not covered by the present index: porcelain, bricks and tiles, hats, barrelmaking, glass and soap.

10 Villaverde Cabral, *Portugal na Alvorada do Século XX*, pp. 27, 145-46. It is also much less than the rates achieved after the 1930s, something which confirms the view that the second half of the nineteenth century, though a period of change, was comparatively less impressive in terms of actual growth. On this point, see J. Reis, "O Atraso Ecónomico Português em Perspectiva Histórica, 1860-1913," *Análise Social*, 20 (1984), p. 8. According to Pereira de Moura, *Por Onde Vai a Economia Portuguesa?*, pp. 135 and 154, industrial rates of growth were 4.9% for 1933-53 and 8.3% for 1953-70.

11 According to the recent calculations of M. E. Mata, real national income increased at about 1% per annum during 1891-1913. See her "A Unidade Monetária Portuguesa face à Libra, 1891-1913" (Faculdade de Economia, Universidade Nova de Lisboa, Lisbon, working paper no. 22, 1984), pp. 22-26. There is a wealth of information on agricultural production in M. Mateus and A. Mateus, "The Agrarian Revolution in Nineteenth Century Portugal–Technological Change, Trade Regimes and the Response of Agriculture" (Faculdade de Economia, Universidade Nova de Lisboa, Lisbon, working paper, forthcoming). Between the 1860s and the early twentieth century, their estimate of agricultural output increased at an annual rate of 0.8%.

[12] In this respect, Portugal can be likened to Bulgaria, as described in A. Gerschenkron, *Economic Backwardness in Historical Perspective* (Cambridge, Mass., 1962), p. 213, a fact which is brought out also by P. Bairoch's, "International Industrialization Levels from 1750 to 1980," *Journal of European Economic History*, 11, no. 2 (1982), p. 330, which shows Portugal, in 1913, as very similar to Balkan levels of industrial development and about 50% below Spain and Italy.

[13] See the convenient summary in A. Carreras, "La Producción Industrial Española, 1842-1981," *Revista de História Económica*, 2 (1984), p. 140.

[14] There is a discussion of this question in Villaverde Cabral, *Portugal na Alvorada do Século XX*, p. 327.

[15] Gerschenkron, *Economic Backwardness*, p. 78.

[16] Ibid., p. 213; D. F. Good, "Stagnation and Take-Off in Austria, 1873-1913," *Economic History Review*, 28 (1974), pp. 84-85.

[17] P. Lains, "O Sector Exportador Português e a Economia Nacional (1842-1914)" (mimeo, 1984). I am grateful to the author for his permission to use these results.

[18] See, for example, J. M. Amado Mendes, "Sobre as Relações entre a Indústria Portuguesa e a Estrangeira no Século XIX," *Análise Social*, 16 (1980), p. 48, and A. Lopes Vieira, "The Role of Britain and France in the Finance of Portuguese Railways, 1850-1890. A Comparative Study in Speculation, Corruption and Inefficiency" (unpublished Ph.D. diss., University of Leicester, 1983), p. 18.

[19] The latter series of years is drawn from a study of the Portuguese banking sector during the second half of the nineteenth century by G. Gomes and which is part of a project on the history of the Bank of Portugal, 1846-1914.

[20] Information drawn from the research project mentioned in the previous note. The annual reports of the Bank of Portugal made constant reference to the help provided to industrial concerns, large and small.

[21] See A. Maddison, *Phases of Capitalist Development* (Oxford, 1982), pp. 74-75. In a pioneering study first published in 1947, de Castro drew attention to what he called the "speculative" as opposed to the "productive" nature of the Portuguese crises. See his *A Revolução Industrial*, chap. 4.

[22] There is a considerable literature on these two crises but see, for example, J. P. de Oliveira Martins, *Estudos de Economia e Finanças* (Lisbon, 1956) and J. P. da Costa Leite Lumbralles, *Organização Bancária Portuguesa* (Coimbra, 1927).

[23] A brief discussion of this argument is to be found in Reis, "O Atraso Económico Português," pp. 12-13.

[24] I. T. Berend and G. Ranki, *Economic Development in East-Central Europe in the 19th and 20th Centuries* (New York, 1974), pp. 86-91. For a strong defense of how important State intervention could be in such countries, see R. Vaccaro, "Industrialization in Spain and Italy (1860-1914)," *Journal of European Economic History*, 11 (1982), pp. 731-38.

[25] Gerschenkron, *Economic Backwardness*, pp. 87-88. In 1913, 15% of all industrial capital in Hungary was foreign owned, a small but nevertheless significant proportion. See J. Komlos, "Economic Growth and Industrialization in Hungary, 1830-1913," *Journal of European Economic History*, 10 (1981), p. 27. In Rumania, 80% of industrial share capital was foreign owned, according to Berend and Ranki, *Economic Development*, p. 106. Curiously enough, in Portugal the only manufacturing sectors which attracted foreign capital and entrepreneurship to any extent were the two fastest growing as well as the most strongly export-oriented, i.e. fish canning and cork.

[26] This is calculated on the basis of production data found in the 1881 and 1890 industrial enquiries but is confirmed by other evidence for later years derived from the reports on various districts published from time to time in the *Boletim do Trabalho Industrial*.

[27] F. P. de Moura *et al.*, "Estrutura da Economia Portuguesa," *Revista do Centro de Estudos Económicos*, no. 14 (1954), p. 206. Much is also made of this argument by Halpern Pereira in her *Política e Economia*, pp. 16-49.

[28] Steam engines were predominantly foreign in their origin, as can be seen even as late as the 1920s from "Os Motores que a Indústria Nacional Utiliza," *Boletim do Trabalho Industrial*, no. 143 (1934). The same was true and for the same reasons of the more complex types of farm machinery e.g. threshers and their engines. See J. Reis, "Latifundio e Progresso Técnico: A Difusão da Debulha Mecânica no Alentejo, 1860-1913," *Análise Social*, 18 (1982), pp. 371-433.

[29] A. Monteiro, "Relatório dos Serviços de 4ª Circunscrição dos Serviços Técnicos da Indústria," *Boletim do Trabalho Industrial*, no. 18 (1905).

[30] *Relatórios da Exposição Industrial Portuguesa em 1891 no Palácio de Crystal Portuense* (Lisbon, 1893), pp. 292-304. For a clear instance of the exhaustion of opportunities for import-substitution in the cotton industry, by the end of the nineteenth century, see L. F. de Oliveira, *Indústria Algodoeira. Estudo. 1887-1903* (Oporto, 1904), pp. 29-30.

[31] Villaverde Cabral, *Portugal na Alvorada do Século xx*, p. 349; G. Clarence-Smith, *The Third Portuguese Empire, 1825-1975. A Study in Economic Imperialism* (Manchester, 1985), p. 90.

[32] Reis, "O Atraso Económico Português," pp. 19-20.

[33] For an illustration of this, the flour milling industry provides good material. See *Inquérito às Fábricas de Moagem Ordenado pela Portaria do Ministério da Fazenda em Data de 20 de Fevereiro de 1890* (Lisbon, 1890).

[34] One example of this is the problem of water scarcity leading to the overequipment of plants which relied on hydraulic power. The 1881 and 1890 enquiries show a surprising number of firms owning expensive steam engines as sources of back-up power for their hydraulic motors during the dry season, which meant that their use was confined to less than half the year.

[35] E. H. Phelps Brown and M. H. Browne, *A Century of Pay* (London, 1968), appendix 3. Since only average yearly earnings are given in this source, I assumed a working year of 250 days, in order to calculate the daily wage. The data for Germany is corroborated by the evidence from the town of Bochum, where in 1910 unskilled workers were earning the equivalent of 0.9 milreis, according to D. F. Crew, *Town in the Rhur. A Social History of Bochum, 1860-1914* (New York, 1974), p. 177. For Germany overall, the daily wage in industry, in 1913 and assuming a six-day week, was 0.937 milreis. See G. Bry, *Wages in Germany 1871-1945* (Princeton, 1960), p. 51.

[36] de Oliveira Simões, *Indústria Fabril Portuguesa*, p. 31. In some sectors, however, the wage differential relative to Germany was not large. In Germany, males earned the equivalent of 0.643 milreis per day in cotton spinning and 0.801 in hosiery, whilst in Lisbon they earned 0.650 and 0.600, respectively. Bry, *Wages in Germany*, p. 351.

[37] The situation in Portugal was more like that portrayed for the United States at the end of the nineteenth century than that of the Italian engineering industry at the same time. See G. R. Hawke, "The United States Tariff and Industrial Protection in the Late Nineteenth Century," *Economic History Review*, 28 (1975), pp. 84-99; and G. Toniolo, "Effective Protection and Industrial Growth: The Case of Italian Engineering, 1898-1913," *Journal of European Economic History*, 6 (1977), pp. 659-74.

[38] See, respectively, *Relatório Apresentado ao Exmo. Sr. Governador Civil do Distrito do Porto pela Sub-Comissão Encarregada das Visitas aos Estabelecimentos Industriais* (Oporto, 1881), pp. 167-224; and R. Church, ed., *The Dynamics of Victorian Business. Problems and Perspectives to the 1870s* (London, 1980), p. 173.

[39] "Inquérito sobre as Indústrias Texteis," *Boletim do Trabalho Industrial*, no. 105 (1916). In the case of vertically integrated firms, we calculated the number of

employees that the spinning and the weaving sections would have required, on the basis of the British average. The sum of these two figures was generally half of the existing labor force in the mill.

[40] The British data are taken from D. A. Farnie, *The English Cotton Industry and the World Market, 1815-1896* (Oxford, 1979), p. 199.

[41] See L. Sandberg, "Ignorance, Poverty and Economic Backwardness in the Early Stages of European Industrialization," *Journal of European Economic History*, 11 (1982), pp. 675-97, for a recent and illuminating comparative analysis of the problem. Also I. T. Berend and G. Ranki, *The European Periphery and Industrialization, 1780-1914* (Cambridge, 1982), pp. 56-59.

[42] *Inquérito Industrial de 1881* (Lisbon, 1881), part 3, p. 69.

[43] For data on literacy, schools and teaching staff, see A. H. de Oliveira Marques, *História de Portugal desde os Tempos Mais Antigos até ao Governo do Sr. Pinheiro de Azevedo* (7th ed., Lisbon, 1977), vol. 2, p. 228.

[44] P. Mathias and M. M. Postan, eds., *The Cambridge Economic History of Europe* (Cambridge, 1978), vol. 7, p. 459.

[45] *Inquérito Industrial de 1881*, part 3, p. 70.

[46] D. S. Landes, *The Unbound Prometheus. Technological Change and Industrial Development in Western Europe from 1750 to the Present* (Cambridge, 1972), p. 183, note 2.

[47] C. Geral da Alfândegas, *Documentos sobre a Renovação do Tratado de Comércio com a França* (Lisbon, 1881), p. 52.

[48] S. B. Saul, "The Economic Development of Small Nations: The Experience of North West Europe in the Nineteenth Century," in C. P. Kindleberger and G. di Tella, eds., *Economics in the Long View. Essays in Honour of W. W. Rostow* (New York, 1982), vol. 2, p. 118.

[49] Companhia da Real Fábrica da Fiação de Tomar, *Relatório da Direção. Balanço. Parecer da Comissão Fiscal* (Lisbon, 1886), p. 27.

[50] Associação Industrial Portuguesa, *Catálogo da Exposição Nacional das Indústrias Fabris Realizada na Avenida da Liberdade em 1888* (Lisbon, 1889), vol. 2, pp. 125-130; *Catálogo da Exposição Industrial Portuguesa em 1891*, p. 183.

[51] A very rough indicator of these possible differences in terms of the intensity with which the railway network was used is given by the freight tonnage carried per kilometre of track in a given year. The following are the figures in thousands of tons/Km for 1900 in several countries: Austria-Hungary, 6.18; Belgium, 12.08; Denmark, 1.42; Germany, 6.96; Portugal, 1.24; Sweden, 1.91; U.K., 14.18. See B. R. Mitchell, *European Historical Statistics 1750-1970* (London, 1978), pp. 315-43.

[52] J. d'Azambuja Proença, *A Indústria de Tecidos de Lã em Portugal. Algumas Considerações sobre o seu Atraso e Estacionamento* (Belem, 1884), p. 11.

[53] See Landes, *Unbound Prometheus*, pp. 131-33, for a case in point.

[54] *Inquérito Industrial de 1881*, part 3, p. 17.

Uneven Development in the Nineteenth Century: A Comparison of the Habsburg Empire and the United States

DAVID F. GOOD

Unlike most studies of uneven development before World War I, this paper uses the region (not the nation) as the unit of study. Weak market links with the national market partially explain persisting relative backwardness in the Habsburg Empire's eastern hinterland and in the American South. Even if product and factor markets had been perfectly integrated, institutional rigidities would have greatly retarded development. In the Empire, growth emerged in the west where serfdom was weakening and spread slowly as feudal institutions decayed. In America, capitalistic institutions promoted development in the North more thoroughly than did slavery and postbellum institutions in the South.

UNEVEN development is likely to threaten the world well into the twenty-first century. Social scientists devote much effort to the problem, but disagree still on the question posed by Richard Easterlin in 1981: "Why Isn't the Whole World Developed?"[1]

The answers reflect two competing traditions—one focusing on the internal structure of less developed countries, the other on their economic links to the outside world, both tracing back to classical economics. The first draws on Smith's basic premise that economic growth requires laws and institutions that guarantee "natural liberty."[2] Marx put Smith's premise at the heart of his analysis, identifying the mode of production as the driving force of history and the spatial diffusion of capitalism as a prerequisite for the worldwide transformation to socialism.[3] Yet non-Marxists are today the advocates of theories

Journal of Economic History, Vol. XLVI, No. 1 (March 1986). ISSN 0022-0507.

The author is Associate Professor of Economics at Temple University, Philadelphia, Pennsylvania 19122.

I would like to thank Michael Bernstein, Scott Eddie, Michael Edelstein, Stanley Engerman, John R. Hanson III, Carol Heim, William Hutchinson, John Komlos, Donald McCloskey, Joel Mokyr, Larry Neal, William Stull, and workshop participants at the University of Chicago and Indiana University for helpful suggestions and criticism on earlier drafts.

[1] Richard Easterlin, "Why Isn't the Whole World Developed?" this JOURNAL, 41 (March 1981), pp. 1–17.

[2] See Andrew Skinner, "Adam Smith: An Economic Interpretation of History," in Andrew Skinner and T. Wilson, eds., *Essays on Adam Smith* (Oxford, 1975), pp. 154–78; and Ronald Meek, "Smith, Turgot and the 'Four Stages' Theory," *History of Political Economy*, 3 (Spring 1971), pp. 9–27.

[3] On the classical Marxist view see Anthony Brewer, *Marxist Theories of Imperialism* (London, 1980), pp. 15–17.

stressing the bad effects of local conditions, arguing that poor economies need to create incentives that reward individual decision makers.[4]

The second, externalist tradition draws on a different thread of classical analysis. Smith pointed to expanding markets as the source of wealth, and Ricardo refined the insight as comparative advantage. After World War II the neoclassical economists used the Smith-Ricardian framework to argue for worldwide economic integration. In their view flows of trade and investment provide the chief mechanism for promoting growth.[5] Outside the developed world, however, the neoclassical vision has little influence. Like their neoclassical counterparts, neo-Marxists and Third World nationalists focus on the external links of less developed countries to the world economy, but see the world economy as a game in which the developed world wins, exploiting its power through the very markets praised by neoclassical theorists.[6]

W. Arthur Lewis recently attacked the economics profession for educating "a generation of economists with no historical background" who fail to look "to history for enlightenment on the process of development."[7] The history of development in the nineteenth century is rich, but most of it uses the nation as the unit of study.[8] Here I draw on a smaller but equally illuminating tradition that examines diffusion at the national level, using the region as the unit.[9]

The key is the comparative method, of which Marc Bloch wrote in 1928: "I consider its perfection and general use one of the most pressing needs of present-day historical science."[10] The comparative method has not achieved the high status foreseen by Bloch, but it has become an important tool.[11] Its strength is its ability to generate something like

[4] See, for example, Peter T. Bauer, *Dissent on Development* (London, 1971), pp. 74–82 and 158–61, and the analysis of imperialism's role in Third World economic history by Tony Smith, *The Pattern of Imperialism* (New York, 1981), pp. 50–84.

[5] The vast literature on economic integration is surveyed in Fritz Machlup, *A History of Thought on Economic Integration* (New York, 1977).

[6] The views of Andre Frank, Immanuel Wallerstein, and other neo-Marxists are summarized in Brewer, *Marxist Theories*, pp. 17–24.

[7] W. Arthur Lewis, "The State of Development Theory," *American Economic Review*, 74 (March 1984), p. 7.

[8] See, for example, Rondo Cameron, *France and the Economic Development of Europe, 1800–1914* (Chicago, 1965); A. G. Kenwood and A. L. Lougheed, *The Growth of the International Economy, 1820–1980* (London, 1983); and David Landes, *The Unbound Prometheus* (Cambridge, 1969).

[9] See, for example, several contributions in Paul Bairoch and Maurice Lévy-Leboyer, eds., *Disparities in Economic Development Since the Industrial Revolution* (London, 1981); and Jeffrey Williamson, "Regional Inequality and the Process of National Development," *Economic Development and Cultural Change*, 13, Supplement (July 1965).

[10] Marc Bloch, "Toward a Comparative History of European Societies," in Frederic C. Lane and Jelle C. Riemersma, eds., *Enterprise and Secular Change* (Homewood, Ill., 1953), pp. 494–95.

[11] On the comparative method see Sylvia L. Thrupp, "The Role of Comparison in the Development of Economic Theory," this JOURNAL, 17 (Dec. 1957), pp. 554–70; W. T. Easterbrook, "Long-Period Comparative Study: Some Historical Cases," ibid., pp. 571–95; and Rondo Cameron, "Comparative Economic History," in Paul Uselding, ed., *Research in Economic History*, Supplement 1 (Greenwich, Conn., 1977), pp. 287–305.

controlled experiments, and to provide standards of significance beyond a theoretical or statistical ideal.[12] Comparative history provides standards which are explicit and historical rather than implicit and ideal.[13]

The United States and the Habsburg Empire seem at first too different for a useful comparison. By the eighteenth century the Empire had declined from its dominant position in Europe two centuries earlier, and over the two centuries to follow its status eroded further. America went approximately the other way. The older view of the Empire as one of Europe's sickest economies has given way to a more balanced view, but a picture of sluggish growth and increasing relative backwardness still dominates. The American economy, by contrast, is typically held up as the paragon of success, an engine of growth.[14]

Yet the two cases show many similarities. Compared to other nations, both had large internal markets. In both, a coerced labor force survived well into the nineteenth century. Both experienced debilitating conflict along sectional lines. Above all, uneven development has long been a theme in both American and Habsburg economic history. In the Habsburg case the theme has been central to interpretations of the Empire's political demise, and served indeed to nurture national ideologies in post-1919 East-Central Europe. The same theme appears in American history, but as a relatively minor matter. A comparative study shows that such differential treatment is not justified.

PATTERNS OF UNEVEN DEVELOPMENT

Uneven development was pronounced late in the nineteenth century in both economies, and in both reflected the delayed diffusion of growth. On the eve of World War I regional income disparities were very similar in the two countries. In the Habsburg Empire a large gap distinguished the most advanced, western regions from the less developed areas to the east and southeast (Table 1). Levels of income and industrialization in the Alpine and Bohemian lands were one and a half times those of the southern lands, and twice those of the Carpathian

[12] Donald N. McCloskey, "The Rhetoric of Economics," *Journal of Economic Literature*, 21 (June 1983), pp. 496–99.

[13] Charles Kindleberger, *Economic Response: Comparative Studies in Trade, Finance and Growth* (Cambridge, Mass., 1978), p. 2.

[14] Available data support these contrasting pictures. In 1774 GNP per capita was roughly the same in the American colonies and the United Kingdom. Alice Hanson Jones, *Wealth of a Nation to Be: The American Colonies on the Eve of the Revolution* (New York, 1980), p. 68. By 1913 the level in the United States was 25 percent above that of the United Kingdom. Based on Angus Maddison, *Phases of Capitalist Development* (New York, 1982), Tables 1.4, A2, and B2. For the Empire, GNP per capita was at best 30 percent below the United Kingdom in 1830 and barely 50 percent of the United Kingdom in 1913. Based on the controversial figures in Paul Bairoch, "Europe's Gross National Product: 1800–1975," *Journal of European Economic History*, 5 (Fall 1976), pp. 283, 286.

TABLE 1
REGIONAL INCOME PER CAPITA AND TWO PROXIES: HABSBURG EMPIRE, 1890–1913 (Empire = 100)

Region	Income Per Capita, 1911/13	Literacy Rate 1890	Literacy Rate 1900	Literacy Rate 1910	Savings Bank Deposits Per Capita 1890	Savings Bank Deposits Per Capita 1900	Savings Bank Deposits Per Capita 1910
Austria							
Alpine Lands	130	172	153	136	318	304	284
Bohemian Lands	134	176	155	133	169	171	179
Southern Lands	77	76	86	86	61	67	62
Carpathian Lands	59	42	58	68	24	27	28
Hungary							
Inner Hungary-West	—	124	120	124	120	116	108
Inner Hungary-East	—	86	87	94	48	52	61
Transylvania	—	62	69	79	29	33	38
Croatia-Slavonia	—	61	73	81	31	31	39

Note: For the literacy and deposit variables the relatives for each region were estimated using the unweighted mean of the 8 regional units. For income per capita the relatives were estimated using the unweighted mean of the 4 Austrian regions only. The relatives have been rounded. For 14 provincial units of Austria regressions of these proxies on 1911/13 income per capita yield $r^2 = .77$ for the literacy rate and $r^2 = .51$ for savings bank deposits per capita. David F. Good, *The Economic Rise of The Habsburg Empire, 1750–1914* (Berkeley, 1984), p. 51.
Source: Estimated from official Austrian and Hungarian government statistics. For complete description see Good, *Economic Rise*, Appendix B.

lands. Hungary, too, lagged behind the advancing regions of Austria, and had its own pattern of unevenness—higher levels of development prevailed in western Inner Hungary than in eastern Inner Hungary, Transylvania, and Croatia-Slavonia.[15] In the United States income per capita in the most developed region, the Northeast, was one and a half times that of the North Central and twice that of the South: the lag of the South within America was on par with that of the eastern periphery within the Empire.[16]

Regional disparities were not always so great. For the Empire the scattered evidence on trade, urbanization, and industrial activity re-

[15] Income per capita in Hungary was roughly 70 percent that of Austria on the eve of World War I. Based on income data from Friedrich Fellner, "Das Volkseinkommen Österreichs und Ungarns," *Statistische Monatschrift*, 21 (Vienna, 1917), p. 594 and population statistics from B. R. Mitchell, *European Historical Statistics, 1750–1970* (New York, 1975), Table B1, pp. 19 and 21.

[16] Richard Easterlin, "Regional Income Trends, 1840–1950," in Seymour E. Harris, ed., *American Economic History* (New York, 1961), p. 528. Although poor by American standards, the South was rich compared to the rest of the world. Robert Fogel and Stanley Engerman, *Time on the Cross: The Economics of American Negro Slavery* (Boston, 1974), p. 250. On the eve of World War I its per capita income was perhaps 15–20 percent *above* the richest regions of the Habsburg Empire, the Alpine and Bohemian lands. Based on real GDP estimates for 1913 reported in Maddison, *Phases*, Table A3, converted to a per capita basis with population figures in ibid., Table B2 (for the United States) and in Mitchell, *Historical Statistics*, Table B1, p. 19 (for Austria). My estimate assumes that southern income per capita was 55 percent of the American average and that Alpine and Bohemian land income per capita was 130 percent of the Austrian average. See Table 1 above and Easterlin, "Regional Incomes," p. 528.

TABLE 2
REGIONAL PER CAPITA INCOME AS PERCENT OF UNITED STATES AVERAGE, 1774–1860

Region	1774	1790A	1790B	1840	1860
New England	97	100	105	117	124
Middle Atlantic	107	113	113	121	118
South Atlantic	97	87	82	61	58
United States[a]	100	100	100	100	100

[a] Simple mean of three regions.
Sources: *Regions*: As defined by Richard Easterlin, "Regional Income Trends, 1840–1950," in Seymour E. Harris, ed., *American Economic History* (New York, 1961), p. 526. For 1774, Middle Atlantic consists of Middle Colonies including New York as defined by Alice Hanson Jones, *Wealth of a Nation to Be: The American Colonies on the Eve of the Revolution* (New York, 1980), pp. xxxiv–xxxv. New England consists of the New England colonies, and South Atlantic consists of the Southern colonies as defined by Jones. For 1790A Maryland is included in the South, following Gary M. Walton and James F. Shepherd, *The Economic Rise of Early America* (Cambridge, 1979), p. 196. For 1790B, Maryland is included in Middle Atlantic, following Easterlin.
1774 Income Estimates: Based on Jones, *Wealth of a Nation*, p. 63. Jones converted her per capita wealth estimates to per capita income estimates using capital-output ratios of 3.5 and 3.0. I used the average of these for each region and computed an index number where the simple mean of all three regions = 100.0.
1790 Income Estimate: Based on the per capita real export estimates for each colony (state) for 1768–1772 and 1791–1792 in Walton and Shepherd, *Early America*, p. 196. The change in real income per capita between the two dates was estimated from the change in real exports per capita assuming a multiplier of 2.0. The 1774 level of income per capita was adjusted by this change to get 1790 income per capita. The index numbers are based on the simple mean of the three regions = 100.0. For the United States as a whole my procedure yields a decline in per capita income between 1774 and 1790 of 5.5 percent. This estimate is in accord with the 6.6 percent drop in per capita income between 1774 and 1793 as reported in Claudia Goldin and Frank Lewis, "The Role of Exports in American Economic Growth during the Napoleonic Wars, 1793 to 1807," *Explorations in Economic History*, 17 (Jan. 1980), pp. 22–23.
1840 and 1860 Income Estimates: Uses the estimates of Easterlin, "Regional Income," as adjusted in Stanley Engerman, "Some Economic Factors in Southern Backwardness in the Nineteenth Century," in John F. Kain and John R. Meyer, eds., *Essays in Regional Economics* (Cambridge, Mass., 1971), Table 2. I allocated Engerman's estimates of Northeast income per capita to New England and Middle Atlantic using population weights.

veals a modest gradient stretching from west to east in the late eighteenth century.[17] Habsburg mercantilism indirectly confirms such evidence since it aimed at promoting regional specialization between an industrial west and an agrarian east. Yet compared to 1913 the regional disparities in the eighteenth-century Empire were small. Estimates of American wealth in the eighteenth century suggest virtual equality in regional income per capita (Table 2). The northern colonies were slightly more urban and less agricultural than the southern colonies, but agriculture largely dominated life anyway.

In both cases regional disparities intensified in the nineteenth century. The post-Napoleonic decades mark the emergence of sustained

[17] Summarized in David F. Good, *The Economic Rise of the Habsburg Empire, 1750–1914* (Berkeley, 1984), pp. 14–24.

TABLE 3
AGRICULTURAL LABOR FORCE IN THE HABSBURG ECONOMY BY REGION, 1756–1910

Year	Alpine Lands	Bohemia[b]	Southern Lands[c]	Carpathian Lands[a]	Hungary[b]
1756	—	.780	—	—	—
1771	—	—	.890	—	—
1846	—	.644	—	—	—
1857	—	—	.833	—	—
1869	.542	.486	.815	.871	.750
1880	—	—	.811[b]	—	.740
1890	.486	.406	.758	.838	.725
1900	.427	.356	.733	—	.684
1910	.387	.322	.667	.786	.645

[a] Share of agricultural labor force in economically active population.
[b] Share of agricultural population in total population.
[c] Slovenia

Sources: *Alpine Lands*: Gustav Otruba, "Wachstumsverschiebungen in den Wirtschaftssektoren Österreichs 1869–1961," *Vierteljahresschrift für Sozial-und Wirtschaftsgeschichte*, 62 (1975), Table 8.
Bohemia: František Lom, "Die Arbeitsproduktivität in der Geschichte der tschechoslovakischen Landwirtschaft," *Zeitschrift für Agrargeschichte und Agrarsoziologie*, 19 (1971), Table 5.
Southern Lands: Toussaint Hočevar, *The Structure of the Slovenian Economy 1848–1963* (New York, 1965), p. 86.
Carpathian Lands: Herbert Matis, *Österreichs Wirtschaft: Konjunkturelle Dynamik und gesellschaftlicher Wandel im Zeitalter Franz Josephs I* (Berlin, 1972), p. 389.
Hungary: Scott Eddie, "Agricultural Production and Output per Worker in Hungary, 1870–1913," this JOURNAL, 28 (1968), p. 210.

growth in the Habsburg Empire.[18] An irreversible acceleration in output per capita occurred and the Empire began a slow structural transformation. The growth was centered in the Alpine and Bohemian lands experiencing rapid growth and technological change, in industries such as textiles and sugar refining. Table 3 shows that in the early nineteenth century the share of agriculture in the labor force dropped steadily in these regions. It remained at high, preindustrial levels elsewhere. In America sustained growth also emerged after 1820 when the growth of output per person began to dwarf that of the previous century. The accompanying structural transformation, however, was not uniform. Growth in the North Central and South regions was tied to regional specialization in agriculture. Although all regions experienced some urban and industrial growth prior to 1860, the rate of structural change was substantially higher in the Northeast during the antebellum period.[19]

[18] For a survey of the issues and evidence see Good, *Economic Rise*, chap. 2. For the underlying quantitative research see John Komlos, *The Habsburg Monarchy as a Customs Union* (Princeton, 1983), chap. 3; and Richard Rudolph, "The Pattern of Austrian Industrial Growth from the Eighteenth to the Early Nineteenth Century," *Austrian History Yearbook*, 11 (1975), pp. 3–43.

[19] Based on time series data for urbanization in *Historical Statistics of the United States: Colonial Times to 1970*, Part 1 (Washington, D. C., 1975), Series A172-94, p. 22, and for the agricultural labor force in Simon Kuznets, Ann Miller, and Richard Easterlin, *Population Redistribution and Economic Growth: United States, 1870–1950* (Philadelphia, 1960), vol. 2, Table A2.4, p. 82.

Development diffused with a delay into more backward regions. With increasing intensity after mid-nineteenth century, development diffused eastward in the Empire and by the 1870s sustained growth had emerged in much of the Hungarian lands.[20] The income per capita proxies of Table 1 suggest that the eastward spread may have been strong enough to begin moderating the gap between the Alpine and Bohemian lands and the less developed areas to the southeast and east. The narrowing came earliest for western Inner Hungary, somewhat later for eastern Inner Hungary, last for the Carpathian lands of Austria and for Transylvania and Croatia-Slavonia in Hungary. Other evidence confirms the eastward diffusion. Hungary's large agricultural sector declined in relative terms during the last pre-World War I decades, reaching the level of the Alpine and Bohemian lands roughly a half-century before (see Table 3). The mining and metallurgy sectors received new technology from abroad, and flour milling became Hungary's single most important industry with substantial linkages to the rest of the economy. Between 1885 and 1910 processed foodstuffs, semi-manufactures, and manufactures accounted for a growing percentage of Hungarian exports (Table 4). By 1900 some limited industrialization was apparent even in the far reaches of the Hungarian lands, and in the Carpathian lands of eastern Austria.

In America the lag was much longer. In the two pre-Civil War decades the Easterlin-Engerman income data show the North Central states at roughly half the level of the Northeast, with a narrowing of the gap soon after the Civil War. The same data show that by 1840 the South held an income position relative to the Northeast not much different from that of the North Central region. Some speculation on relative income levels in the pre-1840 South is possible from Table 3. The southern lag most likely developed early, during the restructuring of the post-Revolutionary American economy, and intensified after 1820. The Southern income relative deteriorated sharply after the Civil War and began narrowing only after 1900.

The South was not, of course, devoid of cities and factories.[21] But by 1850 the North Central region was already more urbanized than the South, and by 1920 it was twice so. After 1860 the agricultural sector of the North Central region shrank in relative terms but stayed large in the South. Even in 1910 the share of the southern labor force in agriculture was at the level of the North Central region four decades before.[22] In the

[20] See Iván Berend and György Ránki, *Hungary: A Century of Economic Development* (Newton Abbot, England, and New York, 1974), pp. 40–68; Komlos, *Customs Union*, pp. 112–13, 132–47; Iván Berend and György Ránki, *Economic Development in East-Central Europe in the 19th Centuries* (New York, 1974), pp. 122–29; and John Lampe and Marvin Jackson, *Balkan Economic History, 1550–1950* (Bloomington, Ind., 1982), pp. 287–94, 310–22.

[21] For details see Fred Bateman and Thomas Weiss, *A Deplorable Scarcity: The Failure of Industrialization in the Slave Economy* (Chapel Hill, N.C., 1981), pp. 7–14.

[22] See sources in footnote 19. The South's industrial development appears retarded compared to the North, but not compared to much of Europe. Fogel and Engerman, *Time on the Cross*, p. 256.

TABLE 4
DISTRIBUTION OF EXPORTS BY COMMODITY CLASS FOR AUSTRIA AND HUNGARY, 1885 AND 1910

		Exports (percent)			
		Abroad		Interregional	
Region	Class	1885	1910	1885	1910
Austria	All agricultural commodities	37.7	29.0	9.5	7.8
	Processed foodstuffs	4.1	1.9	6.1	4.5
	Raw materials, fuels	12.8	19.0	4.0	5.9
	Semi-manufactures and manufactures	45.4	50.2	80.5	81.9
Hungary	All agricultural commodities	47.4	38.3	64.5	60.4
	Processed foodstuffs	19.6	20.6	14.1	19.5
	Raw materials, fuels	14.8	14.6	5.5	3.9
	Semi-manufactures and manufactures	18.1	26.9	15.9	16.3

Sources: For each region and year the columns add to 100.0 percent except for rounding errors.
Hungary: Exports abroad and interregional exports, that is, Hungarian exports to Austria, are from Scott Eddie. "The Terms and Patterns of Hungarian Foreign Trade, 1882–1913," this JOURNAL, 37 (June 1977), Table 7, p. 349.
Austria: Exports abroad from Scott Eddie, "Austria in the Dual Monarchy: Her Trade Within and Without the Customs Union," *East Central Europe*, 7 (1980), Table 2 plus official government statistics supplied to me by Scott Eddie. Interregional exports, that is, Austrian exports to Hungary, available as Hungarian imports from Austria in Eddie, "Terms and Patterns."

plantation states the lag was even greater: roughly two-thirds of the labor force was employed in agriculture in 1910.[23] Hutchinson's trade data (Table 5) exhibit the structural lag of the South. Already by 1870 a sizable share of North Central exports (interregional and foreign) was in processed foodstuffs, semi-manufactures, and manufactures, and these shares rose further to 1910. Only at the turn of the century was there any significant increase in the share of such industrial categories in southern trade.

THE ROLE OF ECONOMIC INTEGRATION

In both America and Austria-Hungary economic integration began well before 1800 but was far from complete on the eve of World War I. By the eighteenth century legal and natural barriers to trade were eroding. Yet barriers were still high. Producers and resource owners were not strongly tied into international and interregional markets. Major military events jolted both economies toward closer integration. In the Empire during the 1740s a major policy shift occurred after the Habsburg defeat by Prussia and the loss of an economically advanced Silesia. Implementing mercantilist principles, the government promoted

[23] Jay R. Mandle, *The Roots of Black Poverty* (Durham, N.C., 1978), p. 59.

TABLE 5
DISTRIBUTION OF EXPORTS BY COMMODITY CLASS FOR REGIONS OF THE UNITED STATES, 1870–1910

		Exports (percent)			
		Abroad		Interregional	
Region	Class	1870	1910	1870	1910
Northeast	Crude food	—	—	—	—
	Processed food	30.4	0.8	25.0	13.2
	Crude materials	8.7	17.3	14.0	7.8
	Semi-manufactures	18.9	33.4	6.6	2.3
	Manufactures	42.0	46.5	54.4	76.7
South	Crude food	1.0	—	9.4	—
	Processed food	—	9.0	0.5	30.8
	Crude materials	98.5	85.6	90.0	50.9
	Semi-manufactures	—	4.1	—	10.0
	Manufactures	1.1	1.2	0.1	8.3
North Central	Crude food	42.4	26.5	67.9	32.6
	Processed food	38.5	55.0	29.1	55.5
	Crude materials	5.0	2.2	0.5	2.8
	Semi-manufactures	1.6	3.1	—	—
	Manufactures	12.6	13.3	2.5	9.1
West	Crude food	69.4	12.0	65.2	32.5
	Processed food	23.9	13.4	34.8	0.4
	Crude materials	—	3.8	—	36.6
	Semi-manufactures	4.2	12.2	—	28.5
	Manufactures	2.5	58.5	—	2.0

Sources: For each region and year the columns add to 100.0 percent except for rounding errors.
Exports Abroad: From William Hutchinson, "Regional Exports of the United States to Foreign Countries: A Structural Analysis, 1870–1910," in Paul Uselding, ed., *Research in Ecomonic History*, 10 (Greenwich, Conn., forthcoming), Table 2.
Interregional Exports: Uses residual values of per capita output for individual commodities *not* exported abroad or consumed within the region as estimated by William Hutchinson, "Regional Exports to Foreign Countries: United States, 1870–1914," in Paul Uselding, ed., *Research in Economic History*, 7 (Greenwich, Conn., 1982), Tables 4 and B.1.

unity with river and road improvements and attacks on the fiscal privileges of the nobility and local government bodies. In America the key event was the Constitution, creating in another way the institutional prerequisites for a full-fledged monetary and customs union.

The transportation revolution accelerated the process of economic integration. In the Empire the railroad gradually linked the eastern lands to the more advanced Alpine and Bohemian lands. An increasingly integrated economy emerged, with Vienna as its focal point and Prague and Budapest as secondary. Capital flowed from the surplus western regions to the deficit eastern regions.[24] Waves of peasants migrated from the east, pulled by higher wage levels and pushed by the widely

[24] David F. Good, "Financial Integration in Late Nineteenth-Century Austria," this JOURNAL, 37 (Dec. 1977), pp. 890–910; and David F. Good, "National Bias in the Austrian Capital Market before World War I," *Explorations in Economic History*, 14 (April 1977), pp. 141–66.

TABLE 6
TRADE RATIOS FOR FOUR REGIONS OF THE UNITED STATES, 1870–1910

	Ratio of Exports to Regional Output			
	Exports Abroad		Interregional Exports	
Region	1870	1910	1870	1910
Northeast	.013	.024	.053	.108
South	.175	.109	.106	.108
North Central	.022	.034	.145	.205
West	.112	.032	.153	.160

Sources: Ratios were estimated from data in Hutchinson, "Exports to Foreign Countries," as follows: *Regional Exports Abroad*: As indicated in Hutchinson, Table 1, p. 145.
Interregional Exports: Estimated using per capita interregional exports in Hutchinson, Table 4, pp. 158–60, and regional population figures in Hutchinson, Table B.3, p. 193.
Regional Output: Uses Gross Regional Product estimates of Hutchinson as reported in Table B.6, p. 200, converting these from nominal to real terms using the Federal Reserve Consumer Price Index for the relevant years.

discussed "land hunger."[25] The Empire's low export-to-GNP ratio reflected a large internal market built on specialization and trade between the industrial areas of Austria and the agricultural areas of Hungary (Table 6).[26] In the United States, too, a national capital market evolved to channel funds from the surplus east to the deficit west.[27] Waves of settlers moved westward, attracted by abundant land and higher real wages.[28] As in the Empire, America's low export-to-GNP ratio reflected a large internal market based on growing specialization and trade between the industrial Northeast and agricultural West and North Central regions (see Table 4).

The spatial spread of development was closely related to the drawn-out character of integration. In both cases regions lagged behind when

[25] Michael Mesch, *Arbeiterexistenz in der Spätgründerzeit-Gewerkschaften und Lohnentwicklung in Österreich, 1890–1914* (Vienna, 1984), pp. 195–228; Scott Eddie, "The Changing Pattern of Landownership in Hungary, 1867–1914," *Economic History Review*, 2nd ser., 20 (Aug. 1967), pp. 297–98, 306–9.

[26] For 1913 the following ratios of exports to GNP can be computed: United States (.043); Habsburg Empire (.108); Germany (.193); and United Kingdom (.193). For the United States exports are from Hutchinson, "Exports to Foreign Countries," Table 1, p. 145, and GNP is the sum of Gross Regional Product estimates for U.S. regions in Hutchinson, "Exports to Foreign Countries," Table B.6, p. 200. For the Habsburg Empire exports are from official government statistics as compiled by Scott Eddie and transmitted privately; GNP for Hungary is Fellner's estimate reported in Komlos, *Customs Union*, pp. 210–11; GNP for Austria from Anton Kausel, "Österreichs Volkseinkommen 1830 bis 1913," in *Geschichte und Ergebnisse der Zentralen amtlichen Statistik in Österreich* (Vienna, 1979), p. 718. GNP for the Empire equals Austrian GNP plus Hungarian GNP. For Germany and the United Kingdom exports are from B. R. Mitchell, *European Historical Statistics, 1750–1970* (New York, 1975), Table K1, pp. 783–90.

[27] Lance Davis, "The Investment Market, 1870–1914: The Evolution of a National Market," this JOURNAL, 25 (Sept. 1965), pp. 355–93; and John James, "The Development of the National Money Market," this JOURNAL, 36 (Dec. 1976), pp. 878–97.

[28] Lowell E. Gallaway and Richard K. Vedder, "Mobility of Native Americans," this JOURNAL, 31 (Sept. 1971), pp. 613–49; and Philip R. Coelho and James F. Shepherd, "Regional Differences in Real Wages: The United States, 1851–1880," *Explorations in Economic History*, 13 (April 1976), pp. 203–30.

they were poorly integrated into the larger national economy. In the Habsburg Empire the delayed spread of development reflected the lengthy, sequential integration of the eastern lands into the western lands.[29] By World War I many of the most backward Habsburg territories in the east and southeast were still poorly integrated into the larger economy. Although densely settled, they resembled the parts of the American West which had not been touched by the railroad.

No lagging region of the Empire stood apart from integration in the same sense as did the American South. In the late nineteenth century regional interest rates converged throughout America, but much more slowly in the South than in the North Central and Western regions.[30] Up and down the Atlantic seaboard Americans moved west, but they did so in highly stratified fashion.[31] Compared to other American regions the South was more dependent on foreign than interregional trade (Table 6), and a single commodity class, of course, dominated its exports (see Table 4). Unlike the Northeast it remained virtually self-sufficient in foodstuffs, which sharply limited its potential interregional links with the North Central region.[32] Since a large percentage of output was produced and consumed on plantations, the market links within the South were relatively weak, too.[33] In the Empire regional differentials persisted in peripheral areas, but began to close once these areas were attached to the Empire-wide capital market. Migration was not stratified: Vienna was a hub for migration waves from the east and southeast. Tables 4, 5, 6, and 7 show that Hungary's role in regional specialization and trade was more like that of the North Central and Western regions than the South. Like them, Hungary relied more heavily on interregional than foreign trade and exported a diversified bundle of agricultural commodities. The western regions of Austria, on the other hand, resembled the American South in export orientation. But Austria's exports were highly diversified and heavily slanted towards processed foodstuffs, semi-manufactures, and manufactures.

THE ROLE OF INSTITUTIONS

A mix of capitalistic and noncapitalistic forms provided the framework for growth in the United States and the Empire. Clear-cut regional

[29] See Good, *Economic Rise*, pp. 104–8, 116–19, 121–22.

[30] Davis, "Investment Market," pp. 388–92; and James, "National Money Market," p. 897.

[31] Richard Steckel, "The Economic Foundations of East-West Migration during the 19th Century," *Explorations in Economic History*, 20 (Jan. 1983), pp. 14–19.

[32] Albert Fishlow, "Antebellum Interregional Trade Reconsidered," in Ralph Andreano, ed., *New Views of American Economic Development* (Cambridge, Mass., 1965), pp. 187–200; William K. Hutchinson and Samuel Williamson, "The Self-Sufficiency of the Antebellum South: Estimates of the Food Supply," this JOURNAL, 31 (Sept. 1971), pp. 591–612; and Diane Linastrom, "Southern Dependence upon Grain Supplies: A View of the Trade Flows, 1840–1860," in William Parker, ed., *The Structure of the Cotton Economy of the Antebellum South* (Washington, D. C., 1970), pp. 101–13.

[33] Bateman and Weiss, *Deplorable Scarcity*, p. 97.

TABLE 7
TRADE RATIOS FOR AUSTRIA AND HUNGARY, 1882 AND 1913

	Ratio of Exports to Regional Output			
	Exports Abroad		Interregional Exports[b]	
	1880[a]	1913	1880[a]	1913
Austria	.176	.123	.100	.081
Hungary	.111	.069	.256	.214

[a] 1882 data for exports.
[b] Austria's exports to Hungary and Hungary's exports to Austria.
Sources: *Exports*: From official government statistics as compiled by Scott Eddie and transmitted privately. Available in the official data are the Empire's exports abroad, Hungary's exports abroad, and Hungary's imports from Austria. Austria's exports abroad are estimated by Eddie as the difference between the Empire's exports abroad and Hungary's exports abroad. Austria's exports to Hungary are available as Hungary's imports from Austria. According to Eddie the data include a considerable amount of throughput that shows up as Hungarian exports to Austria, and again as Austrian exports abroad. This implies that the Hungarian ratio for trade abroad is understated and for interregional trade is overstated. For the same reason the Austrian ratio for trade abroad is overstated.
GNP: Austrian data from Kausel, *Volkseinkommen*, p. 718; Hungarian data based on Komlos, *Customs Union*, pp. 210–11. Komlos cites Fellner's estimate for 1913 and makes a "back-of-the-envelope" estimate for 1870. These estimates of nominal GNP were converted to real terms with the price index in Vera Mühlpeck, Roman Sandgruber, and Hannelore Woitek, "Index der Verbraucherpreise 1800–1914," in *Geschichte und Ergebnisse*. The implied growth rate from 1870 to 1913 was used to estimate real GNP for 1880 which was converted back to nominal GNP with the price index of Mühlpeck et al.

differences in economic organization are apparent in the Habsburg case.[34] By the mid-eighteenth century serfdom was already weakening in the western regions. In the mountains of Alpine Austria and the hills of northern Bohemia the large-estate agriculture had died out. Peasants supplemented their income by wage labor and part-time employment in the domestic system. Large noble estates were often significant centers of industrial activity, especially in Bohemia, and city-based industries sprouted up in the more urbanized west. Serfdom remained solidly entrenched in Galicia and Hungary. Agricultural reforms improved the status of serfs and facilitated the long-term conversion of agriculture to capitalism, but were applied less vigorously in the Hungarian lands. The institutional differences between west and east persisted. In the nineteenth century many nobles voluntarily commuted labor services into money payments and agitated for far-reaching reforms of the landlord-

[34] The following relies on Richard Rudolph, "Social Structure and Austrian Economic Growth," *East Central Europe*, 7 (1980), pp. 212–20; Herman Freudenberger, "Progressive Bohemian and Moravian Aristocrats," in Stanley Winters and Joseph Held, eds., *Intellectual and Social Developments in the Habsburg Empire from Maria Theresa to World War I* (Boulder, Colo., 1975), pp. 115–30; Roman Rozdolski, *Die Grosse Steuer-und Agrarreform Josephs II* (Warsaw, 1961), chap. 2; Jerome Blum, *Noble Landowners and Agriculture in Austria 1815–1848: A Study in the Origins of the Peasant Emancipation of 1848* (Baltimore, 1948), pp. 87, 203–21; Berend and Ránki *East-Central Europe*, pp. 32–33; Eddie, "Landownership in Hungary," pp. 296–300; and Herbert Matis, *Österreichs Wirtschaft 1848–1913: Konjunkturelle Dynamik und gesellschaftlicher Wandel im Zeitalter Franz Josephs I* (Berlin, 1972), p. 145.

peasant tie. The mid-century Emancipation abolished all remaining vestiges of serfdom, but did not undermine the economic and political power of the aristocracy, especially in Hungary. The aristocracy's post-1848 power base was secured in economic terms by the strengthening of large-estate agriculture and in political terms by the Dual Settlement between Austria and Hungary. The transformation to capitalism remained incomplete throughout the Empire, but more so in the eastern than in the western lands.

In contrast, institutional forms in America were much more sharply polarized along regional lines, especially after the Revolutionary War. In the North, capitalist institutions spread westward as the North Central and Northeast regions became integrated. In the South the slave system expanded inexorably. The Civil War brought no end to institutional polarization between North and South. In face of labor shortages many former slaveholders subdivided plantations and turned to sharecropping and tenancy to recruit a labor force. Country store merchants dominated the postbellum capital market. Capitalistic agriculture built on wage labor or on the market-oriented family farm failed to emerge in the South.

In both cases the regional pattern of institutional change determined the spatial spread of economic development. Regions lagged so long as they were burdened with growth-inhibiting forms of economic organization. In the Empire modern economic growth emerged first in the west where serfdom was already decaying and gradually spread eastward in response to its continued erosion. Some studies have challenged the thesis that growth required the dissolution of precapitalist values and institutions.[35] Industrialization seemed to emerge early as an integral part of serfdom, not something alien to it, and Emancipation was not associated with any observable surge in output.[36] Late in the century Hungarian agriculture prospered under the guidance of a market-oriented aristocracy, and the aristocracy-dominated Hungarian government used its newly won autonomy to promote economic development.[37]

These findings are an antidote to the widely held but erroneous view that the Empire's institutional rigidity precluded economic development along Western European lines. Yet it is hard to imagine the emergence and diffusion of modern economic growth in the Empire *within the*

[35] For a particularly strong version of the feudal institutions-as-burden thesis see Oscar Jaszi, *The Dissolution of the Habsburg Monarchy* (Chicago, 1961), pp. 222–39.

[36] Rudolph, "Social Structure," p. 224; and Komlos, *Customs Union*, pp. 236–37.

[37] Scott Eddie, "Farmer's Response to Price in Large-Estate Agriculture: Hungary and Germany, 1870–1913," *Economic History Review*, 2nd ser., 24 (Nov. 1971), pp. 585–86; and his "Agricultural Production and Output per Worker in Hungary, 1870–1913," this JOURNAL, 28 (June 1968), pp. 215–17; Alexander Eckstein, "National Income and Capital Formation in Hungary, 1900–1950," in Paul Horecky, ed., *International Association for Research in Income and Wealth*, Series 5 (London, 1955), p. 175; Berend and Ránki, *Hungary*, p. 55; and Komlos, *Customs Union*, p. 207.

institutional framework existing before 1750. In this sense the standard arguments about the negative consequences of Habsburg serfdom retain some force.[38] Serfdom was inefficient because serfs tended to shirk and use inferior capital during forced labor; it immobilized labor and retarded the growth of a consumer market; its incentive structure discouraged capital formation and technological change; and its value system sharply retarded the development of the entrepreneurial spirit. Of course, modern economic growth emerged in the Empire before serfdom was abolished—serfdom was not incompatible with growth. But growth in the pre-1848 period is really testimony to how much serfdom had eroded over the previous century.[39] The marginal impact of the formal Emancipation was perhaps small, but the likely gains from serfdom's decline between 1750 and 1847 were undoubtedly large, and the eastward diffusion of development in the Empire depended on serfdom's continuing decay.

In America institutional forms provided more fertile ground for the emergence and spread of modern economic growth in the North than in the South. Early cliometric research, however, challenged this view. The level and growth of southern per capita income proved to have been surprisingly high before 1840; investment in slaves was profitable; slave agriculture was more efficient than northern farming; and slaves possessed a strong work ethic.[40] Yet slavery was not good for growth. Southern planters seemed to spurn profitable investment opportunities outside slavery; their plantations isolated them from outside forces; and compared to northern farmers they conducted operations on a larger scale but with lower capital-to-labor ratios and weaker incentives for innovation.[41] The impact of postbellum institutions on southern development has of course been a hotly contested issue, and southern agriculture did not completely stagnate after the Civil War. Yet many observers agree that tenancy, sharecropping, and debt peonage limited

[38] See Blum, *Noble Landowners*, pp. 192–94; Berend and Ránki, *East-Central Europe*, pp. 28–30; Karl Dinklage, "Die landwirtschaftliche Entwicklung," in Adam Wandruszka and Peter Urbanitsch, eds., *Die Habsburgermonarchie, 1848–1918* (Vienna, 1973), pp. 410–17; Arnošt Klíma, "Industrial Growth and Entrepreneurship in the Early Stages of Industrialization in the Czech Lands," *Journal of European Economic History*, 6 (Winter 1977), pp. 564, 573–74; Eduard März, "Comments," *Austrian History Yearbook*, 11 (1975), p. 31; Matis, *Österreichs Wirtschaft*, p. 44; and Ferdinand Tremel, *Wirtschafts-und Sozialgeschichte Österreichs. Von den Anfängen bis 1955* (Vienna, 1969), pp. 341–46.

[39] On the eve of the 1848 Emancipation forced labor in the form of *robot* was only a small portion of total labor applied in the agricultural sector—about 9.0 percent in Austria and 4.5 percent in Hungary. Komlos, *Customs Union*, p. 236.

[40] Fogel and Engerman, *Time on the Cross*, pp. 4–5.

[41] Douglass North, *The Economic Growth of the United States, 1790–1860* (New York, 1966), p. 52; Bateman and Weiss, *Deplorable Scarcity*, pp. 113–21; Heywood Fleisig, "Slavery, the Supply of Agricultural Labor, and the Industrialization of the South," this JOURNAL, 36 (Sept. 1976), pp. 572–97; and Robert Gallman, "Slavery and Southern Economic Growth," *Southern Economic Journal*, 45 (April 1979), pp. 1007–22.

flexibility in the southern economy by immobilizing resources and discouraging technological change.[42] Taken as a whole the evidence supports Wright's view that nineteenth-century prosperity depended less on the South's internal dynamic and more on external conditions governing the demand for cotton.[43]

CONCLUSION

These cases support classical Marxists and present-day non-Marxists who stress the critical role of local institutions in determining the spatial spread of economic development. In the Habsburg Empire growth emerged early in the west where serfdom was already weakening and spread eastward with a lag as serfdom continued its long-term decay. In the American North, capitalist institutions nurtured the emergence and diffusion of growth more thoroughly than did the growth-inhibiting institutions of the South.

The external linkages of lagging regions played an important but secondary role in the two cases. Here the evidence supports the classical rather than the neo-Marxist interpretation. In the Empire the weak links to the rest of the Habsburg economic union promoted increasing relative backwardness in the eastern hinterland into the twentieth century. In America the weak market links within the South, and the South's relative isolation from the rest of the economy, kept the South backward. If institutions had not been rigid and if product and factor markets had been perfectly integrated, economic development would have spread instantaneously. In the real world, such frictionless states do not exist. The American and Austrian cases, so disparate, demonstate how powerful such rigidities and imperfections can be. The spatial spread of development required most of the nineteenth century, and remained incomplete at its end.

[42] William Parker, "The South in the National Economy, 1865–1970," *Southern Economic Journal*, 46 (April 1980), pp. 1038–39, 1045–46; Mandle, *Black Poverty*, chap. 5; Roger Ransom and Richard Sutch, *One Kind of Freedom: The Economic Consequences of Emancipation* (New York, 1977); Warren Whatley, "Institutional Change and Mechanization in the Cotton South" (Ph.D. diss., Stanford University, 1983); and the recent textbook treatment in Jonathan Hughes, *American Economic History* (Glenview, Ill., 1983) pp. 272–77. For a more positive assessment of postbellum Southern institutions see Joseph Reid, Jr., "White Land, Black Labor, and Agricultural Stagnation," *Explorations in Economic History*, 16 (Jan. 1979), pp. 31–55; and Robert Higgs, *Competition and Coercion: Blacks in the American Economy, 1865–1914* (New York, 1977), chap. 6.

[43] Gavin Wright, "Slavery and the Cotton Boom," *Explorations in Economic History*, 12 (Oct. 1975), p. 450.

Economic Growth and Industrialization in Hungary 1830-1913

John Komlos *
Aurora University

Throughout the nineteenth century and into the twentieth, Hungary was predominantly an agricultural country. In the middle of the century manufacturing contributed 7-8 percent to her national product,[1] and by 1913 this share was close to 25 percent.[2] This was far from being a spectacular achievement. Nonetheless during most of the three and a half decades prior to World War I, her industrial sector was developing rapidly. In 1841 Hungary's manufacturing equalled 16 percent [3] of Austria's output. By World War I it was 30 percent.[4] Although in many respects Hungary remained an underdeveloped country, relative to her

---

* My gratitude is extended to the International Research and Exchanges Board for subsidizing the research for this project and to Professors Donald McCloskey, Arcadius Kahan, Emile Karafiol, William McNeill, Bert Hoselitz and Ira Glazier for commenting on an earlier version of this paper and editorial assistance.

1 Without mining. LÁSZLÓ KATUS, " A Kelet-europai iparosodás és az önalló tökés fejlödés kerdéséhez," *Történelmi Szemle*, 10, no. 1 (1967): 1-43, p. 9.

2 FRIEDRICH FELLNER, *Das Volkseinkommen Österreichs und Ungarns* (offprint from the *Statistischen Minatschrift*, Vol. 21, September-October, 1917), p. 123.

3 NACHUM GROSS, " Industrialization in Austria in the Nineteenth Century " (Unpublished Ph. D. dissertation, University of California, Berkeley, 1966), p. 107.

4 FELLNER, *Das Volkseinkommen*, p. 133.

custom union partner her growth was creditable; she had ameliorated to a considerable extent her relative backwardness.

The first widely diffused spurt of industrial activity in Hungary took place in the late 1870s. Perhaps the most important finding of this paper is that the inflow of Austrian capital was instrumental in the beginning and in sustaining industrialization in Hungary. The Hungarian government played an important role in attracting Austrian funds by open market operations. At the same time the outflow of capital from Austria contributed to, and prolonged, the effects of the post-1873 depression there. The conclusions emerge that the Hungarian economy would have functioned less effectively without unencumbered access to the Austrian market, and that the economic ties that bound Austria to Hungary throughout the nineteenth century were less beneficial, and in the case of capital exports, even detrimental, to the Austrian half of the union.

Due to the small size of the manufacturing sector in Hungary before the 1870s, the accurate measurement of its rate of growth eludes us at this time. However, scattered evidence of modernization after 1825 does exist. For instance, the growth of Hungary's iron production (Table 1) and export of manufactures in the 1840s were positive signs of development. Yet these sectors cannot be assumed to have been indicative of the overall behavior of

TABLE 1

ANNUAL GROWTH RATES OF THE HUNGARIAN IRON INDUSTRY 1830-1913

Years	Percent
1830-1847	4.5
1947-1864	8.6
1864-1873	3.6
1873-1886	3.5
1886-1898	5.9
1898-1907	1.5
1907-1913	8.1

Source: Table 19.

industry or the economy. Mining, perhaps the most significant non-agricultural endeavor during the first half of the century, was growing slowly, at a rate of 2.1 percent per annum (between 1831 and 1847), and much of this growth took place in the late 1840s. The cotton, textile and sugar industries were too small to be measured meaningfully. Flour production, also part of the industrial output index (Table 18), was extrapolated linearly between 1841 and 1867 at the same rate as wheat and rye consumption in Hungary. Hence, ferrous metals is the only sizeable industry with reliable annual statistics in the index in this period. From this industry alone, one ought not draw definitive conclusions about the overall behavior of the industrial sector.

In the 1830s and 1840s the growth of the ferrous metals branch exceeded 4 percent per annum. Production was exclusively for domestic consumption until 1837. Thereafter trade developed with Galicia and Moravia in pig iron due largely to the demand for rails manufactured in Moravia. The increased export demand had a negligible impact on iron production until the elimination of the 10 percent export tariff to Austria in 1843. Whereas the net output of the iron industry was growing at 4.0 per cent per annum between 1830 and 1843, in the four years after the lifting of this tariff it averaged 6.3 per cent. Under the impact of the spread of railroads in the 1850s iron production expanded enormously. (See Figure 1.)

Despite the well known predominance of agricultural exports in Hungary's foreign trade, Hungary was exporting manufactured goods not only to the Austrian provinces but to a limited extent outside the monarchy as well, These exports increased by a third between 1841 and 1847, from 2.4 to 3.2 million gulden. This is another indication that manufacturing was perhaps not stagnant.

In spite of the growth of manufactured exports, and of the ferrous metal industry, one cannot formulate a thesis with respect to the development of manufacturing in Hungary prior to 1850

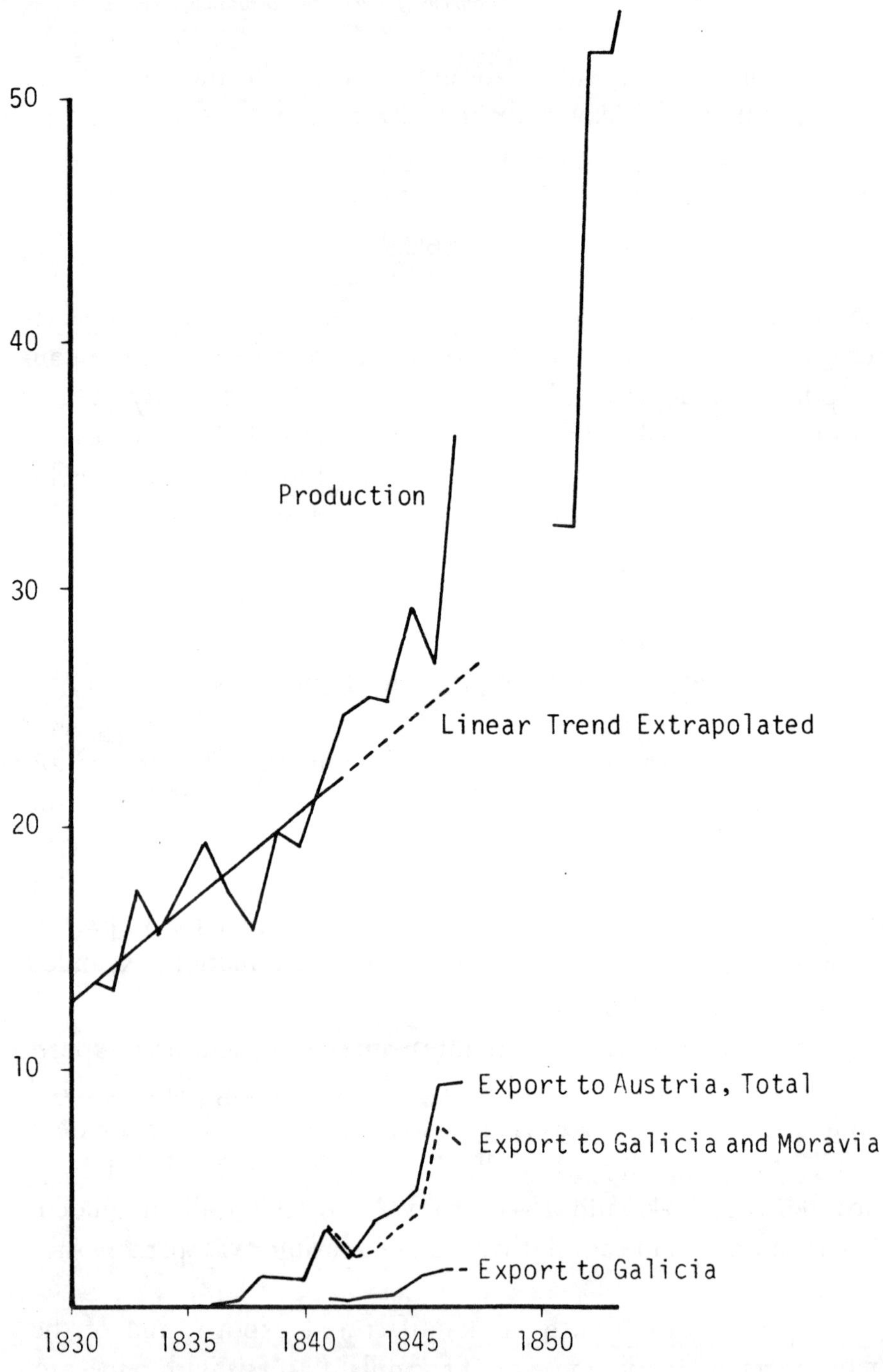

Figure 1 - Hungarian Production and Exports of Pig Iron, 1830 - 1854 (thousands of tons).

SOURCE: *Ausweise über den Handel von Österreich.*

except to intimate that the 1.3 per cent annual growth rate in manufacturing between 1830 and 1847, shown in Table 2, appears reasonable. In the 1850s, manufacturing accelerated somewhat, possibly as a response to railroad construction. Data from the iron sector do point in this direction (growth was 8.6 percent per annum), but the poor performance of mining, and the absence of Hungarian trade statistics in the 1850s hampers overall generalization.

One can be less equivocal about the behavior of manufacturing during the 1860s and early 1870s as the several indicators (iron, sugar, cotton textiles, flour) begin to move more closely together, and because the flour index becomes more precise. The worldwide upswing in economic activity of the early 1870s does not appear to have had a significant impact on Hungarian manufacturing. Though there was increased industrial activity compared to the slump of 1865 and 1866, production did not exceed greatly the previous peak of the early 1860s. Expansion of the iron industry even decelerated. Flour output, the most important manufactured good, could not have increased significantly because large amounts of wheat were exported from Hungary. (The average export of wheat in 1867 and 1868 was not only at an all time high, but was, in fact never to be surpassed.) The failure of both the flour and the ferrous metal industries to increase production substantially during this period renders the rapid advance of manufacturing in the late 1860s-early 1870s improbable.

The upswing and boom between 1867 and 1873 was nonetheless of considerable intensity, but was confined to expanding the real and financial infrastructure.[5] Trasportation and banking grew rapidly. Railroad mileage tripled,[6] and the number of financial institutions increased by sixfold within the six years after

---

[5] Alexander Matlekovits, *Die Zollpolitik der Österreichisch-Ungarischen Monarchie und des Deutschen Reiches seit 1868 und deren nächste Zukunft* (Leipzig: Duncker and Humblot, 1891), p. 9.

[6] From 2,160 to 6,253 kilometers, *Ungarisches Statistisches Jahrbuch*, 1912, p. 245.

TABLE 2

GROWTH OF HUNGARIAN INDUSTRY, 1830-1913

Business Cycle	Manufacturing	Manufacturing Mining	Manufacturing Mining Construction
1) 1830-1847	1.30	1.93	2.38
2) 1847-1864	2.18	1.72	1.34
3) 1864-1874	1.37	1.38	2.16 [b]
4) 1874-1883	5.47	5.10	6.07 [b]
5) 1883-1896	3.45	3.69	3.77
6) 1896-1906	2.24	2.41	1.78
7) 1906-1912 [a]	5.95	5.33	4.74
1867-1913 [a]	3.18	3.16	3.18

[a] includes woolens textiles.
[b] The peaks of the construction cycles are different in this period. Because of lumpiness of the construction series the growth rates differ substantially if measured from the alternative peaks: 1861-1873: 3.28; 1873-1883: 3.42, and perhaps reflect better the intensity of railroad construction.

Source: Table 19.

1867.[7] Large sums were expended by municipalities on enlarging and renovating urban centers.[8] Although the number of corporations increased rapidly most of them were in banking and transportation,[9] and those manufacturing enterprises which were incorporated often represented merely legal changes.[10] The absence of contemporary reports pointing to significant advances in manufacturing during this upswing is a further substantiation, though indirect, that the advances were confined to enlarging the extent of social overhead capital.

Thereafter the Hungarian manufacturing index becomes less impressionistic since flour production can be estimated anually.

[7] From 107 to 637 institutions, " A Magyar Szent Korona Országainak Hitelintézetei az 1894-1909. Években " *Magyar Statisztikai Közlemények*, Uj Sorozat, 35, 73.

[8] *Ungarns Handel und Industry, 1870-1872* (Budapest: Kereskedelmi és Iparkamara, 1873), p. 13.

[9] Of the 1965 incorporations in 1872 only 15 were manufacturing enterprises. *Ibid*, p. 17.

[10] MATLEKOVITS, *Die Zollpolitik*, p. 9. The Ganz works is but one example.

Economic Growth and Industrialization in Hungary

TABLE 3

GROWTH OF AUSTRIAN INDUSTRY, 1830-1913

Business Cycle	Manufacturing	Manufacturing Mining	Manufacturing Mining Construction
1) 1830-1846	2.64		3.03
2) 1846-1861	2.62	3.08	3.09
3) 1861-1871	3.25	3.88	4.69
4) 1871-1884	1.84	2.42	2.12
5) 1884-1898	2.25	2.80	2.63
6) 1898-1907	3.89	3.77	3.38
7) 1907-1912	1.74	1.57	1.58
1867-1913	2.79	3.21	3.13

Source: Table 19.

During the next two business cycles, 1879-1883 and 1883-1896, Hungarian manufacturing production grew at a rapid pace. That is, while Austria and much of Western Europe were in the throes of the great depression, Hungary was undergoing its first widespread wave of industrialization. Value-added in industrial production is estimated to have grown at a rate of 5.5 per cent annually between 1883 and at a rate of 3.5 per cent between 1883-1896.[11] At the same time, Austria's industrial growth rate was extremely slow: in the neighborhood of 2 per cent. After 1896, the growth of industrial production in Hungary and Austria continued to be out of phase with one another. Hungary's industry slowed to a rate just over 2 percent per annum between 1896 and 1906 while Austria's accelerated to close to 4 per cent per annum. The-

[11] According to *Gyáripara*, Vol. 14, p. 85, " Our whole economic life, including agriculture, entered the road of intensive development only a few decades ago." Though the language of this assertion is quite imprecise it does nonetheless belittle the importance of the upswing which began in 1887 and which has been characterized as Hungary's " great spurt " by Katus in László KATUS, " Economic Growth in Hungary During the Age of Dualism 1867-1913. A Quantitative Analysis," in E. PEMLÉNYI (ed.), *Social-Economic Researches on the History of East-Central Europe* (Budapest: Akadèmiai Kiadó, 1970), p. 56.

John Komlos

reafter Hungary's industry grew rapidly once again, reaching almost 6 per cent per annum, while Austria's decelerated to below 2 per cent (Table 4). The mechanism for this pattern is to be sought in the flow of capital between Austria and Hungary. Before discussion of that aspect of the relationship, however, the significance of the late 1870s in Hungarian economic development should be explored further.

Since Hungary was predominantly an agricultural country, her greatest comparative advantage lay in food processing. This was her largest manufacturing sector until World War I. The sector expanded enormously at the end of the 1870s. The production of spirits and beer (not included in the index) almost doubled in the early 1880s. Sugar production increased and flour milling was booming. This is not likely to be a statistical quirk: annual data reported by the mills situated in Budapest beginning in 1866 show a pronounced upsurge in flour output at the end of the 1870s. Alexander Matlekovits, a respected authority on the Hungarian economy, intimated that the flour industry had acquired worldwide significance by the late 1870s.[12]

Prior to the late 1870s, there was insufficient growth in grain production to allow a substantial development of flour production and little grain was imported into Hungary. Domestic consumption of wheat and rye grew at a rate of 1-½ per cent per annum and flour production could not have exceeded this rate in spite of the the 12 per cent annual growth of flour exports between the 1830s and 1870s.

The rapid growth of flour production in the late 1870s was based on the comparative advantage provided by the good quality of Hungarian wheat and its low relative price. In the mid-1870s, demand for flour in Austria and Germany increased sufficiently so that the exclusive reliance on domestic supplies of raw materials in Hungary for producing flour had become a constraining

---

[12] MATLEKOVITS, *Die Zollpolitik*, p. 531.

Economic Growth and Industrialization in Hungary

TABLE 4

HUNGARIAN AND AUSTRIAN MANUFACTURING GROWTH RATES COMPARED[a]

Business Cycle	Hungarian	Austrian
3) 1864–74	1.37	3.25
4) 1874–83	5.47	1.84
5) 1883–96	3.45	2.25
6) 1896–06	2.24	3.89
7) 1906–12	5.95	1.74

[a] The growth rates over a business cycle are measured from peak to peak. Since the peaks in Hungary may not correspond precisely to the peaks in Austria the periods compared are not exactly identical. The growth during the business cycle are the rates compared in this table.

Source: Table 2 and Table 3.

force. As a consequence, the importation of wheat commenced and the acceleration of domestic grain production was also induced. In order to increase grain output, investments were undertaken to enlarge the land under cultivation by 1.1 million hectares or by 13 percent between 1873 and 1880.[13]

Although the agricultural sector did respond to the stimulus of increased demand for grain by the milling industry, the latter was no longer constrained by the ability of the former to increase production after 1875 because a trade treaty was signed with Romania which allowed duty-free entry of Romanian grain. This increased the competitiveness of Hungarian flour on foreign markets not only because of the cheaper price of Romanian grain, but also because the mixture of the different qualities of wheat resulted in a grade of flour that was better suited to the tastes of Englishmen and Germans. Thus the access to Romanian raw materials was important to the flour industry as evinced by the increase in imports between 1876 and 1880 from 16 thousand to 130 thousand tons.[14]

[13] " A Magyar Korona Országainak 1891. és 1892. évi mezögazdasági Termelése," *Magyar Statisztikal Közlemények*, Új Folyam, Vol. 6, 1893.

[14] MATLEKOVITS, *Die Zollpolitik*, pp. 48, 573.

In 1878 grain duties were eliminated entirely until 1882. Thereafter the mills could import wheat duty-free irrespective of country of origin as long as its flour-equivalent was exported within a year.[15] Though the significance of the special treaty with Romania was therefore brief, Hungary continued to import large quantities of wheat from Romania until trade war erupted in 1886.[16] By then, however, the mills were able to substitute for Romanian wheat because of the special grain import provision of 1882.

The pattern of Germany's flour imports shows well the increase in Hungary's competitiveness in this sector. In the late 1870s half of Austria-Hungary's exports were destined for Germany. Between 1874 and 1878 exports to Germany quadrupled to 125,000 tons. Soon thereafter Germany lost its short-lived significance as a consumer of Hungarian flour because increasingly burdensome tariffs were imposed. Outside the customs union England remained the sole important consumer of Hungarian flour, at least until the turn of the century. Thereafter, even tiny Bosnia consumed more Hungarian flour than England did.

In short, high profits (26% return on capital), brought about by increased demand for wheat flour in Austria and Germany and access to inexpensive wheat from Romania, induced an unprecedented advance of this branch in the late 1870s. The process was accompanied by technological breakthroughs: milling machinery invented and improved by András Mechwart became world famous in the 1870.[17]

Hungary's flour industry prospered until 1896 as exports to Austria were growing at a rate of 10 per cent per annum. There-

---

[15] JOSEF GRUNZEL, *Handelspolitik und Ausgleich in Österreich-Ungarn* (Vienna and Leipzig: Alfred Holder, 1912), p. 50.

[16] The trade war lasted 5 years. LUDWIG LANG, *Hundert Jahre Zollpolitik* (Vienna and Leipzig: Carl Fromme, 1906), p. 220. MATLEKOVITS, *Die Zollpolitik*, pp. 114, 115.

[17] Ganz began producing an improved milling "stool" in 1874. "Magyarország Malomipara 1894-ben." *Magyar Statisztikai Közlemények*, Új Folyam, 13.

after, both exports and production stagnated until the outbreak of World War I. After 1896, exports to the world fell rapidly, caused to some extent by internal forces: the privilege to import wheat duty-free into Hungary was limited and then abolished.[18] In addition, external forces also contributed to a decline in exports, viz, the low price of American flour. As a consequence of the decrease in foreign demand, and the stagnation of Austrian demand, the development of the flour industry in Hungary ceased, and in fact, never regained its vitality (Figure 2).

In spite of the relatively low value-added proportions of the flour industry, it is a paradigm of the development of manufacturing during the three decades after 1875. It was the largest industry by gross output. (In 1898, it accounted for one-fourth of total industrial production). Although the labor force employed in the flour industry ranked only sixth in Hungary, after the ferrous metal, machine and the (labour intensive) tobacco, brick and wood-working branches, it was the second largest industry in the amount of horse-power utilized, surpassed only by the iron industry. Even in value-added terms, the flour industry was equalled only by iron and machine production.[19]

The industrial production index which encompasses the flour, iron, sugar and cotton textile sector confirms this pattern. Although few in number, these sectors together employed one half of the power and one fourth of the labor force in Hungary's manufacturing industry in 1898 and accounted for 43 percent of the value of gross production.[20] The 5.5 per cent annual growth rate

[18] "Die Kreditanstalten der Länder der ungarischen Heil. Krone in den Jahren 1894–1909," *Ungarische Statistische Mitteilungen* (Budapest: Kön. Ungarischen Statistischen Zentralamt, 1913), Neue Serie, 35, p. 9 (Hereafter referred to as *Kreditanstalten.*) In 1896 the percentage of flour that had to be exported in place of the wheat imported duty-free was increased from 70 percent to 100 percent and the time limit on re-export was decreased to six months. Four years later the privilege was abolished entirely. Offergeld, p. 12.

[19] *Gyáripara*, p. 245.

[20] *Ungarischen Statistisches Jahrbuch*, 1900, p. 146.

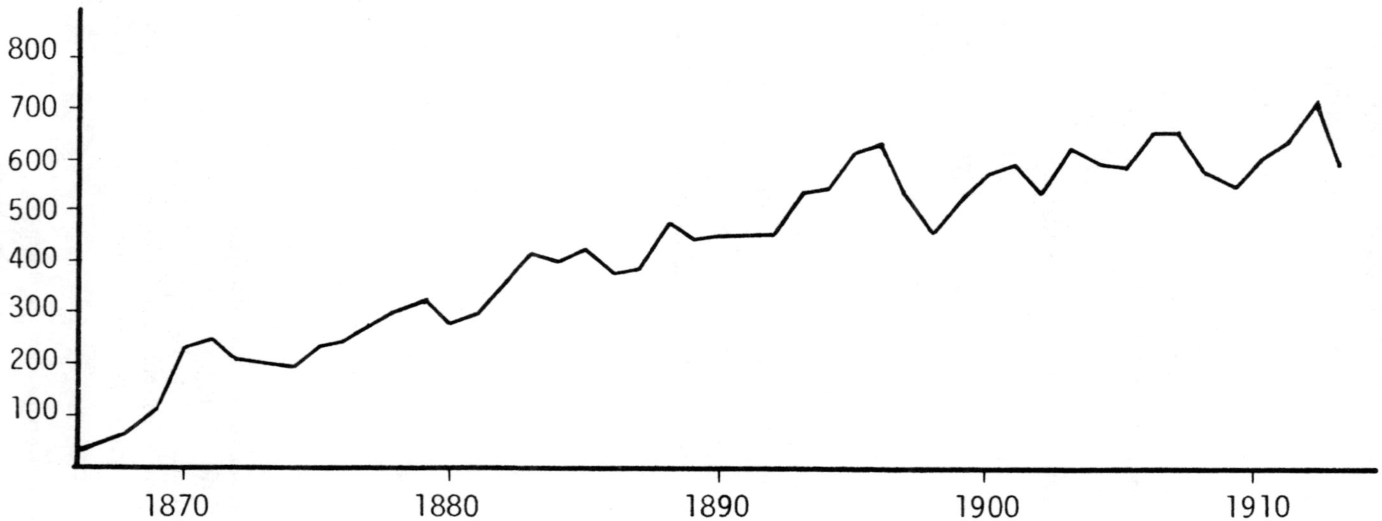

Figure 2 - Flour Production in Budapest, 1866 - 1913 (thousands of tons).

SOURCE: « A Magyar Szent Korona Orszagainak Malomipara az 1906. évban » *Magyar Statisztikai Közlemények* (Budapest: Statisztikai Hivatal, 1909) Vol. 26, p. 75. Alexander Mathekovits, *Das Monigreich Ungarn* (Leipzig: Duncker and Humblot, 1900), p. 108. Wilhelm Offergeld, *Grundlagen und Ursachen der industriellen Entwicklung Ungarns* (Jena: Gustav Fischer, 1914), p. 11.

Magyar Statisztikai Evkönyv, 1910, p. 186.

of these sectors during the decade after 1874, even if it were an overestimate of the performance of total manufacturing, is nevertheless indicative of the general advance in industrial production during this business cycle.

Similar processes of modernization are detectable in many other aspects of economic life, even in such unlikely areas as ocean shipping: seven steam vessels were added to the fleet in 1882.[21] The growth of the mining sector in the 1880s surpassed 3 per cent per annum, which was faster than at any time previously in the nineteenth century. The increase in the volume of internal trade might be indicated by the growth in the value of postal money orders processed in Hungary: 86 million gulden in 1877 and 135 million by 1882.[22] The accounts of several steel mills and engineering firms evinced a steep increase in the production in the early 1880s.[23]

Moreover, Hungary's terms-of-trade (prices of exports/prices of imports) vis-a-vis Austria improved until 1898, perhaps as a consequence of the large transfer of purchasing power to Hungary.[24] In fact, Hungary's terms-of-trade continued to correlate positively with the index of manufacturing production and the transfer of funds from Austria. After improving until 1898, the terms-of-trade worsened until 1906 and improved again thereafter (Table 5). During periods in which Hungary's terms-of trade were improving, her rate of industrial growth was relatively high. The favorable change in Hungary's terms-of-trade caused a rise in real income there and a fall in real income in Austria.

---

21 "Fiume hajo-és Áruforgalma az 1906. es 1907. években," *Magyar Statisztikai Közlemények*, Uj Sorozat, 24, p. 24. (Hereafter referred to as *Fiume, 1906–0197*.)

22 Austria. K.K. Finanz-Ministerium, *Statistischen Tabellen zur Währungsfrage der österreichisch-ungarischen Monarchie* (Vienna: Hof- und Staatsdruckerei, 1892), p. 267.

23 ALEXANDER MATLEKOVITS, "Die wirtschaftliche Entwicklung Ungarns seit 1867' *Zeitschrift für Volkswirtschaft, Sozial politik und Verwaltung*, Vol. 7, (1898), p. 545.

24 This would have been the case if the sums of Hungary's and Austria's marginal propensities to import had been less than unity. RICHARD E. CAVES and RONALD W. JONES, *World Trade and Payments* (Boston: Little, Brown and Co., 1973), p. 457.

TABLE 5

HUNGARY'S NET BARTER TERMS OF TRADE WITH AUSTRIA IN DOMESTIC PRICES, 1882-1913

Year	
1882	82.7
1898	105.0
1906	89.5
1913	100.0

Source: SCOTT M. EDDIE, "The Terms and Patterns of Hungarian Trade, 1882-1913," *Journal of Economic History* 37, no. 1 (1977): 358.

Hence, if the transfer of funds from Austria on a relatively long term basis was the reason for the change in the terms-of-trade.[25] the transfer induced a "secondary burden" on Austria as the "price changes created an international redistribution of income additional to the initial loan."[26]

Simultaneously, the state expanded its direct promotion of economic development and the formation of social overhead capital. The value of the infrastructure (public buildings, roads, state railroads, bridges, harbors) grew at a rate of 10 per cent per annum between 1874 and 1896 (Table 6). (In the 1870s, the

TABLE 6

THE VALUE OF SOCIAL OVERHEAD CAPITAL IN HUNGARY,[a] 1872, 1878, 1896
(millions of crown)

Year	
1872	248.8
1878	581.6
1896	2,442.6

[a] The valuation is at historical cost without allowance for depreciation. Note that the period was one of secular deflation.

Sources: *Statistisches Jahrbach für Ungarn*, 1872, p. 701; 1879, p. 23. MATLEKOVITS, *Des Königreich Ungarn*, 2, 869-70.

---

25 This conclusion may be warranted all the more since Hungary's terms of trade in world prices diverged significantly from this pattern.

26 Caves and Jones, p. 457.

growth rate exceeded 15 per cent per annum). Total value of state property (which in addition to infrastructure, included forests, mines, estates, and industrial enterprises) increased more slowly than infrastructure alone, but it too reached a growth of 5.6 per cent per annum between 1873 and 1896 (Table 7). (During the 1870s, its growth was faster: 7.7 per cent per annum.)[27]

TABLE 7

VALUE OF GOVERNMENT PROPERTY 1872, 1878, 1896
(Current prices, millions of crowns)

	1872	1878	1896
Railroads	150.0	417.8	1,885.2
Bridges	17.0	29.8	33.6
Roads, harbors [a]	1.1	26.9	346.5
Public buildings [b]	40.8	40.1	221.7
Estates	98.8	139.2	271.2 (Estates and Forests)
Forests	50.2	207.0	
Mines	418.0	391.0	423.7
Inventory [c]	135.4	167.0	331.2
Total	951.2	1,485.8	3,535.0

[a] Includes river embankments.
[b] Includes other real estate and in 1896 industrial enterprises.
[c] Includes other movable property.

Source: See Table 6.

The growth of government investment between 1874 and 1896 was accompanied by an increase in operating expenditures of 3.6 per cent per year. The extensive governmental spending was not accomplished by a proportional increase in taxation. Tax revenue increased only at a rate of 2.9 percent per annum (Table 8). The deficits were financed to a small extent by the profits from government enterprises but primarily by the expansion of the debt. The latter not only covered the servicing of the debt until the 1890s, but also a large portion of the investments under-

27 MATLEKOVITS, *Das Königreich Ungarn*, p. 700.

John Komlos

TABLE 8

ANNUAL GROWTH RATE OF GOVERNMENT EXPENDITURES AND REVENUES, 1878-1912

	State Debt	Expenditures [c]	Investments	Taxes [d]
1878-1896 [a]	5.0%	3.0%	9.7%	2.9%
1896-1906 [b]	0.8%	2.0%	— 0.2%	3.0%
1906-1912 [b]	2.5%	10.9%	11.7%	3.7%

[a] For investments 1879 is the initial terminal point.
[b] For investments 1905 is the final terminal point.
[c] Does not include servicing the debt.
[d] Does not include revenues from the post and telephone offices and government enterprises.

Source: Ungarisches Statistisches Jahrbuch.

taken by the state. Since most of the debt was foreign-owned in the 1870s and 1880s, government investments were not undertaken at the expense of private domestic investments. That is, the crowding-out effect must have been unimportant.

Some of the Austrian funds put at the disposal of the Hungarian government found their way into the agricultural sector. Agricultural production accelerated in Hungary after 1878. This was achieved to a large extent by increasing the amount af land under cultivation. Plow land in Hungary, and Transylvania (without Croatia) grew between 1873 and 1877 from 8.3 to 9.2 million hectares.[28] The necessary capital for this investment was provided by the government through the purchase of forests from the large estates. Between 1872 and 1878, government-owned forests increased fourfold and their purchase provided 150 million crowns to the large landowners, which could in turn be used by them to undertake the necessary investments and to increase the seed to land ratio.

Another beneficial aspect of the increase in government debt was that the influx of Austrian funds was tantamount to increasing

[28] "A Magyar Korona Országainak, 1891. és 1892. évi Mezögazdasági Termelése," *Magyar Statisztikal Közlomenyek*, Új Folyam, VI, p. 16*.

the supply of money in Hungary. Prices in Hungary were, on the other hand, determined not exclusively in Hungary but in the common market with Austria, and were even in secular decline due to worldwide deflation. Consequently, the increase in the quantity of money circulating in Hungary (assuming that the velocity of circulation did not decrease proportionately) put upward pressure on output and imports. Hungary's trade deficit in the 1880s (Table 9), financed by the transfer of funds from Austria, conforms to this pattern. Hence, even without an autonomous monetary policy the government, perhaps inadvertently, but no less effectively, did increase the quantity of money circulating in Hungary in the 1870s and 1880s. This, in turn, induced an increase in production and imports.

The expansionary governmental activity was reversed between 1896 and 1906. Fiscal policy was curtailed (Figure 3). The national debt expanded at a rate below 1 per cent per annum and was no longer underwritten by foreigners. Hence, government revenue was increasingly consumed in servicing the debt and government expenditures expanded more slowly than the increase in taxes. Government investments stagnated utterly. Consequently, the rate of growth of industrial production decelerated. During the next cycle (1906-1912), the growth of government expenditures exceeded 10 per cent per annum as did government investments. These growth rates are in nominal

HUNGARY'S BALANCE OF TRADE
ON MERCHANDISE ACCOUNT, 1882-1913
(millions of crowns, current prices)

	Import Surplus	Export Surplus
1882-1898	363	
1899-1905		700
1906-1913	973	

Source: Magyar Statisztikai Közlemények, LVIII, 3.

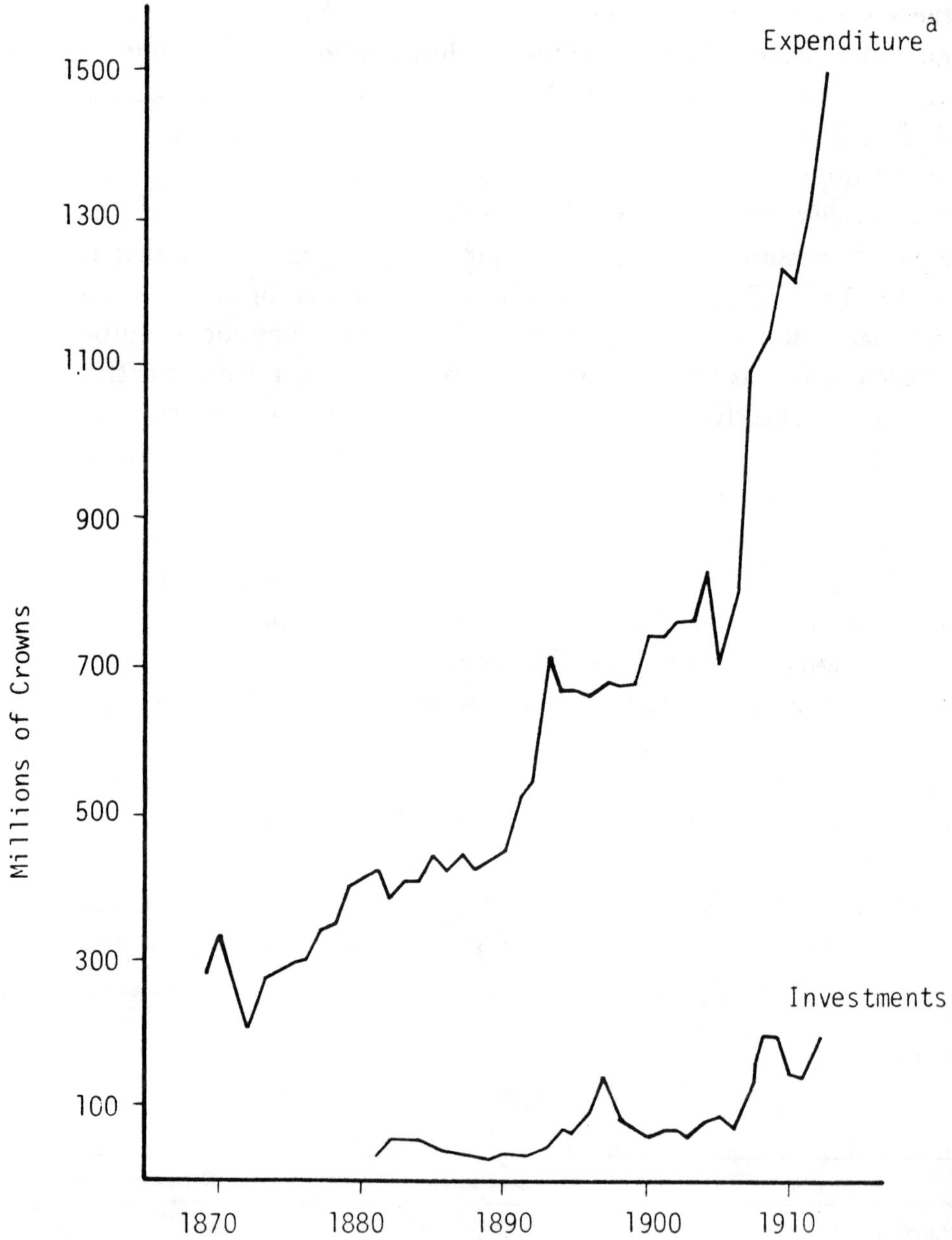

[a] Does not include investments or service of the debt.

Figure 3 - Hungarian Government Expenditures and Investments 1869 - 1913.
SOURCE: *Ungarisches Statistisches Jahrbuch.*

terms, but the increase in real terms was still substantial as the inflation rate was only about 2.5 per cent per annum.[29] Though the national debt grew in this period, it was underwritten only to a small extent by foreigners who were more interested in Hungarian municipal and mortgage bonds. Taxes increased more slowly than government expenditures; between 1907 and 1912 only 70 per cent of government spending was financed through taxes. This, too, put upward pressure on output.

In sum, government fiscal policy correlates well with the industrial production index. In the 1870s and 1880s, the ability of the government to attract foreign investments shifted the burden of the debt on to foreigners and enabled the state to pursue an active fiscal policy without stifling the private sector through excessive taxation. The pattern was reversed between 1896 and 1906 when the debt weighed heavily on the domestic economy.

The influx of foreign funds between 1878-1892 and again after 1906 enabled the government to finance a significant proportion of its day-to-day activities, its industrial promotion programs, and its investments in social overhead capital with foreign funds. Since the deficit did not have to be financed exclusively in the domestic money market, the private sector was not drained of capital. In addition, disposable income grew faster than it would otherwise have grown since the government did not have to resort to excessive taxation as a substitute for open market operations. Hence the growth of consumption was not restrained by the growth of taxes. As Table 8 shows, taxes grew faster than government expenditures and investments only between 1896-1906 a period of slow industrial growth in Hungary which was accompanied by a cautious attitude toward the Hungarian money market on the part of foreign investors. This suggests that to a large extent Hungary's ability to industrialize depended on unencumbered access to Austria's capital market, which, gi-

[29] David F. Good, "The Cost of Living in Austria 1874–1913," *The Journal of European Economic History*, 5 (1976): 395.

John Komlos

ven the psychological milieu of the post-1873 depression, found Hungarian state securities quite attractive.

After the stock market crash of 1873, which was superimposed on the downturn of the business cycle in 1872, the confidence of Austrian capitalists in their domestic economy, particularly in its industrial sector, wavered.[30] This led to a decrease in investments in Austria relative to trend which lasted well into the 1880s. The funds that accumulated in Austrian savings institutions and mortgage banks [31] increased, in effect, the demand for safe government securities.[32] Both governments responded readily.[33]

After the crash, the general banks in Austria developed into a new type of institution in which " regular bank business was the underlying basis of the bank's existence. " [34] They also divested themselves of industrial equity as much as possible. Since the stocks had to be absorbed by an unwilling public prices were further depressed. New issues were seldom undertaken: " Insofar as promotional activity initiative in establishing new enterprises was concerned, not a thought was given to it in this

---

30 Referring to the causes of the depression Rudolph states that " ... an important key is the aversion to joint stock companies and industrial shares on the part of the government, the public, and the banks, which came to be part of the fact and folklore of the post depression period." RICHARD RUDOLPH, *Banking and Industrialization in Austria-Hungary. The Role of Banks in the Industrialization of the Czech Crownlands, 1873-1914* (Cambridge: Cambridge University Press 1976), p. 70.

31 The deposits at savings institutions in Austria increased by 45 percent between the end of 1873 and 1879. Austria. *Statistisches Jahrbuch für das Jahr 1879* (Vienna: k.k. Hof- und Staatsdruckerei, 1882), IX, p. 43.

32 Outside the banking system, government bonds were bought partly by " the propertied middle class who want to invest their savings without risk at constant interest, and partly by the very wealthy for whom the possible capital gains of other securities are not enticing." EUGEN PHILIPPOVICH, *Grundriss der Politischen Oekonomie* (5th Ed.; Tübingen: J.C.B., Mohr, 1915), II, Part II, 219.

33 EDUARD MÄRZ, *Österreichische Industrie- und Bankpolitik in der Zeit Franz Josephs I. Am Beispiel der k.k. priv. Österreichischen Credit-Anstalt für Handel und Gewerbe* (Vienna: Europa, 1968), p. 220.

34 FRITZ STEINER, *Die Entwicklung des Mobilbankwesen in Österreich* (Vienna, n.p., 1913), p. 265, as cited by Rudolph, *Banking and Industrialization*, p. 78.

time." [35] This attitude predominated in spite of the apparent profitability of some industries.[36]

Until 1878, the majority of Hungary's creditors were in Western Europe. This pattern changed quickly thereafter and Austrian capital predominated, particularly until 1893. Fully 61 per cent of Hungarian state securities outstanding was in Austrian hands and less than a fourth was owned by Hungarians. Austrians effectively injected at least two billion crowns into Hungary's economy, which was responsible for the absence of signs of depression in Hungary.

Toward the end of the 1880s, Austrian banks as well as the general public began to turn their attention to industrial and bank stocks once again.[37] The shock of 1873 was beginning to wear off and the previous upswing of the early 1880s convinced investors that the depression they had been experiencing was a cyclical ra-

---

[35] JOSEF HORAK, *Přehld vývoje českých obchodnich bank* (Prague, 1913), p. 92, as cited by Rudolph, *Banking and Industrialization*, p. 78.

[36] In 1881, the profit rate of banks exceeded 8 percent, natural gas utilities 13 percent, machine producers 8 percent, distilleries 15 percent, sugar refineries 6 percent, shipping companies 7 percent, insurance companies 10 percent, and miscellaneous industrial enterprises 9 percent. These returns were characteristic of the 1880s in general for these branches except for the distilleries. One should also consider that the profits of some of the enterprises in branches that were not doing well in the 1880s surpassed the average return in their branch by quite a margin. For instance, in 1879, 55 corporations in Austria disbursed dividends of over 10 percent of their paid-up capital. In spite of these investment opportunities, the paid-in capital of enterprises in the above named branches increased by only 14.2 million crowns during 1881 (to 840.9 million crowns), while about 50 million crowns (71 million crowns nominal) were invested in Hungarian securities during the same year even though their return decreased to 5.6 percent. Contemporaries in fact noted that real investments lagged far behind savings and that the liquid position worked to Hungary's advantage as large amounts of funds were placed at the Hungarian government's disposal. Austria, k.k. Statistisches Central Commission, *Statistisches Jahrbuch für das Jahr 1879*, VIII, p. 67. Austria, k.k. Finanzministerium, *Tabellen zur Währungs-Statistik* (2d ed., Vienna: K.K. Hof- und Staatsdruckerei, 1900-1914), II, 651 ff. MATLEKOVITS, *Das Königreich Ungarn*, II, 888. Westphal, pp. 2, 11, 12. März, p. 368.

[37] *Kreditanstalten*, p. 12. Rudolph writes that after 1873 the banks "... lost most of their interest in industrial enterprises for years to come. It was only in the 1890s that such interest returned, ..." RUDOLPH, *Banking and Industrialization*, p. 70. Also see MÄRZ, pp. 295-297.

ther than a permanent attribute of the economy. Thus, by the early 1890s, the biases against industrial investment disappeared for the most part but not soon enough for Austria to undertake the investments that would have been necessary to allow manufacturing to reach its long-run growth rate during the cycle that peaked in 1896. But the repatriation of capital became crucial during the expansionary phase that followed (1896-1907) when Austrian industrial growth reached over 3 per cent per annum per capita, a rate unsurpassed in any cycle throughout the nineteenth century.

Between 1896-1907, the performance of Hungarian and Austrian industry reversed. While between 1873 and 1896 Hungary's industry grew relatively fast and Austria's did not, between 1896 and 1907 Austrian industry advanced rapidly while Hungarian industry stagnated. The growth of the national debt slowed and Hungary had to bear a much larger share of it than previously. Since the foreign holdings of the debt could no longer be increased (though German capital did replace Austrian capital to some

TABLE 10

DISTRIBUTION OF HUNGARIAN STATE SECURITIES AND TOTAL NOMINAL VALUE OUTSTANDING, 1878-1913
(millions of crowns)

	A	B	C	D	Total W. Europe	Austria	Hungary	Total Value
1878	17.8	4.8	9.2	15.7	47.6	14.6	37.9	1,780
1893	0.3	—	2.5	13.8	16.6	61.0	22.5	4,380
1896	1.4	—	4.8	25.1	31.3	31.6	37.1	4,526
1905	1.5	0.3	7.4	30.5	39.7	20.9	39.6	5,147
1913	1.3	0.2	5.7	25.6	32.8	21.8	45.2	6,229

A = England
B = Holland
C = France
D = Germany

Source: FELLNER, *Das Volkseinkommen*, p. 135 and FRIGYES FELLNER, *Die Zahlungsbilanz Ungarns. Ein Beitrag zur Lehre von der internationalen Zahlungsbilanz im allgemeinen* (Vienna and Leipzig: F. Deuticke, 1908), Table III.

extent), this meant that the interest on the debt weighed heavily on the economy and necessitated a contraction in government spending (Figure 4). Until 1893, debt servicing was accomplished through expansion of the foreign holdings of government securities. After 1893, this was no longer possible, and the Hungarian economy had to bear the full burden of the previous expansionary epoch.

After 1906, the economic situation reversed once again: Hungary's industrial growth rate exceeded Austria's. This, too, was accompanied by an increasing inflow of capital; in the form of mortgage and municipal bonds and direct investments in industrial enterprises rather than in the form of state securities.

Table 11 shows the extraordinary increase in foreign, predominantly Austrian (including Czech), investment in Hungarian industry and banking between 1907-1913, the final business cycle in the monarchy's history. (However, in spite of the growth of foreign investment, foreign holdings of industrial and bank stocks in 1913 accounted for only 15 per cent of the stocks outstanding in this category). After the lull in Austrian investments in Hungary around the turn of the century, and the coincidental stagnation of Hungarian industry, the flow of capital between Austria and Hungary correlates well with the rate and pattern of industriali-

TABLE 11

FOREIGN INVESTMENT IN HUNGARY, 1905-1913
(million of crowns)

	1905	1913	Increase	% Increase
A	4,321.6	4,712.5	390.9	9
B	1,429.9	2,135.9	706.0	49
C	76.1	416.1	340.0	447
Total	5,827.6	7.264.5	1,436.9	25

A = State securities, railroad stocks and obligations.
B = Municipal obligations and mortgage bonds.
C = Industrial and bank stocks.

Source: FELLNER, *Das Volkseinkommen*, p. 122.

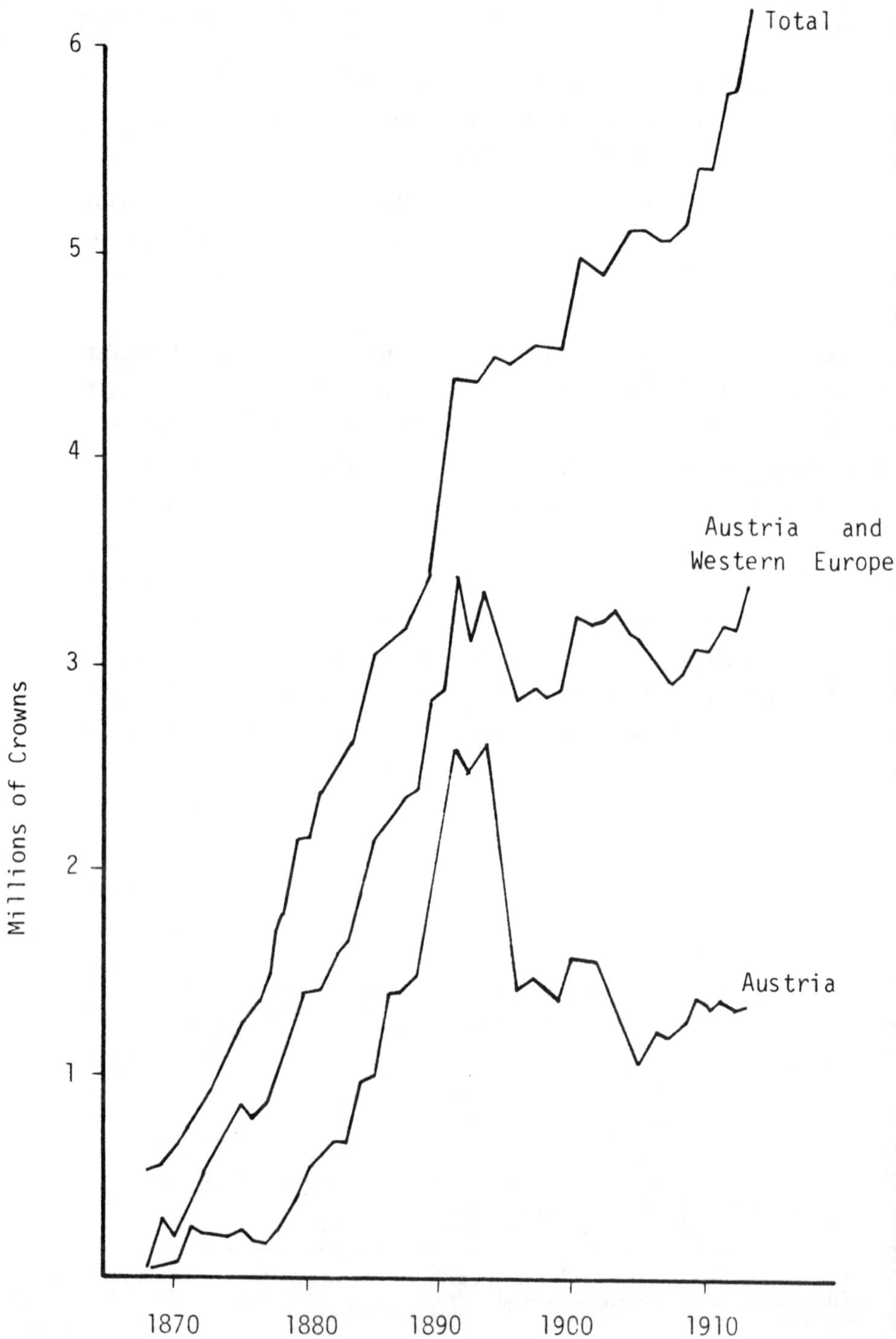

Figure 4 - The Nominal Value of Total Hungarian State Securities Issued and Those Held Abroad, 1869 - 1913.

SOURCE: See Table 10.

zation in the two halves of the monarchy. The influx of Austrian capital between 1907 and 1913 was accompanied by an acceleration of industrial growth in Hungary and a deceleration in Austria, similar to the pattern observed between 1873 and 1896.[38]

The export of capital from Austria hurt its economic development between 1873 and 1898. The reason is not only that the terms-of-trade had turned against her, that the quantity of money circulating in Austria decreased, but also that her industrial sector was not in long run competitive equilibrium. Profits for a number

---

[38] Rough calculations show that Austrian investments in Hungarian bonds were large enough to have accounted for the relatively slower growth of Austrian industry during the 1873-1896 depression. Austria invested 600 million crowns in Hungarian state securities between 1871 and 1884. In 1885, the output of Austrian industry was 1.246 billion crowns. Extrapolating this backward at a rate of 1.8 percent per annum yields a net output of 1.148 billion crowns in 1871 (in 1885 prices). The value of capital invested in Austrian industry in 1885 can be estimated by assuming that wages absorbed 60 percent and capital 40 percent of total product. Assuming, moreover, a marginal return on capital of 10 percent suggests that Austrian capital stocks was on the order of: $K = \frac{4\,(1.246)}{1} = 5$ billion crowns in 1885. If we were to assume that the initial 600 million crowns (746 million crowns assuming 80 percent discount.) had been retained at home and invested in the industrial sector, the accumulated industrial capital stock would have been 5.6 billion. If, as a first approximation, the marginal product of capital and the share of capital in total output were constant, industrial output in 1885 would have been about $Y = \frac{1\,(5.60)}{4} = 1.4$ billion crowns. The rate of growth then between 1871 and 1884 would have been 2.7 percent, or 50 percent higher than the rate actually realized (and closer to the rate achieved between 1861-1871). The calculation is not sensitive to either the factor shares or the return of capital. (Since the industrial sector was not in long run competitive equilibrium, some investment opportunities existed which yielded a return well above this marginal rate. Consequently the marginal return on capital need not have decreased with the repatriation of funds from Hungary.) Given a growth rate of 2.25 percent in the following period (1884-1898) and assuming the same parameters, implies a hypothetical Austrian industrial capital stock in 1898 of approximately $K = \frac{.4\,(1.972)}{1} = 7.89$ billion crowns. Adding 1.33 billion crowns (1.66 billion crowns assuming 80 percent discount rate) that Austrians invested in Hungarian securities yields an output of 2.30 billion and consequently a rate of growth between 1884 and 1896 of 3.4 percent. Thus the rough calculations suggest that Austrian investments in Hungarian government bonds were large enough to have been capable of accounting for the relatively slower growth of Austria's industry during the depression of 1873-1897.

of industries were above normal and they persisted. Hence, there were economic rents to be earned which were foregone as a consequence of the flight of capital into Hungary. To be sure, measured in terms of utility the Austrians were better off holding Hungarian bonds than industrial equities, but a substantial part of this income consisted of non-pecuniary services of Hungarian bonds. Measured in physical units, the growth of national income was undoubtedly slower between 1873 and 1893 than it would have been if the Hungarian bonds had not been bought to the extent they were by Austrians.[39]

It is readily admitted that even without the availability of Hungarian issues not all of the Austrian capital would have been placed in industrial enterprises. No doubt more would have been placed in industrial ventures however, and sooner. Though the capital flow into Hungary was sufficiently large to account fully for the slow industrial growth during the depression (1873-1896), the capital drain was also responsible for prolonging the depression in Austria. Therefore, Hungary's detrimental impact on the Austrian economy was probably more crucial during the later years (1884-1896) of the depression than during the earlier years (1873-1884) since Austrians would probably not have undertaken large scale domestic investments immediately after 1873 in any event.

Austrian investments were indispensable to the mobilization of Hungary's industrial sector. In 1893, the Austrian holdings of Hungarian securities amounted to 2.7 billion crowns. In contrast, the gross output of Hungarian industry in 1898 was only half this amount, and total fixed capital invested in industry other than buildings was one tenth of this sum as late as 1910.[40] Without

[39] Gerschenkron's suggestion that "... the government's demand for assistance from the banks in its credit operations was deflecting the banks from industrial financing" holds true for both the Austrian and Hungarian governments. ALEXANDER, GERSCHENKRON, *The Economic Spurt that Failed* (Princeton: The Princeton University Press, 1977), p. 50.

[40] FELLNER, *Die Zahlungsbilanz*, p. 71.

access to such huge foreign sums, Hungary could not have begun to industrialize in the late 1870s, and industrialization would undoubtedly have been delayed.

How easily Hungarians could have substituted foreign capital for Austrian capital in the absence of the common market is a question worthy of consideration. The more isolated Austria-Hungary was in the 1870s and 1880s from the world capital market the harder would it have been for Hungary to acquire foreign funds in order to engineer the beginning of its industrialization.

Table 12 provides some evidence that support Austria-Hungary's isolation from the world capital market. If the capital market had been fragmented, the movements in the interest rates in Austria would not correlate with the fluctuations of interest rates in the rest of the world, after allowing for the fluctuations in the exchange rate. More precisely, the domestic return on bonds (r) would be equal to the foreign yield (R) plus the expected percentage depreciation of the Austrian currency with respect to the foreign currency (e*). Since the forward rate did nor diverge appreciably from the spot rate,[41] e* might be replaced by the actual future rate (e). The correlation between Austrian yields on state securities not denominated in gold and R + e shows no correlation with any of the major developed countries of Western Europe prior to the adoption of the gold standard in 1892 (Table 12). (Yet R + e correlated highly amoung the contries as shown in Table 13). After 1892 the correlation between Austrian and foreign rates increased enormously. This pattern holds also for the correlation between r and R (Table 14) but not for securities denominated in gold (Table 15). The correlations in Table 15 do not increase as greatly after 1892. This suggests that the stability of the Austro-Hungarian currency blocked the

[41] LELAND B. YEAGER, " Fluctuating Exchange Rates in the Nineteenth Century: the Experiences of Austria and Russia " in ROBERT A. MUNDELL and ALEXANDER K. SVOBODA (ed.), Monetary Problems and the International Economy (Chicago: The University of Chicago Press, 1969).

John Komlos

TABLE 12

CORRELATION BETWEEN r AND R + e, 1872-1900

		Germany	France	England
Austria	A	—.18	.10	.00
	B	.55	.93	.96

Germany: A = 4% Imperial bonds 1878-1891
B = 3-1/2% Imperial bonds 1892-1900

France: 3% bonds, A = 1872-1891
B = 1892-1900

England: 3% consols until 1887 and 2-3/4 consols thereafter.
A = 1872-1891
B = 1892-1900

Sources: Währungs-Statistik, II, pp. 127, 130; *Währungsfrage*, p. 215; SIDNEY HOMER, *A History of Interest Rates* (New Brunswick: Rutgers University Press, 1963(, p. 260. Austria. k.k. Finanz-Ministerium, *Statistische Tabellen zur Währungsfrage der österreichisch-ungarischen Monarchie* (Vienna: Hof- und Staatsdruckerei, 1892), p. 215. Austria, k.k. Finanz-Ministerium, *Tabellen zur Währungs-Statistik* (2d ed., Vienna: Hof- und Staatsdruckerei, 1900-1904) Vol. II, pp. 127, 130.

TABLE 13

CORRELATION OF R + e FOR GERMANY, FRANCE AND ENGLAND, 1872-1900

	France	England
Germany	.74	.99
France		.97

Source: See Table 12 .

TABLE 14

CORRELATION OF AUSTRIAN (r) AND FOREIGN (R) INTEREST RATES, 1866-1900

		Germany	France	England
Austria	A	.42	.20	.01
	B	.82	.82	.91

Germany: A = 1878-1892
B = 1892-1900

France: A = 1866-1892
B = 1890-1900

England: A = 1866-1900
B = 1893-1900

The first differences of the series were correlated in order to adjust for trend.

Source: See Table 12.

TABLE 15

CORRELATION BETWEEN YIELD OF AUSTRIAN GOLD BONDS AND YIELD ON FOREIGN SECURITIES, 1872-1900

		Germany	France	England
Austria	A	.75	.44	.66
	B	.82	.80	.72

A = prior to 1892
B = 1892-1900
The first differences of the series were correlated in order to adjust for trend.

Source: See Table 12.

inflow of foreign capital into the monarchy prior to 1892,[42] but there was a foreign market for the securities denominated in real terms.

Government securities guaranteed in gold served as the only investment on which foreigners did not expect to sustain large capital losses that would have offset the relatively higher return of Austrian bonds. As these securities were a greater burden on the government prior to 1892 their issue was limited and they could not have easily replaced the securities denominated in crowns. The low correlations reported in Table 12 prior to 1892 and the higher correlations thereafter suggest a fragmented capital market for securities in the 1870s and 1880s due to the instability of the florin.[43] With the adoption of the gold standard in 1892, con-

[42] Milton Friedman and Anne Schwartz suggest that " highly variable inflation,... even if the average rate is largely anticipated... tends to destroy financial intermediation, rendering the capital market both limited and inefficient,... The result is that the real yield on physical capital may differ widely from one enterprise or industry to another, in addition to the yield to the holder of securities deviating widely from the yield on the physical capital to which the securities are linked," This holds for the fluctuations in the exchange rates as well. M. FRIEDMAN and A.J. SCHWARTZ, " Money and Interest Rates," paper presented to the Money and Banking workshop of the University of Chicago, January 13, 1976, p. 25. They mention lags of some to thirty years in the formation of anticipations, p. 32.

[43] *Kreditanstalten*, 1909, p. 12.

fidence increased and capital markets became less imperfect. The fragmented capital markets are but one more indication that Hungary could not have substituted easily for Austrian capital in the absence of the customs union.

The unencumbered access to the Austrian capital market was crucial to the beginning of industrialization in Hungary in the late 1870s. Until recently the first upsurge of industrial activity in Hungary was believed to have occurred after 1867 and to have come to an abrupt halt in 1873 after the crash of the stock market, only to begin anew after 1887.[44] From this pattern, especially from the upswing after 1867, scholars have concluded that the political consolidation that followed the compromise with Austria in 1867 was a significant element in Hungary's industrialization. No evidence was found to support this view. On the contrary, although transportation and banking did expand considerably during the post-1867 upswing, industrial production appears not to have been significantly affected. In this regard researchers might have been misled by the formation of corporations after 1867, which is not indicative of industrial production, especially during the early stages of development. The appearance of numerous large steam flour mills was also deceptive as they were initially substitutes for flour mills conventionally powered.[45]

In contrast to the view that relegates the period between 1873-1887 to minor importance, the finding presented in this paper is that within a few years after the beginning of depression in Austria in 1873 the influx of foreign, predominantly Austrian, capital into Hungary was channeled, through the purchase of government securities, into investment in the infrastructure, and into increasing government expenditures. Without Austrian capital as a catalyst the beginning of industrialization in Hungary would have been unthinkable in the 1870s because the rate of

[44] Katus, " Economic Growth in Hungary," pp. 55-56.
[45] *Ibid.*, pp. 56, 58.

domestic accumulation was insufficient. In 1896 the advances of industrial production practically came to a halt as a consequence of the recession of Austrian capital from Hungary. After a decade of stagnation the influx of new funds from Austria correlated well once again with the acceleration in industrial production between 1906 and 1913.

TABLE 16

MINING PRODUCTION IN HUNGARY, 1830-1913
(ten thousand ton)

Year	Coal: Anthracite	Coal: Lignite	Gold	Silver	Copper	Lead
1830	.27	.37	.000012	.00017	.016	.0012
31	.16	.22	.000015	.00021	.020	.0011
32	.21	.29	.000014	.00020	.021	.0014
33	.15	.21	.000014	.00013	.021	.0017
34	.16	.22	.000014	.00019	.019	.0022
35	.12	.16	.000015	.00019	.019	.0031
36	.20	.27	.000016	.00020	.018	.0023
37	.16	.22	.000016	.00020	.022	.0030
38	.15	.20	.000017	.00020	.023	.0028
39	.33	.45	.000016	.00019	.021	.0018
40	.32	.44	.000018	.00021	.021	.0035
41	.34	.47	.000017	.00020	.020	.0035
42	.20	.28	.000017	.00020	.022	.0038
43	.21	.29	.000018	.00019	.022	.0050
44	.44	.61	.000019	.00021	.022	.0051
45	.21	.28	.000019	.00021	.022	.0051
46	.15	.20	.000021	.00021	.024	.0023
47	.39	.54	.000020	.00020	.028	.0048
48	.0	.0	.0	.0	.0	.0
49	.0	.0	.0	.0	.0	.0
50	.0	.0	.0	.0	.0	.0
51	.61	.98	.000017	.00022	.013	.0018
52	.69	1.40	.000016	.00020	.019	.0016
53	.86	1.01	.000017	.00022	.020	.0040
54	1.04	1.15	.000017	.00022	.021	.0038
55	1.12	1.35	.000017	.00020	.022	.0061

John Komlos

TABLE 16 (*continued*)

1856	1.25	1.27	.000013	.00018	.020	.003
57	1.26	1.81	.000015	.00017	.020	.008
58	1.00	1.00	.000010	.00001	.020	.005
59	2.06	1.93	.000016	.00021	.020	.013
60	2.38	2.38	.000016	.00020	.019	.023
61	2.58	2.37	.000016	.00019	.020	.022
62	3.24	2.58	.000017	.00018	.021	.013
63	3.40	2.65	.000015	.00020	.020	.017
64	3.43	2.44	.000018	.00026	.022	.016
65	3.62	2.57	.000018	.00026	.022	.013
66	4.13	2.87	.000016	.00027	.022	.012
67	4.02	3.34	.000018	.00027	.024	.016
68	4.87	4.43	.000017	.00027	.021	.011
69	4.94	5.68	.000016	.00026	.017	.015
70	5.37	6.02	.000015	.00020	.012	.011
71	6.17	8.56	.000014	.00020	.013	.016
72	6.41	9.45	.000014	.00017	.012	.017
73	6.84	9.50	.000012	.00019	.011	.015
74	6.25	7.74	.000013	.00017	.010	.015
75	6.36	8.16	.000016	.00021	.010	.017
76	6.67	8.84	.000019	.00023	.010	.022
77	6.82	9.07	.000017	.00021	.010	.021
78	6.87	9.09	.000018	.00020	.011	.019
79	6.74	9.32	.000016	.00019	.010	.020
80	8.05	10.13	.000016	.00017	.008	.017
81	8.49	11.13	.000016	.00018	.008	.018
82	7.99	12.60	.000017	.00017	.007	.017
83	8.93	14.74	.000016	.00017	.008	.021
1884	9.40	15.84	.000068	.00015	.0061	.018
85	9.56	15.87	.000017	.00017	.0050	.040
86	8.59	15.68	.000018	.00016	.0037	.021
87	7.86	17.23	.000019	.00018	.0034	.018
88	8.51	18.74	.000018	.00017	.0038	.020
89	9.37	19.35	.000022	.00017	.0031	.024
90	9.95	22.52	.000021	.00017	.0028	.013
91	10.19	24.28	.000015	.00017	.0024	.022
92	10.52	25.54	.000016	.00018	.0032	.023
93	9.83	29.18	.000017	.00024	.0034	.025
94	10.37	31.75	.000027	.00020	.0027	.021
95	10.68	34.75	.000032	.00020	.0029	.023
96	11.33	37.62	.000032	.00020	.0016	.019
97	11.18	38.71	.000031	.00027	.0021	.025
98	12.40	42.07	.000028	.00019	.0015	.023
99	12.39	42.93	.000031	.00021	.0016	.022

Economic Growth and Industrialization in Hungary

TABLE 16 (*continued*)

1900	14.47	51.28	.000033	.00020	.0018	.020
01	13.65	51.80	.000033	.00024	.0016	.020
02	11.83	51.32	.000034	.00023	.0008	.022
03	12.33	52.72	.000034	.00019	.0004	.021
04	11.55	55.19	.000037	.00016	.0006	.021
05	10.88	60.89	.000037	.00016	.0007	.021
06	12.38	63.65	.000037	.00014	.0006	.019
07	12.74	64.91	.000035	.00013	.0008	.014
08	12.10	71.51	.000033	.00013	.0017	.015
09	13.97	76.59	.000027	.00011	.0027	.016
10	13.02	77.34	.000030	.00013	.0021	.021
11	12.90	81.55	.000032	.00011	.0021	.016
12	13.02	82.85	.000028	.00011	.0024	.016
13	13.20	89.54	.000029	.00009	.0040	.011

Year	Salt	Total Value Added in Mining in 1913 Prices (Millions of Crowns)
1830	.64	14.39
31	.94	19.07
32	.76	16.88
33	.89	18.26
34	.91	18.22
35	.82	17.60
36	.84	18.01
37	1.03	20.68
38	1.17	22.52
39	1.11	21.83
40	1.09	22.39
41	1.07	21.65
42	1.04	21.04
43	.88	19.66
44	.90	20.70
45	.98	21.26
46	1.09	22.86
47	1.34	26.60
48	1.00	11.30
49	.90	10.17
50	.82	16.87
51	1.11	22.42
52	1.21	23.82
53	1.15	23.67
54	1.18	24.47

TABLE 16 (*continued*)

55	1.37	36.67
56	1.95	31.95
57	1.43	27.05
58	1.49	24.86
59	1.30	27.36
60	1.15	36.58
61	1.25	27.81
62	1.10	27.18
63	1.10	27.06
64	1.16	29.06
65	1.10	29.69
66	1.00	27.74
67	1.46	34.20
68	1.37	34.00
69	1.48	35.78
70	1.55	35.88
71	1.69	40.41
73	1.46	38.47
74	1.21	33.49
1875	1.11	34.04
76	1.20	37.13
77	1.19	36.58
78	1.46	39.87
79	1.51	39.94
80	1.58	42.37
81	1.50	42.76
82	1.64	45.06
83	1.60	47.47
84	1.60	48.53
85	1.53	48.85
86	1.52	46.86
87	1.60	48.35
88	1.64	50.65
89	1.65	53.33
90	1.60	55.35
91	1.63	56.02
92	1.76	59.48
93	1.67	61.43
94	1.69	66.57
95	1.69	70.82
96	1.80	74.78
97	1.72	75.02
98	1.79	78.32
99	1.83	80.42

Economic Growth and Industrialization in Hungary

Table 16 (*continued*)

1900	1.89	90.79
01	1.84	90.03
02	1.75	86.77
03	1.83	88.95
04	1.88	91.22
05	1.95	96.00
06	2.01	100.53
07	2.17	102.93
08	2.24	107.87
09	2.31	113.47
10	2.30	114.00
11	2.75	122.42
12	2.96	125.13
13	2.95	130.72

Table 17

INDEX OF INDUSTRIAL PRODUCTION OF AUSTRIA-HUNGARY, 1830–1913 in 1913 PRICES (millions of crowns)

	Hungary		Austria	
Year	Manufacturing	Manufacturing Mining Construction	Manufacturing	Manufacturing Mining Construction
1830	48.44	62.83	163.60	165.08
31	48.92	62.83	163.60	165.08
32	49.36	66.24	174.29	175.85
33	50.65	68.91	171.53	173.00
33	50.65	68.91	171.53	173.00
34	50.79	69.01	165.98	167.80
35	51.63	69.23	181.13	183.05
36	52.48	70.49	181.44	183.50
37	52.57	73.25	191.92	194.82
38	52.66	75.18	200.98	204.54
39	53.88	75.71	192.49	201.73
40	54.32	76.71	210.15	213.74
41	55.13	76.78	205.64	220.71
42	56.23	77.27	217.79	223.30
43	56.62	76.28	220.17	224.11
44	57.06	77.76	225.80	235.82
45	60.04	81.30	221.24	240.44
46	58.94	83.66	248.09	263.61
47	60.36	93.68	234.45	248.91

TABLE 17 (*continued*)

48	75.57	69.77	216.37	225.16
49	56.36	66.53	236.30	253.22
50	55.05	74.27	244.37	257.77
51	55.83	85.40	269.79	282.58
52	57.00	80.82	292.60	305.70
53	61.79	88.60	282.03	296.61
54	62.66	90.82	315.38	334.39
55	64.87	95.95	303.79	331.25
56	67.54	105.13	309.44	342.38
57	70.89	113.67	330.25	364.89
58	71.13	112.76	323.79	371.83
59	74.84	110.00	294.48	336.17
60	76.10	115.36	342.93	391.18
61	79.70	119.96	365.79	416.34
62	82.74	114.68	276.83	327.40
63	82.79	112.50	260.65	310.81
64	87.06	117.41	255.61	299.93
65	84.97	127.84	262.96	317.61
66	87.23	116.65	268.94	327.91
67	97.52	140.59	363.26	429.37
68	104.01	159.08	392.97	477.99
69	90.00	133.96	429.32	537.76
1870	81.96	160.04	426.36	546.12
71	84.80	177.26	503.51	658.56
70	76.36	171.64	488.76	641.31
73	68.33	176.60	409.26	556.64
74	99.73	145.35	429.58	554.82
75	93.20	130.81	428.70	577.06
76	77.96	133.85	418.04	560.66
77	91.06	135.67	440.99	587.46
78	125.19	178.40	477.65	606.07
79	110.52	165.98	480.95	621.35
80	101.16	153.10	450.57	600.84
81	123.10	189.25	529.10	701.19
82	151.98	239.59	568.56	741.71
83	161.04	247.08	629.36	821.10
84	152.94	230.68	638.58	864.79
85	157.27	231.94	549.39	742.55
86	159.30	233.04	571.12	777.19
87	165.33	264.99	613.18	840.99
88	180.05	254.53	619.11	869.66
89	165.34	254.61	656.92	901.58
90	184.13	274.97	689.77	944.15
91	204.98	313.82	702.97	976.22

Economic Growth and Industrialization in Hungary

TABLE 17 (*continued*)

92	208.52	289.86	722.24	984.43
93	233.59	330.83	737.25	1019.83
94	240.64	259.68	804.20	1096.78
95	239.27	374.66	825.40	1118.89
96	250.16	399.58	781.61	1094.50
97	224.15	370.41	806.12	1154.26
98	234.82	372.55	872.17	1243.86
99	260.23	395.95	874.46	1249.43
1900	245.73	371.84	848.02	1206.80
01	245.75	375.83	872.95	1242.72
02	265.16	396.07	927.85	1297.77
03	283.48	420.71	901.01	1283.90
04	265.20	392.49	918.69	1377.71
05	276.29	430.05	970.61	1374.27
06	312.32	476.47	1039.51	1474.96
07	305.54	455.82	1229.78	1677.19
08	318.83	496.81	1199.86	1682.10
09	348.33	529.24	1205.95	1708.95
10	344.77	513.50	1168.86	1642.37
11	379.11	552.98	1250.10	1711.96
12	419.94	606.14	1340.67	1814.33
13	415.26	598.63	1285.51	1773.54

Source for Table 16-20: JOHN KOMLOS, " The Habsburg Monarchy as a Customs Union: Economic Development in Austria-Hungary in the Nineteenth Century " (Umpublished Ph. O. Dissertation, The Uinversity of Chicago, 1978), pp. 233-246.

John Komlos

TABLE 18

SECTORAL BREAKDOWN OF AUSTRIA'S MANUFACTURING PRODUCTION IN 1913 PRICES, 1830-1913

Year	Iron	Sugar	Cotton Textiles	Woolen Textiles	Flour	Electricity
1830	11.21	0.700	18.420	81.270	52.00	
31	12.14	0.630	13.730	85.970	52.50	
32	11.93	0.720	20.790	87.850	53.00	
33	12.28	0.650	19.800	85.300	52.50	
34	12.88	0.700	21.570	76.830	54.00	
35	14.10	0.920	27.280	84.330	54.50	
36	14.03	0.920	34.160	77.330	55.00	
37	15.57	0.950	35.680	84.220	55.50	
38	15.54	0.960	37.240	91.00	56.00	
39	15.93	1.040	35.840	83.180	56.50	
40	16.84	1.110	45.550	89.650	57.00	
41	18.05	1.090	37.040	91.570	57.89	
42	18.40	1.240	42.380	97.770	58.00	
43	19.03	1.360	45.969	95.320	58.50	
44	20.93	1.420	47.510	96.940	59.00	
45	21.19	1.420	51.740	87.390	59.50	
46	24.67	1.580	66.980	95.869	59.00	
47	25.71	1.750	57.290	90.700	59.00	
48	24.16	1.330	45.570	86.310	59.00	
49	22.96	1.370	59.390	93.580	59.00	
50	27.08	1.730	66.060	90.500	59.00	
51	26.97	1.710	65.240	116.870	59.00	
52	30.93	1.990	84.130	116.050	59.50	
53	34.80	2.370	91.040	93.820	60.00	
54	35.84	1.920	76.250	141.370	60.00	
55	38.84	2.420	84.480	118.050	60.00	
56	41.81	2.400	100.170	105.060	60.00	
57	47.95	2.450	92.540	127.310	60.00	
58	49.26	2.920	101.250	110.360	60.00	
59	40.44	3.160	93.480	97.400	60.00	
60	41.53	4.020	123.44.	113.940	60.00	
61	53.15	2.910	129.110	120.620	60.00	
62	59.11	3.070	57.820	96.830	60.00	
63	58.17	3.700	48.480	90.300	60.00	
64	46.35	3.360	52.160	93.740	60.00	
65	44.01	4.250	68.100	86.600	60.00	
66	37.60	3.630	71.260	96.450	60.00	
67	50.78	4.850	111.800	135.740	60.00	
68	81.81	4.150	117.330	129.390	60.29	
69	90.56	3.370	113.890	157.090	64.41	
70	91.13	5.520	119.880	135.750	74.08	

Economic Growth and Industrialization in Hungary

TABLE 18 (*continued*)

Year	Iron	Sugar	Cotton Textiles	Woolen Textiles	Flour	Electricity
1871	99.98	1.030	163.420	160.070	79.01	
72	117.16	6.170	141.280	149.080	75.07	
73	126.50	8.310	125.420	83.570	65.47	
74	82.55	6.760	135.120	129.010	76.14	
75	74.67	4.700	148.390	118.760	82.18	
76	64.58	6.600	154.790	119.840	72.23	
77	64.59	7.620	159.280	131.200	78.30	
78	78	11.950	160.000	139.560	87.30	
79	71.76	14.790	172.520	140.760	81.12	
80	76.80	14.140	163.110	116.690	79.83	
81	97.14	23.820	181.210	135.850	91.08	0.00
82	116.29	24.120	171.340	164.890	91.76	0.17
83	138.74	27.630	237.130	138.330	87.13	0.40
84	171.19	23.740	212.410	142.450	88.12	0.67
85	110.90	25.130	182.190	134.920	95.42	0.8
86	11.86	15.130	201.650	146.800	94.58	1.10
87	115.70	25.340	228.710	144.280	97.81	1.33
88	139.96	19.290	202.900	153.860	101.43	1.67
89	151.18	28.690	210.410	177.310	87.13	2.20
90	150.51	37.050	250.740	159.630	88.77	3.07
91	163.33	39.490	248.870	153.650	93.80	3.84
92	167.77	38.950	254.770	163.480	92.83	4.44
93	176.10	42.620	253.540	165.390	94.63	4.97
94	203.10	38.990	290.100	172.050	94.15	5.80
95	218.43	52.980	296.020	158.280	93.22	6.47
96	228.57	33.390	268.200	151.330	92.92	7.20
97	256.85	46.180	271.250	134.580	88.85	8.41
98	277.75	40.470	321.650	134.230	88.73	9.34
99	287.48	44.880	292.340	140.750	98.67	10.34
1900	287.84	50.110	284.520	119.730	94.48	11.34
01	276.71	42.820	306.080	141.430	93.30	12.61
02	255.46	51.150	324.090	180.290	103.65	13.21
03	249.75	37.370	329.160	163.310	106.95	14.48
04	250.78	44.500	336.200	162.060	109.01	16.14
05	291.88	39.800	350.180	153.950	116.86	17.94
06	322.92	58.140	350.930	164.220	123.49	19.81
07	375.99	51.460	469.930	189.450	119.93	23.01
08	401.86	52.730	395.810	200.890	121.62	26.95
09	368.85	49.140	413.040	212.120	130.85	31.95
10	379.62	48.180	367.350	206.830	129.89	36.99
11	392.38	61.360	420.630	204.750	129.05	41.93
12	450.79	40.840	455.570	210.040	135.67	47.76
13	445.22	68.660	406.620	176.920	134.69	53.40

John Komlos

TABLE 19

SECTORAL BREAKDOWN OF HUNGARY'S MANUFACTURING PRODUCTION IN 1913 PRICES, 1830-1913
(millions of crowns)

Year	Iron	Sugar	Cotton Textiles	Flour	Woolen Textiles	Electricity
183.	2.44			46.00		
31	2.42			46.50		
32	2.36			47.00		
33	3.15			47.50		
34	2.79			48.00		
35	3.13			48.50		
36	3.48			49.00		
37	3.07			49.50		
38	2.66			50.00		
39	3.38			50.50		
40	3.32	0.00		51.00		
41	3.45	0.04		51.64		
42	4.19	0.04		52.00		
43	4.08	0.04		52.50		
44	4.02	0.04	0.00	53.00		
45	4.44	0.04	2.06	53.50		
46	3.52	0.05	2.37	53.00		
47	5.20	0.05	2.61	52.50		
48	3.36	0.05	2.16	52.00		
49	3.45	0.05	1.36	51.50		
50	3.29	0.05	1.71	50.00		
51	5.10	0.10	1.63	1.63		
52	5.19	0.16	1.65	50.00		
53	8.74	0.13	1.92	51.00		
54	8.69	0.20	1.77	52.00		
55	9.57	0.20	2.10	53.00		
56	11.09	0.30	2.15	54.00		
57	13.48	0.30	2.11	55.00		
58	12.35	0.40	2.38	56.00		
59	15.11	0.40	2.33	57.00		
60	13.70	0.50	2.90	59.00		
61	15.06	0.53	3.11	61.00		
62	17.47	0.61	2.66	62.00		
63	19.55	0.40	2.84	60.00		
64	21.18	0.79	3.09	62.00		
65	18.12	0.47	3.38	63.00		
66	19.02	0.57	3.64	61.00		
67	16.77	0.70	4.05	76.00		
69	24.26	0.80	6.29	76.33		
70	23.79	1.38	1.21	61.96		

Economic Growth and Industrialization in Hungary

TABLE 19 (*continued*)

Year	Iron	Sugar	Cotton Textiles	Flour	Woolen Textiles	Electricity
1871	25.74	1.19	5.10	55.58		
72	25.94	1.32	3.70	52.77		
73	29.17	0.86	2.67	45.40		
74	28.57	1.64	4.55	55.63		
75	29.74	1.34	3.81	64.97		
76	22.98	1.65	5.88	58.31		
77	23.61	1.72	7.69	47.45		
78	25.73	3.08	9.25	58.04		
79	21.46	2.87	10.82	87.13		
80	26.17	3.79	12.37	75.37		
81	29.57	3.33	13.89	58.83		
82	32.30	3.80	14.67	76.31		
83	35.65	4.20	18.55	101.21		
84	40.95	3.85	14.71	102.64		
85	39.78	3.37	12.33	93.43		
86	45.72	3.56	14.46	101.79		
87	36.27	2.90	15.35	95.56		
88	37.53	4.74	15.85	110.81		
89	45.12	6.72	16.81	121.93		
90	57.90	5.95	16.69	96.69		0.00
91	58.85	8.13	19.14	103.59		0.85
92	60.36	8.35	20.67	118.01		1.16
93	63.36	9.83	22.12	117.98		1.59
94	67.80	9.35	19.64	136.69		2.05
95	72.12	8.15	20.33	141.80		2.71
96	81.01	9.92	24.32	135.97		2.91
97	83.31	9.36	26.11	132.00		3.10
98	91.01	9.96	27.22	102.27		3.77
99	87.75	11.29	26.35	102.86		4.18
1900	78.57	13.77	25.34	130.66		4.60
01	87.42	13.70	28.49	123.45		5.08
02	83.70	14.03	34.64	111.06		5.67
03	80.00	16.60	37.49	127.12		6.59
04	74.57	12.20	39.09	142.80		7.63
05	80.57	14.78	39.93	131.71		8.24
06	81.15	20.60	49.70	132.78	10.0	9.79
07	103.94	18.55	42.22	151.00	18.0	11.54
08	130.73	16.26	43.26	129.29	21.0	12.96
09	135.96	20.15	48.32	115.63	29.0	12.77
10	124.22	22.02	43.19	129.13	33.0	16.91
11	133.23	22.51	51.96	138.44	38.0	19.57
12	152.75	37.34	58.72	151.84	36.0	21.91
13	165.93	37.48	46.87	140.68	38.0	24.30

TABLE 20

VALUE ADDED IN RAILROAD CONSTRUCTION
AND REPAIR IN AUSTRIA AND HUNGARY IN 1913 PRICES, 1830-1913

Year	Hungary	Austria	Year	Hungary	Austria
1830	0.00	0.00	1871	52.02	71.98
31	0.00	0.00	72	54.87	66.84
32	0.00	0.00	73	49.80	50.01
33	0.00	0.00	74	12.13	22.93
34	0.00	0.00	75	3.58	41.55
35	0.00	0.00	76	18.77	30.86
36	0.00	0.00	77	8.03	33.66
37	0.00	0.75	78	13.35	12.52
38	0.00	0.96	79	15.52	15.70
39	0.00	5.98	80	9.56	15.66
40	0.00	0.00	81	15.61	28.00
41	0.00	11.05	82	39.68	26.34
42	0.00	1.44	83	38.57	30.84
43	0.00	0.00	84	29.21	64.05
44	0.00	5.09	85	25.82	24.77
45	0.00	13.65	86	26.88	31.71
46	1.87	9.38	87	51.30	43.90
47	6.73	8.10	88	23.83	50.42
48	0.91	1.70	80	35.93	32.31
49	0.00	10.07	90	31.40	25.30
50	2.35	6.22	91	49.95	34.45
51	7.15	2.51	92	21.87	22.78
52	0.00	0.99	93	35.81	32.27
53	3.14	1.22	94	43.86	39.13
54	3.69	3.61	95	55.96	26.61
55	4.41	9.73	96	65.63	36.95
56	5.63	12.48	97	61.82	52.16
57	15.73	12.09	98	49.18	64.38
58	16.76	24.26	99	46.70	56.33
59	7.80	14.70	1900	26.31	45.63
60	12.68	17.16	01	26.54	38.29
61	12.45	15.50	02	26.54	44.12
62	4.76	11.23	03	36.00	45.71
63	2.66	11.24	04	27.47	37.00
64	1.29	4.72	05	38.51	44.99
65	13.18	10.95	06	52.57	57.14
66	1.89	17.84	07	37.53	31.86
67	8.87	13.58	08	60.28	38.31
68	21.07	25.03	09	56.79	51.46
69	8.10	44.04	10	44.49	41.85
70	42.20	49.56	11	42.46	33.76
			12	49.60	35.50
			13	44.05	34.81

Source for Table 16-20: JOHN KOMLOS, "The Habsburg Monarchy as a Customs Union: Economic Development in Austria-Hungary in the XIXth Century" (Unpublished Ph. D. Dissertation, The University of Chicago, 1978, pp. 233-246).

ARTICLE 24

JOHN R. LAMPE

Varieties of Unsuccessful Industrialization: The Balkan States Before 1914

THIS paper should begin with a brief defense of its title. "Variety" and "unsuccessful" are doubtful if not dirty words to most economists and many economic historians. The "success stories" of rapid development in Western Europe, Russia and Japan have been the most frequent subject of this JOURNAL's articles on non-American topics. And the discovery of uniformity in the past, rather than variety, is admittedly essential to the development economist's search for predictability in the future that has informed so much of the economic history written since the Second World War.

Yet the variety in the often stereotyped history of the Balkan states is too great to be ignored. It is my strongest single impression after seven years of studying the pre-1914 economic development of Serbia, Bulgaria, Romania and Greece. Clearly missing is the comprehensive uniformity that would permit use of John Stuart Mill's classic list of methods for comparative study.[1] There is all the same enough uniformity to attempt the "intensive comparison" of both similarities and differences that Sylvia Thrupp urged upon us in 1957.[2]

Industrial stirrings (see Table 1 below) that fell short of a sustained spurt and did not absorb significant factors of production from the rest of the economy are one experience shared by these four states before the First World War. But there are other common points of reference that recommend the economic history of Serbia, Bulgaria, Romania and Greece to comparative analysis.

This paper was prepared with the assistance of grants to Romania in 1974, to Bulgaria in 1972 and to Yugoslavia in 1969-70 from the International Research and Exchanges Board, New York, N.Y. Any comment or criticism will assist the author and Marvin R. Jackson with their book in progress, *An Economic History of the Balkan States: The First Developing Nations.* Translations of titles of works in the various Balkan languages are omitted only where the meaning is self-evident.

[1] E.g., to isolate the cause or effect of two contrasting situations by discovering that they share every circumstance in common save one. On Mill's six methods of comparison, see Elias H. Tuma, *Economic History and the Social Sciences: Problems of Methodology* (Berkeley: University of California Press, 1971), pp. 38-9.

[2] Sylvia Thrupp, "The Role of Comparison in the Development of Economic Theory," JOURNAL OF ECONOMIC HISTORY, XVII (December 1957), 555.

Balkan States 57

Most obvious but rarely well defined are the uniformities in the Ottoman economic legacy. Imperial domination from Constantinople lasted from roughly 250 years for the Romanian Principalities (and cut them off from largely Romanian Transylvania until 1918) up to nearly 500 years for the Bulgarian lands. In the process, the Turkish word for mountains gave the "Balkan" peninsula its nineteenth-century name.

Only the last complete century of Ottoman domination concerns us here. During this eighteenth century both Balkan agriculture and urban commerce were marked for the future, but by the weakness rather than the strength of the Porte's once centralized system.

The most striking feature of the rural landscape throughout the peninsula during the eighteenth century was the scarcity of population. Crude available estimates indicate amazingly low totals of under half a million people for Serbia and also the Romanian Principalities, with the Greek and Bulgarian lands somewhat higher only because of a higher urban population.[3] The same sources make it clear that this was a major decrease from earlier centuries under Ottoman rule. The reason was undoubtedly the increasing disorder and actual warfare that accompanied the Porte's weakening central control. Fighting and epidemics took their toll. Surviving peasants sometimes reacted by fleeing to the Habsburg lands and in any case by deserting open country for foothills and forests. Raising livestock predictably became the principal agricultural activity. Peasant mobility and scarcity generally prevented the establishment of estate agriculture, except in Macedonia and northern Greece, neither of which belonged to a Balkan state before 1912.[4] The result was a heritage of scattered smallholdings and livestock-raising that left the sizeable Balkan grain lands almost virgin territory for nineteenth-century expansion, even under existing methods of cultivation.

[3] John R. Lampe, "Financial Structure and the Economic Development of Serbia, 1878-1912" (Unpublished Ph.D. Dissertation, University of Wisconsin, 1971), pp. 68-75; Jack Tucker, "The Rumanian Peasant Revolt of 1907: Three Regional Studies" (Unpublished Ph.D. Dissertation, University of Chicago, 1972), 1-6; Zhak Natan et al., *Ikonomika na Bulgariia*, v. I (Sofia, 1969), pp. 197-207; Hariton Korisis, *Die politischen Parteien Griechenlands, 1821-1910* (Harsbruck/Nürnberg, 1966), pp. 112-21.

[4] Even in the Romanian Principalities, the powerful native boyars were lucky to collect the semi-nomadic, livestock-raising peasantry for the full twelve days of annual demesne labor (*clacă*) that they were authorized to impose. On eighteenth-century Romanian agriculture, see V. Mihordea, *Relaţiile agrare din secolul al XVIII în Moldova* (Bucharest, 1968) and F. Constantiniu, *Relaţiile agrare din Ţara Românească în secolul al XVIII-lea* (Bucharest, 1972).

Ottoman political control of the towns denied the native population the chance to develop its language and culture, but not its commerce.[5] The Ottoman practice of loose, ill-defined commercial control plus the relative paucity of Turkish traders gave other nationalities the chance to supply the growing needs of the Sultan's army and the million mouths in Constantinople. Every Balkan ethnic group and a few foreign ones (Jews, Germans or Austrians and Armenians mainly) seized this opportunity. Yet this initial pattern of mixed urban populations would pose a problem that each of the ethnically self-conscious Balkan states would have to face sooner or later.

The emergence of these four Balkan states from the Congress of Berlin in 1878 with complete or nearly complete political independence from the Ottoman Empire is another important uniformity. Serbia, Romania and Greece, it is true, had started on this road from different directions almost fifty years before. But 1878 brought them all to its end and launched them, along with an autonomous Bulgaria, as modern nation-states at the same time. Native political leaders committed to some form of "modernization," in the phrase of a noted American scholar born in Bulgaria,[6] were subsequently able to struggle for power with monarchs drawn from native stock in the Serbian case and elsewhere from princely European families not directly tied to the ruling dynasties in the Great Powers. The outcomes of these various struggles would have fateful consequences for the economic development of each Balkan state.

Several common parameters in the data for the pre-1914 Balkan states also facilitate comparison. From 1886 to 1912 the borders of all four did not change. Neither did any of them go to war with each other, nor with anyone else, excepting the brief Greek war with the Ottoman Empire in 1897. The economic confusions of war are thus absent. From 1881 to 1914, the currencies of all four nations (the Serbian dinar, the Bulgarian lev, the Romanian leu and the Greek drachma) were tied formally or informally under the rules of the Latin Monetary Union to a nominal unit value equal to the

[5] On these commercial activities, see Traian Stoianovich, "The Conquering Balkan Orthodox Merchant," JOURNAL OF ECONOMIC HISTORY, XX (June 1960), 234-313. On urban economic life in the Balkans, the definitive work, soon to be published in English, is Nikolai Todorov, *Balkanskiiat grad, XV-XIX vek* [The Balkan Town, 15th-19th Centuries] (Sofia, 1972).

[6] Cyril Black, *The Dynamics of Modernization* (New York: Harper, 1967).

TABLE 1
GROSS OUTPUT OF SELECTED ECONOMIC SECTORS, 1911-13
(lei, leva, and dinars)

	Total in Millions			Per Capita		
Sector	Romania (1912-13)	Bulgaria (1911)	Serbia (1911, 1913)[a]	Romania (1912-13)	Bulgaria (1911)	Serbia (1911, 1913)
Large-scale[b] private industry	519	112	95	71	25	33
Small-scale[c] private industry	185	88	116	25	20	39
Agriculture[d]	1,501	611	590	206	139	197
Forestry	206	97	95	28	22	32
Livestock products	292	213	273	40	48	91
Mining and oil	106	1	16	15	.3	5

[a] The figures for large-scale industry and mining are for 1911. The others are for 1913, after Serbia had been at war for two years and had experienced a marked but not runaway inflation of 10-20 percent.
[b] Generally defined as an enterprise using some mechanical horsepower and employing at least 20 workers and 20,000 lei, leva, or dinars in capital.
[c] Generally defined as artisan shops or rural household industry.
[d] Grain for flour milling included here; other double-counting not eliminated.
Sources: See Table 2.

French gold franc, or 19.3 cents. By 1910, moreover, the market value of each currency had achieved par with the gold franc, although only Romania had ever converted (in 1890) to the gold standard.

Any comparison of the industrial statistics in Tables 1, 2, and 3, available only for Serbia, Bulgaria and Romania,[7] to the production of the other sectors presented there makes it plain that the mini-spurts of industrial growth in the Balkan states before the First World War do not meet the criteria for sustained industrialization. Gross output of large-scale industry, with the value of grain for flour milling excluded, simply accounts for too little of total gross output, that is, social product (under ten percent of the sum of all sectors listed for Serbia and Bulgaria, with Romania's eighteen percent cut in half when we insert the value-added industrial output

[7] For Greece, comprehensive industrial statistics are not available until 1917. By then the country had almost doubled its prewar territory and prices were badly inflated after five years of mobilization (the Romanian economy did not mobilize fully until 1916). Large-scale industrial production, as defined in Table 1, did match the Romanian total for 1912-13. But when flour output (over 300 million of the 500 million total) is adjusted for a 250 percent inflation since 1912 and textiles (73 million) for an inflation and increased military output that had quadrupled its value, the prewar Greek total reduces to little more than the Serbian and Bulgarian totals. Available Greek data are collected in E.-J. Tsouderos, *La relèvement économique de la Grèce* (Paris/Nancy, 1919).

TABLE 2
COMPARATIVE REAL GROWTH RATES, 1900-1915
(Annual Average Rate by Geometric Method)

	Romania		*Bulgaria*		*Serbia*	
	Gross Output	*Per Capita*	*Gross Output*	*Per Capita*	*Gross Output*	*Per Capita*
Large-scale private industry	(1901/02-1915) 7.0%	5.3%	(1904-1911) 14.3%	12.7%	(1901-1911) 12.5%	10.5%
Grain crops	(1901/05-1911/15) 1.7%	0.0%	(1894/1904-1909/12) 2.3%	0.5%	n.a.	
Mining and oil	(1901/02-1912/13) 13.4%	12.0%	n.a.		(1901-1911) 21.5%	19.3%
Population (1901-1911)	1.6%		0.5%		1.9%	

Sources: John R. Lampe and Marvin R. Jackson, "The Genesis of the State Sector in the Balkans," paper presented to the VIth International Congress on Economic History, Copenhagen, August 19-23, 1974, Table I. For a further discussion of Balkan data problems for the period before 1914, with emphasis on Romania, see Marvin R. Jackson, "Quantitative Economic History in the Balkans: Observations on the Pre-1914 Period," paper presented at the 3rd International Congress on Southeastern Europe, Bucharest, September 4-10, 1974.

even for 1915, in Table 3, into Table 1). In addition, the length of these mini-spurts is too short to assume them to be self-sustaining. I have reversed my initial impression, moreover, that the respectable growth rates of the last prewar decade (see Table 2) could have turned into sustained industrialization on either a rapid, German-style or slow, French-style pattern, had the First World War not intervened. I will argue that the long-term prospects of these industrial stirrings were very limited.

In this brief paper perhaps the best approach is to pose the most obvious question that appears from an inspection of Table 1.[8] Why did pre-1914 Romania achieve totals of gross industrial output per capita that were well over twice those of Serbia and Bulgaria and probably Greece? Nicholas Spulber, has suggested several answers without further comment: "Its [Romania's] better endowment in raw materials—notably oil; its somewhat better access to natural transportation facilities; its larger volume of foreign trade, due

[8] I have considered elsewhere the other question about comparative industrial development that is suggested by Table 1: namely, why was the per capita output of large-scale industry and mining for Serbia about half again the Bulgarian total? See John R. Lampe, "Finance and the Pre-1914 Industrial Stirrings in Bulgaria and Serbia," forthcoming in *Southeastern Europe.*

TABLE 3
COMPOSITION AND REAL GROWTH OF ROMANIAN LARGE-SCALE INDUSTRY
(million lei)

Branch		*Nominal Gross Output in 1915*	*Nominal Net Output in 1915*	*Price Index of Gross Output (1901/02 = 100)*	*Average Annual Real Growth of Gross Output, 1901/02-1915*
Mineral Materials and Glass		21.4	13.3	89	10.6%
Metal Processing		42.4	14.6	102	11.0%
Wood Processing		56.7	26.3	177	4.4%
Textiles		45.8	20.9	151	7.1%
Food		215.1	67.1	155	4.5%
Of Which:	Flour	115.5	17.2	117	2.3%
	Sugar	37.8	20.3	143	6.6%
Chemicals		154.7	53.0	n.a.	n.a.
Of Which:	Petroleum	141.3	48.7	243	14.9%
Leather		24.6	3.5	98	9.0%
Paper and Printing		20.9	9.5	73	8.4%
Total		584.1	210.2	145	7.0%
Total without Petroleum		413.4	161.5	n.a.	6.2%

Sources: An unpublished study, "Economic Change in Romania, 1862-1914," by Professor Marvin R. Jackson, Department of Economics, Arizona State University, Tempe. Basic data for 1915 are found in *Anuarul statistic al României, 1915-16*, (Bucharest, 1919), pp. 194-95. Real growth rates are based on data for 1901/02 in the *Ancheta industrială, 1901/02* (Bucharest, 1904), adjusted to conform to 1915 data and reflated by the above price indices for individual industries to reflect the price changes between 1901/02 and 1915.

partly to depressed consumption levels in the countryside . . . its relatively larger domestic market."[9] Each of these advantages deserves further comment. So do other possible answers more directly concerned with the process of industrial production in Romania and the other Balkan states: the various means of borrowing more advanced European technology, the effect of Romanian adherence to the gold standard on its financial system and on its access to foreign funds, the forms of industrial organization, and the Romanian government's initiation of protective tariffs and industrial legislation ahead of its Balkan neighbors.

A second question deserves to be asked, although it can be less thoroughly answered. What external limits or internal weaknesses kept the greater Romanian industrial stirrings from doing better

[9] Nicholas Spulber, *The State and Economic Development in Eastern Europe* (New York: Random House, 1966), pp. 61-9.

and achieving a pattern that in Czech fashion would have survived both the World War and the Great Depression?[10]

These inquiries can be usefully organized around a comparison of the factors of production (land, labor, capital and a residual factor for productivity) in the four economies. Further quantitative evidence will be offered where possible, but a good deal of intensive comparison will often take the place of the comprehensive, fully reliable body of statistical data that is missing for all these countries.

I

The neoclassical tendency to emphasize capital and the Marxist tendency to emphasize labor have too often led both sets of scholars to exclude land as a factor of production. In the overwhelmingly agricultural economies of the pre-1914 Balkan states such an omission is dangerous. And here the broadest possible definition of land, including all natural resources and the geo-political location of the respective Balkan states proves instructive.

The distribution of grain-growing land clearly favored Romania and hurt Greece, while leaving Serbia and Bulgaria in an intermediate position. Yields per hectare for Romanian wheat, rye, barley and oats, but not corn, were fifteen to thirty percent higher than for Serbia and Bulgaria and up to fifty percent more than for Greece during the last prewar years.[11] This was in large part a testimony to the rich flatlands north from the Danube along the Black Sea that distinguish the Romanian provinces of Wallachia and Moldavia from the sharply rolling, often mountainous countryside that generally prevails south of the river and that doubtless explains the Turkish name for the peninsula. The large Romanian boyar estates, unique in the nineteenth-century Balkan states except for northern Greece, admittedly achieved higher yields than peasant

[10] For evidence that the Romanian economy lost rather than gained ground during the interwar period, see the still standard work by the late Henry L. Roberts, *Rumania, Political Problems of an Agrarian State* (New Haven: Yale University Press, 1951). On the accomplishments of the interwar Czech economy see the chapter by Zora P. Pryor in Victor Mamatey and Radomir Luža, eds., *A History of the Czechoslovak Republic, 1918-1948* (Princeton: Princeton University Press, 1973), pp. 188-215.

[11] I. Adam and M. Marcu, *Studii despre dezvoltarea capitalismului în agricultură Romîniei* [Studies Concerning the Development of Capitalism in Romanian Agriculture], v. II (Bucharest, 1959), pp. 204-07; Dresdener Bank, *The Economic Forces of the World* (Berlin, 1930), p. 19.

TABLE 4
TOTAL EXPORT VALUE AND TRADE SURPLUS OR DEFICIT, 1886-1911
(million lei, leva, dinars or drachma)

	Romania		*Bulgaria*		*Serbia*		*Greece*	
Annual Average	*Total Exports*	*Surplus or Def.*	*Total Exports*	*Surplus or Def.*	*Total Exports*	*Surplus or Def.*	*Total Exports*	*Surplus or Def.*
1886-90	265.6	−64.8	62.4	−8.2	40.3	+1.1	96.1	−26.0
1891-95	298.0	−96.9	77.5	−6.0	47.4	+10.6	84.8	−29.3
1896-00	251.2	−74.7	68.5	+.5	59.7	+15.7	87.8	−38.9
1901-05	360.6	+61.7	120.0	+25.0	66.4	+13.7	86.7	−52.0
1906-10	499.3	+90.3	118.6	−21.6	84.5	+14.1	119.6	−29.6
1911	691.7	+122.0	184.6	−14.7	116.9	+1.5	141.0	−32.5

Sources: See Table 6.

smallholdings. But this was more the result of their retaining better quality land in the limited land reform of 1864 that it was of using better methods of cultivation.[12] The typical share-cropping arrangement based on an annual contract between the head of the individual peasant household and the boyar owner, or increasingly his lessee manager, was hardly conducive to modern, large-scale agriculture.

The accelerating increases in the value of Romanian exports during the last prewar decades did depend almost entirely on cereals, with oil exports reaching even twelve percent of total value only after the fighting had started. By 1910 Romanian wheat export value had climbed past that of the U.S. to fourth place in the world. Even when the expansion of cultivated land slowed down after 1895, this grain-led increase was not prompted by rising productivity, in the manner described for pre-1914 Hungary by Scott Eddie.[13] It was instead the result of a sharp, probably cyclical increase in world grain prices and subsequent pressure on the peasantry to turn over a larger marketable surplus to the boyar estates. Calculations by Marvin Jackson reveal that the annual rise in real crop output per capita after 1900 was literally nil.[14] Table 6 indi-

[12] Neither Romanian yields per hectare nor productivity per capita showed significant progess after the increases of the 1870's and 1880's. These were in any case led by a rapid rise in cultivated land, from 2 to 4 million hectares between 1866 and 1891. From that date, only a million new hectares were brought into cultivation by 1914, and no significant increase in cereal productivity is recorded in the index calculated by Tucker, *Romanian Peasant Revolt*, pp. 75-78.

[13] Scott M. Eddie, "Agricultural Production and Output Per Worker in Hungary, 1870-1913," JOURNAL OF ECONOMIC HISTORY, XXVIII (June 1968), 197-222.

[14] In his unpublished paper cited in Table 2, "Quantitative Economic History in the Balkans: Observations on the Pre-1914 Period," Jackson also calculates that the

TABLE 5
PERCENTAGE VALUE OF PRINCIPAL EXPORTS, 1886-1911

Annual Average	*Romania*		*Bulgaria*	*Serbia*		*Greece*
	Cereals	*Oil*	*Cereals*	*Cereals*	*Livestock & Proc.Meat*[b]	*Currants*
1886-90	76.8%	—	72.2%	17.8%	38.0%	54.7%
1891-95	79.5%	—	78.0%	30.4%	42.0%	45.2%
1896-00	74.2%	.8%	72.7%	27.7%	29.9%	42.0%
1901-05	75.1%	2.2%	68.8%	20.5%	44.6%	34.5%
1906-10	79.7%	7.3%	64.4%	43.5%	11.8%	30.8%
1911	71.4%	11.8%[a]	70.1%	31.7%	19.8%	33.1%

[a] 1912-13 average; 1911 not available.
[b] Processed meat exceeds one percent of total export value only after 1895 and averages 4-8 percent until rising to 17 percent in 1911.
Sources: See Table 6.

cates the decisive part that price increases for wheat and corn played in the fast-rising export totals for Serbia and Bulgaria as well as Romania before 1914.

In contrast to its neighbors, Greece was forced to import about one quarter of its own grain needs during these last prewar decades. For the largest single part of its exports, Greece had to rely on a current crop whose price had declined with the recovery of French cultivation from phylloxera by the early 1890's.

The Serbian economy was in the fortunate position of being less dependent on grain for her export earnings than either Romania or Bulgaria, as noted in Table 5. Only Serbia remained tied to the animal husbandry that had dominated Balkan agriculture in the late Ottoman period. This was surely an advantage, cushioning Serbian export earnings from the strong cyclical swings of the international grain market and the risk of a bad harvest.[15] These

real value of *total* crop output rose 2.2 percent a year from 1862-66 to 1896-1900 but only 1.7 percent a year from 1901-05 to 1911-15. The rapidly rising value of total exports after 1900, stressed by Professor Hohenberg in his comment on this paper, were the nearly equal result of rising grain export prices and stiffer share-cropping contracts that extracted a higher percentage of peasant production. (See Adam and Marcu, *Studii*, 268-69.) That this heavier extraction was more important as a burden for future Romanian development than as an accumulation of new investment capital may be judged from the seriousness of the 1907 peasant revolt. See Tucker, *Romanian Peasant Revolt* and Philip G. Eidelberg, *The Great Romanian Peasant Revolt of 1907: Origins of a Modern Jacquerie* (Leiden: E. J. Brill, 1974) for evidence that the Romanian rural economy was not nearly comparable enough to the grain-growing regions of recent European settlement like Canada to permit much inference from comparable statistical trends in export growth.

[15] By the end of the 1906-1911 tariff war with Austria-Hungary, moreover, the Serbian economy had converted over three quarters of its meat export value into

TABLE 6
PRICE INDEX FOR PRINCIPAL EXPORTS, 1896-1911

1886-1905 =100	*Romania*		*Bulgaria*		*Serbia*			
	Wheat[a]	*Corn*[a]	*Wheat*[a]	*Corn*[a]	*Wheat*[a]	*Corn*[a]	*Live Cattle*	*Proc. Meat*
1896	92	77	90	87	105	74	127	100
1897	123	82	117	98	124	74	127	139
1898	115	87	125	109	133	107	124	138
1899	115	97	126	103	118	81	129	145
1900	115	107	119	110	95	96	129	168
1901	100	100	113	103	107	97	142	155
1902	108	112	110	111	93	108	130	171
1903	n.a.	n.a.	104	107	100	90	115	143
1904	n.a.	n.a.	103	101	125	97	97	107
1905	n.a.	n.a.	110	116	116	97	95	143
1906	n.a.	n.a.	108	108	105	97	71	155
1907	n.a.	n.a.	134	128	171	119	101	159
1908	n.a.	n.a.	152	151	156	124	95	161
1909	160	n.a.	158	148	168	138	120	180
1910	136	119	140	130	149	132	135	216
1911	136	123	129	124	123	128	177	189

[a] Wheat and corn consistently made up over 80 percent of the cereal exports of these countries. The similar behavior of the prices for rye, barley and oats, the other cereal exports from the Balkan states, further justifies the use of wheat and corn prices.

Sources: *Anuarul statistic al României, 1901-02* (Bucharest, 1904), pp. 291, 298; Ibid., *1909* (Bucharest, 1910), pp. 471, 478, 486; Ibid., *1912* (Bucharest, 1914), pp. 303, 321, 334, 338; Ibid., *1915-16* (Bucharest, 1919), pp. 161, 172-73; C. I. Băicoianu, *Istoria politicei noastre vamale şi comerciale* [A History of Our Tariffs and Commerce], v. I (Bucharest, 1906), pp. 356-57; *Bulletinul statistic al României*, v. XII (Bucharest, 1915), nos. 38-39; Maurice Pearton, *Oil and the Romanian State, 1895-1948* (Oxford: Clarendon Press, 1971), p. 43; Kiril Popoff, *La Bulgarie économique, 1879-1911* (Paris, 1920), pp. 169, 404; John R. Lampe, "Financial Structure and the Economic Development of Serbia, 1878-1912" (Unpublished Ph.D. Dissertation, University of Wisconsin, 1971), pp. 38-41; Douglas Dakin, *The Unification of Greece, 1770-1923* (London: Ernest Benn, 1972), p. 321.

threats continued to hang over the potential accumulation of industrial capital in the impressive Romanian export totals, nearly two and a half times the per capita figures for the other Balkan states.

There are two additional problems for industrial development that might have followed from Romanian foreign trade's heavy dependence on cereal cultivation. First, there is the possibility, insufficiently studied at this writing, that the rising prices for agricultural exports diverted venture capital away from industry. The large number of Jewish lease-holders, originally town merchants,

processed sales that could anticipate a secure future now immune from the threat of border closings by neighboring Austria-Hungary on veterinary pretexts. See Lampe, "Financial Structure," 21-29.

who were managing boyar estates by the turn of the century points in this direction.

Second and more certainly, had the favorable world trend in grain prices continued to give Romania a solid base of export surpluses from which to finance industrial development, that development would almost certainly have operated under the handicap of rising domestic food prices.[16] Table 7 indicates that increases of over thirty percent in the price of daily necessities had occurred in the last prewar decade in Serbia and Bulgaria as well as Romania.[17] Given the slow growth of towns, one result was successful upward pressure on urban wages, also noted in Table 7. Another was the first appearance of serious labor unrest in the Balkan towns. Neither result boded well for the continued growth of private industry. The latter one would of course have momentous consequences for the rise of socialist movements, at least in Serbia, Bulgaria and Romania, in time for the further stimulus of the Russian Revolution.[18] The failure of prices for most manufactures made for domestic use, as noted for Romania in Table 3, to keep pace with the cost of food and sometimes wages was of course a further source of discouragement to private investment in industry.

Relevant to the utilization of labor and capital as well as land were several Romanian advantages in geo-political location. As noted by Professor Spulber, the country enjoyed "somewhat better access to natural transportation facilities." This access took the form of an eastern border on the Black Sea, a southern border consisting largely of the Danube, and a western border with Austria-Hungary and its railway network leading to Central Europe. Bulgaria shared only the first two borders and Serbia only the last two.

[16] During the last two prewar decades a rapidly rising rural population was cancelling out the increase in per capita production that additions to cultivated land would have afforded (see Table 2). Crop lands in turn were rapidly approaching the arable limits, about half of total acreage. In the absence of higher productivity, lower birthrates, or more urban migration, this situation would make interwar Romanian agriculture synonymous with overpopulation and underemployment.

[17] It should be noted that in the Romanian and Bulgarian cases there was a less prominent rise in the price of bread but instead sharp increases in the price of meat and vegetables. We may speculate that the greater Romanian and Bulgarian emphasis on grain production did keep bread prices down but encouraged neglect of other kinds of agricultural activity and thus left these goods in short supply.

[18] The Balkan labor situation may well be comparable to the Italian experience with aggressive trade unionism and developing industry before the First World War, as observed by Alexander Gerschenkron in "Notes on the Rate of Industrial Growth in Italy, 1881-1913," in his *Economic Backwardness in Historical Perspective* (New York: Praeger, 1965), pp. 86-7.

TABLE 7
INDEX OF URBAN FOOD PRICES AND DAILY WAGES, 1900-1912

	Romania[a]			Bulgaria[b]			Serbia[c]		
1900—100	*Bread*	*Pork*	*Beans*	*Bread*	*Pork*	*Beans*	*Bread*	*Pork*	*Beans*
1901	104	100	122	96	102	92	105	97	110
1902	96	100	117	94	106	95	111	103	130
1903	100	106	130	89	123	95	105	117	155
1904	108	113	152	88	108	107	95	118	184
1905	108	115	156	94	117	121	95	130	251
1906	104	130	130	91	123	116	105	116	203
1907	117	136	139	115	129	132	125	106	163
1908	133	140	113	130	148	149	140	115	155
1909	133	149	135	135	160	121	n.a.	n.a.	n.a.
1910	121	164	152	119	158	106	136	132	246
1911	113	166	152	110	154	133	136	146	299
1912	108	168	165	123	151	147	n.a.	n.a.	n.a.

	Romania[a]	Bulgaria[b]		Serbia[c]	
	1905=100	*1891-1900=100*		*1892-1902=100*	
	Railway Worker	*Day Laborer*	*Bricklayer*	*Day Laborer*	*Bricklayer*
1900	n.a.	83	84	104	100
1901	n.a.	n.a.	n.a.	n.a.	100
1902	n.a.	n.a.	n.a.	102	100
1903	n.a.	n.a.	n.a.	103	104
1904	n.a.	87	87	106	104
1905	100	95	96	105	109
1906	104	101	107	110	113
1907	124	103	118	111	118
1908	128	104	121	114	121
1909	135	109	124	n.a.	n.a.
1910	128	115	131	n.a.	n.a.
1911	134	125	138	n.a.	n.a.
1912	140	n.a.	n.a.	n.a.	n.a.

[a] Bucharest only.
[b] Sofia only.
[c] All towns.

Sources: *Anuarul statistic al României, 1915-16* (Bucharest, 1919), pp. 39, 160; I. Mollev and A. U. Totev, *Tseni na zemedelskite produkti u nas, 1881-1934* [Prices of Our Agricultural Products, 1881-1934] (Sofia, 1935), pp. 90-98; *Iubileena kniga na grad Sofiia, 1878-1928* [Jubilee Book of the City of Sofia, 1878-1928] (Sofia, 1928), p. 377; John R. Lampe, "Financial Structure and the Economic Development of Serbia, 1878-1912" (Unpublished Ph.D. Dissertation, University of Wisconsin, 1971), pp. 405-06; *Statistički godišnjak Kr. Srbije, 1906* [Statistical Yearbook of the Kingdom of Serbia, 1906] (Belgrade, 1908), p. 39; Ibid., *1907-08* (Belgrade, 1913), p. 391.

Greece extended into the Mediterranean, it is true, but was barred from rail access to the European mainland by Ottoman refusal to allow a link-up in northern Greek lands that the Porte continued to control until 1912. And when we consider all of the energy and expense that the pre-1914 Balkan states concentrated in their centers of government, the location of the Romanian capital in Bucharest helped to develop the above advantages. Although not on the Danube like Belgrade, the city was less than fifty miles away and was linked by rail to the river port of Giurgiu in 1869. It was less than one hundred miles from the Habsburg border.[19] Athens had the port of Piraeus only a few miles away but no other connections. I have commented elsewhere on the location of the Bulgarian capital in Sofia, far from either the Danube or the Black Sea and largely surrounded by mountains.[20] Here was the one Balkan seat of government in a patently poor position to develop the economic advantages of its borders.

II

Turning to the supply of industrial labor, I wish to emphasize the tremendous advantages offered to Balkan industrial enterprises by immigration from more developed economies. Such workers, trained in manufacturing or at least accustomed to the disciplines of urban life were a profitable alternative to a labor force of local peasants with perhaps a limited primary education. Both Romania and Serbia benefited from this kind of immigration, and had little emigration to reduce the net gain.

During the growing disorders late in the Ottoman period, sizeable numbers of Romanian and Serbian peasants had moved across the nearby Habsburg border to safer ground. The Romanians migrated west to join their fellows in Transylvania, and the Serbians north into the Vojvodina. There each group found most of the arable land in the hands of the ruling Hungarian magnates but urban commerce open to their migrants' initiative. As their homelands secured internal order and external independence during the nineteenth century, some of them returned, usually to the capital city.

Prećani Serbs (from "across" the Danube and Sava rivers in Habsburg territory) started to come back in the 1830's and accounted

[19] Lines to the Black Sea and the Danube delta soon followed. The rail connection with Austria-Hungary was completed in 1879. *Istoria Romîniei*, v. IV (Bucharest, 1965), pp. 463-66; G. M. Dobrovici, *Istoricul dezvoltării economice şi financiare a Romîniei, 1823-1933* (Bucharest, 1934), p. 101.

[20] It was placed there on purpose, to guard against Ottoman attack and to strengthen the Bulgarian claim to the Macedonian lands lost in the Treaty of Berlin. Lampe, "Finance," forthcoming in *Southeastern Europe*.

for about a third of Belgrade's population by 1900. They were joined by Austrian, Czech and Jewish immigrants who made up only five percent of the city's population but were heavily represented among artisans and traders. It was recent, unnaturalized arrivals among these groups that accounted for almost half of the skilled labor in Serbian industry by 1910.[21]

The Romanian migration from Transylvania started later but by 1900 accounted for almost 40,000 of Bucharest's 276,000 residents. They were typically employed in artisan shops or manufacturing enterprises. So were the relatively fewer ethnic German immigrants from Transylvania, who already totalled 4,000 by 1824 and whose very name became a Romanian word for master-craftsman or technician. Bucharest's impressive Jewish population of 50,000 by the turn of the century had come more from Russia than Central Europe but brought commercial if not manufacturing skills along with it.[22]

Bulgaria had no emigrées outside its border in sizeable numbers except for certain areas in the Ottoman Empire and on the Romanian side of the Danube, including Bucharest. Few immigrants from the Habsburg lands of any nationality found their way to pre-1914 Bulgaria. In addition, there was an outflow of educated Bulgarians, mainly teachers, to the Ottoman-controlled Macedonian lands.[23]

The Greek economy of course gained by the return of an undetermined number of widely dispersed emigrées, from Western as well as Eastern Europe, throughout the nineteenth century. After 1900 this gain must be balanced against two burdens. First, the new immigrants were now mainly impoverished refugees from the Ottoman lands. This total of about 25,000 only anticipated the problem of absorbing over a million refugees from Turkish territory that proved so serious for the interwar Greek economy. Second and more significant for the prewar Greek economy, the one outward emigration of any size in an independent Balkan state gathered momentum rapidly.[24] This drain damaged both industrial and agri-

21 Lampe, "Financial Structure," 86-88, 260; Industrijska Komora Kr. Srbije, *Izveštaj* [Report], Tables 9 and 10.

22 C. C. Giurescu, *Istoria Bucureştilor* (Bucharest, 1966), pp. 159, 172.

23 Lampe, "Finance," forthcoming in *Southeastern Europe*.

24 From an annual emigration, principally to the United States, of 1,000 to 2,000 in the 1890's, the yearly average jumped to 10,700 for 1901-05 and then to 31,500 for 1906-10. British Parliamentary Papers, *Sessional Papers*, 1909, v. 95, no. 4208; Korisis, *Die politischen parteien Griechenlands*, p. 119. On the interwar problems of

cultural development. Gone were people mobile and enterprising enough to move to new jobs. The departures created a shortage of urban labor that forced up industrial wages during the period 1906 to 1910 by as much as sixty percent in some areas. British consular reports of the period regarded these impositions as serious enough to outweigh the rising remissions of overseas income, which did cancel out the Greek trade deficit during the last prewar decade.[25]

The quality of industrial labor remaining in Greece is also open to question. As in the other three Balkan states, the Greek system of public education paid very little attention to technical or practical training that might have prepared pupils for factory labor. A few small secondary schools for such purposes were opened throughout the Balkans before 1914 but had few graduates. The primary and other secondary schools spent virtually all of their time teaching the national language, literature and history. The fact that Romania and Serbia had lower levels of literacy than Bulgaria and Greece[26] and yet made better industrial beginnings only reinforces the emphasis laid at the start of this section on the importance of already trained immigrants.

Concerning labor's role as a source of demand for manufactured goods, we can broaden our vision beyond the relatively small industrial labor force. But how far? A good approximation for the non-agricultural labor force, fully absorbed in the money economy and capable of rapid accommodation to industrial employment, would seem to be urban population. However much the Balkan peasantry had been drawn into selling in the money economy to pay for supplies and taxes, its surplus over subsistence would more likely be saved or spent on buying more land than on manufactures.[27]

---

the Greek economy with refugees from Turkish territory, see Dimitri Pentzopopoulos, *The Balkan Exchange of Minorities and Its Impact on Greece* (Paris, 1962), pp. 143-67.

25 British Parliamentary Papers, *Sessional Papers*, 1907, v. 90, no. 3818, p. 6 and 1911, v. 93, no. 4649, p. 9.

26 Literacy among army recruits for the Balkan Wars in 1912, the one common measure available, was 70 percent for Greece and 75 percent for Bulgaria. Lagging behind were Romania with 59 percent and Serbia with 50 percent. Douglas Dakin, *The Unification of Greece, 1770-1923* (London: Ernest Benn, 1972), Appendix VI; Popoff, *La Bulgarie économique*, pp. 27, 31; L. Colescu, *Statistica ştiutorilor de carte din România din 1912* [Statistics of Literacy for Romania in 1912] (Bucharest, 1915), p. 63.

27 For Serbia, see the statistical sampling of peasant income, consumption and saving in M. Avramović, *Naše seljačko gazdinstvo* [Our Peasant Economy] (Sarajevo,

At first glance, however, this urban standard is of little help in identifying the "relatively larger domestic market" that Professor Spulber rightly speaks of as one reason for Romania's more significant prewar industrial development. The far larger industrial production per capita for Romania may to some extent be the result of stimulus to economies of scale from the largest total population in the Balkans (7.2 million by 1912, versus 4.4 million in Bulgaria and 3 million in Serbia and Greece). Yet when we look at the proportions of urban population in these totals, recorded in Table 8, we find Romania's seventeen percent in a middle position, just behind Bulgaria at nineteen percent, ahead of Serbia's fourteen percent but well behind Greece's twenty-four percent. These aggregate percentages only mask the concentration of the Romanian total in a few towns; Bucharest, Iași, Brăila, Galați and Ploești accounted for over half. Table 8 reflects the unchallenged position of Bucharest as the largest Balkan capital. Surely this concentration counted for more than total urban numbers in offering Romanian industrial enterprises a market that promised to justify large-scale, mechanized production.[28]

A growing urban market for industry typically extends past clothing and other consumer goods to construction materials. In this last regard the Romanian capital city enjoyed an advantage beyond its size. I do not refer to the extensive construction of impressive public buildings and European-style facilities like sewer systems and tram lines. All this was done in the other Balkan capitals as well.[29] What appeared only in Bucharest was the steady demand for impressive private houses. A brief tour of the four capital cities even today makes the contrast clear. The villas of pre-1914 Bucharest easily compare to some of the most lavish residences built

1927). The judgment of a Russian scholar on his own country in 1911 probably fits the Balkan states just as well: "The town population . . . consumes far more than the rural, of some products 10 to 20 times as much . . . the growth of consumption can to a considerable extent be explained by the growth of town population." Cited by Olga Crisp, "The Pattern of Russia's Industrialization Up To 1914," in Pierre Leon, François Crouzet, Richard Gascon et al., *Industrialization en Europe en XIXm siècle* (Paris, 1972), pp. 370-71.

28 As evidence we may note the location in Bucharest of almost half the Romanian industrial enterprises listed for 1901-02, a much higher fraction than in Belgrade or Sofia. Giurescu, *Istoria Bucureştilor*, p. 291; Lampe, "Finance," forthcoming in *Southeastern Europe*.

29 See Frederic Damé, *Bucharest en 1906* (Paris, 1907), *Iubileena kniga na grad Sofiia, 1878-1928* [Jubilee Book of the City of Sofia, 1878-1928] (Sofia, 1928); *Istorija Beograda*, 3 vols. (Belgrade, 1974).

TABLE 8
POPULATION OF TOWNS[a] AND CAPITAL CITIES, 1860-1910
(thousands)

	Romania		*Bulgaria*		*Serbia*		*Greece*	
Year	*Total Pop.*	*Urban %*	*Total Pop.*	*Urban %*	*Total Pop.*	*Urban %*	*Total Pop.*	*Urban %*
1860-61	3,918	8%	—	—	—	—	—	—
1874	—	—	—	—	1,354	11%	—	—
1877-79	4,486	—	—	—	1,700	—	1,654	18%
1884	4,862	—	3,024	18%	1,902	12%	—	—
1889-90	5,318	—	3,232	19%	2,162	13%	2,187	—
1895-96	5,710	—	3,441	—	2,312	14%	2,434	—
1899-1900	5,957	19%	3,716	19%	2,493	14%	—	—
1907	6,684	—	—	—	—	—	2,632	24%
1910	6,996	17%	4,306	19%	2,912	14%	—	—
Year	*Urban Pop.*	*Bucharest*	*Urban Pop.*	*Sofia*	*Urban Pop.*	*Belgrade*	*Urban Pop.*	*Athens*
1860-61	313	122	—	—	—	—	—	—
1874	—	—	—	—	174	27	—	—
1877-79	—	177	543	20	—	—	293	—
1884	—	—	—	—	236	35	—	—
1889-90	—	—	610	45	286	54	465	—
1895-96	—	—	—	—	319	59	526	—
1899-1900	1,120	276	700	67	351	70	—	—
1907	—	—	—	—	—	—	628	111
1910	1,195	293	829	103	413	90	—	—

[a] Minimum population of 2,000.

Sources: *Anuarul statistic al României, 1912,* (Bucharest, 1914), p. 18; G. D. Cioriceanu, *La Roumanie économique, 1860-1915* (Paris, 1928), p. 312; *Statisticheski godishnik na Bulgarskoto Tsarstvo, 1911* [Statistical Yearbook of the Bulgarian Kingdom, 1911] (Sofia, 1912), pp. 60-61; Zhak Natan et al., *Ikonomikata na Bulgariia,* v. I (Sofia, 1969), p. 374; *Durzhaven vustnik* [Official Gazette] (Sofia), Feb. 5, 1911, p. 15; John R. Lampe, "Financial Structure and the Economic Development of Serbia, 1878-1912" (Unpublished Ph.D. Dissertation, University of Wisconsin, 1971), p. 15; Hariton Korisis, *Die politischen Parteien Griechenlands, 1866-1910* (Harsbruck/Nürnberg, 1966), p. 118.

in the United States during that era, although the typical owner was a wealthy boyar or merchant rather than an industrialist. The production of construction materials (cement, glass and lumber) that grew with this expanding market achieved nearly ten times the value of comparable production in Serbia or Bulgaria by the start of the First World War.[30]

[30] N. P. Arcadian, *Industrializarea României* (Bucharest, 1936), pp. 125-26; John R. Lampe, "Serbia, 1878-1912," in Rondo Cameron, ed., *Banking and Economic Development: Some Lessons of History* (New York: Oxford University Press, 1972), p. 127; *Statisticheski godishnik na Tsarstvo Bulgariia, 1911* [Statistical Yearbook of the Kingdom of Bulgaria, 1911] (Sofia, 1912), p. 267.

At the same time, the promising growth of this sector of Romanian industry should not obscure the small base from which it began. Its share of net industrial output rose in real terms from 7.4 to 10.0 percent between 1901 and 1915, according to data in Tables 1, 2, and 3 which do not include lumber. This limited production is partly explained by the fact that as big as Bucharest was compared to the other Balkan capitals, almost three times larger, it was not big enough. Also, there was no export market near enough for such bulky goods. The rest of the explanation had best await the final section of this paper and its inquiry into state policy and the private organization of industry.

III

First, the relative unimportance for this comparative study of industrial capital, the last of the three classic factors of production, needs explaining. For this brief paper, our focus can be narrowed to the question of capital formation under the gold standard, recently well explored for Tsarist Russia by Haim Barkai.[31] Romania had formally adopted the gold standard in 1890, seven years before Count Witte's Russia, and the subsequent history of its financial structure is no less instructive if not as well documented. In the absence of data on the behavior of the Romanian national product and stock of money over time, the experience of the other Balkan experience under bimetallism can be consulted.

In Romania, as in Serbia and Bulgaria, the creation of a national money supply had to await the cutting of sufficient ties with the Ottoman Empire. When this occurred in 1878, all three governments quickly approved the founding of national banks with new and exclusive powers of note issue. The Banca Națională a României opened on a permanent basis ahead of the other two, in 1880.[32] By 1911 its total assets had reached a figure, as noted in Table 9, that was half again the combined total for the Serbian and Bulgarian central banks. Hence a larger base of loanable funds from which to support industrial development and at interest rates after 1900 that were about half the twelve percent prevailing in 1880.

[31] Haim Barkai, "The Macro-Economic Development of Tsarist Russia in the Industrialization Era: Monetary Developments, the Balance of Payments and the Gold Standard," JOURNAL OF ECONOMIC HISTORY, XXXIII (June 1973), 339-71.

[32] For details, see V. Slavescu, *Istoricul Băncii Naționale a României, 1880-1924* (Bucharest, 1924) and Lampe, "Finance," forthcoming in *Southeastern Europe.*

TABLE 9
DISTRIBUTION OF BANK ASSETS IN 1911
(millions of lei, lev, dinar and drachma)

Country	*Central* Bank	*Other State Banks* (Number of Banks)	*Private Native Banks* (Number of Banks)	*Foreign Banks* (Number of Banks)	*Total*[a]
Romania	760	463 (2)	512 (144)	295 (4)	2,030
Bulgaria	310	235 (1)	100 (53)	90 (5)	735
Serbia	210	164 (1)	320 (172)	70 (3)	764
Greece	494	n.a.	n.a.	335 (2)	n.a.

[a] These totals should be reduced by about five percent for layering of interbank transactions, if the Serbian situation is any guide.

Sources: *Anuarul statistic al României, 1912,* (Bucharest, 1914), pp. 176, 184, 444; V. Slavescu, *Marile băncii comerciale din România* [The Large Commercial Banks of Romania] (Bucharest, 1915), pp. 14, 27; *Banca Românească, 1911-1920* (Bucharest, 1920), p. 37; *Banca Marmarosch-Blank & Co., 1848-1923* (Bucharest, 1923), pp. 62-64; British Parliamentary Papers, *Sessional Papers,* 1914, v. 92, no. 5224, pp. 40-41; John R. Lampe, "Finance and the Pre-1914 Industrial Stirrings in Bulgaria and Serbia," forthcoming in *Southeastern Europe,* Table II.

Yet the financial history of the last two prewar decades suggests that none of these banks was able to offer special assistance to native industry.

The supply of currency and central bank credit was held back throughout the Balkans by the interaction of government demand for credit and the European gold standard. Ironically, it was the National Bank of Greece, already in successful operation for thirty-five years by 1878, that suffered most and longest from this combination of forces.[33] The bank was released from three decades of rising state demands for loans of unsecured note issue in 1898 only to run afoul of the gold standard. A European financial commission, empowered to reorder Greek finances when the government could no longer meet annual payments on over 600 million francs worth of foreign debts, obliged the bank to do its part in reducing to par the sixty-eight percent premium payable for European gold-backed currency in paper drachma. This part consisted of putting a moratorium on new note issue and withdrawing two million drachmas a year from circulation; the explicit goal was Greek adoption of the gold standard.[34] Only when par had been achieved in 1910

[33] The best secondary source on this complex relationship is Dimitrios Tourpalis, *Das Bankwesen Griechenlands* (Würtzberg, 1933), pp. 27-51.

[34] British Parliamentary Papers, *Sessional Papers,* 1911, v. 93, no. 4285, 20-27. A

was the bank permitted to pump 70 million drachmas into a badly strained money supply (outstanding note issue had been about 140 million). In the meantime the factories founded during the 1890's to take advantage of the implicit import tariff and export subsidy (furnished by the domestic premium paid for European currencies and goods) had been badly squeezed. The gradual reduction of the premium had eliminated the hidden tariff and the subsidy at the same time that the limit on note issue had restricted domestic credit.

The Serbian and Bulgarian central banks faced lesser versions of the same problem in the 1890's because of rising state demands for silver-backed rather than entirely unsecured loan issues of new currency. The *agio* on their currencies never exceeded twenty-five percent, however, and after 1900 both banks held their note issue in check to maintain par with the gold franc almost continuously until the Balkan Wars.[35]

Despite its early adoption of the gold standard, Romania did not escape similar restrictions on the note issue of its central bank. From the start the country adhered to what one Romanian observer has rightly called a cereal standard.[36] The Banca Națională a României was never prepared to export its gold reserves to support the leu internationally and would exchange only small amounts of currency for gold on demand in Bucharest. Large export surpluses, of the sort in 1888 to 1889 that had first permitted adherence, were counted on to keep the leu at par. In their frequent absence, the bank was able to export foreign exchange reserves but was unable to prevent an annual average premium of two and a half percent on francs bought with lei. This premium actually passed the gold export point only once (in 1899, the year of Romania's worst prewar harvest). The recurring threat of higher premiums was nonetheless sufficient to restrain the bank's note issue even after the reserve requirement was cut back to thirty-three from forty percent,

good brief history of the pre-1914 Greek foreign debt is found in William H. Wynne, *State Insolvency and Foreign Bondholders*, v. II (New Haven: Yale University Press, 1951), pp. 283-348.

35 Lampe, "Finance," forthcoming in *Southeastern Europe*.

36 A. I. Popescu, *Variațiile sezonale ale leului* [Seasonal Variations of the Leu] (Bucharest, 1927), p. 7. For an extended discussion of Romania's early experience with the gold standard, see C. C. Kirițescu, *Sistemul banesc al leului și precursorii lui* [The Banking System of the Leu and Its Forerunners], v. I (Bucharest, 1964), pp. 327-36.

to which it had been raised as part of going on the gold standard. The bank's note issue never exceeded the rate of increase for the Serbian and Bulgarian central banks until the large export surpluses of the last few prewar years.[37] In addition, the attempt to adhere to the gold standard repeatedly frustrated the Romanian bank's announced intention of aiding commerce and especially industry by adjusting its rates countercyclically.[38]

What about the flow of foreign capital that the gold standard was supposed to bring into Romania? The first place to look would logically be the state loans that have long been recognized as the principal form of European investment in the pre-1914 Balkans. What advantages did Romania enjoy in borrowing European capital after 1890, as opposed to before, and also as compared to the other Balkan states? None of significance. For the period 1891-1911, the Romanian government borrowed a nominal total of 1,286 million francs at an average effective interest rate of five and a half percent. For the shorter period 1881-1890, a total of 798 million francs nominal was contracted at five and three quarters percent effective. Leaving Greece and her special problems with state borrowing aside, Serbia and Bulgaria were obliged to pay effective rates averaging six and a half percent until after 1900. But during the last prewar decade the former received five and a half percent effective and the latter five and three quarters percent. Also, during the 1891 to 1911 period, each had borrowed about 700 million francs, comparable to the Romanian total when we consider their smaller size.[39] The Romanian fraction of effective borrowing spent on the economy, as opposed to debt service and military or administrative

37 The note issues of the Romanian and Serbian central banks roughly doubled between 1891 and 1911. After a late start the Bulgarska Narodna Banka raised its note issue more rapidly than any Balkan central bank after 1900. Kirițescu, *Sistemul banesc*, v. II, Annex III; Lampe, "Financial Structure," 199; *Iubileen sbornik na Bulgarska Narodna Banka, 1879-1929* [Jubilee Book of the Bulgarian National Bank, 1879-1929] (Sofia, 1929), p. 184.

38 Discount rates shot up from 5 to 7-8 percent during the monetary crises of 1894, 1899-1900, 1904 and 1907, as the bank's preoccupation with protecting its reserves won out every time. C. I. Băicoianu, *Istoria politicei noastre monetare și Băncii Naționale*, v. II (Bucharest, 1932), pp. 392-94, 623-28; *Enciclopedea României*, v. IV (Bucharest, 1941), pp. 683-84, 718. The regular rules of the gold standard would have required contraction of the bank's note issue, an even harsher penalty.

39 I. Tutuc, *Studiul valorilor mobilare* [The Study of Movable Assets] (Bucharest, 1927), pp. 34-35. L. Aleksić-Pejković, *Odnosi Srbije ca Francuskom i Engleskom, 1903-1914* [Serbia's Relations With France and England, 1903-1914] (Belgrade 1965), p. 808; *Doklad do Ferdinand I . . . 1887-1912* [Report to Ferdinand I . . . 1887-1912] (Sofia, 1912), pp. 254-5.

expenses, was little more than the one fifth allocated by the Serbian and Bulgarian governments. Romania did enjoy substantial advantages in infrastructure over Serbia and Bulgaria, but these were the result of railway, bridge and harbor construction before 1890.[40]

The second place to look for an influx of foreign capital after the adoption of the gold standard is in the activities of the four large European banks that established themselves in Bucharest.[41] Together their assets made up half of the impressive total for the nine large Romanian banks in 1911, noted in Table 9, a figure that in turn exceeded the combined assets of all private banks, native and foreign, in Serbia and Bulgaria combined.

Despite this potential reservoir for investment in domestic manufacturing, the four banks behaved exactly as their counterparts in the other Balkan states did before 1914. They provided trifling long-term assistance to industry.[42] Furthermore, the one foreign bank for which we have a record of the distribution of its short-term credits gave industry, principally textiles, just 12.5 percent of the total.[43] While the one domestic bank to provide important assis-

40 Romania's pre-1914 railway network consisted of 3,552 kilometers, and 2,491 had been built by 1890. So had twenty bridges across the Danube and other rivers, plus modern harbor facilities at Galaţi and Brăila near the mouth of the Danube. Over half of the pre-1890 railway construction was already in place by 1879, when Serbia and Bulgaria had to start virtually from scratch. *Istoria Romîniei,* v. IV, pp. 460-74; I. Cârtână and D. Stafteric, *Dunărea în istoria poporului român* [The Danube in the History of the Romanian People] (Bucharest, 1972), p. 102.

41 These four banks were the Bank of Romania, Ltd., founded in 1865 with English capital and by 1912 having assets exceeding 50 million lei, the Banca Generală Româna, founded in 1895 with German capital from the Disconto-Gesellschaft and Bleichroder banks and by 1912 having assets approaching 150 million lei, the Banca de Credit Român, founded in 1904 with Austrian capital from Vienna's Länderbank and having assets exceeding 100 million lei by 1912, and the Banca Comerciala Româna, founded in 1906 by a long list of European banks and having assets by 1912 approaching 100 million lei. P. M. Sitescu, *Die Kreditbanken Rumäniens* (Bucharest, 1915), pp. 62-3.

42 About a million lei apiece in direct industrial participation came from three of the foreign banks and none from the fourth. C. D. Creangă, *Raport asupra activitate Băncii Generale Române* (Bucharest, 1919); V. Slavescu, *Marile băncii comerciale din România* [The Large Commercial Banks of Romania] (Bucharest, 1915), p. 42. An unknown and undoubtedly important fraction of their short-term credit went to help European oil companies. This was the one exceptional case where capital from large European banks flowed into the private sector of the Romanian economy in large amounts. See Maurice Pearton, *Oil and the Romanian State, 1895-1948* (Oxford: Clarendon Press, 1971), pp. 1-69.

43 This was the Länderbank's Banca de Credit Român, with 10.6 of its 87 million lei in short-term credits for 1915 granted to industry. A. Vijoli, *Cercetare asupru capitalului financiar in ţare noastre* [Research Concerning Finance Capital in Our Country] (Bucharest, 1949), p. 120.

tance to Romanian industry, Banca Marmarosch-Blanc, did have major German and Austrian banks as partners, its Jewish director and most of its management had been born in Romania. Its aid, moreover, took the form of entrepreneurial energy in placing new stock issues and technical expertise in setting up new production processes rather than large investments (its total direct participation in industry was, like the large foreign banks, about one million lei in 1912). Its contribution can be better assessed, therefore, as one of the residual sources of productivity.

So can the private foreign capital that collected in several branches of manufacturing per se, as opposed to the essentially extractive oil industry. Much more European capital unmistakably was assembled in prewar manufacturing in Romania than in any other Balkan country. But why? Surely not because adherence to the gold standard had created internal credit conditions or an external exchange rate for the leu that was noticeably different from the situation in Serbia and Bulgaria after 1900 and in Greece after 1910. Instead, the sources of this European capital must be traced to the 168 joint-stock companies that comprised one third of the almost 500 Romanian industrial enterprises by 1910, in contrast to about 80 of 345 for Bulgaria in 1911, and just 51 of 465 for Serbia in 1910 and 26 of 282 for Greece in 1917.[44] The proportion of capital invested in Romanian industry by these joint-stock enterprises goes well beyond half, although exact figures are lacking. We do know that eighty percent of this joint-stock industrial capital came from outside the country, in contrast to perhaps twenty-five percent for other industrial firms. When we leave aside a petroleum investment of which large European banks and oil companies furnished ninety-two percent, we are still left with foreign shares ranging from forty-six to ninety-four percent of the stock in the timber, metallurgical, paper and sugar industries. Widely scattered sources give the impression that this was mainly capital from

[44] N. N. Constantinescu, *Din istoricul formării şi dezvoltarii clasei muncitoare din România* [Concerning the Formation and the Development of the Working Class in Romania] (Bucharest, 1959), pp. 318-19; *Anuarul statistic al României, 1915-1916* (Bucharest, 1919), pp. 232, 238; G. Toshev, in the *Bulgarsko Ikonomichesko Druzhestvo* [Bulgarian Economics Society journal], XVII (1915), 431; Industrijska Komora Kr. Srbije, *Izveštaj*, 1910, Table 1; Tsouderos, *La relèvement économique de la Grèce*, p. 175. On the obstacles to joint stock incorporation posed by the Byzantine legal code adopted by the Greek state, see A. A. Pepelasis, "The Legal System and the Economic Development of Greece," Journal of Economic History, XIX (June 1959), 173-98.

smaller Central European, especially Austro-Hungarian investors assembled in Romania by local entrepreneurs.

IV

The entrepreneurial skills and commercial connections of the men who founded and managed Romanian joint-stock firms appear to have been one of the two important residual sources of productivity. Here the variety of Balkan history asserts itself again. Setting industrial management in Romania apart from counterparts in the other Balkan states, including Serbia after 1900, was the large number of foreign immigrants. Unlike the industrial labor force, more had come early rather than late in the nineteenth century. More of them were not ethnically Romanian. Jewish and German or Austrian names frequently appear on enterprises and lists of stockholders. With these immigrants came technical knowledge, typically diffused before 1914 by individuals rather than international companies as J. D. Gould has observed.[45] Behind stayed relatives and acquaintances who offered continuing access to new technology and more capital. While the extent of their participation elsewhere in Romanian industry is unclear, such immigrants dominated three of the four branches, aside from oil, which European capital appeared to dominate.[46] This seems significant because equal contri-

[45] J. D. Gould, *Economic Growth in History* (London: Methuen, 1972), pp. 354-55.

[46] For timber with joint-stock capital seventy percent foreign, the Goetz company was admittedly set up by Austrian interests in Vienna in 1873 to exploit forests elsewhere in the Dual Monarchy. But the crash of 1873 turned Goetz and his associates away from the Monarchy to accept the proposal of the Banca Marmarosch-Blank to buy timber lands in Romania. By 1883 the firm had incorporated with its head office in Bucharest and only a branch in Vienna. By 1902 Romanian investors, urged on by the Bucharest bank, had bought up over a third of the company's stock, now totaling 15 million lei, and enjoyed equal representation with the Austrians on the board of directors. In the last prewar years the firm's sawmill, the largest and most modern in Romania, employed 3,000 workers, one third of whom were immigrants from Austria-Hungary. Most of the staff and technicians were Austrian.

For metallurgy with joint-stock capital seventy-four percent foreign, all three of the combined iron foundries and machine shops operating in 1912 in Craiova, the country's chief railway head west of Bucharest and an industrial center today, bore Austrian or German names. Only one of seven joint-stock metallurgical firms did not bear such a name. And this was the nail factory of the young Romanian doctor turned industrialist, Emil Costinescu, almost half of whose three million lei in paid-capital was held by a scattering of Austrian investors and one third of whose labor force were non-Romanian immigrants from Austria-Hungary.

For paper with joint-stock shares forty-six percent foreign, the picture is more mixed because the largest firm, the Letea factory, was founded in 1881 on state ini-

butions from these four branches accounted together for one half of the real net output for Romanian industry in 1915. Also, the real growth rate for their net output between 1901-02 and 1915 nicely matches the average rate of seven percent a year for all industry noted in Table 3 above.

For sugar, however, with ninety-four percent foreign stock shares, there is no denying that established firms from France, Belgium and England furnished most of the capital for the five Romanian refineries. It is worth noting that the first of these firms reopened with modern equipment in 1898, after operating unsuccessfully between 1876 and 1883, only because of the entrepreneurial initiative of the Banca Marmarosch-Blanc.[47] Remaining to be explained is the arrival of Western European capital and management around 1900 to revive or establish the four firms that produced about ninety percent of the sugar consumed in Romania in 1910. We need not look far for the reason. The sugar industry was the only one in Romania or anywhere in the Balkans to receive a production subsidy from the state on the Hungarian model. The legislation dated back to Austrian export subsidies for its own sugar in 1882 and the Romanian desire to retaliate. The result was a 16 lei subsidy for every 100 kilograms of domestic sugar (about fifteen percent of the retail price) that attracted European attention only after being renewed in 1897.

Enter the issue of state aid to industry and the second important residual element of productivity that contributed to the higher industrial output per capita in Romania. Exit after the sugar subsidy the easy answers promised by some aggregate measure of industrial

tiative with King Carol I as the largest stockholder. But typical of the other dozen joint-stock firms in the field was the Schiel Brothers enterprise. They came from Braşov, for centuries a center of German settlement and just across the Romanian border in Austria-Hungary. With them they brought capital and skilled labor. See *Banca Marmarosch-Blanc & Co., 1848-1923* (Bucharest, 1923), p. 93; K. Und K. Handelsministerium, *Rumänien: Landes und Wirtschafts Statistische und Topographische Übersichten* (Vienna, 1917), pp. 155-60, 171-72; N. I. Paianu, *Industria mare, 1866-1906* [Large-scale Industry, 1866-1906] (Bucharest, 1907), pp. 82, 90; N. Costache, *Industria hârtiei în România* [The Paper Industry in Romania] (Bucharest, 1929), p. 6.

47 The bank contributed one-third of the small founding capital of 1.2 million lei and, more important, the services of board member I. G. Cantacuzino, a trained engineer who had already set up the country's first Portland cement works in 1890. *Banca Marmarosch-Blanc*, pp. 28-29. There is a striking parallel between Cantacuzino's career with the bank and that of Miloš Savčić, the Serbian engineer who was director of the Serbian bank most active in industrial promotion. See Lampe in Cameron, ed., *Banking and Economic Development*, p. 151.

tax exemption and tariff protection. All the prewar Balkan states, save Greece, passed industrial laws offering indirect tax and import tariff exemption, free land, railway rebates and some state purchase of production to manufacturers with over twenty employees and over 20,000 lei, leva or dinars of capital. Again with the exception of Greece, they all revised their tariff structures to allow industrial raw materials and machinery almost duty-free entry and to allow what were heralded in the respective National Assemblies as protective rates on imported maufactures, comparable to France's Méline Tariff of 1892. Yet the aggregate benefits accorded Romanian firms under the industrial legislation, sugar aside, never exceeded 4-5 million lei a year before 1905 and dropped to 2.1 million by 1910, well below the 3.7 million leva granted in 1912 under Bulgarian legislation.[48] Nor do Romanian tariff rates appear to offer us a clue. During the period 1907-1911 total tariff revenues averaged 11.3 percent of import value for Romania, but 15.4 percent for Serbia, 20.8 percent for Bulgaria and 26.8 percent for Greece.[49]

Despite the apparently inverse correlation between the levels of industrial production and government assistance, state policy was still a greater stimulus to manufacturing in Romania than elsewhere for several reasons. Romanian industrial and tariff legislation, plus a new commercial code recognizing joint-stock incorporation, were adopted in 1886-87, at the start of a successful six-year tariff war with Austria-Hungary. Bulgaria did not follow suit comparably until 1897 nor Serbia until 1898. Thus the prosperous decade of good harvests before the agricultural slump of 1897-1900 gave Romanian industry a real head start.[50] Beyond this legislation, the three

48 Although figures are lacking on the total benefits from the Serbian legislation, the fact that parliamentary infighting kept it from being applied to more than one fifth of the qualified enterprises suggests a lesser impact. The Bulgarian legislation applied automatically to all qualified firms from the start, and the Romanian legislation first to 70 percent and then to all of them.

49 *Anuarul statistic al României, 1915-1916,* 256; Lampe, "Financial Structure," 44, 186; *Doklad do Ferdinand I,* p. 195. The most obvious danger of taking these percentages as conclusive evidence is of course the risk that truly protective tariffs for particular goods would cut off their import entirely. A comprehensive presentation of ad valorem rates must await their calculation from the French-style tariff legislation throughout the Balkans that listed most duties by quantity. A start for converting these specific rates to ad valorem has been made for Romania in John R. Lampe and Marvin R. Jackson, "The Genesis of the State Sector in the Balkans," a paper presented to the VIth International Congress on Economic History, Copenhagen, Aug. 19-23, 1974.

50 Some 110 industrial firms with a total capital of 30 million lei had received

branches of manufacturing other than sugar also enjoyed a special relationship with the Romanian state. The lumber and paper industries both benefited from access to state timber land at low rents. The latter found most of its production bought up for the offices of the burgeoning state bureaucracy. Most metallurgical enterprises were located near the railheads of Bucharest and Craiova and in the ports of Brăila and Galați, not around the late-blooming oil fields of Ploești. The earlier, more extensive railway and dock construction promoted by the Romanian state thus spawned the growth of the largest metallurgical sector in the Balkans.

Finally, the Liberal party's dominance of Romanian political life for most of the prewar period counted for more than its introduction of industrial and tariff legislation in 1887. Unlike any other major and non-Marxist party in the prewar Balkans, the Romanian liberals were ideologically committed to the rapid but non-foreign industrial development of their country along Listian lines. They could more accurately be called National Liberals on the Central European pattern. Their ready acceptance of the idea that "manufacturing for Romania is a matter of life and death" marks them as the party of the rising bourgeosie.[51] But the base of their great power would seem, in the absence of the collective biography of their leadership that is needed, to lie with wealthy boyar estate owners educated in Paris. Generally Francophile and always opposed to the idea of Habsburg hegemony, they were reluctant to share political power with their German king or the boyar leaders of the Conservative party who followed his royal lead. The only native aristocracy in the Balkans used their equally unparalleled wealth and influence to stand up to their foreign king as no Bulgarian or Greek party had the resources to do. Some liberals invested personally in manufacturing as a new, more independent base of economic power, in the above-noted fashion of Emil Costinescu or I. G. Cantacuzino. Most were determined to break Romania free of the German financiers and Austrian commercial

encouragement by 1894, a year when total capital in Serbian and Bulgarian industry combined was less than 20 million dinars and leva. N. N. Constantinescu, *Din istoricul formării*, p. 192; Lampe, *"Finance,"* forthcoming in *Southeastern Europe*, Table I.

[51] So the Liberal party is consistently identified in postwar Romanian historiography. At the same time, the liberals' championing of balanced industrial growth has won increasing sympathy from scholars in a society pursuing the same strategy, albeit along socialist lines. See the late Olga Constantinescu, *Critica teoriei: România țara eminamenta agricolă* [A Criticism of the Theory of Romania As a Predominantly Agricultural Country] (Bucharest, 1973.)

interests who supported King Carol I from his accession to the throne in 1866. Industrialization was a means to that end.

Given the nature of the private sector, however, the motives of the Romanian liberals did not hold much promise for moving the country into sustained industrial growth had the First World War not intervened. Liberal antipathy to foreign interests unfortunately extended to the very immigrants whose entrepreneurship and other skills had been at the heart of the advantages Romanian industry enjoyed over that of the other Balkan states before 1914. Indeed the party's famous slogan of the 1920's, *prin noi înșine* (through ourselves alone), which led to the national take-over of Austrian and German firms and sustained pressure on immigrant interests, had already appeared in a prewar pamphlet.[52]

Let us not leave this comparative survey of pre-1914 Balkan industry with the impression that the aversion of the ruling Romanian policial party to foreign influence, understandable against Austria-Hungary and lamentable against non-Romanian immigrants, was the only barrier to further development. The indifference of the centers of European capital to manufacturing, if not extractive investment in Romania merits final mention. Probably more important were the obvious limits to growth in branches of manufacturing that did attract European capital but were largely dependent on government consumption, inelastic in the long-run and unpredictable in the context of pre-1914 European capitalism. Throughout the prewar Balkans, the state sector in fact diverted more financial and human resources away from productive investment than into it.[53]

The weakness of the mainly agricultural foundations of the pre-1914 Romanian economy also augured badly for the future. The nearly virgin lands of a century before had now filled up with a peasant population tied to ever smaller plots (there was no primogeniture) and to traditional methods even on the large estates. Any prolonged downturn in world grain prices would have spelled disaster for agriculture and capital accumulation. Any attempt to

[52] The introduction to C. Hălăceanu, *Proiectul de lege industriala* (Bucharest, 1912), pp. 8-11, is by Vintilă Brătianu, one of the famous brothers who dominated Liberal party leadership before and after the war.

[53] Sir John Hicks dates the Administrative Revolution, i.e., the first comprehensive state programs to promote economic growth, even among the major European governments no earlier than the First World War. See his *A Theory of Economic History* (New York: Oxford University Press, 1969), pp. 99, 162.

extract a larger exportable surplus or to increase productivity within the existing system of land tenure ran the risk of a peasant revolt still more widespread than the uprising of 1907. In addition, this peasant population was untrained for other work and by this time largely immobile. After 1900 the urban market for manufactures afforded by Bucharest, the biggest city in the Balkans, ceased to grow anywhere as rapidly as total population, and trained industrial labor was scarce.

The response to all of this by the joint-stock companies that were the vanguard of Romanian industry was to form sales cartels of the sort that were discussed elsewhere in the Balkans but never formally established. During the fifty years of European economic history before the First World War such cartels were a defensive response to the poor market conditions that accompanied cyclical downturns in the developed economies. They discouraged technological innovation. They were hardly the symptom of aggressive capitalist growth that sometimes sees them confused with trusts, but rather the most passive form of industrial organization imaginable for private industry trying to grow in a developing economy.[54] For Romania, the most firmly established cartels were in cement, flour and sugar. Conditions for the sugar industry were perhaps the best of any branch of Romanian manufacturing. Competing sugar imports had been eliminated by 1902, and afterwards the state subsidy helped keep yearly profits at twice the five to ten percent proportion of capital common in other branches.[55] By the same year of 1902, however, a sugar cartel had been formed. By 1906 its complaints of over-capacity for the domestic market and insufficient opportunities for export outside the limited Ottoman market culminated in legislation that barred any new sugar producers from all state privileges until 1914. Only the lumber industry was able to export an important amount of its production, over

[54] A Romanian case of this confusion is C. Cassasovici, *Trustul zaharului* [The Sugar Trust] (Bucharest, 1915). A good antidote to confusion is Eric Maschke, "An Outline of the History of German Cartels from 1875 to 1914," in François Crouzet, W. H. Chaloner and W. M. Stern, eds., *Essays in European Economic History, 1789-1914* (New York: St. Martin's Press, 1969), pp. 226-58.

[55] The less than conclusive evidence on Romanian rates of industrial profit during the last prewar decade may be found in Hălăceanu, *Proiectul de lege industriale,* pp. 17-18, and Arcadian, *Industrializarea României,* pp. 125-26. The comprehensive enterprise records needed to pin down this point have been lost during the two world wars.

one half in the last prewar decade, but this averaged just over five percent of the total export value recorded in Table 4.

The strategy of import substitution around which the majority of pre-1914 Romanian (and other Balkan) manufacturing had been organized also appeared to be a dead-end for sustained industrial development. Greek industry admittedly operated under the special burdens of skimpy agricultural resources, rising emigration and a European-enforced appreciation of the drachma. Bulgarian manufacturing was hampered too, by a lack of trained immigrants and an economically isolated capital city. But import-competing industry in Serbia, while free from these handicaps, could not match the growth of processed meat exports. Finally, Romanian production of domestically consumed goods showed signs after 1900 of saturating the home market in every major area save textiles, where imports continued to swamp domestic output.[56] Romania's experience in the golden age of unprocessed exports before the First World War was the best of any Balkan state. All the more significant, therefore, were the limited prospects generated for modern manufacture in this biggest prewar Balkan state.

University of Maryland

[56] The evidence of domestic over-capacity and the inability to develop exports in most branches of Romanian industry has been assembled in an unpublished paper by Professor Marvin R. Jackson, Department of Economics, Arizona State University, Tempe, "Economic Change in Romania, 1862-1914."

The Growth of Finnish Industry in 1860 - 1913. Causes and Linkages

Sakari Heikkinen - Riitta Hjerppe
University of Helsinki

The years 1860-1913 are usually referred to as Finland's first phase of industrialization. The Grand Duchy of Finland, an autonomous part of the Russian Empire, was a sparsely populated country with a population of only 1.7 million in 1860 which decreased because of the famines in the 1860s. Population growth was, however, rapid after 1870 up to World War I, about 1.3 per cent annually, and the number of inhabitants was 3.0 million in 1913.[1] Finland was an agrarian country with about 60 per cent of its Gross Domestic Product coming from agriculture and forestry in the 1860s and still over 40 per cent in the early 1910s.[2] So, even immediately before World War I, society was still quite agrarian, although the change from the conditions in the 1860s was pronounced.

In European terms, industrialization started late in Finland. In 1860, 4 per cent of the Finnish population was engaged in industry

[1] *Suomen taloushistoria 3. Historiallinen tilasto* (Economic History of Finland 3. Historical Statistics). Helsinki 1983. See also Yrjö Kaukiainen Finland 1860-1913, *Handbuch der europäischen Wirtschafts - und Sozialgeschichte*, Band 5, Hrsg. Wolfram Fischer, Stuttgart 1985.

[2] Riitta Hjerppe - Erkki Pihkala The Gross Domestic Product of Finland in 1860-1913. A Preliminary Estimate, *Economy and History* 1977:2.

and industrial handicraft, and they produced 7-8 per cent of GDP. Half a century later, the labour force engaged in industry and industrial handicraft represented one-tenth of the economically active population. Industry and industrial handicraft accounted for about one-fifth of GDP.

The importance of foreign trade – trade with Russia included – increased considerably between 1860 and 1913. The share of imported goods in GDP was well over 10 per cent in the 1860s and the share of exported goods almost 10 per cent. Before World War I, these shares had grown to 31 and 25 per cent respectively. The share of industrial products in all exports varied between 60 and 70 per cent in 1860-1913.[3]

International prices were declining in the late XIXth century and in Finland the prices of imported goods fell by one-third from the mid-1870s to the early 1880s. Import prices remained then at about this level until nearly World War I. With the exception of the timber boom which raised prices temporarily in the 1870s, the prices of Finnish export products stayed fairly constant from the 1860s up to 1913, with the result that Finland's terms of trade improved by about 50 per cent between the 1860s and World War I.[4]

The following essay will examine industry and industrial handicraft during the first phase of Finland's industrialization.[5] In

[3] ERKKI PIHKALA *Suomen ulkomaankauppa 1860-1917* (Finland's Foreign Trade, 1860-1917), Bank of Finland Publications, Studies in Finland's Economic Growth II, Helsinki 1970; HJERPPE - PIHKALA loc. cit.

[4] HEIKKI OKSANEN - ERKKI PIHKALA *Suomen ulkomaankauppa 1917-1949* (Finland's Foreign Trade, 1919-1949), Bank of Finland Publications, Studies in Finland's Economic Growth VI, Helsinki 1975, pp. 116-119.

[5] Data on industry are based on SAKARI HEIKKINEN, RIITTA HJERPPE et. al. *Suomen teollisuus ja teollinen käsityö 1860-1913* (Industry and Industrial Handicraft in Finland, 1860-1913), Bank of Finland Publications, Studies in Finland's Economic Growth XII, Helsinki 1986. See also PER SCHYBERGSON *Hantverk och fabriker I-III*, Finlands konsumtionsvaruindustri 1815-1870, Bidrag till kännedom av Finlands natur och folk, H. 114, 116, 117, Helsingfors 1973-1974 and KAI HOFFMAN *Suomen sahateollisuuden kasvu, rakenne ja rahoitus 1800-luvun jälkipuoliskolla* (Growth, structure and financing

addition to industrial growth and structural change, the importance of export and home market demand is discussed. Furthermore, attention is paid to supply factors and the role of the central government. Finally, industry's backward linkages are examined.

Growth

The volume of industrial and handicraft output in Finland grew by more than 5 per cent annually in 1860-1913. The growth in output can be divided into seven cyclical periods, using the troughs of the business cycles as benchmarks: 1860-1869, 1869-1879, 1879-1886, 1886-1893, 1893-1902, 1902-1908, 1908-1913 (See Chart 1). Hence the rhythm of cyclical variations in Finnish industry was approximately the same as elsewhere in Europe.[6]

In the 1860s, growth in production was slower than the average for the entire period, a fact which could be ascribed to crop failures at the end of the decade and to the international depression. The upswing in the first half of the 1870s was the most vigorous in the entire 50-year period. In particular, it was a period of expansion for export industries. In the middle of the 1870s, the growth in production came to a halt. Industrial output levelled off for a few years, then fell sharply in 1878-1879. In the decade 1876-1886, growth was slow, averaging only less than one per cent per year.

After the middle of the 1880s, output started to rise rapidly, largely in response to home market demand. In 1892 and 1893, output declined. The world economy was in depression, in addition to which a crop failure in agriculture in 1892 reduced the demand for manufactured goods. The depression was followed by

of Finland's sawmill industry in the latter half of the 19th century), Bidrag till kännedom av Finlands natur och folk, H. 124, Tammisaari 1980.
Industry is here defined as industrial plants with at least five persons working in the plant, industrial handicraft as plants smaller than that.

[6] W. ARTHUR LEWIS *Growth and Fluctuations, 1870-1914*, Cambridge 1978, pp. 17-24.

Chart 1.
THE VOLUME INDEX OF PRODUCTION IN INDUSTRY AND INDUSTRIAL HANDICRAFT
(1900=100) AND ITS DEVIATIONS FROM TREND (PER CENT) IN 1860-1913.

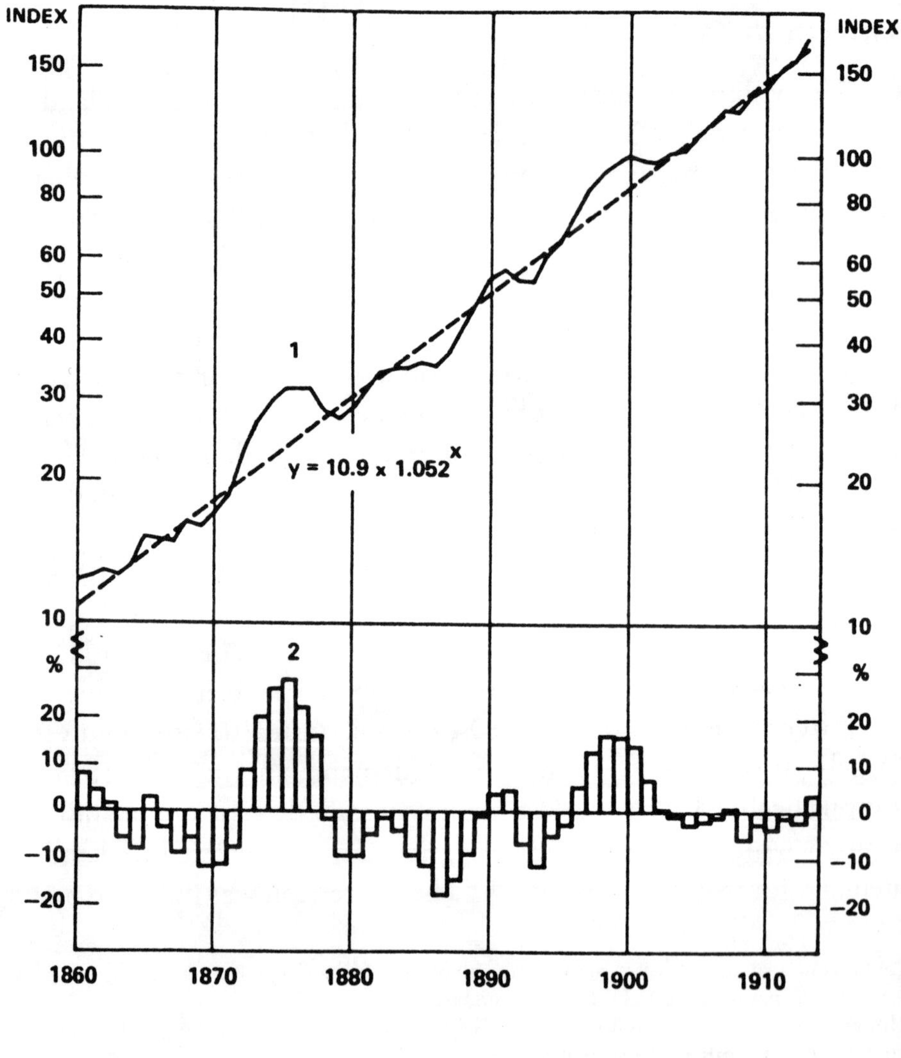

a vigorous upswing. The years 1894-1900 were second only to the early 1870s in terms of the pace of industrial growth during the period examined. From the turn of the century until World War I, industrial output grew at a fairly even rate. Output did fall slightly in 1901-1902, and again in 1908.

In 1860, the number of persons engaged in industry and industrial handicraft was 31,000, rising to 148,000 by 1913. The annual increase averaged 3 per cent. Comparison of the growth figures for employment and the volume of production shows an annual increase in productivity of about 2 per cent. According to this fairly rough comparison, Finnish industrial growth in 1860-1913 was based on an increase in the labour force rather than in the productivity of labour. During the first half of the period examined, the growth rates of employment and productivity were approximately the same, whereas during the latter half of the period, employment grew distinctly faster than productivity (Table 1).

TABLE 1

ANNUAL GROWTH OF PRODUCTION, EMPLOYMENT AND LABOUR PRODUCTIVITY IN INDUSTRY AND INDUSTRIAL HANDICRAFT IN 1860-1912, PER CENT

Period	Volume of production	Employment	Labour productivity
1860/62-1910/12	5.1	2.9	2.1
1860/62-1890/92	5.1	2.6	2.5
1890/92-1910/12	5.1	3.1	1.5

There were significant differences in the development of productivity by industrial sector. In the sawmill industry, labour productivity grew by only 0.3 per cent annually between 1860-1900. Steam did not produce any significant change in the sawmill technology itself,[7] but steam sawmills could be operated for a longer period annually than water sawmills. In the other major

[7] HOFFMAN op. cit. pp. 100-103.

industrial sector of the wood-processing industries, the paper industry, the development of productivity was substantially faster than the average for industry.

Structural changes

In the early 1860s, the largest branch of Finnish industry was mining and quarrying and the production of metals and metal products, including the manufacture of transport equipment (e.g. wooden ships). It accounted for one-third of the value added of total industry (Table. 2). The second-largest sector was the manufacture of textiles, clothing and footwear (27 per cent of the value added). The manufacture of textiles consisted mainly of a few forerunners of mechanical manufacturing, i.e., modern cotton mills, which were among the largest of Finnish industrial plants. A roughly equal proportion of the value added in this sector was produced by shoemakers and tailors and other artisans of the clothing industry. The remainder of the overall total was represented by three virtually equal branches: the food, beverage and tobacco industries, the wood-processing industries, and other, mainly handicraft-type production (12-14 per cent each).

In the period leading up to World War I, the sectoral structure of industry and industrial handicraft underwent a substantial change. The sawmill and other wood industries grew to become the largest sector (20 per cent of the value added). The output of the sawmill industries expanded particularly rapidly in the 1870s (1870-1877), when the volume of production quadrupled. The vigorous upswing was followed by a stagnation of more than ten years, for the level of output reached in 1877 was not exceeded until 1889. Thereafter the output of the sawmill industries grew at the same rate as other industrial production.

From the 1860s onwards, the manufacture of paper and paper products also started to become part of the wood-processing industries. The first mechanical pulpwood mill was established in

TABLE 2

THE BREAKDOWN OF VALUE ADDED IN INDUSTRY AND INDUSTRIAL HANDICRAFT IN 1860-1913, PER CENT, BY BRANCH

Branch	Breakdown of value added			
	1860-64	1880-84	1890-94	1909-13
Food beverage and tobacco industries	13.7	13.6	15.5	15.8
Manufacture of textiles, footwear, other wearing apparel and made-up textile goods	26.7	19.8	23.1	15.7
Manufacture of wood, cork, furniture and fixtures (except metal furniture)	12.2	23.4	14.0	20.1
Manufacture of paper and paper products	1.7	6.4	6.4	14.1
Mining and quarrying, Manufacture of metal and metal products, Manufacture of transport equipment	33.2	25.3	24.1	19.0
Other manufacturing industries	12.5.	11.5	16.9	15.3
Total industry	100.0	100.0	100.0	100.0

1860, and by the turn of the 1860s and 1870s it had been joined by a number of others. Even after the outset of the production of wood pulp, the traditional raw material, rag pulp, was used abundantly in the manufacture of paper, as mechanical pulp was not suited to the manufacture of finer qualities of paper. The raw material problem was eased after a chemical manufacturing method for wood pulp was developed. The manufacture of chemical pulp started in Finland in the 1880s, although production expanded into large-scale industry only during the decade preceding World War I. The manufacture of paper and paper products grew from being an insignificant branch to account for 14 per cent of total industry. Whereas the paper industry employed only 400 workers in 1860, the number had reached more then 12,000 by 1913, and output grew even more rapidly and uninterruptedly, while cutbacks in the volume of output were temporary and insignificant.

Mining and quarrying and the manufacture of metals and metal products fell to 19 per cent. The former major metal industry, the manufacture of iron, originally and mainly based on Swedish ore, shrank almost into obscurity. In 1913, the most important sector in the manufacture of metals and metal products was the engineering industry, which also included the manufacture of steel ships and rolling stock.

The manufacture of textiles, footwear, and clothing was also relatively smaller in 1913. The share of cotton mills in the output of the textile industry slightly declined, whereas that of woollen mills increased. The food, beverage and tobacco industries as well as the miscellaneous sector of other industries and handicrafts had maintained their positions.

In the 1860s, industrial handicraft (establishments with less than five employees) accounted for 30 per cent of the value added of total industry and handicraft, and for half of the overall employment. As industrialization proceeded, the share of handicraft declined. Immediately prior to World War I, handicraft accounted for one-fifth of the total industrial labour force and for less than one-tenth of the value added. Although handicraft expanded, it did so much more slowly than industry. In 1860-1913, employment in handicraft increased from 16,000 to 26,000, whereas industrial employment grew from 15,000 to 121,000 during the same period.

The growth in industry certainly did not cause the disappearance of handicraft. By contrast, industrialization and related growth in income, the expansion of the monetary economy and urbanization increased the demand for many handicraft products. Handicraft partly offset home production (e.g., tailors, bakers); it partly satisfied new needs (repairs of agricultural machinery and tools). True, in some industrial sectors, the position of handicraft was already shaken by mechanized large-scale production - this was also the case with the footwear and leather industries. However, many industrial products did not compete with handicraft. They substituted home production, such as yarns, textile goods and tobacco or they were exported. Moreover, new industry manufactured entirely new products, created and satisfied new needs (cigarettes, wallpapers, matches), and consumer goods, which had previously been rare (sugar, cotton materials, glassware) gained ground.

Exports and domestic demand

Throughout the period examined, the rapidly-expanding sawmill industry was an export industry, as was the new, wood-consuming paper industry from the late 1860s. Exports were of significance in the manufacture of textiles, industry and in several minor branches during the first half of the period examined. At the beginning of the second decade of the XXth century, the bipartite structure of Finnish industry was, however, already distinct: export industries consisted of the wood and paper industries, and domestic industries of other branches (Table 3).

TABLE 3

THE COMPOSITION OF EXPORTS OF INDUSTRIAL PRODUCTS IN 1860-1913, PER CENT

Branch	1860-1864 A	1860-1864 B	1880-1884 A	1880-1884 B	1890-1894 A	1890-1894 B	1909-1913 A	1909-1913 B
Manufacture of textiles and wearing apparel	17.1	95.7	12.9	97.9	9.4	97.3	3.9	96.8
Manufacture of wood, cork, furniture and fixtures (except metal furniture)*	39.3	5.4	53.9	2.1	56.8	2.0	63.3	4.9
Manufacture of paper and paper products	1.7	99.3	13.1	99.3	15.6	72.3	25.9	60.3
Manufacture of metal and metal products, Manufacture of transport equipment	32.4	99.1	13.6	90.2	9.8	96.0	2.9	83.9
Other manufacturing industries**	9.5	68.3	6.5	88.4	8.4	96.1	4.0	81.8
Total industry	100.0	58.6	100.0	44.7	100.0	37.3	100.0	28.2

A = Share in total exports of industrial products.
B = Russia's share in the exports of the respective industrial branch.
* For the period 1860-1864, it was assumed that the exports of the sawmill industry equalled its production; the export data for 1880-1884 and 1890-1894 are taken from Kai Hoffman's study.
** Excluding tar and butter.
Sources: Pihkala, *op. cit.*, Hoffman, *op. cit.*

Manufactured goods were exported to Russia and Western Europe, where the United Kingdom was the major export market. In the early 1860s, Russia received more than half of the manufactures exported and in the 1880s more than 40 per cent, whereas in the first decade of the XXth century no more than about 30 per cent. This change was attributable to a downswing in textile and metal industry exports to Russia after the mid-1880s. At the same time, paper industry exports to Western Europe started to expand. They accounted for 25-30 per cent of total paper industry exports in the late XIXth century and increased to 40 per cent by World War I. While it was mainly paper that was exported to Russia, Western Europe primarily imported mechanical and chemical pulp, and paperboard. Sawn goods were exported almost exclusively to Western Europe, where the United Kingdom, Germany and France were the major recipient countries. The purchasing area for timber had gradually extended further afield from the economic centres of Western Europe. A fall in freight rates made transport over longer distances more attractive than previously.

Other industrial exports went to Russia, and from the 1860s to the 1880s, exports to Russia stimulated an expansion in the cotton and metal industries. Many Finnish manufactures enjoyed exemption from customs duties in the Russian market, being thus in a favourable competitive position compared with actual «foreign» exports. During the peak export period of the 1870s, two-thirds of the manufactured cotton materials had been exported to Russia.

From the 1880s, Russia started, like many other countries, to increase the customs protection of her own industry. The rise in customs duties naturally weakened the export prospects of Finnish industry, even though Finland's position remained favourable compared to other nations. However, the increase in customs duties was not the only reason for the less favourable trend in Finnish exports eastwards. From the 1880s, Russia's own industry began to expand rapidly and satisfy domestic demand. Obviously, Finnish industry did not fare well in competition with the new Russian production. The paper industry was an exception, however, and Finnish paper accounted for one-third of the Russian paper consumption in the 1910s.

From the early 1860s until the mid-1880s, the growth in Finnish industry was, on the whole, highly export-led. Exports of manufactures to both Western Europe and Russia grew rapidly, and the share of exports in manufacturing output reached its peak in the early 1880s (see Chart 2). In the 1880s, the course of development changed. The period of export-led growth was followed by a decade during which the expansion in manufacturing output was fuelled by an increase in domestic demand. The share of exports in manufacturing output fell to a level that was then maintained until the first World War. This change could be ascribed on the one hand to the prolonged international recession

Chart 2.
THE SHARE OF EXPORTS IN INDUSTRIAL PRODUCTION AND THE SHARE OF IMPORTS IN DOMESTIC MARKETS FOR INDUSTRIAL PRODUCTS IN 1860-1912, PER CENT

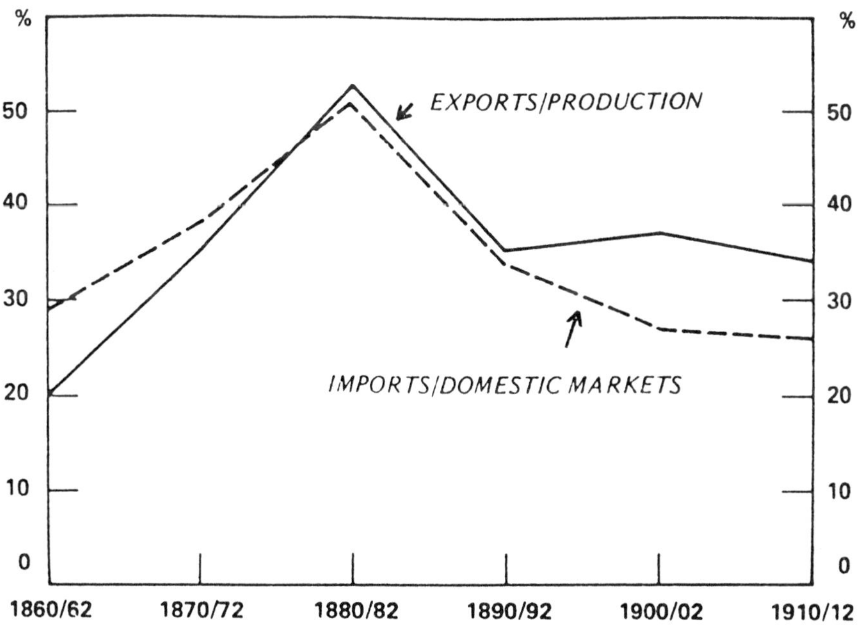

and Russia's protective measures, and on the other hand to the growth in domestic demand.

In the Finland of 1860, the level of income was relatively low; agriculture and forestry were the sectors dominating the economy, and the networks of transport and communication and of commerce were undeveloped. At this stage, domestic demand could not be a significant stimulus to industrial growth. After the years of famine in the 1860s, the level of income began to rise and the population growth continued at a rapid rate. A growth in forestry brought money to the countryside. This was channelled into consumption and agricultural investment. Urban expansion reduced the area of subsistence economy and enlarged the scale of local market demand. As the real prices of textiles and other manufactures fell, home production was replaced by manufacturing. The economic integration of the country was fostered by the construction of railways and the increased density of other transport networks. This also fuelled demand for railway equipment and steamships.

The growth in domestic demand did not automatically increase domestic industry, which was forced to compete for the markets with imported goods. In the 1860s and 1870s, the share of imported goods in the market for manufactures increased (see Chart 2). The 1880s saw a change, and the market share of domestic industry increased, with the same trend continuing into the 1890s. However, the domestic metal industry was not able to increase its market share from the level to which it had fallen in the 1860s and 1870s.

Domestic manufactures also enjoyed customs protection, the degree of which varied by sector. As customs duties were collected on the basis of imported volumes, a fall in product prices meant an increase in customs protection and a rise in prices a fall therein. Hence, the relative import duties on manufactures rose when, for example, prices fell in the 1880s, whereas the duties fell after the middle of the 1890s when prices generally rose. However, the customs duties were primarily fiscal: they were the central government's major source of income. Moreover, it should also be recalled that no import duties were levied on imports of Russian manufactures.

Supply factors, finance and the central government

In the short term, industrial development depended mainly on change in demand. In the long term, changes in the growth and structure of production were markedly affected by supply factors, which in Finland were mainly natural resources and labour force. Finland's major natural resources from the point of view of industrial development were forests and hydropower, which were easy to utilize even by using fairly simple technology. In other respects, the impact of natural resources on industrial growth was insignificant.

The availability of labour depended naturally on demographic developments and the structure of the economy. In 1860-1913, the increase in the resident population averaged one per cent annually, and in 1870-1913 1.3 per cent per year. In 1870, the population of working age accounted for 62 per cent of the total population and in 1913, for 59 per cent. The supply of labour was fundamentally influenced by the economic and social structure of the rural areas. A rapid growth of population in the countryside led to an excess of population in rural areas, which provided a labour reserve for industry. The excess population in the agrarian sector tended to maintain a low wage level, a trend which was also bolstered by the fact than no customs duties were levied on imported corn. The difference in the wage level was appreciable as compared with that in Sweden: in 1900-1913, the average wage in the Finnish sawmill industries was two-thirds of the average wage of a Swedish sawmill worker.

It is more difficult to estimate how industrial development was affected by the quality of the labour force. While technical knowledge had no traditions, the technology needed by industry was adopted mainly from abroad. It was there that machinery was acquired and technicians, foremen, skilled workers and some entrepreneurs immigrated. In the major industrial branches — for example, the wood-processing industries — the requirements set for workers' professional qualifications were modest, so that the quality of the labour force should not have impeded the growth of output. On the other hand, as a consequence of the low level of technical knowledge, the industries which demanded more developed technology were insignificant. Literacy rates were high, but it is difficult to assess their importance for industrial development.

There are very few studies on the importance of capital for the industrialization of Finland. On the whole, researchers have been content with explaining the lack of capital as a barrier to industrialization or as the reason for the postponement of industrialization.[8] It was not until the 1860s that the Finnish banking system began to take shape. Subsequently, the role of banks in the channelling of monetary flows increased rapidly. Previously, the money needed for industrial investments had come from private persons, such as civil servants and farmers, state funds, the Bank of Finland, the central bank founded in 1812, the commercial sector, pension funds, and savings banks. Merchant houses played a central role in financing and setting up the wood-processing industries. Direct investment by foreign investors in Finnish industry was insignificant. Capital from St. Petersburg was invested in Finlayson's, the largest cotton mill, and the iron industry in Eastern Finland received capital from Russia, while a few wood-processing firms obtained capital from Western Europe. All in all, however, the role of actual foreign investment capital remained small.

When examining Finland's industrialization, it has been common to emphasize the legal reforms which were implemented

[8] EINO JUTIKKALA Industrialization as a Factor in Economic Growth in Finland, *First International Conference of Economic History*, Paris 1960.

mainly between the late 1850s and the late 1870s.[9] That period saw the abolition of a number of restraints on economic activity and the creation of a legal framework for the development of a market economy (freedom of trade, the ensuring of the free mobility of labour, company legislation, abolition of the restraints on sawmill activities, etc.).

Nevertheless, the central government interfered even more directly in the economy and industrial development. During the first half of the XIXth century, the central government supported particularly mining and quarrying and the manufacture of metal and metal products. The first cotton mills and certain other industrial plants — not sawmills — also received loans or subsidies for starting up their activities. Central government lending to industry played a significant role until the beginning of the 1860s, during the very earliest phase of industrialization when the banking system was still undeveloped. Subsequently, the role of the central government's direct financing declined and the emphasis was shifted over to the construction of the infrastructure, e.g. railroads, other transport, and education.

Using Alexander Gerschenkron's famous tripartition, it can be claimed that Finland by no means represented an "English" model of industrialization nor even a "German" one. On the other hand, the central government's direct interference in industrial development was, with the exception of the middle of the XIXth century, so insignificant that there is no reason for describing it as an "East European" model either.[10]

[9] EINO JUTIKKALA Suomen teollistuminen (Industrialization of Finland), *Suomen talous - ja sosiaalihistorian kehityslinjoja*, Porvoo 1968. See also K.O. ALHO *Suomen uudenaikaisen teollisuuden synty ja kehitys 1860-1914* (Origins and development of the Finnish modern industry, 1860-1914), Suomen Pankin taloustieteellisen tutkimuslaitoksen julkaisuja, Sarja B: 11, Helsinki 1949, p. 49-52.

[10] ALEXANDER GERSCHENKRON *Economic Backwardness in Historical Perspective. A Book of Essays*, Harvard 1962.

Backward Linkages

The macroeconomic significance of industry cannot be described merely in terms of its share of total output, as the growth of industry also affected the other sectors of the economy and its dynamics. Using Alfred Hirschman's terminology, we can talk about the backward and forward linkages of the industrial sector. Here we shall deal only with the backward linkage which refers to an increase in the domestic supply of production inputs by a growth in output.[11]

If backward linkage is measured in terms of the contribution of domestic raw materials and semi-finished products to the gross value of output of each industrial branch, the backward linkage was high in the timber industry (63 per cent) and the paper industry (49 per cent) in 1910-1913. In other branches it was low: the food, beverage and tobacco industries 33 per cent; the mining, quarrying and metal industries 23 per cent; the textiles and clothing industries 14 per cent, and other industries 16 per cent.

If we examine the total of the value added and domestic raw materials, the contribution to industry of the timber and paper industries is much larger than when measured in terms of employment or only the value added. During the second decade of the XXth century, the woodprocessing industries represented almost half of the macroeconomic weight of total industry (Table 4).

We can hardly agree with Lennart Jörberg's claim that "the Finnish export industry was an 'enclave' within the economy".[12] Although true for the cotton industry of the 1850s to 1870s, it does not apply to the major export industry, the wood-processing industries. The multiplier effects of the wood-processing industries on other industries were fairly insignificant, although the sawmill industry and the pulp industry utilized domestic machin-

[11] Albert O. Hirschman *The Strategy of Economic Development*, Forge Village 1965 (1958), p. 100.

[12] Lennart Jörberg The Industrial Revolution in the Nordic Countries, *The Fontana Economic History of Europe 4*, The Emergence of Industrial Societies 2, p. 463.

TABLE 4

EMPLOYMENT, VALUE ADDED AND VALUE ADDED PLUS DOMESTIC RAW MATERIALS IN 1910-1913 (per cent)

Branch	Employment	Value added	Value added plus domestic raw materials
Food, beverage and tobacco industries	13.3	15.3	19.5
Manufacture of textiles and clothing	20.4	15.3	11.0
Manufacture of wood, cork, furniture and fixtures (except metal furniture)	24.6	20.6	29.4
Manufacture of paper and paper products	8.8	14.1	16.1
Mining and quarrying industries, Manufacture of metal and metal products	18.7	19.4	13.6
Other industries	14.2	15.3	10.4
Total industry	100.0	100.0	100.0

ery to a greater extent than the average. By contrast, the growth in the wood-processing industries was reflected outside industry, in forestry, agriculture and the transport and communication sector. The expansion of forestry brought capital and labour income to the countryside. Capital income spread fairly widely, because the ownership of forests was not very concentrated.[13]

Conclusions

The development of Finnish industry in 1860-1913 can be divided into three main phases:

1) From the 1860s to the 1880s, export demand was the major growth factor. Exports to both Western Europe (sawmill industry) and Russia (paper, textiles and metal industries) accelerated the growth of industry.

[13] See also Dieter Senghaas *The European Experience. A Historical Critique of Development Theory*, Dover, New Hampshire 1985, p. 72-80.

2) From about the mid-1880s up to the mid-1890s, export developments were unfavourable. The growth of domestic demand and an expansion in the market share of domestic industry fuelled the growth of industry.

3) From the 1890s up to World War I, both exports and the domestic market were significant growth factors. Paper exports to Western Europe now grew considerably. At the same time, the bipartite nature of industry became increasingly pronounced: the wood-processing industries were export industries, whereas the other sectors were for the most part home market industries consuming foreign raw materials. Exports of manufactures were fostered by a marked improvement in the terms of trade.

Finland was part of the Russian Empire and in the Russian market she enjoyed a customs benefit which the paper industry, the textiles and metal industries were able to utilize. On the other hand, whereas Russian products were duty-free in the Finnish market, Finnish industry fared well in this competition.

The major supply factors of industrial growth were forest and hydropower resources and the labour reserve in the rural areas. Industry was labour-intensive, and the degree of processing relatively low. The growth of output was based on an increase in the labour force rather than on a rise in productivity. By West European standards, the level of industrial wages was low. From the point of view of the dynamics of the economy, the wood-processing industries as a whole was the crucially expanding branch: it was an export industry, whose multiplier effects on the economy (forestry, agriculture, transport and communication) were substantial and direct.

Growth in a Dual Economy – The Norwegian Experience 1814–1914

By FRITZ HODNE

For a majority of the advanced nations today, the key to economic growth has been the foreign sector. By entering the international trade routes and international markets, state after state has escaped the constraints of its traditional home economy. When home markets lacked infrastructure and buying power, and when the home economy was too undifferentiated to justify risk-taking in new technology, the world market, in contrast, gave each state the chance to utilize its comparative factor advantages, and specialize in areas where each had an advantage over its competitors. Moreover, with larger markets came the opportunity to realize economies of scale, since, in turn, risks became more tolerable.

The present article will discuss how far the economic growth that Norway experienced in the period 1814—1914 can be explained as a result of the country's successful integration into the world economy. It seeks to answer the following question: To what extent was the foreign sector the proximate cause of economic growth in Norway? [1]

The discussion will proceed within the framework of the dual economy model. Section I offers a brief, descriptive sketch of the model, section II considers the period 1814—1850 and the last section deals with the period 1850—1914. The first period shows the relation between stagnant foreign demand and deflation in the Norwegian economy. The last period illustrates the interrelationship between expansion in world commerce and Norway's specialization in the export field. The interaction between rising costs and technical innovation is considered, and also the performance gains realized in the transfer of labour from the agricultural into the industrial sector.

I

In the present context the dual economy model [2] consists of two sectors, the agrarian sector and the industrial sector. Agriculture has two factors,

[1] The article is prompted by the discussion stirred up by the appearance of two recent books on Norwegian growth.
Lennart Jörberg, *The Industrial Revolution in Scandinavia 1840—1914*, London 1970.
Sima Lieberman, *The Industrialization of Norway 1850—1920*, Oslo 1970.

[2] Arthur Lewis, Development with Unlimited Supplies of Labor, *The Manchester School*, XXII, May 1954, pp. 139—191.

labour and land; the industrial sector capital and labour. Through the industrial sector, the economy is linked to the world market. Only the open version is of interest here.

The agricultural sector is overpopulated in relation to resources and technology. The marginal product of labour, therefore, is very low, approaching zero. The wage rate, paid out in goods, is higher than the marginal product of labour. Wages may be said to be dictated more by social convention than economic maximizing behaviour. In the latter case, wages, of course, would be equal to, or higher than the marginal product of labour. This was Sir Arthur Lewis' original assumption, later elaborated in several publications by the scholars John C. H. Fei and Gustav Ranis. Dale Jorgensen, criticizing Fei and Ranis, has argued that the assumption of zero marginal product for labour is unnecessarily strict. All that is necessary, is to assume a productivity gap between the two sectors. This is the approach adopted here. For the agricultural sector the assumptions are:

1. hidden unemployment or under-employment,
2. marginal product of labour lower than the traditional or conventional wage,
3. wage-rate determined by tradition, not by the competitive market mechanism.

The industrial sector, by contrast, is urban and organized on rational economic principles. That is, productive factors, including labour, are

John C. H. Fei & Gustav Ranis, *Development of the Labor Surplus Economy: Theory and Policy*, Homewood, Ill. 1964.

— — Agrarianism, Dualism, and Economic Development in I. Adelman & E. Thorbecke, eds., *The Theory and Design of Economic Development*, Johns Hopkins Press, Baltimore & London, 1966, pp. 3—43.

H. Myint, *The Economics of the Developing Countries*, London 1964.

Dale Jorgensen, *Theories of the Development of a Dual Economy*, in Adelman & Thorbecke above, pp. 45—66.

D. M. G. Newbery, Public Policy in the Dual Economy, *The Economic Journal*, vol. 82, June 1972, pp. 567—590.

Taxation issues are dealt with in Richard Musgrave, *Fiscal Systems*, New Haven & London, Yale Univ. Press, 1969, pp. 69—167.

For the cultural aspects of dualism, rather than detailing a long list here, the reader interested in these matters, is referred to a recent scholarly summary, see

Y. Hayami & V. W. Ruttan, *Agricultural Development: An international perspective*, Baltimore, Md. 1971, especially pp. 1—66.

I have especially profited from Charles Kindleberger's books, see

Charles P. Kindleberger, *Europe's Postwar Growth. The Role of Labor Supply*, Harvard Univ. Press, Cambridge, Mass., 1967.

— *Foreign Trade and the National Economy*, New Haven, Yale Univ. Press, 1962.

paid the equivalent of their marginal product. Employment is thus determined by demand for labour, its price, both dependent on the state of technology, and the demand for industrial products. Income distribution is equally given as in the competitive model.

The term "industrial" must be taken to mean different things at different points of time. In what has been termed "the proto-industrial period" real capital per worker was low or non-existent, productivity per time unit low, but high per worker, since workers were paid their marginal product. "Industrial" then, refers to a mode of factor payment. With the advent of the factory system, machinery, and new forms of mechanical energy, productivity (in both senses mentioned above), has shown a sustained increase. "Industrial" now, thus means both a mode of factor payment and a form of production that involves large-scale capital investments in production plants and eqiupment. The assumptions for the industrial sector, in turn, are:

1. wage rate is equal to the marginal product of labour, (a subsistence wage in Ricardian sense),
2. free mobility of labour, implying an infinitely elastic supply of labour in the industrial sector,
3. "normal" price- and income elasticities,
4. for the dual economy as a whole, the rate of increase of population is a given, independent variable.

The above sketch suggests the co-existence of two economies,—hence the term "dual"—one is linked to the demand patterns of the world market, the other is largely unaffected by the forms of marginal factor payment that obtain in the market sector.

Dualism points beyond the cleavage between a market- and a non-market sector. A group of writers has stressed other dimensions.[3] Their materials are largely drawn from the tropical, under-developed states today, that have had contact with various forms of European colonialism in the past. The aim of these studies is not historical, they seek to illuminate obstacles that have to be overcome in order to set these young nations on the path of progress. However, so far from reducing the model's relevance to a country like Norway, now a mature economy, these aspects, in fact, enhances its usefulness.

The dual economy model brings out precisely those social, ethnic and cultural cleavages that characterized Norwegian society in the period

[3] See Hayami & Ruttan, op. cit. pp. 31—52.

under consideration, and in many ways, those cleavages still remain in force. In contrast to neighbouring Sweden and Denmark, Norway, it must be remembered, has a colonial past. It was a colony under Denmark for 400 years. From 1815 till 1905 it was allied to the Kingdom of Sweden in a personal union. Sovereignty is clearly not a significant part of the Norwegian people's heritage. Thus, as commonly observed in present-day enclave economies, foreign trade in Norway was likewise dominated by foreigners, first by the German Hansa, then (after 1560s), by Danish, Dutch, Scottish and German civil servants and entrepreneurs. These formed a political, cultural and ethnic elite, apart from the natives. Again, as in present-day developing countries, Norway never had the opportunity of setting up its own credit institutions till after 1814—it got its first university 3 years earlier—and only after 1814 did it control its own customs tariff, its central bank, credit, and money supply.

Rural Norway was a traditional peasant culture. Urban Norway was an outward-looking commercial culture, closely in touch with European trends. This can be seen from such things as laws regarding property and inheritance, voting rights and representation in the national assembly, tax structure, poor laws and school laws. It is not usually realized, but nevertheless a pervasive fact, that economic dualism in Norway was closely matched by far-reaching legal and cultural dualism. Similarly, the main contents of Norwegian politics in the period under consideration was dominated by the urban-rural conflict, and the political mobilization of the rural periphery against the urban centre. This aspect will not be explored further here.[4]

However, parallels to dual economies today should not be carried too far. For one thing, Norway was always a free-holding peasant society. This feature reflects the combined effect of climate, geography and the extreme smallness of its population. Thus, l'ancien regime, with large-scale agrarian estates, a feudal nobility and a population of peasant serfs never took hold in Norway. Hence, the problem of economic development in Norway was largely a matter of acquiring skills, knowledge and capital, and this could only be acquired abroad.[5]

[4] Steinar Rokkan, Geography, Religion, and Social Class: Crosscutting Cleavages in Norwegian Politics, in Seymour Lipset & S. Rokkan, eds., *Party Systems and Voter Alignments: Cross-National Perspectives,* New York & London 1967, pp. 367—444.

[5] Cfr. Alan S. Milward, *The Fascist Economy in Norway,* Oxford 1972, pp. 30—31. This thoughful, elegantly written book on Hitler's plans for integrating the Norwegian economy in a Pan German superstate, has, like Jörberg's book, the added merit of viewing Norway's

This shows that the dual economy model has explanatory value going beyond the under-developed countries of today. A recent survey of the post-war growth in Europe, for example, by Charles P. Kindleberger, employs the Lewis model with remarkable success.

We now turn to consider the interaction between the two sectors in the dual economy. Initially, a dis-equilibrium is set up in the industrial sector. This may be due, either to increased foreign demand for its products, new technology that raises efficiency, or to capital investments from outside. The increased activity results in a stepped-up demand for labour. Normally wages would rise, but with labour supply infinitely elastic, wages will remain stable while free hands move in from the agricultural sector. In this situation, unlimited supplies of labour means increased employment, stable wages, high profits and savings. With re-investment of the savings out of current production, a new round of the cycle is set up, wages going to consumption, profits to re-investments. This is, in fact, the classical Ricardian case of expansion, sustained by buoyant demand.

What happens when labour supply begins to run out? When supply turns inelastic, wages will begin to rise, and cut into profits. Wages will probably increase its share of total income, and the process of expansion will come to a halt. The capital-widening period ends, and the capital-deepening phase begins. Faced with increasing labour costs, industry will begin to cast about for new technology, which alone offers an alternative to total lay-off. New technology holds the promise to save profits in the face of increased wage-rates, even compensating for a fall in the market price of the relevant products.

Turning now to the agricultural sector, we expect the process here to match the chain of events in industry. When labour begins to move over to industry, the marginal product of labour, originally somewhat above zero, begins to rise. To the extent that total agricultural product is stable, this means that fewer hands produce more on the average than before; productivity is raised, or, in money terms, per capita income is increased.[6] As yet, labour may still be regarded as a sunken cost, with no alternative value, since its marginal product is below the institutional

economy as part of Europe, with the corollary of new view-points and fresh parallels, usually absent in works written by Norwegian scholars.

[6] See H. Myint, *The Economics of the Developing Countries*, London 1964, repr. 1969, pp. 86—90, who poses the pertinent question whether "zero marginal product" exists at all, hence whether "redundant" labour can be removed from the agricultural sector without a fall in production. Obviously, with unchanged techniques, those who remain behind will have to work harder. The issue is not material to the position taken here.

wage. Therefore, little or no incentive exists for substituting capital for labour. Labour emigration is as yet unable to topple the traditional agricultural system. But with continued emigration of labour out of agriculture, the productivity of those remaining behind will rise correspondingly, until the marginal product of labour matches the going wage in the market sector of the economy.

This is the crucial point, for from here on, agriculture will treat its labour as an alternative cost factor; this opens the way for the substitution of machinery for labour, a shift that is normally self-sustaining, because it is rewarded in the market. The necessity of a change in mental outlook is mentioned only in passing here. Also, when the interaction between the two sectors has gained momentum, productivity levels will be raised, of course, in both sectors. What sustains the process, therefore, is the differentials in productivity—hence factor payments—between the two sectors, that ensures the continued transfer of labour from inefficient to efficient sectors. To this extent it may be said that no advanced nation today has achieved economic growth without simultaneously improving efficiency levels in agriculture.

The Lewis model assumes growth to be achieved with unlimited labour supplies. In one sense it is a supply theory for growth. In the context of the dual economy, however, labour is seen only as a necessary condition. What triggers off the process is foreign demand, which should be regarded as the sufficient, or proximate, cause of growth. To this extent we have a demand theory.

Now, whether we call it a supply- or a demand theory, the model will assign a central role to the components of the foreign sector. Specifically, we should expect the size of the foreign sector to be at least 20 % of the GNP, preferably a good deal higher, in order to act as a pump mechanism for a massive resource transfer into the growth sector. Secondly, we should expect buoyant foreign demand for export goods and services to be related to boom conditions for the economy as a whole. Conversely, a depressed export sector would be linked with stagnation in the rest of the economy. In both cases, economic growth will be a function of foreign demand.

II

If the foreign sector acts as a trigger mechanism for the transfer of labour into the industrial sector, the size of the foreign sector as a share of GNP is of considerable interest. Norwegian national accounts series

Table 1. The Mercantile fleet, production and export of major commodities, 1791—1850.

Year	Export timber	Production iron	Production copper	Export dried & smoked fish	Export klippfish	Export salted herring & fish	Mercantile fleet
	1000 m³	tons	tons	tons	tons	tons	1000 net tons
1791	—	—	498	—	—	—	—
1792	—	5,775	—	—	—	—	—
1805	994	—	—	—	—	—	157
1817	—	3,900	—	—	—	—	138
1815—1819	597	—	—	7,784	1,463	15,521	—
1825—1829	708	—	—	15,551	5,941	33,958	113
1829	—	5,633	346	—	—	—	127
1832	671	—	—	15,748	6,093	57,584	140
1830—1835	—	6,248	—	—	—	—	143
1835	837	—	—	17,091	7,765	48,845	150
1838	894	—	—	13,467	10,264	37,706	168
1836—1840	—	5,811	673	—	—	—	175
1841—1845	—	6,337	619	12,246	8,655	58,259	220
1845	1,060	—	—	—	—	—	240
1850	1,137	—	—	15,174	11,810	56,822	284

Source: A. M. Schweigaard: *Norges Statistik*, Christiania 1840, pp. 89—91, 100—101, 114—115.
M. Braun Tvethe: *Norges Statistik*, Christiania 1848, pp. 66—67, 59, 78—79, 82.
Historisk Statistikk, 1968, Table 175.

As conversion factors are used: "1 trelast-lest"=120 cu.ft.=3.7 m³, "1 skippund"=159.4 kg. "1 fisketønne"=115.8 litres, 1 litre salted herring=0.86 kg.

go back to 1865. For estimates prior to this we are forced to rely on deductions from scattered evidence. Now, if we knew the export volume through these years, including shipping, and if we could measure the GNP percentage of exports of goods and services for one or two years in the period, we would have an estimate of the foreign sector for the entire period. This conclusion rests on the assumption that the structure of the economy remained stable. Thus formulated, the problem can be solved, at least provisionally. On the one hand, we do have fairly accurate volume figures for production and exports, on the other hand, we have "national income" estimates for two years, viz. the years 1835 and 1845.

The production figures are given in table 1.

For the moment we are interested only in the long-term trends. Consider the figures for timber exports: In 1805 they were almost as high as in 1845, 40 years later. There is thus no long term growth, but a sharp fall around 1815, a long depression, followed by a slow recovery, until the 1805-level is finally overtaken in the 1840s. The same movement is apparent in the figures for the mercantile fleet, and a similar picture of stagnation after 1815 can be seen up in the production figures for copper and iron. Only fish exports break the pattern, showing a steady rise throughout the period. We note in passing that the winter herring returned to the south-western coastal waters of Norway in 1808, after an absence of 80 years, giving the Norwegians a virtual monopoly for a cheap, nourishing mass-consumption commodity.

We now turn to professor Schweigaard and Braun Tvethe for the national income estimates in 1835 and 1845.

	1835	1845
Agricultural production	14,283	19,500
Forestry	3,000	3,750
Fisheries	3,000	4,000
Mining	611	1,000
Freight earnings in ocean transport	1,595	2,300
	22,489	
33 % addition to agricultural production to compensate for underreporting	4,285	
Total	26,774	30,550

The figures are in 1000 speciedaler, 1 spd. being equal to 4 gold kroner.

These totals do not represent the aggregate GNP for the respective years, since they do not include estimates for manufacturing, inland trade, private services and public sector services. However, assuming stability in the structure of the economy, the latter sectors may be ignored, since they can be taken to have been of the same relative size throughout the period. We know that the entire agricultural production was consumed at home. According to Schweigaard, 60 % of the forestry output was exported, and 80 % of the yields from fisheries. Braun Tvethe found that 48 % of the mining products was sold abroad in 1835, and 41 % in 1845, and he assumed the same percentages as Schweigaard for the other sectors.

The relative size of the foreign sector in the Norwegian economy in the two years may now be summarized, (table 2).

Table 2. The relative size of the Export Sector of the Norwegian Economy in 1835 and 1845.

	1835			1845		
	Total production	of which exported		Total production	of which exported	
	1000 spd.	1000 spd	%	1000 spd.	1000 spd	%
Total production	26,774	—	—	30,550	—	—
Forestry	3,000	1,800	60	3,750	2,250	60
Fisheries	3,000	2,400	80	4,000	3,200	80
Mining	611	294	48	1,000	410	41
Ocean freights	1,595	1,595	100	2,300	2,300	100
	26,774	6,089		30,550	8,160	
Foreign sector in percentage of total		22.8 %			26.7 %	

Of the estimate for total production used here, 22.8 % was exported in 1835, and 26.7 % a decade later. Compare the export figures for 1835 and 1845 with those for 1805 in table 1 above. Note further that total population was 895,000 in 1805 against 1.2 million in 1835. Further, we may assume that the labour force remained relatively stable, in keeping with the general assumption of structural stability in the economy. On this basis, it appears safe to assume that the foreign sector was larger in 1805 than in 1835, and even 1845, when the population stood at 1.3 million. It does not seem unreasonable to conclude that the foreign sector made up 30 % of the national income already by 1800. For the decade 1865—1874 the national accounts figures of Bjerke show that the export sector had increased to 28.5 % of GNP.

As a provisional conclusion we may say that the foreign sector of the Norwegian economy was of a magnitude which could act as a trigger for economic growth, in the sense that it permitted the transfer of labour from inefficient to efficient sectors. The Norwegian economy thus appears to have been very much an open economy, even in the proto-industrial period.

We now turn to some descriptive remarks, that will afford a background to the trade figures discussed so far.

The international background is well known. From 1799 to 1815 the Napoleonic wars were fought in Europe. Small neutral states were drawn into the maelstrom. Peace and economic depression followed from 1815. In the international economy, stagnation lasted well into the 1830s, broken by a few good years. Real advance was delayed until the early

1840s. In 1807 it meant disaster for Norway to be on the French side, and find itself at war with Great Britain, its biggest customer.[7] Timber exports from the towns around the Oslofjord had traditionally found their best markets in Britain. With the war this market was virtually closed, and British men-of-war blockaded the Norwegian coastal waters. From 1809 Britain, moreover, levied prohibitive import duties on the Norwegian timber, partly as a measure of retaliation against the Danish-Norwegian kingdom, partly to stimulate the export of colonial timber from Canada. British duties were ten times higher for Norwegian timber than for timber from British Canada. As late as 1846 timber shipments from Norway to England were less than half of what they had been in 1805.[8] A compensatory market was found in France from the early 1830s, but prices were low. In aggregate terms, the 1805-export level was not matched till 1842, when the old record of about 900,000 m^3 was overtaken. Right up to 1830, timber merchants in traditional export towns like Christiania, Moss, and Drammen continued to export timber to clear off debts that dated back to the years before 1815. The towns simply dozed off in apathy, particularly Kristiansand and Halden. With falling prices, fixed costs and timber bought at pre-depression prices, the members of the merchant aristocracy succumbed, one by one. The place of the patricians was, with few exceptions, taken by newcomers, who were less burdened by overheads in the form of debts and social commitments.[9] An added difficulty was the heavy export duties paid to the government. Timber duties were, in fact, the largest single government revenue up to 1830.

7 For this period, see

Jacob S. Worm-Müller, *Byen og tiden. Christiania sparebank 1822—1922*, Oslo 1922.

Nicolai Rygg, *Norges Banks historie*, I—III (Oslo 1918—1954), vol. 1 contains masses of descriptive material.

Sverre Steen, *Det frie Norge*, I—V, Oslo 1951—1962, especially vol. *Det gamle samfunn* (vol. 1) and *På fallittens rand* (vol. 2) which offers the best general survey of the post-Napoleonic depression.

Wilhelm Keilhau, *Det norske folks liv og historie*, I—XI, Oslo 1929—1938, has written vols. VIII through XI, offering in vol. VIII, a general account of this period.

8 See in addition to titles under note 7 above:

Francis Sejersted, Aspects of the Norwegian Timber Trade in the 1840's and 1850's. *The Scandinavian Economic History Review*, vol. XVI, nr. 2, 1968.

9 Ada Polak, *Wolffs & Dorville. Et norsk-engelsk handelshus i London under Napoleons-krigene*, Oslo 1968, the crisis in the timber trade seen from the London side.

A major source for the economic situation are the quintennial county governors' reports, beginning in 1829, issued up to 1912. The years 1830—50 are covered in *Beretning om Kongeriget Norges økonomiske Tilstand i Aarene 1830—1835*, Christiania 1837, do. for the period 1836—1840, 1841—1845, and 1846—1850, hereafter cited "Amtmennenes beretninger".

The traditional iron export also had hard times after 1815. Norway broke away from Denmark in 1814, at the same time leaving a cozy, duty-free area for its iron and glass. Denmark, in turn, imposed heavy import duties on Norwegian iron, at the same time as prices took a downward turn. This was the beginning of the end. As the English coke iron began to compete, the Norwegian charcoal-based iron-foundries, at the time the most capital-intensive ventures in the country, were forced out, one by one, though, as the caroer of Fritzøe works shows, the end was postponed into the 1860s by cost-saving innovations and product specialization. Norway, it must be remembered, had no coal deposits, and its iron deposits were small, scattered, and of low-grade quality.

A third major source of market incomes was freight earnings. (Up to this time mostly associated with the carrying trade between Norway and her foreign customers). When the bottom fell out of the timber trade in 1808, the sailing fleet was hit immediately. British seizures of Norwegian mercantile tonnage before 1814 added to the difficulties. Thus, while the sailing fleet amounted to 121,000 tons in 1800, and approached 150,000 tons in 1815, it dropped to 113,000 tons by 1825.[10] Even as late as 1835 the fleet only totalled 151,000 tons, that is, only slightly more than 40 years earlier. Jacob Aall, a central figure in the economic and political affairs of the nation at the time, looking back on this period, later declared it to have been the worst period in the entire history of Norwegian shipping. Tonnage duties to the government worsened the situation, but taxes only reduced profits, they did not determine whether or not the ships sailed or lay tied up in the buoys. This was ultimately determined by foreign demand for timber.

There is thus a direct relation of cause and effect between reduced foreign demand and depression in the industrial sector of the economy, also in the proto-industrial period.

The outstanding exception in these years was, of course, the fisheries, which provided a growing volume for export, indeed, as table 2 above shows, fish products were the biggest single earner of foreign currency right up to 1850. Again, the expansion in fishing underscores the conclusion reached above: herring and other fish were caught and marketed precisely because foreign demand remained firm throughout the stagnation years. For one thing, fish is a consumption article, not an investment input, which goes a long way to explain why demand abroad was stable.

[10] *Historisk Statistikk 1968*, Central Bureau of Statistics of Norway, Oslo 1969, table 175.

Table 3. The population of Norway 1800—1855.

Year	Total population in 1000 inh.	Total Growth %	Rural Pop. 1000	% of total	Rural Growth %	Urban Pop. 1000	% of total	Urban in-crease
1801	883	—	806	91.2	—	77	8.8	—
	—	+ .8	—	—	− .8	—	—	+11.5
1815	885	—	799	90.2	—	87	9.8	—
	—	+18.3	—	—	+17.3	—	—	+30.9
1825	1,051	—	937	89.1	—	114	10.9	—
	—	+13.7	—	—	+14.2	—	—	+13.0
1835	1,195	—	1,066	89.2	—	129	10.8	—
	—	+11.2	—	—	+ 9.5	—	—	+25.5
1845	1,328	—	1,167	87.8	—	162	12.2	—
	—	+12.2	—	—	+10.8	—	—	+22.2
1855	1,490	—	1,292	86.7	—	198	13.3	—

Source: *Historisk Statistikk 1968*, table 13.

Moreover, the population increase in Europe accelerated after 1815, with the advent of peace and the return of the soldiers. A growing population needed more food, and fish is a cheap type of nourishment for the poor, which is why demand for fish was inelastic. Finally, new markets for klippfish, that is, dried and salted cod fish, were found in Spain and Portugal during these years. Since the Newfoundland codfisheries were disrupted during the war years, the Norwegians were quick to step in to fill the vacuum.

Admittedly, the post-war depression hit all the countries in the Atlantic economy, but no nation at the time, or since, for that matter, seems to have had such a large foreign sector as Norway. The presumption is, therefore, that the stagnation in the foreign sector had adverse linkage effects that were relatively stronger in Norway than elsewhere.

What happened to the other sectors of the economy after 1815?

Only a few of these linkage effects can be dealt with here. As the model anticipates, stagnation in the market sector of the economy meant that the population increase would have to be absorbed largely in the non-market sector. The exception again would be the towns and areas associated with the fisheries. These expectations are largely confirmed by the actual population figures for the period. Consider first the aggregate population figures in table 3.

From 1801 to 1815 the urban population, reflecting the boom conditions in the market sector up to 1808, grew by 11.5 %, while rural population actually dropped by .8 %. After 1815, the urban population continued to grow, but now more in step with the increase in the rural districts. More detailed evidence would show that the urban growth was now largely confined to the towns serving the fisheries. Relative urban stagnation was especially marked in the decade 1825—35, when the rural districts showed a growth higher than that of the towns, 14.2 % versus 13 %. Note that there are no statistical pitfalls here, such as, changes in existing town boundaries or the creation of new towns. These trends were to be dramatically reversed in the latter half of the century.

The aggregate population grew markedly after 1815. This was due to the elimination of periodic famines, the introduction of vaccination and better nourishment, following the systematic cultivation of the potato. Total population increase 1815—1845 amounted to 443,000 inhabitants, of this the rural districts absorbed 368,000 or 75 %, the remaining 75,000 fell on the towns.

Parallel to the shift away from the market sector—fisheries excepted — cultivation of new land took a sharp upward turn. With less money for grain imports, people were forced to cover a higher percentage of their food requirements from their own soil. Similarly, the rural labour force expanded more quickly than the non-agricultural labour force, that is to say, stagnation in the foreign sector forced second-best resource allocation upon the country. The potato became a strategic element of the diet in these years. Between 1820 and 1865 total farm acreage just about doubled, that of the potato quadrupled. No general statement can be made on productivity developments in agriculture, but, in the best areas around the Oslofjord, the counties of Akershus, Vestfold and Hedmark, evidence suggests that crops improved considerably. No such evidence exists regarding agriculture in the west and north.[11]

With more people on the land, the "husmenn", a rural class that emerged in the 16th century, as the rural proletariat, now swelled to uncomfortable proportions. The husmenn were farm-workers, or tenant small-holders, with a family and a plot of land, rented on good, and sometimes on bad terms, in return for labour services to the farmer. With more applicants for vacant husmanns-plots, rent conditions were tightened. Their number increased from 48,000 in 1825, to 67,000 in 1855. With an average family of 6, the husmenn-class thus made up

[11] Stein Tveite, *Jord og Gjerning*, Kristiansand 1959, passim.
Fartein Valen-Sendstad, *Norske Landbruksredskaper 1800—1850-årene*, Lillehammer 1964, English summary pp. 303—316.

more than a quarter of the total population. Still, they were not the bottom stratum. Below the husmenn were the growing number of really landless rural workers, mostly without a family, the so-called "drenger" and "innerster", whose prospects were truly Ricardian in the 1840s. We touch here on the social problem of pauperism, the dominant social issue of the day. Compare the fact that the Storting passed draconic poor-law bills twice in less than twenty years, in 1845 and 1863. In 1825 the number of paupers on the public dole, which was organized in rural areas by passing the old and decrepit dependents from farm to farm, stood at 21,000. By 1845 the figure had doubled to 39,000. The corresponding figures for the towns were 3,500 and 7,000 persons. The provisions in the poor-laws against vagrancy and begging were understandably harsh.[12]

As mentioned above, the fisheries represented the exception in this period. Like the American prairies later, the coastal waters of the Lofoten islands were now drawn more closely into the Atlantic market economy. The frontier of the Atlantic economy was pushed, not only westwards along the great American plains, but also northwards along the semi-arctic fjord territories of Norway. The Lofoten fisheries were remarkably stable; the products found an inelastic, but stable demand and a steadily growing market in the Latin parts of Europe, and from the 1850s, also in Brazil. The herring fisheries, by contrast, were more unstable, but catches were truly enormous and markets were readily found in the Baltic and in the inland districts of Scandinavia. Something akin to the Klondyke atmosphere hung over the coastal towns and townlets engaged in the fisheries during these years, Stavanger, Haugesund, Ålesund, Kristiansund and Molde. Especially the three northernmost counties, Nordland, Troms, and Finnmark, received an astonishing stream of settlers from outside. These outlying districts, now largely regarded as declining, population-losing areas, had a population growth up to 1900 that surpassed in relative terms, all the other rural counties.[13] An exception, of course, was the county of Akershus, that surrounded the capital Oslo. Thus, while most of the 18 rural counties of the kingdom doubled their population during the century, the three northernmost counties tripled theirs, and Finnmark alone quadrupled its population. Towns like Tromsø, and Bodø, shot up from next to nothing to substantial

[12] Helge Seip, *Kommunenes økonomi,* Oslo 1949, pp. 28, 40.
M. Braun-Tvethe, *Norges Statistik,* Christiania 1848, p. 375.

[13] *Historisk Statistikk 1968,* table 5.

towns by the standards of the time.[14] The interplay between international demand and expansion of market sector-activities is clear and direct.

Added light on the character of the interaction is shed by a consideration of the balance of payments-situation after 1815.[15] With peace and depression, export incomes dwindled. In the long run imports would be adjusted accordingly, but meanwhile the country faced a balance of payments-deficit. The dis-equilibrium in the foreign sector resulted in a series of devaluations over the years 1813—1822. Import bans, government borrowing abroad, or export of precious metal offered no satisfactory solution to the problems in those days, any more than they do today. In consequence the nation's money had to be devalued, though several political figures expressed apprehensions on behalf of the national honour. With independence in home affairs in 1814, the country was in a position to create its own national bank and its own separate national currency. The lengthy debates within the government and in the Storting demonstrate that those who had to make the decisions interpreted the situation in the terms sketched here, that is to say, they understood that the nation faced a trade deficit, adverse terms of trade, hence, that it was impossible to introduce convertibility for the new money at once. In 1818 convertibility was, therefore, postponed indefinitely. This decision was, in effect, a ban on metal export. Instead the groundwork was laid for a policy of long-term deflation,—along the familiar pattern: the devaluations would raise the price of foreign goods at home, while Norwegian exports became less expensive abroad. In the long run, the detoriation of the exchange-rate was bound to improve the country's terms of trade, and ultimately lift the exchange-rate for the Norwegian speciedaler towards par. This occurred finally in 1842, when par exchange was introduced by law.[16]

[14] *Tabeller over Folkemengdens Bevegelser i Aarene 1856—1865,* NOS Christania 1868, pp. 191—192.

Tabeller over Folkemengdens Bevegelser i Aarene 1886—1900, NOS Christiania 1906, appendix nr. 4, pp. 79—81.

[15] For the difficult monetary questions, see

N. Rygg, op. cit., vol. 1, passim.

W. Keilhau, *Den norske pengehistorie,* Oslo 1952, pp. 48—97.

S. Steen, *Det frie Norge,* vol. II, pp. 126—206, vol. III, pp. 24—70, 271—288.

Gunnar Jahn et. al., *Norges Bank gjennom 150 år,* Oslo 1966, pp. 1—58.

F. Sejersted, Review of Jahn op. cit., in *Historisk Tidsskrift,* 1968, pp. 149—172.

[16] Partial convertibility had been in effect from January 1st 1823, but at a very heavy discount. In modern terms, the country's Central Bank authorities were allowed to "float" the Speciedaler within a specified band, beginning with the band of 190—175, the actual

The currency story 1815—1842 shows to what extent international demand and tariff policies influenced an open economy like Norway's. Clearly, the currency decisions reflected concern for the external, rather than the internal value of the speciedaler. Internally, the money, far from depreciating in value, underwent a revaluation, which accounted for the deflationary pressure right up to 1842. This was possible in part because the home producers, again except the fishermen, worked without imported materials. (Salt, of course, had to be imported.) To the extent that best technology at all existed abroad, the prospects for profits for Norwegian entrepreneurs were less sanguine than they otherwise would have been.

It does not follow that the authorities, with a different set of currency and credit policies could have removed the more deep-seated causes of the depression. These must be sought fundamentally in the failure of foreign demand after 1815, that hit the major export goods of the country directly, and indirectly hit the carrying trade of the sailing fleet.

On the whole, the survey above does reveal the sensitivity of the Norwegian economy to the international business cycle, and how far exports determined the allocation of labour. A measure of the influence exerted by the foreign sector is its size. As shown, already by 1805 export of goods and services probably accounted for a third of the Norwegian total income. By 1825 it had dropped below 20 %; it hovered around 20 % in the 1830s, and around 25 % a decade later. Labour was pushed out of the marker sector. The post-Napoleonic depression, in short, meant a reallocation of inputs away from the export sector over to agriculture. The one bright spot was represented by the fisheries, but apart from that, unfavourable trading conditions forced the adoption of various second-best allocations of resources, with a consequent loss of potential income.

III

Turning to the period 1850—1914, our concern will be to show the interaction,—beginning in the 1850s,—between world demand and the expansion of market activities in the dual economy. The discussion will be limited mainly to the development of Norwegian shipping,

exchange rate of the paper money to be adjusted passively to the rate recorded on the stock exchange. This was in effect a de facto devaluation, the fourth in 9 years. What happened up to 1842, was that the external balance, through deflationary measures, was regained, step by step. Pari passu the exchange rate improved till it was pegged at par level in 1842.

which affords the classical example. Since, as will be shown, the foreign sector in this period came to represent about a third of total GNP, and since shipping made up 39 to 45 % of the foreign sector, it appears warranted to consider shipping as an indicator of the shifting forms of the interaction. Shipping, in fact, will directly comprise 13 to 15 % of the total Norwegian economy.

The international setting was favourable in the 1850s. By virtue of its industrial lead in coal, iron, textiles, chemicals and machinery, Britain by 1850 had become the hub of the world economy. Its pound acted as reserve currency for international settlements; the discount rate of the Bank of England was a standard followed by the other national banks. In 1846 and 1849 Britain opted for a future as an industrial nation by repealing old restrictions on trade (the Corn Laws), and sea-borne transport (the Navigation Acts). Free trade became the declared policy of Britain. The program was taken up by other European states in the years 1850—1870. By 1862 the great powers of Europe had turned the Continent into a single low-tariff area.

Population growth, international migration, gold discoveries, railways and the telegraph, extensive foreign lending and borrowing at low rates of interest, gave a massive boost to international trade in the 1850s. In per capita terms, world trade grew at a decennial rate of 33 % between 1800 and 1913, in the period 1840—1870 it reached a peak rate of growth of 53 % per decade.[17]

An important factor facilitating the international expansion was improved technology in the construction of ships and shipping machinery, in the use of fuels and in navigation. In consequence, freights dropped steadily, especially in two periods, viz. the years 1815—1851 and 1875—1909. In the first period the freight declines took place on outward voyages in the Baltic, Mediterranean routes and on the North-Atlantic run. In the latter period the drop in freights occurred in the long hauls, helped, of course, by the opening of the Suez Canal in 1869 and the Panama Canal in 1901. Reflected in the freight declines were a range of improvements, including the sailing-ship designs, more intensive use of the ship per year, resulting from improvements in cargo handling

[17] For a modern text, see

A. G. Kenwood & A. L. Loughead, *The Growth of the International Economy 1820—1960*, London, 1971.

Simon Kuznets, Quantitative Aspects of the Economic Growth of Nations: x-Levels and Structure of Foreign Trade: Long-term Trends, in *Economic Development and Cultural Change*, Part II. January 1967.

C. P. Kindleberger, *Foreign Trade and the National Economy*, New Haven, 1962.

and dock facilities. Advance also stemmed from the use of the double and triple compound expansion engine, which cut coal consumption from 10 lbs per hp per hour, to nearly 1 lb after 1900, while the power of the marine engine increased about forty times. The introduction of iron, and later steel, in ship-building made possible a steady increase in carrying capacity and continuous opportunities for cost-cutting.[18]

Reduced freights had far-reaching significance for world trade. For one thing, bulky goods whose value is low in relation to volume, could now enter world trade. Goods that only in exceptional times had been traded internationally, now came to dominate the carrying trade, viz. coal, grain, timber, iron ores—and emigrants.[19]

The point to note here is that free trade and expansion in world commerce cleared the way for national specialization and international division of labour and resource use. In turn, the small peripheral states were given an opportunity for growth they otherwise would not have had. Poor internal factor endowments could now be overcome through international exchange of a few goods for a wide variety of inputs and consumer goods. These observations seem especially pertinent to a country like Norway.[20]

It possessed a very small range of resources, hence economic growth was, in the logic of things, destined to take place through specialization in the international economy. Therefore, it was, perhaps, only natural that the shift to economic liberalism met with very little opposition. Matching trends abroad, the home authorities removed most of the obstacles, legal and otherwise, to the free movement of labour and capital in the period 1839—1872. Thus, in 1839 the guilds were abolished, internal commerce was thrown open by a series of laws beginning in 1842, saw-mill privileges were likewise repealed by law in 1860, and the customs tariff was given a free-trade profile over the years 1842—1872. However, rural land, the largest single capital asset of the country, remained an exception to the free resource use, as the institution of the "odel"-right was retained, due to the unanimous insistence on the part of the country's farm population. "The odel" functioned, in effect, as the legal equivalent to entail in Britain and the fideicommiss in contem-

[18] H. J. Dyos & D. H. Aldcroft, *British Transport. An Economic Survey from the Seventeenth Century to the Twentieth*, Leicester University Press 1969, pp. 231—299.

[19] Douglass C. North, Ocean Freight Rates and Economic Development 1750—1913, *Journal of Economic History* vol. XVIII, no. 4, 1958, pp. 537—555.

[20] Douglass C. North, *The Economic Growth of the United States 1790—1860*, Englewood Cliffs, N. J.: Prentice Hall, 1961, a perceptive study of the US in the period, using postulates and materials similar to those used here for Norway.

porary Prussia; both represented the determination on the part of the old society to incapsulate itself from the hardships implicit in the unhampered market economy.

As in England, government revenue became, restricted to a few import duties on the consumption of sugar, coffee, tobacco and alcohol, that is to say, government revenue was autonomously determined by the income elasticities for these goods. When a third of GNP depended on the proceeds from the export sector, it followed that government revenue was in fact, directly determined by the international business cycle.

Norway, as said, found its field of specialization in the international carrying trade. Consider the key figures. In the years 1850—1880 the country sailed up from 8th place to becoming the third largest shipping nation of the world—after Britain and Germany. Its sailing fleet jumped from 288,000 net tons to 1,500,000 tons by 1880, the number of sailors swelled from just under 19,000 to 62,000 by 1878.[21]

These figures point to the fact that the fleet during the period outgrew its old function of carrying goods to and from Norwegian harbours. Already by 1872 80 % of the freight earnings stemmed from international carrying trade. While the fleet that served the carrying demands of the Norwegian economy doubled during the years 1860—1880, the part that sailed exclusively in international trade quadrupled. This shift meant that henceforth the fortunes of Norwegian shipping were almost exclusively determined by the international business cycle.[22]

In consequence, the Norwegian economy became more open, for with the expansion in foreign demand for shipping services, the relative size of the foreign sector climbed upwards. This was a repetition of a pattern observed for the last boom period in the years 1800—1808. Table 4 summarizes the development up to 1914.

The table shows that the decade averages, both on the export and the import side, represented close to a third of the entire GNP during most of the period after 1865. By 1905 the export sector had passed beyond even this level. There was a slowing-down or a halt in the decades 1875—1884 and 1885—1894, which points to the general stagnation in the international economy, and again in the decade 1895—1905.

[21] Historisk Statistikk 1968, table 175.

[22] The standard work on the history of Norwegian shipping is Jacob S. Worm-Müller, et. al., *Den norske sjøfarts historie I—III*, in 6 separate parts, (Oslo 1923—51). Of relevance for this present period, vol. II, part 1 and 2.

Table 4. Exports and imports of goods and services at current prices. Moving decade averages. 1865—1914.[23]

Decade	Absolute figures. Mill. kroner			Percentage shares of gross domestic product at current prices		
	Exports	Imports	Export surplus	Exports	Imports	Export surplus
1865—1874	166	157	+ 9	28.5	27.0	+1.5
1870—1879	202	200	+ 2	28.8	28.6	+0.2
1875—1884	214	214	—	28.8	28.8	—
1880—1889	217	209	+ 8	30.3	29.1	+1.2
1885—1894	226	236	−10	30.1	31.5	−1.4
1890—1899	258	297	−39	29.7	34.1	−4.4
1895—1904	297	346	−49	29.3	34.0	−4.7
1900—1909	354	396	−42	30.5	34.0	−3.6
1905—1914	490	520	−30	33.6	35.6	−2.0

Table 5. Composition of exports at current prices. Percentage figures, total export of goods and services=100. Selected years 1865—1915.[24]

	1865	1875	1885	1895	1905	1915
Fish, whale-oil and other products from hunting	21.8	20.1	16.4	17.4	15.6	14.6
Timber, wood, pulp, paper and paper products	23.6	19.0	19.7	20.4	21.9	14.0
Products from mining; metals and chemical products	2.9	3.3	3.2	2.4	3.4	11.7
Other industrial products	5.3	6.7	9.2	11.0	9.7	13.7
Gross freight earnings by ocean shipping	41.4	45.2	43.3	38.9	32.5	39.9
Other exports	5.0	5.7	8.2	9.9	16.9	6.1
Total exports of goods and services	100.0	100.0	100.0	100.0	100.0	100.0

We turn to the composition of exports, to appraise the relative importance of shipping in the national economy (table 5).

The table brings out the dominant position of shipping and shipping services in the country's export sector. Above 45 % of total exports originated in the ocean-borne carrying-trade around the peak year of 1875. It thereafter decreased till 1900 when it again picked up momentum and by 1915 freight earnings had reached about 40 % of total export incomes.

[23] Juul-Bjerke, *Langtidslinjer i norsk økonomi 1865—1960,* Statistisk Sentralbyrå, Oslo 1966, table 27.

[24] Ibid., table 30.

If we view the development in terms of the anticipations generated by the dual economy model, we note that the initial stimulus came from abroad. During the capital-widening phase 1850—1880, growth of the fleet depended on high international demand for freight services, and at home, on cheap, elastic supply of labour, moving in from the rural sector. The situation permitted high profits, savings and reinvestments of current profits in conventional technology, i.e. sailing vessels. Gross freight earnings increased from 53.2 million kroner in 1865 to 71.1 million kroner in 1870. Six years later they had soared to 103.3 million kroner. Fairly reliable evidence indicates that net returns on capital averaged 16 % over the years 1866—1870.[25] As noted above, the labour force in shipping trebled while total tonnage quintupled. Ton/man ratio improved somewhat, up from 15 tons per man in 1850 to 23 in 1876, still, the figure underscores the capital-widening rather than the capital-deepening character. Compare the fact that of the total tonnage in 1876 only three per cent represented steamships, while the rest was sail tonnage. Business management was altogether the captain's domain. He negotiated contracts, independent of home base, picking up cargoes wherever they could be found. That is to say, incomes represented voyage freights rather than time freights. The company form was loose, with a partnership company being set up separately for each ship contracted, which underscores the haphazard, amateurish operation methods of newcomers.

It was a matter of course—and concern—that the Norwegian assets were the low wages for crew and officers, the poor fare for both, and the second-best technology that the Norwegian sailing ships represented. English and American ships could not compete on the routes that the Norwegians entered, and gradually left the field, first the long-haul routes, involving bulky and durable goods, then the short-haul routes involving non-durables, such as fruit from the West-Indies to the U.S.

World price trends slumped after 1873.[26] It was marked in consumption goods and industrial products, and was severely felt in shipping freights. Industrial wages on the other hand, having risen especially in

[25] A. N. Kiær, *Den norske Skibsfarts Økonomi*, Christiania 1871, Scattered figures also found in Worm-Müller, op. cit. and the Reports of the County Governors.

[26] Douglass North, *Ocean Freight Rates*, above.
Dyos & Aldcroft, op. cit. 265—269.
E. A. V. Angier, *Fifty Years of Freight 1869—1919*, London 1920, provides a precise record and commentary on the developments in the freight markets.
J. Worm-Müller, op. cit. II, 2. passim.

Table 6. Average tonnage increase or decrease for selected countries 1860—1886.

	Great Britain	USA	Germany	Norway	France
Steam					
1860—1870	8.2	8.4	10.7	9.7	8.2
1870—1880	10.2	−2.0	9.7	16.5	4.2
1880—1886	8.0	−4.5	13.5	13.0	11.5
Sail					
1860—1870	1.3	−4.7	1.7	5.7	−0.33
1870—1880	−1.6	0.7	0.6	4.4	−2.8
1880—1886	−2.8	−0.4	−2.1	0.0	−4.9

Source: *Tabeller vedk. Handelsflaaderne 1850—1886,* NOS, Christiania 1887, p. 68.

1871—1875, failed to adjust downwards to the same degree. The resulting profit squeeze in shipping posed the typical challenge: How to compensate for permanently higher variable costs in the face of continually falling freight rates. We will limit ourselves to some analytical remarks. The British, with easier access to capital, predictably switched to steam. At the same time they concentrated on liner company routes, which were subsidized by the British Government in return for mail carrying. The Americans, who in the fifties had been unrivalled on the clipper runs, now packed up and left the field altogether, leaving the US coastal trade, and even the Great Lakes traffic, almost exclusively to foreigners. Comparative costs made shipping no longer a profitable venture for American capital. The Germans, like the British, sought refuge in liner companies, and the French and Italian governments began subsidizing their ship-yards. In short, the capital-deepening phase had begun in international shipping.

The Norwegians, on the other hand, continued to make money undisturbed for while, since they operated with lower variable costs than their competitors. This is indicated in table 6.

The table shows that while the foreigners were selling off their sailing tonnage, the Norwegians continued to make money on this type of tonnage well into the eighties, although they also bought steam-ships.

In the stagnant eighties, freights kept sinking under the impact of the growing use of steamships. These boasted larger carrying capacity per gross ton than the sailing vessels, a greater independence of winds, and hence a better utilization of the capital. They also represented greater speed and reliability, which can easily be seen from contemporary

Table 7. The Norwegian mercantile fleet 1850—1910. Selected years, in 1,000 net tons, and Gross freight earnings in ocean transport.

Year	Sail 1000 t	Steam 1000 t	Total 1000 t	in percent		Gross freights 1000 kroner
				Steam	Sail	
1850	284	—	284	100	0	—
1855	405	—	405	100	0	—
1860	532	—	532	100	0	—
1865	706	—	706	100	0	53,200
1870	961	13	974	99	1	71,135
1875	1,310	42	1,352	97	3	90,538
1880	1,461	58	1,519	96	4	97,800
1885	1,449	114	1,563	93	7	82,779
1890	1,503	203	1,706	88	12	112,800
1895	1,284	321	1,605	80	20	93,021
1900	1,003	505	1,508	67	33	143,911
1905	814	664	1,478	55	45	115,523
1910	63	893	1,526	41	59	139,941

Source: *Historisk Statistikk 1968*, table 175, 195.

insurance rates and ship-wreck statistics. Finally, they offered economies of scale, above all, because of lower manpower requirements.

One or two examples will suffice to show the downward freight trend. Ships sailing with ballast up to Archangel to load board and props for England in 1862 received 77 sh per standard, in 1900 50 sh. For coal to Rio de Janeiro in 1860, the rate was 30 to 40 sh. per ton, by 1900 it had slumped to 15/20 sh. Coffee in bags on the run from Rio to Europe in 1861 paid from 50 to 75 sh. per ton, against 25 sh. in 1875 and 15 to 20 sh. by 1900. On the cost side, wages decreased somewhat in nominal terms, but so did food prices, hence, real wages for the sailors held up fairly well.

Experience in the eighties convinced the most obdurate that the profit squeeze was of a permanent nature. Competition from steamships simply intensified, while the easier travel opportunities to the US set a floor to wages on Norwegian ships. The contours of the gradual conversion to steam is seen in table 7, which gives the overall figures.

One interesting feature in the table is that only around 1890 did the sailing fleet reach its peak level of 1,503,000 tons net, and not till 1907 did steam tonnage outstrip sail, (viz. 814,000 against 751,000 tons). Nor was the transition to steam a direct exchange—ton equivalent by ton equivalent—of one type of tonnage for another, for as the table suggests, the fleet in fact diminished somewhat during these years, 1890—1910, down from about 1.7 million tons to 1.5 million tons. Allowance must, of course, be made for the higher cargo capacity of steamships, but, even so, it is clear that the rate of growth approached zero level after 1885. All in all, the transition from the second best to the best technology was an uncertain and slow process.

This is a general observation; a more detailed review of the matter will show a marked contrast between developments in Bergen and Oslo on the one hand, and Sørlandet on the other, the three shipping regions of Norway. Oslo, for example, offered more credit facilities, entrepreneurship, and the whole economy was more diversified there. Outward-looking Bergen had sufficient amounts of capital, entrepreneurship and forged ahead as pace-setter in steam shipping; also because Bergen in contrast to Oslo, apparently lacked alternative outlets for its pool of risk capital. Sørlandet, in comparison, lacked the necessary entrepreneurship, and its risk capital was scattered. Sørlandet, in short, seems to have lacked the resources for the new and risky venture, both in terms of organization, capital and technology. We cannot go into a detailed discussion here about the causes for success or failure. The fact remains that the economy of Sørlandet went into a decline during the period 1890—1910, following a series of bank failures in the preceeding decade. In terms of the dual economy model, our concern here is what happened to the stream of labour that, up to now, had been flowing from the rural into the market sector.

With the gradual introduction of steam, the mercantile fleet was no longer able to absorb rural labour at the same rate as before 1876. Compare for example, the decennial growth rates of the labour force on board. For four straight decades the labour force showed a growth above thirty per cent per decade, from 11,279 men in 1835 to 60,281 men in 1875. In the following three decades the growth stopped and turned negative, down .1 % 1875—1884, and in the following two decades, down 6 % and 8.9 % respectively. In 1904 the number of men on board had dropped to about 50,000 sailors.

In other words, labour now being barred from expanding employment in the home industrial sector, was forced into seasonal unemployment, underemployment, or simply pushed back into the agrarian sector. But in contrast to the years after 1815, when a similar process occurred, there was another alternative in the 1880s: America. As the home industrial sector failed to expand in step with the growth in the work force, Norwegians emigrated to the USA in waves of growing proportions, in the 1860s, above all in the 1880s, and again in the decade after 1900.[27] The causal relationship between stagnation in shipping and emigration in the eighties is revealed by the fact that in the nineties, the counties of Rogaland, Vest Agder and Øst Agder delivered the greatest flow of rural emigrants. In order to evaluate this result, it should be emphasized that the economy of Sørlandet offered no job alternatives in manufacturing, and even the herring fisheries failed during this period. Since the Oslo region and Bergen converted to steam, the labour surplus of these areas could be absorbed, partly in shipping as before, partly in industrial employment, whereas the population of the Sørlandet had no alternatives to shipping when sailing ships were laid up for good. It is against this background that the record-high emigration rate from Sørlandet should be viewed. (It was double the national average in the decade 1895—1904).

In retrospect, the rise of Norwegian shipping would seem to confirm the anticipations inherent in the dual economy model. Surplus labour may sustain a high level of demand once the growth process has got under way, it cannot initiate a growth process. Labour is a permissive rather than an initiating factor. The determining factor is foreign demand. As long as cheap labour gave Norway a comparative advantage, growth was buoyant. When freight-rates dropped below the critical point at which the sailing fleet, even with subsistence wages, could no longer be profitably used, stagnation set in, first in shipping, and from there the ripples spread to the entire economy. This goes to show that growth in the Norwegian economy, as illustrated by Norwegian shipping, was a function of foreign demand. This is a bald statement that needs qualifications. We shall consider two.

First, look at the national income figures. According to Bjerke GNP per capita tailed off from 763 kroner to 743 kroner, that is .3 % annually, during the decade 1877—1887.[28] A similar set-back occurred in the years 1899—1905, when per capita GNP eased downwards from 933 to 916 kroner, or .2 % annually. In comparison, gross freight earnings in 1896 stood at the same level as they had done 20 years before, around 103 million kroner.[29] In view of the dominant position of the

[27] For Norwegian emigration, see the standard work:
Ingrid Semmingsen, *Veien mot vest,* 2 vols. Oslo 1941—50, cfr. a recent demographic study. T. Moe, *Demographic Developments and Economic Growth in Norway, 1740—1940,* Ann Arbor, Mich., USA, 1970, (mimeograph).

[28] Juul Bjerke, op. cit., p. 39.

[29] *Historisk Statistikk,* table 195.

Table 8. Total labour force and its distribution by industry, 1865—1910. Absolute figures and per cent.

Sector	1865	1875	1890	1900	1910
Agricultural					
Agriculture, forestry, fishing and whaling	59.8 %	51.8 %	49.2 %	40.7 %	39.0 %
Non-agricultural					
a) manufacturing	13.6 %	18.1 %	21.9 %	26.3 %	25.0 %
b) services	20.5 %	27.5 %	27.5 %	30.3 %	32.3 %
Total labour force, thousands	654	731	781	883	920

Source: *Bjerke,* op cit., table 21. Note that the figures do not add up to 100 % because of a fraction whose occupation is unknown.

Table 9. Gross domestic product at current prices by industry. Absolute figures and percentage distribution for selected years 1865—1910. (Figures before 1910 calculated on the assumption of constant ratios between final output and gross product in each industry).

Sector	1865	1875	1890	1900	1910
Agriculture, forestry, fishing and whaling	33.8 %	32.7 %	27.2 %	22.4 %	23.7 %
Manufacturing	21.0 %	21.2 %	22.5 %	24.4 %	26.2 %
Services	45.2 %	46.1 %	50.3 %	53.2 %	50.1 %
Total GNP in millions kroner	480	771	780	1,115	1,435

Source: *Bjerke,* op. cit., table 24.

carrying trade in the Norwegian economy at the time, there can be no doubt that there is a direct causal link between stagnation in gross freight earnings and the parallel levelling-off in national income.

Since shipping's labour absorbing power up to 1880 mostly reflected foreign demand, and since economic growth can be achieved simply by moving labour from low- to high-productivity sectors, it is of considerable interest to appraise how much of the advance in the Norwegian economy can be explained as a result of sectoral re-allocations within the labour force, for example moving Sørlandets labour force from agriculture into shipping. We need to know total labour force, total GNP, percentage figures for the labour force employed in each sector, and the percentages of overall GNP stemming from each sector. To begin with compare table 8, which gives the figures for the labour force. Bjerke has computed the contribution to total GNP by the various

Table 10.[31] GNP per worker, by industrial sectors. Current prices in kroner.

Sector	1865	1875	1890	1900	1910
Agriculture, forestry, fishing and whaling	415	666	552	695	948
Manufacturing	1,133	1,235	1,026	1,172	1,635
Services	1,618	1,914	1,827	2,217	2,419
Average GNP per worker	734	1,055	999	1,263	1,560

sectors for the same period. The computation is based on the kind of figures that was used for the years 1835 and 1845 in table 2 above, that is, they represent gross value of final deliveries from each sector. Major structural shifts in the composition of the labour force are discernible in tables 8 and 9. One inference is that the average gross productivity—as measured in money units—was four times higher in the service sector than in agriculture in 1865.[30] The recorded shift towards employment in the market sector, in itself implies a shift towards higher productivity levels, since labour, in the competitive market setting, is paid the value of its marginal product. Part of the economic advance undoubtedly reflects better technology, skills, and organization. But how much of the increase in GNP per worker mirrors changes in the sectoral composition of the labour force? Table 10 thus gives the total productivity growth as 321 kroner in the decade 1865—1875. We now ask, what were the gains due simply to re-allocation of labour?

Now, multiplying the sectoral estimates of GNP per worker in 1865 by the sectoral labour force for 1875, we obtain a measure (partly hypothetical) of the GNP per worker that would have been achieved in each sector in 1875, if labour in each sector had been as productive at that date as it was in 1865. Next, the sectoral figures may be added up to yield the average GNP per worker for the whole economy, and the resulting difference between the GNP averages for the two years will indicate the gain achieved through the sectoral shifts alone. An illustration of this method of computation is offered in table 11.[32]

[30] Juul Bjerke, op. cit., p. 57.

[31] Table 10 is easily seen as based on table 8 and 9, that is overall GNP divided by the work force (by sectors).

[32] The method has been used on Swedish data, see
Y. Åberg, *Produktion och produktivitet i Sverige 1861—1965*, Uppsala 1969.
Cfr. P. Munthe, *Den økonomiske sirkulasjon*, 3. ed., Oslo 1971, pp. 160—164, for transfer gains in the Norwegian economy in the twentieth century. The present author is indebted to Munthe for this section.

Table 11. Transfer gains achieved through re-allocation of labour 1865—1875. (Current kroner).

(Hypothetical) GNP per worker in 1875:	
Agriculture	415×.518=215
Manufacturing	1,133×.181=205
Services	1,618×.275=445
Average GNP per worker 1875	865

Since the average gross productivity level of 1865 was 734 kroner, we may now isolate the gain due to intersectoral labour re-allocation:

	in kroner	
Transfer gain:	(865—734)=131	(40.8 %)
Productivity advance	(1,055—865)=190	(59.2 %)
Total GNP growth per worker, 1865—1875	321	(100 %)

Considering the importance of shipping for the whole economy at the time, it is interesting to note that in the decade when shipping showed its record high absorption of labour, a hefty 40.8 % of the performance gains in the Norwegian economy were due to re-allocation of labour into more productive types of employment. Similarly, when in the following decade, the rate of increase in shipping employment plunged to zero, and even turned negative, the transfer gains shrunk to very modest proportions, as can be seen from the figures in table 12. Of course, we cannot tell for sure how much of the gain prior to

Table 12. Performance gains achieved through shifts in the composition of the labour force, 1865—1910, in current kroner.

Decade	1865—1875	1875—1890	1890—1900	1900—1910
Transfer gain	131	70	52	17
Productivity gain	190	−126	212	280
Total GNP growth per worker	321	−56	264	297

1875, reflects the influence of shipping, but, in view of the importance of freight earnings on the export side, and the labour intensive character of operations, we may conclude, with a good deal of confidence, that almost half of the transfer gains up to 1875 mirrored the fortunes of shipping. They reflect, in other words, the influence of international demand patterns.

After 1875, labour absorption into shipping levelled off, and, at the same time, re-allocation gains dropped steadily. That is, when shipping ceased being a source of transfer gains in the economy, the overall economic growth as measured in terms of GNP per worker, turned negative, down 56 kroner in the years 1875—1890. Although not quantifiable, there is a strong presumption that this kind of correlation expressed a causal link. This is another way of saying that Norwegian shipping ceased to be labour-intensive, and swung over to a more capital-intensive period.

Table 12 indicates that after 1890, economic growth in Norway came to depend on the productivity gains, achieved either through more advanced skills and technology, or better organization. Parallel to this, production gradually became more capital-intensive. This is demonstrated by the conversion to steam in shipping, it is also seen in the new hydro-electric industries established in the decades 1895—1914, which were characterized by massive inputs of capital and relatively little labour. As seen above, gains from labour re-allocations were correspondingly low.

In this connection it must be emphasized that all the investment capital for the electro-chemical industry and the mining industries came from abroad, as did the demand for their products. Excess labour from the rural sector continued to emigrate as before—800,000 in fact, left the country in the period 1840—1920—and here the importance of the free labour market in the US for Europeans right up to 1924 cannot be overestimated. The emigration rate from Norway, it should be added, was surpassed only by that of Ireland.

From the point of view of Norway's economic growth, the role of the external world may now be indicated. The outside world

— absorbed the country's excess labour,
— supplied the largest share of the necessary investment capital, both for the private sector and for government infrastructure, as well as for modernization of the farms,

— provided the markets for the country's very limited range of marketable goods, and
— supplied the major share of the country's food requirements.

These generalizations from shipping to conclusions about the overall economy, should of necessity be regarded as exploratory, i.e. as suggested lines for further inquiries.

We would like to see, for example, how the dual model fits the developments in the other major exports businesses, e.g. the fisheries and the timber and pulp industry.

Was there a similar relationship between home costs and changing foreign demand after 1875, and was there a parallel swing to capital-deepening product functions? The implications are that, at least for the pulp & paper industry, such an interplay between fixed wage rates and cost-reducing technology can easily be established.[33] To that extent the dual economy model would prove a fruitful framework both for explanation and prediction.

[33] Cfr. F. Sejersted, *Historisk introduksjon til økonomien,* Cappelen A/S, Oslo, 1972, stensil, pp. 11—13.

Norway's School of Business administration, Bergen.

Structural Change and Economic Growth: Sweden in the 19th Century

By LENNART JÖRBERG

DURING the 19th century, Sweden, similar to other European countries, underwent a drastic economic change. This had many features common to Europe in general. We can also find, however, that the economic expansion in Sweden during the 19th century displays features that differ from what can be considered normal in Europe.

England's economic expansion is usually taken as an example of the general pattern in European development. But it is questionable whether that can help to explain the first stage of development on the Continent of Europe or in Sweden. The development in England is in many ways unique. Her initial position was vastly different from that on the Continent. The economic growth factors which started the industrialization of England were not the same as those that initiated the development on the Continent, although England, as the first industrial country, influenced the economic growth in other countries. The foregoing explains why the development in other countries did not follow the English in all main features.

Summarizing survey

The economic growth rate in Sweden accelerated during three short periods when production had an irregular development.

The first acceleration occurred in the 1850's, the second in the 1870's, and the third in the 1890's. The first greatly influenced agriculture, and the international demand for Swedish grain was a contributing cause of the changes that took place. In the 1870's there occurred the industrial breakthrough, which was strongly affected by the international boom period, whereas the expansion in the 1890's reflected the bigger demand by the home market. Swedish industry began to be self-generating, although the international development was extremely important as an explanation of the irregular tendencies in the Swedish economy.

Similar to all European countries, Sweden was agrarian, perhaps more so than most others at the beginning of the 19th century. Agriculture employed more than 80 per cent of her population, whose growth was relatively slow; the agricultural production was scarcely sufficient to support them. Sweden depended on imported grain, although to a dwindling extent, up to about 1830. At this time, a change occurred. Agriculture developed so greatly that it was able to support an increasing population as well as to produce a surplus for export. The Swedish agricultural exports during a short period after the middle of the 19th century began to compete with the iron exports for its position as the second largest export branch. Timber exports held the lead.

We here face a problem of considerable interest. In the pre-industrial phase, Swedish agriculture achieved a production capacity so great that it could support a strongly increasing population (more than 30 per cent from 1830 to 1860) and could also achieve a substantial export, which during the same period multiplied ten-fold.

We do not know whether the living standard of the population fell, i.e. whether the exports were made at the expense of the home market. General reference is made to the so-called lower classes in agriculture being proletarianized, but no proof of this assertion has been produced. Certain factors, for example the increase in the birth rate and the reduced mortality rate that is notable after 1845, can instead indicate that a positive improvement in the National economy took place after this time.

Coincident with the expanding agricultural yields, signs can be detected of an industrial growth after 1830. Iron mining, the production of pig iron and of bar iron expanded at about 3 per cent per annum between 1830 and 1860, and the textile industry experienced a very rapid upward trend. The production of cotton yarn multiplied more than twenty-fold, whereas the production value of cotton fabric and linen fabric rose by about 20 per cent per annum. The number of workers employed in the factories increased, according to the industrial statistics, by about 3.5 per cent per annum, whereas the production value rose by about 10 per cent per annum, which means that the productivity improved substantially.[1] This may seem to be clear proof that an industrial breakthrough had taken place, but it would be unwise to draw such conclusions. Very large percentage increases show, in this case, that the initial position was very low, and also that very modest absolute changes led to very large relative increase figures. The iron industry, which had for a considerable time been well established, was given a positive jolt. This was intensified by the powerfully increased competition from England and from those countries that began the process of industrialization on the Continent of Europe. At this time the Swedish timber industry also began to develop. Exports multiplied five-fold between 1830 and 1860, and the old water-driven sawmills began to be replaced by steam sawmills whose capacity was larger. This allowed them to change their location from the inland waterfalls to the coast.

The economic expansion of the 1870's, however, can be characterized as an industrial revolution, if by revolution is meant a rapid change in the economic structure. Naturally, the conditions for this rapid change had been created during the previous decades. At an analysis explaining the industrial development, in so far as it concerns the factors contributing to these changes, it is impossible to concentrate on analyzing the changes that took place during the brief period that characterizes a take-off. But if we analyze only this take-off, we can disregard some of the conditions that are believed to be fundamental and concentrate on analyzing the short-period changes.

In the 1870's an extremely rapid expansion took place in most industrial branches. Investments reached new heights. They were concentrated on a few sections, industry and communications, strategic from an economic standpoint.

[1] Report of Sweden's economic and financial development during the years 1834–1860. Stockholm 1863 (Finance committee's report) the table appendices.

TABLE I

Some data concerning the Swedish economic growth, 1870–1914[2]

	Annual percentage growth	Absolute figures (in millions) 1870	1914
Population	0.7	4.2	5.7
Town population	2.3	0.5	1.5
National income	2.8	800	3300
Industrial production	4.4	320	2200
Gross investments	5.0	50	450
Investments in machinery	4.0	10	90
Exports	3.1	140	790
Imports	3.2	150	800

From the 1870's up to the First World War, the Swedish National income rose at the rate of about 3 per cent per annum, i.e. it more than doubled during this period. The industrial growth was even stronger. At the outbreak of the First World War, the part played by industry in the National income was considerably greater than that played by agriculture. More than 30 per cent of the population depended on industry, whereas agriculture employed about 50 per cent.

Here will be discussed some of the factors that have probably contributed to this development, and the consequences for the further economic growth will be analyzed.

Some basic theoretical factors

The intensive studies made during the past decade into the problems of economic growth have led to a large number of factors being considered relevant for study when seeking an explanation for this growth. For a long time the raw material sector was concentrated on. It was believed that access to raw materials was decisive in order to bring about an economic expansion. The availability of capital later played a leading role as the object for

[2] Jörberg, L: Growth and fluctuations of Swedish industry 1869–1912. Studies in the process of industrialization, Lund 1961, p. 27. Percentage growth calculated exponentially by the least squares method.

investigation, and the degree of investment in relation to the national income was considered to be a criterion of a country's ability to initiate and maintain an economic expansion. However, some authors have pointed out that scanty savings are not necessarily an insuperable obstacle to the start of industrial growth. Furthermore, it has been shown that savings and investment are, to an equally high degree, a result as well as a cause of economic development. A theory that has again more recently been debated is Schumpeter's regarding the strategic role played by the innovations and the entrepreneurs in the development process. The innovations lead to structural changes both for enterprises and for branches of economy, i.e. they can result in changed relations between branches of economy or industries. For the under-developed countries the lack of entrepreneurs has been thought to be a big obstacle to economic development. Availability of capital and entrepreneurial activity are sometimes considered complementary, but it has also been suggested that the influence of the entrepreneurs in the development process is more important than capital.

Albert Hirschman in his book "The Strategy of Economic Development" has stressed that the usually long list of "necessary" conditions for an economic development does not justify the conclusion that the difficulties are insurmountable for the under-developed countries of today. Instead, he emphasizes that economic development can be brought about by many different means. It is necessary to break "the vicious circle", but the break-in can occur in many places. When the development has started, it leads upwards as more and more conditions for expansion come into being.

The most debated theory of economic growth has been Walt Rostow's stage by stage exposition, where his "take-off" thesis is the most prominent feature.[3] His aim has been to find a general explanation of how economic growth begins, and he believes his theory to be applicable to all communities, at the same time as he considers it to have a prognosis value. However, he has been

[3] Rostow, W. W.: Stages of economic growth, Cambridge 1960.

subjected to violent criticism and his explanation of economic growth has hardly been accepted.[4] What contributes mainly to the criticism is his somewhat obscure terminology. He discusses thoroughly so-called pre-conditions, but how these can be distinguished from the causes that can be laid down in order to make a take-off, he has not succeeded in explaining. One way of avoiding the Rostow-dilemma was shown by Alexander Gerschenkron, who speaks of the "degree of economic backwardness".[5] An economic growth, a take-off, can be produced on many levels of economic development, and it is the degree of backwardness which will decide the factors to play the leading role.

Other authors have also tried to force the development into a system by speaking of "big push", or "the early phase of modern growth", or "the critical minimum effort", or "basic conditions for economic growth", where they have stressed different factors as decisive for producing a rapid economic growth.[6] In all cases, however, it must be noted that the confusion in terminology has contributed to obscuring the aim of the two basic investigations which must be separated when studying the problems of the economic growth. We either take a long view in studying the economic expansion, i.e. analyze all the factors leading up to a stage where rapid economic expansion is possible, or a short view, i.e. concentrate on analyzing the changes and disregard how the structure develops, which in isolated instances leads to industrial revolution in the real meaning of the word.

Hirschman's grasp of the development problems is for an economic historian more satisfying than, for example, the stage-theories of Rostow, who concentrates solely on the investment

[4] Rostow, W. W. (ed): The take-off into sustained growth. Proceedings of a conference held by the International Economic Association, London 1963.
[5] Gerschenkron, A.: Economic backwardness in historical perspective, Cambridge, Mass. 1963.
[6] Rosenstein–Rodan P. N.: Notes on the theory of the "Big push". Center for international studies MIT (mimeographed) Cambridge, Mass. 1957. Kuznets, S.: see note 4. Leibenstein, H.: Economic backwardness and economic growth, New York 1957. Habakkuk, H. J.: The historical experience on the basic conditions of economic progress (in Economic progress, pubd. by L. H. Dupriez) Louvain 1955.

sector and simplifies reality, and forces it into a system that is far from objection-free.

Another type of explanation is found in the more formalized theories that are based on, for instance, a simple multiplier-accelerator model where, with changes in the parameters, a certain pattern in the development can be arrived at.[7] A model of this type can be very instructive, but we must be certain that we can explain why changes occurred in the production which, in its turn, generated a rise in income or the reverse, because in a multiplier-acclerator model there is a mutual dependence between production and income. It is not enough to argue that a change of this nature has occurred, for if the multiplier-accelerator has once begun to function it leads, according to the construction of the model, to economic growth (or fluctuations).

When we come to study the development in Sweden, it is also difficult to find *one* decisive cause of expansion. Studied more closely, the problems are seen to be more differentiated, the causal connections more complicated, and the conclusions more guarded.

In the following I will concentrate on analyzing the long-term changes and after that will dwell on the special situation in the 1870's and the development up to 1914. This means that in the analysis I must import factors whose significance for the economic growth cannot by themselves be determined or quantified.

Population development

In 1800, Sweden had a population of 2.3-million. Up to 1830, this increased by 23 per cent and during the following 30-year period, 1830 to 1860, by 34 per cent. The population then totalled 3.9-million. The increase was strongly differentiated. In eight counties the population increased by less than 30 per cent; in seven, by more than 40 per cent. It was greatest in the south and the west of Sweden and in Norrland; the eastern and central counties had the lowest growth rate. During the next 40-year period, up to the turn of the century, the increase was approximately as great as

[7] E.g. Duessenberry, J. S.: Business cycles and economic growth, Cambridge, Mass. 1958.

TABLE 2

Percentage change in the size of different age groups, 1800–1910[8]

	1800–1810	1810–1820	1820–1830	1830–1840	1840–1850
0–14	—0.4	+10.4	+22.6	+ 2.9	+ 8.7
15–39	+5.2	+ 8.3	+ 6.6	+14.1	+14.9
40–64	+2.6	+ 2.3	+ 5.6	+11.3	+ 6.9
15–64	+4.3	+ 6.1	+ 6.2	+13.0	+12.0
65–	—7.7	+13.8	+ 9.3	— 2.0	+12.0
Total	+2.1	+ 7.8	+11.7	+ 8.6	+10.9

	1850–1860	1860–1870	1870–1880	1880–1890	1890–1900	1900–1910
0–14	+12.7	+ 9.7	+ 4.8	+ 7.0	+ 4.5	+5.0
15–39	+ 7.6	+ 0.8	+11.2	— 0.4	+ 9.0	+9.8
40–64	+17.4	+17.4	+11.4	+ 2.1	+ 5.5	+6.9
15–64	+ 9.1	+ 6.6	+10.8	+ 0.5	+ 7.7	+8.7
65–	+19.8	+12.6	+18.7	+36.7	+16.9	+8.4
Total	+10.8	+ 8.0	+ 9.5	+ 4.8	+ 7.3	+7.5

during the 30-year period 1830 to 1860, i.e. 33 per cent. Sweden now had a population of over 5-million. It thus had more than doubled during the 19th century. In the latter half of that century, more than a million people had emigrated. The emigration began on a small scale in the 1850's and reached its maximum during the 1880's. In addition, the population showed considerable mobility even before the emigration. The net migration, especially at times of bad harvest, amounted to several thousand per annum in some counties during the years 1830 to 1860. The internal population movements are not, as yet, investigated, but it seems as though the Swedish people were migratory long before the modern movements began. To how large an extent this can have contributed to equalizing the opportunities for employment and to recruiting workers for the new industries, it is difficult to assess, but a mobile population must, reasonably, have been advantageous for raising the level of production and the productivity and for pre-

[8] Sweden's historical statistics, part I, Table A 16.

venting diminishing returns, which would have had an effect of decisive economic importance. The GNP expanded also by moving the labour force from low productive to higher productive industrial sectors. The transfer gains could be considerable.

When appraising the influence of population conditions on economic development, the displacements between different age groups must also be taken into account. A strong increase in the non-productive population in the pre-industrial phase must have contributed to retarding the expansion in the National product. This can have a certain value when appraising the so-called proletarianization. (The population development is only slightly investigated in Sweden, despite our having, from an international standpoint, an unusually complete material available.) A big increase in the population of people of working age, however, does not necessarily lead to greater production. In the pre-industrial phase, that increase could equally well lead to unemployment, more or less hidden, or to a reduction in wages, which could reduce consumption or savings and thereby investments. In an industrialized community, such an effect can also be produced, but there the durability of the depressive effect ought to be shorter. In Sweden, the number of those fit for work—meaning those in the ages 15–64 years—increased more than the total population during the 1830's and the 1840's, during the 1870's, the 1890's, and the first decade of the 20th century. Also during the 1850's, the increase was almost as large as the increase in the total population. We thus find that, during those phases when the economic growth accelerated, the increase in the working population exceeded or almost equalled the increase in the total population. During the 1810's and the 1820's, as well as during the 1860's and the 1880's, when the economic expansion was less accentuated, however, the increase in the number of potential workers was conspicuously slower than the increase in the total population.

But if we consider not only the changes in the size of the age groups but also the proportion of potential workers in the total population, we obtain a rather different picture, although the main features, naturally, remain unchanged.

TABLE 3

The age group 15–64 years in percentage of the total population, 1800–1910[8]

1800	1810	1820	1830	1840	1850	1860	1870	1880	1890	1900	1910
62.0	63.3	62.2	59.2	61.6	62.2	61.2	60.5	61.5	59.0	59.2	59.9

From 1850 up to the 1870's, the proportion of the potential workers decreases, but from 1890 it increases. During the 1880's, the most extensive increase in Swedish manufacturing industry occurs, i.e. the number of workers rises more rapidly than the accumulation of real capital.

It is evident that the changes in the composition of the population can have influenced the rhythm of economic growth, although at present it is not possible to dare to assert in more general terms anything about cause and effect in this process.

If we insist that this increase in population was sufficiently large to constitute an expansive power in the growth process, although not large enough to obstruct it, we touch upon the most important feature of the population problem in the economic growth. In Sweden, as in other European countries, the increase in population was in general no greater than the expansion in production; before 1870 a rise in real income could occur despite a rather modest economic expansion in the beginning.

As far as Sweden is concerned, we find that about 30 per cent of the rise in the National income can be explained by the increased population and the other 70 per cent by greater productivity during the years 1872 to 1889, whereas the productivity expansion was an even more dominant factor up to 1912 and was responsible for over 80 per cent of the measured rise in the National income.[9]

The transformation of agriculture

During the introductory phase before 1870, Swedish society underwent a radical structural change. The relatively large increase

[9] Jörberg, L.: op cit., p. 20.

in population can have resulted in over population, and thereby also in a degree of proletarianization of the lower classes in the rural areas during the period up to 1840, i.e. the rise in National income was slower than the increase in population. The average productivity per person would, in other words, have fallen although it cannot be presumed that the marginal productivity was equal to zero. During the period 1750 to 1840, the number of farmers grew by about 16 per cent, whereas the lower class population increased by approximately 240 per cent.[10]

At the same time, however, an extensive land reclamation took place and thereby agricultural production also expanded. This land reclamation is usually explained by the fact that the land-parcelling activity speeded up after about 1830 when the Act concerning legal land parcelling was issued. This was unquestionably a contributing cause of the land reclamation, but it does not explain why the farmers were so ready to parcel out their villages. For that, the population development must also be taken into account. It was not the parcelling-out movement that led to the population growth increase; it was rather the greater demand on the productive land in the old agrarian community, caused by increased population, that precipitated the parcelling out. In those parts of the country that had a relatively slight increase in population, the parcelling out came later than in the districts with a considerable increase in population. That may have given rise to the land division in Western Europe. In countries that had a rapid increase in population, e.g. England and Germany, land distribution took place; in France, with a stagnating population, this did not happen. The parcelling out also necessitated considerable investments. New buildings, road constructions, and fencing demanded much capital outlay over relatively short periods and, although much of the work was done by the farmers themselves and material for these investments was taken from their own forests, this new investment must have had an accelerating effect on the economy and have resulted in a definite multiplying effect. In addition, agriculture around 1850 became

[10] Höjer, E.: Svensk befolkningsutveckling (Swedish population growth), Stockholm 1959.

more capital demanding. New implements, such as harvester combines, began to come into use and a more capital-intensive production must have led to an increase in the growth of both the investments and consequently the income.

In this connection, the availability of capital and its acquirement in agriculture must be discussed. The work of parcelling out and reclamation of land demanded considerable capital resources. If these had not been mobilized, it is probable that this remoulding of Swedish agriculture would not have taken place during this period and, perhaps, not in the form that it got. What seems surprising is that the availability of capital in the rural areas appears to have been fairly good. The growth of the rural savings banks was, to some extent, a reaction to the savings banks in the towns, which refused to allow the rural population facilities for deposits. In the opinion of the savings banks of the towns, there were no investment possibilities for more money than could be saved by the small number of the population in the towns. That this was a misconception is clear from the rapid growth of the savings banks in the rural districts. The number of savings banks increased between 1834 and 1860 from 31 to 146 and their capital from 2.3-million kronor to 29.0-million kronor.[11]

Another contributing cause of the investment activity was the development of mortgage societies. The first of these for the farming community was founded in 1836, and before 1850 most counties in south and central Sweden had their own. At the end of 1858, these societies had a foreign bond liability of 52-million kronor out of a total bond liability of 72-million kronor.[12] This large import of capital in the pre-industrial phase naturally contributed actively in facilitating the remoulding of agriculture that took place.

A further factor in the discussion concerning the procurement of capital must be considered. Around 1830 an export of grain began on a small scale. This export grew in importance and reached its maximum in the 1870's. The exports consisted mainly of oats, of which England took the major share. There they were

[11] Finance committee's report, Table XXII.
[12] Finance committee's report, p. 83.

used to provide the omnibuses of that time with fuel (in the 1830's, London's largest transport enterprise employed more than 300,000 horses).

The expansion in the production capacity of agriculture thus led not only to its becoming sufficient to support the growing population, but also to a large surplus being made available for export. This development must be considered to have been highly important for the appraisal of Sweden's possibilities to bring about an industrial transformation. Swedish agriculture obtained a large additional income from the exports.

If we assume that farming worked at less than full capacity, the change in demand on the world market must have meant that a larger amount of grain could be disposed of. Thus, the farmers had a primary addition in income, and through the multiplier effect this led to augmented income formation in other sectors of industrial life. It helped to speed up the demand on the home market for industrial goods. This created scope for the beginning of a consumer-goods industry. Furthermore, the possibilities for loans on the foreign market were improved because the import of agricultural products ceased to be of importance in the 1830's. A continued agricultural import would have considerably checked the possibilities of borrowing on the foreign markets.

The early agricultural economic development may have been hindered by the relatively evenly distributed production of different types of grain throughout the country. This prevented production specialization and expansion of the National grain trade. (Some irregularities in the distribution of the resources would have increased trade, raised profits, and made the share of trade in the economic growth more significant.) In actual fact this imbalance in production was made by the exports, but at the same time it created fresh problems in the later development.

Transformation of agriculture was unquestionably one of the more important factors in the economic growth. It had more obvious effects for the people than for the industrialization that was slowly progressing at the same time. It contributed to the growth in capital, and improved the opportunities for rationalization in agriculture. Up to 1875, prices were rising during the greater part of the period—with an inflationary peak during the Crimean War and one at the beginning of the 1870's—which ought to have made the investments more profitable. The credit institutions, created with agriculture as their best customer, cooperated in channeling export profits to help enterprising farmers.

All these factors combined can explain the transformation of Swedish agriculture before the industrial transformation, and the fact that these two changes did not occur simultaneously makes it easier to understand the problems of the growth. The industrial expansion occurred during that epoch when the problems of creating food supplies for the increasing non-agrarian population were largely solved. This is not meant to imply that agriculture had solved all its problems; only that the structural tensions had been reduced. Around 1880, Sweden was again importing grain, and we get the impression that industrial development up to 1914, together with emigration, cooperated in diminishing the agricultural labour force more rapidly than productivity improved. The agricultural crisis during the 1880's and 1890's was perhaps not solely a reflection of changed conditions on the world market.[13]

History can be written in at least three ways: what has happened can be described without trying to explain the reasons for the development; the development can also be analyzed and attempts be made to treat of the underlying factors; we can discuss what might have happened had something else not happened instead. This third method has for long been cherished by the historians because of the possibilities it gives for speculating without running any risk of proof to the contrary arising.

I do not intend to speculate more closely therefore concerning the development that could have been the consequence of the stagnation of Swedish agriculture and its incapacity for exports during the period before 1870. It need only be established that the transformation of Swedish agriculture took place before the Swedish industrial breakthrough. For a capital-poor country, such as Sweden was, this was important, because investments were largely concentrated on different sectors during different periods.

[13] Fridlizius, G.: Swedish corn export in the free trade era, Lund 1957. Cf. also Jörn Svensson: Jordbruk och depression 1870–1900, Lund 1965.

This fact is important if one wishes to compare the development in the under-developed countries of today with that of Sweden in those days—which has often been done. Today's problems are much more complicated, and an obstructive factor in the progress of the under-developed countries is precisely the inability of agriculture to expand production in step with increasing population. For Sweden this seems not to have been a problem. In other matters, too, the resemblances to the problems of the under-developed countries of today are not particularly marked.

Agriculture has always had a low supply elasticity, which has obstructed rapid expansion. However, it is surprising to observe how rapidly the Swedish agriculture responded to a heightened international demand during, for instance, the 1850's and the 1870's. It is not yet clear to what extent distribution of land through the parcelling out reform, land reclamation, or productivity improvement caused this supply elasticity. Nor to what extent it was caused by unrational use of available resources, such as excess feeding of cattle, which at a strongly increased demand (e.g. at the beginning of the 1890's) could be cut down. However, this capacity was an important factor in the economic development. Economic growth, as far as Sweden was concerned, did not depend solely on the industrial capacity of the country, but to an equal degree on its capacity to transform the agricultural sector.

One of the consequences of the development of agriculture was the railway construction, which began during the 1850's. This was, to some extent, inspired by the demand of agriculture for improved transport from inland to the ports. The export industry joined in this demand, and during the 1870's, Sweden experienced a railway boom that has never been surpassed.

International adaptation

The industrial development in Sweden up to the 1870's was to a great extent an adaptation to happenings outside its borders and to a lesser degree to an independent economic expansion. This was the case with the iron industry, which struggled with international competition and had big conversion difficulties in order not to lose contact with the international market. To an even greater extent, the expansion of the sawmill industry was, prior to the 1870's, a clear reflection of a larger demand from England and from those Continental countries that underwent an industrial development. At the beginning of the 1830's the export of sawn timber amounted to 200,000 m^3; around 1860, it reached 1-million m^3. The development of the sawmill exports can be illustrated by a logistic curve. The inflexion point is found in the 1880's, after that the rate of increase slackens up to the First World War.[14]

The growth of the sawmill industry was stronger than almost every other economic expansion in Sweden before 1870. It can also be said to have had a decisive importance for the entire industrial development. The multiplier-accelerator effect of the sawmill industry had importance not only for the capital-goods industry—which furnished the sawmills with machinery and in its turn led to an increased demand by the engineering industry for other products—but also for the boom in the consumer-goods industry, which can to some extent be seen as a reflection of the sawmill industry development.

The export industry thus came to be a driving force in the economic growth. The demand from foreign countries for Swedish products gave the exporters and the producers possibilities for expansion and enabled them to reduce their production costs. The investment activity therefore influenced also other industrial branches, whose production grew as the export industry expanded.

It was easier for the export industry than for the home-market industry to expand in a community such as the Swedish, with its low National income and poorly-developed communications. The expansion in the export sector could take place before the essential basic investments had been developed. An industrial expansion on the home market was delayed, being limited not only by low real income, but also by poor communications between

[14] Söderlund, E.: Svensk trävaruexport under 100 år (Swedish timber exports during 100 years), Stockholm 1951.

the regional markets. Investments in communications were therefore necessary in the initial industrialization phase in order to improve the distribution facilities if an integrated home market was to be developed. However, because the demand for industrial products came from foreign markets, the industry could begin to expand before these investments were made. The industrial breakthrough in the 1870's coincided with the rapidly constructed internal communications.

Because the export industry was first in the industrial development, it had no immediate need to compete with the home-market industries for labour and capital. Later, the competition between the two sectors became a stimulus mainly for the home-market industries, which urged forward innovations suitable for expanding their productivity.

The expansion in the export sector therefore encouraged growth also in other sectors, partly owing to the bigger demand for export products having a multiplier-accelerator effect on other industrial branches and partly on "external economies".

Walter Hoffmann at the beginning of the 1930's investigated the structure of the industry concerning its distribution on capital-goods industry and consumer-goods industry during the industrialization process.[15] As anticipated, he found that at the beginning of the industrialization phase, the consumer-goods industry dominated, but later its share was reduced and exceeded by the capital-goods industry. Hoffmann presumed that this development had general validity. As far as Sweden was concerned this "law" is not entirely correct. Around 1870, before the industrialization had seriously begun, the distribution between the two sectors was equal: both had approximately half of the production value in industry. According to Hoffmann's scheme such a distribution ought not to have been reached until considerably later. But the reason that Sweden fails to fit into the general pattern is easy to find.

The Swedish industrialization was a reflection of and an adaptation to conditions outside Sweden. The export industry was

[15] Hoffmann, W.: Stadien und Typen der Industrialisierung, Jena 1931. (English edition: The growth of industrial economies, Manchester 1958.)

the dominant part in the capital-goods industry, and sawmills and iron industry were responsible for the larger part of the production value of this sector. This enlargement resulted, in its turn, in an expansion in the consumer-goods industry.

The question then arises: why did not the Swedish economy get a structure similar to that of the raw-material producing countries of today? Why did not a "dual economy" arise, with an advanced export sector and an under-developed home-market industry? Apart from historical analogies being of dubious value, although comparisons are made between closer lying countries and periods of time, there are several factors that can be adduced in explanation.

Structure and financing of enterprises

One factor in the development has already been discussed. Not only industry but also agriculture had income from exports. This spreading was advantageous for the origin of a balanced economic expansion. Another factor, also of considerable importance, was the emergence of an entrepreneurial class in Sweden.

Industrial expansion brings with it, by definition, a high degree of investment. But investment is a subject that can be treated from different premises. 1) Who makes the investments? 2) How are the investments used? 3) What decisions lie behind the investments?

All investments can also be classified as either private or public. In Sweden the State was responsible for the large basic investments in communications and education, and both these spheres of investments were essential for the economic growth.

In appraising the industrial growth, however, the interest must be concentrated around the private entrepreneur. The study of "entrepreneurial history" is of importance for the understanding of Swedish industrial expansion. "To study the entrepreneur is to study the central figure in modern economic history, and in my view the central figure in national economy," says, somewhat exaggeratedly, Professor Arthur Cole, who was one of the initiators of the "Research center in entrepreneurial history" at Harvard University.

The Swedish industrial entrepreneurs were recruited from a relatively small group. It is clear that the commercial houses, i.e. the wholesalers, belonged to the most initiative-rich founders of businesses in Sweden.[16] Most of the sawmills that were established, as well as enterprises in other branches, were initiated by merchants. From this small group were trained a large number of the entrepreneurs at the industrial breakthrough. These wholesalers, being both importers and exporters, had an international outlook. They saw the needs of the foreign market and reacted rapidly by founding or cooperating at the founding of the new export enterprises.

The iron and timber industries had, at the beginning, the largest need for capital, because their plants were already of considerable size.

In the iron industry, the promoting of new enterprises was slight. It was rather a modernizing of the old concerns that took place when the commercial houses with their capital resources came in as owners or reorganizers.

In the timber industry, on the other hand, the merchants were also promoters of enterprises; in the engineering industry, too, the trading companies were there from the beginning. Within the latter branch, technicians were often initiators, whereas the commercial houses came in as financiers.

The financing often took the form of short-term credits, for instance by promissory note guarantee, but in actual fact these were made long-term by continual renewal, which was, naturally, a pressure on the liquidity of the enterprises.

The old idea in Sweden was that much of the working capital came from foreign countries. Professor Torsten Gårdlund's investigations of Swedish industrial financing have shown that this was not the case,[17] although foreign capital undoubtedly played a large role because it was probably marginal loans, which were

[16] Gårdlund, T.: Industrialismens samhälle (The society of industrialism), Stockholm 1942.

[17] Gårdlund, T.: Svensk industrifinansiering under genombrottsskedet 1830–1913 (Swedish industrial financing during the breakthrough period 1830–1913), Stockholm 1947.

of greater importance for the enterprises than their actual amount would lead us to suppose.

Another factor in this connection, which Gårdlund does not discuss, is the importance of the foreign short-term credits. It is made clear that, for instance in the sawmill industry and the grain exports, these credits played a large role.[18] The foreign trading houses made advance payments on coming deliveries with the result that the financing of the enterprises was facilitated; these short-term loans could be of considerable size. Without them, Swedish industry would not have been able to finance its operations as successfully as it did.

The reason that the commercial houses were able to engage themselves so strongly in promoting and financing enterprises was that, besides having big capital resources, their own trading yielded large profits. Up to the turn of the century, the trading business was for them still the essential; after that the focal point moved to financial operations. The large profits resulted not so much from the loans they made to industry as from their grip on the sales made by the industry from which they were able to extract large commissions.

It is thus clear that merchants and commercial houses played a decisive role in spreading the entrepreneurial operations in Sweden, and their attitudes towards and opinions concerning these operations were more modern and the planning more far-sighted than that of most other leading entrepreneurs.

Sweden's old association with the European market, which mainly took place through the earlier exports of iron, became highly important because it enabled Swedish enterprises to recognize the possibilities for expansion and to use their foreign connections to relieve the Country's capital shortage by allowing foreign buyers, through advance payments and short-term loans, to be responsible for a part of the operational financing.

When discussing the operational financing, incorporation must also be touched upon. Incorporation reduces the risks for individual financiers and simplifies capital accumulation: a) by

[18] Söderlund, E.: op. cit. and Skandinaviska banken 1864–1914, Stockholm 1964; also Fridlizius, G.: op. cit.

TABLE 4

The number of newly formed limited companies in industry between 1866 and 1910[20]

1866–70	155	1881–85	259	1896–00	976
1871–75	506	1886–90	394	1901–05	1073
1876–80	234	1891–95	545	1906–10	1486

facilitating the procurement of capital in such a manner that those who have capital become willing to invest; b) by making it possible for smaller capital owners to contribute to productive investments; and c) by the limited company form often being a condition for larger bank loans and for the issue of bonds.

From the introduction of the limited company law in 1848 up to the 1860's, the attitude of the State to the limited company form was restrictive and many applications were denied. After the 1860's, the attitude changed, and during the 1870's and 1880's, practically all applications were sanctioned.[19] However, the role of the State was throughout regulating, and it neither encouraged nor supported company formations. The number of new limited companies—whether they were newly formed enterprises or conversions of those already in existence—rose sharply during the first half of the 1870's, and the powerfully accelerated development during the 1890's resulted in the figures of the 1870's being for the first time exceeded.

The increase during the 20th century is due, above all, to the fact that small enterprises now also began to be operated in limited company form. Earlier it was chiefly the large enterprises which utilized the possibilities of safeguarding their owners from unlimited responsibility and which also had need for larger capital in their operations. It is not clear whether it is a rising economic activity which is reflected in these figures during the 20th century or whether a change took place in the attitude towards the limited company form.

[19] Nilsson, C. A.: Business incorporations in Sweden 1849–1896 (in Economy and History, vol. II, Lund 1960).
[20] Jörberg, L.: op. cit., p. 198.

TABLE 5

A comparison between the percentage increase in the number of industrial workers in limited companies and other forms of enterprises, 1872–1912[21]

	Companies	Other forms		Companies	Other forms
1872–80	37.1	8.0	1897–03	29.8	—4.2
1880–89	82.6	49.8	1903–12	24.1	—6.0

It is natural that the growth of the Swedish industry meant not only that production grew and that the number of workers in the industry increased, but also that a change occurred in the production structure and in the ownership structure.

It can be asserted that the limited company form was a driving force in the industrial development, at the same time as, during certain periods, it can be seen to be a consequence of this development. The limited company form created possibilities for more risk taking, and it is obvious that the industrial production, to an increasingly high degree, occurred in enterprises that had the limited company form. In 1872, 45 per cent of all workers were employed in limited companies; in 1912, the figures rose to almost 80 per cent.

The promotion of large-scale enterprises was facilitated also by the relatively new form of ownership. Alexander Gerschenkron has pointed out that the more backward a country is when it begins industrialization, the more likely it is that the first enterprises are large. To some extent, this reasoning also applies to Sweden. The largest enterprises in Sweden had a greater share of the production in their branch at the beginning of the 1870's than later. This does not mean that these enterprises were large from an international standpoint, but that in the main the number of enterprises was few and the range of products small, which thus made it easy for small enterprises to capture a large share of the market. This is in itself natural and not a peculiar characteristic of Swedish industrialization. In 1872, more than two-thirds of the workers in the Swedish industry were employed in enterprises

[21] Jörberg, L.: op cit., p. 201.

TABLE 6

Percentage changes in the number of industrial workers in factories with less than 100, with 100–500, and with over 500 workers, 1872–1912[22]

	Total	-100	100–500	500-
1872–80	13.6	25.4	—1.6	25.9
1880–89	66.0	51.9	67.8	97.0
1897–03	19.1	25.5	13.0	22.8
1903–12	16.3	12.8	17.8	22.1

each of which had more than 100 workers. This proportion fell during the course of the industrialization.

However, the question is what amount of capital was required in the beginning in order to reach effective production. The size of the market naturally depends upon the demand, and the possibilities for development depend upon the slope of the demand curve. If the demand is strong or differentiated, a small capital investment can be sufficient to reach profitable production, and the enterprises can be relatively small without being less competitive than larger enterprises that can take advantage of large-scale operations. On the other hand, investments in, for example, railways must be made on a large scale. The initial costs are great, and not until the railway first comes into operation can income be expected. The fact that this capital-demanding type of investment was provided largely by the State gave the capital market free hands for less cost-demanding enterprises, which yield a more rapid profit.

It can be stated that this enterprise structure was advantageous for the development. Relatively large enterprises were a condition for specialization and many new articles for production were of such a nature that big investments were needed. These could only be provided by the large enterprises. Because these latter were in the beginning often also without any competition worthy of mention, they had the chance to develop by being able to command

[22] Jörberg, L.: op. cit., p. 150. The figures before 1897 do not include the sawmill industry or the iron and steel industry because of lack of statistical information concerning these branches.

TABLE 7

The average number of industrial workers in factories owned by limited companies and in other forms of enterprises, 1872–1912[23]

	Limited companies	Other forms		Limited companies	Other forms
1872	163	32	1897	101	33
1880	108	26	1903	82	29
1889	117	31	1912	75	26

high prices, and the increased profit was to a large extent diverted into new ventures. It is a characteristic feature of the large enterprises of that time that the share of the profits for the owners was often small compared with the gross profit.

At the same time, there was a negative factor in this development. Because the communications were poor and the only available way for the enterprises to expand production was to find their markets nationally and not regionally, the enterprises were not quite able to utilize the advantages of large-scale operations. The optimum production size was, from an international standpoint, small.

Another factor is that the productivity in the larger enterprises was not higher than in the small. In many instances, the opposite was the case, particularly at the beginning of the industrial phase. The reason for this condition appears to have been the difficulty in organizing the enterprises in a rational manner and that trained labour was lacking, which could have been a greater obstacle in the large enterprises than in the small ones. To this was added the lack of technicians.

A high degree of concentration does not necessarily mean that the leading enterprises had a dominant position on the market. The degree of market dominance depends on the size distribution of the enterprises, on the product differentiation, on the occurrence of cartels, on enterprise integration, and finally, on the occurrence of international competition. In order to measure the importance

[23] Jörberg, L.: op. cit., p. 202. Cf. the comment to note 21.

of the concentration, it is not sufficient merely to measure the share of the enterprises in the national production; it is necessary to study the whole market.

A high concentration in the Swedish exports industry, for example, did not mean that the enterprises had a dominant influence on the market. A high concentration has little meaning if the goods produced by the enterprise are easy to substitute. There are, too, other factors. If the demand increases rapidly, or if technical changes take place quickly, or if the enterprises operate on different markets, or if some form of product differentiation occurs, then under such conditions the dominance of the enterprises can easily be nullified.

In this connection, it may be worthwhile to ponder somewhat on the attempts of the Swedish industry by cartelling the production to master what the entrepreneurs considered to be a threatening over-production. The forming of cartels need not first become desirable for the enterprises when over-production has be-

TABLE 8

The three largest enterprises' percentage share of the production value in certain industrial branches, 1872–1912[24]

Branch	1872	1880	1897	1912
Margarine	—	—	89	71
Flour mill	..	..	23	31
Sugar	54	59	43	98
Cotton	29	21	26	27
Woollen	29	26	29	25
Leather	30	24	23	31
Footwear	—	95	36	17
Sawmill	..	..	11	7
Paper & pulp	50	20	15	13
Cement	—	100	84	72
Matches	59	56	57	82
Engineering (not electrical)	40	24	11	10
Electrical engineering	—	100	92	92
Iron and steel	..	..	23	24

[24] Jörberg, L.: op. cit., p. 170.

come a fact; when a balance exists between supply and demand, the enterpreneurs might already feel themselves threatened.

Cartels occurred prominently in the home-market industry, whereas the agreements between the exports enterprises were less extensive. The type of cartels which occurred were predominantly price cartels. However, proposals were made very early for more extensive cartels in, for example, the sugar industry, which already in the 1870's discussed a trust formation, which, however, was not introduced until 1907. Certain special enterprises in the engineering industry tried also to include enterprises in Norway in the cartel. Although this suggestion was not carried through, it was very advanced in form.

The sawmill industry also tried an international link during the economic recession in the 1880's, when an attempt was made to interest both Finnish and Russian enterprises in making an agreement concerning restrictions in production.

It is evident that in the branches where production concentration was greatest, the possibilities for cartelling were more easily accomplished. The home-market industries with a low degree of concentration were, for example, the footwear and the ready-to-wear clothing; to this group belonged also the engineering industry. In the latter, the product differentiation was more important than forming cartels.

At the end of the 1880's, a protectionistic tariff policy was introduced in Sweden. Several authors consider this to be a factor of importance for the growth of Swedish industry. It is possible that the home-market industry had some advantage from this tariff policy. It gave the enterprises a certain monopolistic position in the home market, but counteracted, on the other hand, the formation of a concentration. However, it can not be shown that this tariff policy was a decisive factor in the industrial development. Because the industrialization was rapid and because the demand grew quickly, it was only during certain depression periods that such a pressure on the enterprises arose as to render a tariff protection useful.

Enterprise integration and fusions between enterprises were hardly decisive for the industrial growth. They were instead a consequence of the technical development. Thanks to the introduction of good communications (both railways and post and telegraphs are included here) the management had for the first time possibilities to supervise an enterprise that had several factories, and it was not until the 1890's that such enterprises began to appear more generally on the Swedish market. It was the same with forming cartels. Cartel organizations where the members had no possibilities easily to contact and supervise one another were of no value. Many enterprises experienced this during the 1870's and 1880's, when quite a few cartels were formed, but very soon ceased to function.

The boom of the 1870's

The 1870's mark the beginning of Sweden's industrial breakthrough. As is obvious from the foregoing, by "breakthrough" we do not mean that the industrial development now starts for the first time. Industrialization had increased, although to a limited extent seen comprehensively, since the 1830's. By "breakthrough" we mean a marked acceleration in the development, a break in the previous trend; and furthermore, that the structure changed institutionally, from the enterprise standpoint, and in the production, so that we can speak of a new situation after this acceleration compared with the preceding development. The boom period in the 1870's brought about these changes; therefore we can date the industrial breakthrough to this decade.

During the 1870's, an extensive investment occurred in almost all sectors of the industry, as well as in railway construction and house building.

The agricultural exports increased strongly at the end of the 1870's. The production value of the capital goods industry more than doubled during the boom period up to 1875; industrial investments increased more than 80 per cent and the National income by more than 30 per cent, measured in stable money value, during the first five years of the 1870's: an increase almost twice as great as during any following quinquennium up to 1914.

TABLE 9

Percentage annual changes in the value of exports, imports, and industrial production, 1869–1879[25]

	Export	Import	Industrial production
1869–71	+20	+10	+ 5
1871–74	+14	+27	+17
1874–75	— 9	—13	— 5
1875–77	+ 2	+ 7	0
1877–79	— 8	—14	—10

The economic expansion was encouraged by a powerful foreign demand for Swedish export goods, and the expansion of the export industry was stronger than that of the home-market industry. The inflationary rise in prices at the beginning of the 1870's created large profits for the enterprises, which to a large extent used them for further investments. The money market was, moreover, marked by low interests. This situation was typical of the Swedish capital market during the last decades of the 19th century and is explained by the fact that the expansion was "imported", i.e. the large profits from the export industry were transferred to the capital market.

The extension of the railways was unquestionably important for the industrial development. Transport costs fell considerably, and the development of communications thus changed the costs for the enterprises at the same time as it facilitated the marketing of the products. Lower transport costs resulted in cheaper raw materials, which in turn reduced the prices of the products; this in its turn could contribute to a reduction in the costs of railway construction. Cheaper fuel for the iron works, for example, led to cheaper steel prices, which reduced the costs of rails and other railway material.

The railways in Sweden did not always pass through already densely populated areas. The main line from Scania to Stockholm went through relatively under-developed and thinly-populated districts. Large investments were therefore put into the extension

[25] Jörberg, L.: op. cit., p. 224.

of the new communities that developed around the railway stations, often outside old community centres. These investments were a contributing cause of the acceleration in the economic development, which can be ascribed to the 1870's. They had a strongly income-increasing effect for other sectors of industrial life, and the railways contributed to a considerable re-location of the population from relatively low productive areas to more productive ones. The transfer gains resulting from this can have been just as considerable as the direct effects of the railway investments.

The financing of the railways occurred mostly through State-arranged foreign loans. Thus, the supply of capital for other investments was in nowise reduced. During the period 1855–1860, the State grants towards railway construction amounted to almost 34-million kronor,[26] but it was during the 1870's that the railway construction was given a really considerable scope. In that decade, the railroads in Sweden were extended by an average of 14 per cent annually, i.e. it more than trebled during that period and reached a total of more than 5000 km, whereas the rate of extension after that up to 1914 was about 3 per cent annually.

Thus in the 1870's, decisive changes in communications took place. At the same time, however, structural changes occurred in the investment activity in industry. Production became more capital demanding, and because the investments were especially large in the capital-goods industry, it had repercussions in other sectors of industry. The iron and steel industry made large investments at the beginning of the 1870's. The Bessemer process was successful, and other technical changes were carried out. These technical changes led to cheaper production and had therefore more profound effects on the production structure than price-reducing measures had in the consumer-goods industry. It is true that they demanded relatively large non-recurrent contributions of capital, and often investments that were larger than the immediate demand required; but this in its turn led to cheaper prod-

[26] Finance committee's report, p. 118.

ucts and thereby increased the possibilities of investments in other industrial sectors, which used products from, for instance, the iron and steel industry.

The increasing mechanization within industry is reflected to some degree in the import of coal and coke. This rose by 150 per cent between 1869 and 1878, but by only 66 per cent between 1880 and 1890. Although this import also reflects the rapid railway extension during the 1870's, it none the less gives an idea regarding the scope of investments and thereby the technical changes upon which the expansion was based.

The export industry increased its production almost in step with demand, but to judge from the price development, the production in the iron and steel industry was not quite in step with the demand. The production stagnated, whereas the prices during the years 1871–73 continued to rise in the international market as well as in the Swedish. That this development chiefly became noticeable in the iron industry was due to the fact that the investments there required longer time to be carried out and needed more capital than, for example, the sawmill industry, which, broadly speaking, managed to supply the increased international demand.

The growth in the export sector changed the distribution of the production factors and expanded productivity. The entrepreneurs, to a large extent, used their augmented income for productive purposes.

The depression at the end of the 1870's was the first which obviously influenced the Swedish economy, and it was also the most severe up to 1914. Through its close connection with the international market, we could expect that the business cycles in Sweden would be in agreement with the international ones—as by and large they were—but also that they would show a similar amplitude—which was not the case. The Swedish economic development proceeded unusually quietly with slight fluctuations. Exports naturally fluctuated rather more than did production and had shorter waves than those which marked the home-market industry—shorter than 7–11 years.

It is difficult to establish the extent that a strong increase in

TABLE 10

The annual percentage growth in the production value of the industrial branches, 1867/69–1912[27]

	1867/69–1892/95	1896–1912
1. Paper, pulp, and graphic	6.1	11.0
2. Gravel, stone, clay and allied industries	6.0	5.3
3. Food (inc. tobacco)	5.1	4.0
4. Chemicals	4.5	7.3
5. Metal and engineering	3.7	7.0
6. Timber	3.7	2.0
7. Iron and steel	3.2	2.2
8. Textiles	2.6	5.6
9. Leather, hair, and rubber	1.7	10.0
10. Capital goods	5.2	6.2
11. Consumer goods	3.3	6.1

demand for agricultural products, allied to the decline of industry at the end of the 1870's, supported the initiated growth. However, the industrial expansion that began in the 1870's continued, with slight variations in the percentage growth, up to 1914. This obviously concerns the industrial development as a whole. The various industrial branches showed a more complicated development.

Many countries have passed through an expansion process similar to that of Sweden in the 1870's. Not all, however, were able to take advantage of the expansion to continue the building up of the economy without, after a longer or shorter time, the expansion coming to a standstill.

Sweden's development differs very little in this respect from that common to Western Europe. However, the Swedish expansion was of later date and had a more rapid start than that of, for example, France or Germany. The question is whether any causes can be found for this development in Sweden which were lacking or less obvious in other Western European countries.

In modern discussion concerning the problems of the under-

[27] Jörberg, L.: op. cit., pp. 61, 63.

developed countries it has been pointed out that these countries have an advantage in being late in their economic expansion. They can borrow ideas and technique from the more economically advanced and implant them.[28] Whether this is actually an advantage is open to discussion. It would be just as reasonable to discuss the opposite: the disadvantages of a late economic expansion.

Professor Gerschenkron has pointed out that the more retarded a country is, in economic respects, the more rapid ought the industrialization to be.[29] The development is intermittent. These two points can be discussed together. Sweden was late in starting her industrialization. She experienced a rapid development during the 1870's.

In relation to Western Europe, Sweden was unquestionably a country that was economically less advanced. It would thus appear that she had certain advantages for her further expansion. The capital export from England and France to Sweden can be pointed to as an example. The economic infrastructure and the costly investments associated with it were built up with the aid of foreign capital. It led to the industry having larger possibilities for internal financing by the alleviation of the capital scarcity.

We can also point to the fact that the wages of the Swedish workers were lower than, for instance, the German and considerably lower than the English, which from a competitive standpoint was to the advantage of the enterprises even if it was not so for the workers.

The most discussed advantage of the late economic start is that of being able to utilize the technical knowledge of the advanced countries. It is obvious that Sweden had some advantage from this circumstance. She imported quite a number of technicians, chiefly from England, who were employed in, for instance, the textile industry, engineering, and railway constructions. When the Swedish rubber industry was projected in the 1890's, Russian experts were introduced. But the technical gap between the European countries was much smaller in the 19th century than it is today between the under-developed countries and the advanced

[28] Cf. e.g. Fellner, W.: Trends and cycles in economic activity, New York 1956. [29] Gerschenkron, A.: op. cit.

countries. The advantage Sweden might have had from the opportunity of importing technical knowledge could not have been of decisive importance for her industrialization.

Nor can we argue that the rapid industrial development of the 1870's could imply that Sweden was very much more retarded than any other country in Western Europe. National income calculations for this time are consistently unreliable and say very little about the general living standard. But if Kuznets' calculations can be believed, the difference between Sweden and Western Europe was slight in this respect, even though Sweden had lower National income per capita than most of the Western European countries.[30]

Sweden's rapid development in the 1870's thus could not be ascribed mainly to her backwardness, but to other factors. The simplest available explanation is, again, her ability to satisfy the demand from foreign sources for exports from a limited sector of her economy. But the ability depended on the fact that she had, during the preceding decades, changed and prepared her economic development in, for example, the sphere of agriculture; that teaching and education had been assigned a relatively high status even from an international standpoint; that communications had been developed, although railway construction had its first larger upswing during the 1870's; that her iron industry had been long occupied in channeling production into larger and more rational units; and that the legislation concerning trade had been modernized and the old decrees that regulated industrial life abolished.

Only in the obvious fact that England, as the leading industrial nation, was of considerable help to all other European countries in their economic expansion can we find that Sweden profited by coming later into the industrial development.

The necessary adaptation of Sweden's economic and technical development was such as to render the aid from more advanced countries of secondary importance in the initial phase. These adaptation difficulties were great in the 19th century. They seem, despite everything, to be even more considerable today, and mod-

[30] Kuznets, S.: Quantitative aspects of the economic growth of nations, I–II (Economic development and cultural change, vol. V, 1956/57).

ern opinion concerning the ability of the under-developed countries to start an economic expansion is that this problem is more relevant and unquestionably more difficult to solve.

The development 1880–1914

Sweden, after a rapid start in the 1870's, succeeded in continuing the economic expansion at a high rate. What factors contributed to the development? And is it possible to find that some differed from those that led to the general expansion which occurred in other European countries during the same period?

Between the middle of the 1880's and the start of the First World War, the volume of Swedish exports increased one and a half times. A certain transformation of the structure of exports occurred also. The share of the old staple goods was reduced. The grain exports ceased and were replaced by exports of meat and dairy produce, chiefly butter. Some goods in the exports was replaced by paper pulp and paper, thus a processing of the products of the timber industry. A new raw material came on the market, namely iron ore, which together with the products of the engineering industry was responsible for the most rapid expansion. Iron ore represented the old type of Swedish exports, raw materials, whereas the engineering industry represented a new type in

TABLE II

The distribution of Swedish exports on groups of commodities, 1881–1913, in percentage[31]

	1881/85	1891/95	1901/05	1911/13
Timber	40.4	37.1	38.5	26.1
Iron and steel	16.2	9.5	10.2	9.3
Grain	11.7	4.7	0.4	0.3
Butter	6.3	12.0	8.9	6.0
Paper and pulp	4.6	8.3	12.9	17.6
Iron ore	—	0.4	5.0	8.0
Engineering products	2.6	3.1	6.7	10.5
Others	18.7	22.8	17.6	21.6

[31] Fridlizius, G.: Sweden's exports 1850–1960 (in Economy and History, vol. VI, Lund 1963).

the development, which proved that the Swedish economic development had reached a certain degree of independence and the ability for self renewal.

One factor that to a high degree influenced the Swedish economy was the improvement in her terms of trade for almost the entire period after 1860. During the 1870's, the terms of trade were extremely advantageous for Sweden. Her export prices rose more strongly than the import prices up to about 1875; they then deteriorated up to the depression of 1879. Thereafter a new improvement came about, and for a short period from the end of the 1880's, the improvement was very strong—about 20 per cent. In this phase the cause was not the same as operated during the 1870's, i.e. that the Swedish export prices rose more strongly than the import prices, but that the import prices fell considerably, whereas the prices of export commodities could be maintained.

Arthur Lewis has pointed out that terms of trade swung to the disadvantage of the raw-material producing countries after 1883.[32] He says also that the reason for this was the increased production in these countries, the immigration, the re-allocation of land, and the capital investments, as well as the reduction in the costs of transport. He is probably correct in this argument. However, we should remember that the export prices were reckoned f.o.b. and the import prices c.i.f. If the transport costs fell strongly, as happened at the end of the 19th century, it did not necessarily imply that the prices which the raw-material producing countries received were lower than earlier. Terms of trade can have been improved in the industrialized countries, thanks to lowered transport costs, which reduced the c.i.f. price. Terms of trade can also have been improved for the raw-material producing countries for the same reason. To explain the absence of economic expansion in the raw-material producing countries by referring to the fact that the raw-material prices fell in the industrialized countries is thus only a part, and perhaps not the most important part, in the causal connections.

The Swedish exports consisted of about 50 per cent raw-ma-

[32] Lewis, W. A.: Economic survey 1919–1938, London 1963.

terials, and this increased from the 1890's when iron ore began to be shipped in larger amounts. Terms of trade improvement in the Swedish foreign trade concerned both raw-materials and manufactured goods. The raw-materials that Sweden exported were predominantly of the type that had a big demand on the European market; they were in many cases essential material for the European industrialization. This can explain the development of the terms of trade that took place.

The improved prices for Swedish export goods naturally also led to the fact that the imports could take place to a greater extent than if the improved terms of trade had not come into being. A strongly rising import cost could be paid as well as instalment and interest payment on the large loans that were taken for railway construction and other purposes.

Rising prices contributed to the economic growth both by facilitating the capital accumulation and by solving stagnating social arrangements. Falling prices had the opposite effect. In Sweden, prices were rising throughout the greater part of the 19th century. When prices fell, as happened after the depressions of the 1870's, 1880's, and up to the middle of the 1890's, the decline was less accentuated than in many other European countries. This development might have facilitated the Swedish economic expansion during the period up to 1914. Concerning industry, the profit development was not satisfactory to the same degree during the 1880's as earlier or later, but measured according to volume, the Swedish industry hardly showed any recession during this time. Whereas the investments of the 1870's, and even more so of the 1880's, had an extensive nature, i.e. enlarged the spheres where capital was employed, the development later turned towards a more intensive use of capital per worker. The degree of mechanization is the measure of the capital intensity available. It shows little variation during the period, and it can be argued that the somewhat worse economic climate of the 1880's was a driving force in the increased degree of mechanization. The number of workers increased considerably more slowly than the amount of horsepower used in all sectors of industry, but the increase in horsepower/worker was slower during the 1880's than during

TABLE 13

Percentage changes in the population according to the branch of industry, 1870–1910[33]

	1870–1880	1880–1890	1890–1900	1900–1910
Total	10	5	7	8
Agriculture	3	—4	—5	—5
Industry	30	31	37	24

the previous decade. It is hardly original to state that industrialization implies a substitute of labour by machinery. As a growth factor, mechanization was important, and it was largely facilitated by the fact that the prices for Swedish export products were maintained, and the improved productivity enhanced the competitive power of Swedish industry. Both the mechanization and price development contributed to the rapid economic expansion.

The economic expansion in Sweden occurred at the same time as all the countries in Western Europe strongly increased their industrial potential. It was thus, in part, a reflection of international development, and for a country strongly dependent on its foreign trade, the abolition or reduction of tariffs and other trade barriers was important. (The Swedish timber exports first attained really large proportions when tariff reduction and later free trade was introduced in England and France at the middle of the 19th century.) During the 1880's, a protectionistic tariff policy was initiated in most European countries—including Sweden—as a consequence of, among other things, the poor competitive situation of European agriculture in relation to the USA and other transoceanic countries. The increased tariff protection ought therefore to have been an obstructing factor in the continued expansion during the 1880's and the 1890's. But just as it is difficult to establish the importance of the Swedish tariff protection for Swedish industry, so it is to measure the importance of the increased tariff protection in other countries for the development of Swedish exports. Although it can be established that a certain

[33] Sweden's historical statistics, part I, Stockholm 1955.

TABLE 14

Annual percentage changes in the value of exports, imports, and industrial production, 1893–1900[34]

	Export	Import	Industrial production
1893–95	2.5	3.4	4.0
1895–97	7.4	8.5	13.0
1897–98	—4.8	11.8	9.9
1898–00	6.4	8.5	8.7

redistribution of the exports took place, this need not have been a consequence of increased tariffs, but can be associated with the increasingly reduced costs of transport, which made Swedish commodities competitive also in extra-European markets.

During the third acceleration period for Swedish industry i.e. during the 1890's, the home market had a greater importance than earlier.

The home-market industry expanded at the same rate as the export industry and was responsible for about half of the production value before the First World War—thus it had the same share as it had around 1870. The competitive capacity of Swedish industry was sufficient to withstand the attempts of foreign countries to conquer the Swedish market, and this was shown by, among other things, the ability of the industry before 1914 to begin competing on the foreign markets with typical home-market products such as textiles and footwear. To this is added that an increasing proportion of the production of the capital-goods industry remained within the country.

Erik Dahmén has said in his book "Svensk industriell företagarverksamhet" (Swedish industrial entrepreneurial activity) that limited sales possibilities on the home market were a hindrance to Swedish industrial development prior to the First World War. He points out that the spread of innovations within the industry might have been prevented because of a too-limited market. Certain innovations demanded a bigger market than the Swedish in

[34] Jörberg, L.: op. cit., p. 186.

order to command a reasonable price. This, in his view, was the situation in several branches of the engineering industry during the initial industrial phase.

During the 1890's, however, the engineering industry entered to some extent a new phase. The expansion was based in some cases on Swedish inventions (separators, turbines, internal combustion engines, gas accumulators, and ball bearings). The engineering industry had an irregular structure: a powerfully expanding sector based on Swedish inventions and directed to international competition, and another, slowly growing sector directed to regional markets and only slightly specialized. The latter often consisted of small enterprises with low profitability, partly because of the lack of specialization and partly because of strong mutual competition. Whereas the export of engineering products increased by more than 50 per cent each 5-year period after 1889, the imports increased scarcely half that amount. This means that the expansion of the Swedish engineering industry was strong enough to satisfy to an increasing extent an expanding demand by the home market and to effect an increased export.

A further example of the capacity of the industry to supply the home market is found in the textile industry, especially the woollen branch, which increased its production value from about 20-million kronor around 1890 to 70-million around 1910. Whereas 65 per cent of the Swedish consumption was met by national production at the end of the 1880's, the national production immediately before the First World War was capable of supplying 80 per cent of the home-market demand.

It was mainly during typical boom years that the spread of innovations was greatest; this also led to a considerable establishment of enterprises. There was some difference in the investment activity during the various business cycles. During the 1880's, which was thought to be a "poor" decade because of the then falling prices (although, despite this, the volume increase was considerable—about 5 per cent per annum), the investments were clearly extensive, i.e. the capital/worker rose slowly, the increase being due to the rise in the number of workers, whereas in the boom period during the 1890's the investment activity was intensive, i.e. the capital/worker rose more quickly than the number of workers.

A development similar to that of the 1880's occurred after the accelerated expansion of the 1890's, which was followed by a slowing down in investments within the larger enterprises. Extensive investments were made, which broadened the base of the industrial development by establishing new enterprises.

It is obvious that the variations of the exports were of the utmost importance for explaining the fluctuations of Swedish industry. The income that exports brought in varying extent to industry reacted on the formation of income in the country by the multipiler effect of the export increase. Except for the depression in the 1870's, there were evidently only few primary causes for the turn of the market within the country. The business cycle in the 1870's differed to some extent from the following cycles. The crisis at the end of the 1870's was naturally affected by the international depression. It seems, however, also to have been affected by the investment boom (broken by the economic decline) proving in many instances to have involved a large measure of faulty investments because of the inability of the whole economy to change within a short time from an agrarian to a partly industrial community. The hasty construction of new industries resulted in big balance problems. "The development blocs" were not ready; this resulted in difficulties for the industry to settle those "tensions" which thereby arose. The other cyclical variations could to a large extent be explained by the dominant influence on the exports by the international business cycles and the further influence by the exports on Swedish industry.

The repercussions of the fluctuations were of comparatively slight importance as a hindrance to the industrial expansion. The economic declines, except for the depression in the 1870's, were mild. The explanation of this stability is naturally very complicated. We can only indicate certain points which evidently were moderating and levelling out factors on the fluctuations of industry.

As basis for the whole reasoning concerning the cyclical variations, the expansive and structural changes must be emphasized.

Industrial expansion during the second half of the 19th century and up to the First World War resulted in a change in Sweden's economic structure. The share of industry in the national product increased, a change occurred in the structure of industrial production, and production technique was improved. This is a normal industrial process and does not explain the variations.

The international development facilitated the Swedish industrialization by the fact that the demand for Swedish products, such as iron, timber, and paper, led to increased Swedish exports, and this resulted in increased possibilities for expansion in other parts of industry.

The expansion periods of the 1870's and the 1890's occurred during periods with strong rises in prices, which contributed to producing large profits for the industry. This in turn led to increased investments, which in their turn contributed to an acceleration in the rate of development.

The industrial expansion was also distinguished by increased differentiation. New commodities, new methods and new markets were created, and because its newly added industries were often the driving force in the expansion, they helped to maintain the expansion on a higher level also in those periods when the export industries were affected by the international depressions.

We can point to the breakthrough of the beet sugar industry at the end of the 1880's; to the expansion of the ready-to-wear clothing industry, of the rubber industry, of the carpentry industry, of the paper pulp industry, and of the superphosphate industry in the 1890's; or to the origin of new products in the engineering industry, such as dairy machinery and internal combustion engines; to new ingot steel processes in the iron industry, which led in turn to a rapid expansion of iron mining, and so on.

When the market expanded, industrial production could increase, which naturally resulted in lower costs and also to a certain substituting of imported goods by home production. To this substituting, the protectionistic tariff policy of 1888 and 1892 can also have contributed by creating the basis for a home production for e.g. the footwear industry and the jute industry, although it is difficult to determine the influence of the tariffs on the development of production.

One of the obstacles for the Swedish industrialization was the limited home market at the beginning of the period. This led to difficulties at an overall industrial expansion, and many of the enterprises created in the general optimism that existed during the boom of the 1870's must have discovered that the demand for the products was insufficient to enable them to operate profitably.

Furthermore, poor communications were also an obstacle to an integrated national market, and they limited the production development. Not until these obstacles had been removed, could the industry enter into its self-generating phase, i.e. it reached in the 1890's the stage where its income-producing effect was so large that the fluctuations in other sectors of industrial life, as well as of exports, were of less importance because the demand by Swedish industry on Swedish industrial commodities played an increasingly important role. This demand naturally depended to some extent on the development of the export industries, but their importance for the income formation within the country, aided by the multiplier effect, was diminished because of the stability of the home-market industry. The industrialization resulted in an expanded production at the cost of imports. The expansion of the home-market industry, especially during the 1890's, was stronger than the increase in demand. Because the consumer-goods industry was less sensitive to business cycles than the exports industry, this increase in the degree of economic independence contributed to levelling out the variations in the fluctuations and facilitated the industrial expansion.

The stability which was established for the expansion of the industry can also to some extent be explained by the investments that were put into railway construction and into housing during periods of economic decline. To this must be added that the agricultural exports were to a large extent of a counter cyclical nature, and these exports amounted to about one-fifth of the total exports up to the turn of the century.

Furthermore, it must be remembered that, of the population of Sweden, almost three-quarters around 1870 and almost half around 1910 were supported by agriculture. This sector of the economy can have had a stabilizing effect on industrial production, because the demand from this group of the population was largely unaffected by the fluctuations of industry. Through the introduction of protective tariffs in the 1880's, an income reduction, as far as agriculture is concerned, can also have been prevented.

Industrial growth, however, was concentrated to a few well-identifiable boom periods, during which there occurred an accelerated expansion in production and also in real wages for industrial workers.

TABLE 15

Percentage changes in the value of exports, imports, industrial production, and number of workers in industry, 1871–1909[35]

	1871–1874	1876–1878	1881–1883	1885–1886	1887–1889
Total export	+43	—19	+14	— 8	+20
Total import	+82	—18	+16	—12	+29
Export of industrial production	+51	—18	+12	—10	+27
Import of industrial production	+87	—22	+16	—11	+29
Industrial production	+37	—16	+14	— 1	+13
Number of industrial workers	+15	—10	+ 7	+ 2	+23

	1890–1893	1898–1900	1900–1901	1904–1907	1907–1909
Total export	+10	+13	—10	+26	—10
Total import	—12	+18	—13	+17	— 9
Export of industrial porduction	+ 8	+18	—11	+24	— 8
Import of industrial production	—17	+14	—12	+24	—14
Industrial production	+10	+18	— 2	+27	— 7
Number of industrial workers	— 1	+ 8	— 1	+ 8	— 5

[35] Jörberg, L.: op. cit., p. 338.

In the 40-year period up to 1910, the largest production expansion occurred during the boom periods 1871–1874, 1898–1900, and 1904–1907; and the rise in real wages occurred during three delimited periods: 1869–1875, 1881–1887, and 1892–1896.[36] The latter thus did not altogether follow the schedule for the expansion in production owing to the fact that the real wages did not always rise in step with the nominal wages, which can also help to explain the relatively mild depressions during the 1880's and the 1890's.

These expansion periods also coincided with the international economic development. But the fact that factors outside Sweden facilitated Swedish expansion does not explain how the Swedish economy could change and maintain the possibilities created by this. In the foregoing, the role of agriculture is emphasized as well as the importance of the entrepreneurs; also the fact that the State was responsible for the basic investments—investments made with capital borrowed largely in foreign countries—which facilitated the formation of National capital and relieved scarcity of it. To this is added the technical development leading to the establishment of a number of industries based upon Swedish inventions.

The population development was also a factor which facilitated the change from agrarian community to industrial community. Emigration was probably not solely a negative element in this development but can have contributed to simplifying the change. The conclusion drawn from the above discussion must be that the Swedish community had a large capacity for change and showed a decided ability for creating permanence in the development.

[36] Bagge, G.–Lundberg, E.–Svennilsson, I.: Wages in Sweden 1860–1930, I–II, Stockholm 1935.

Chapter 10

Italy

Giovanni Federico and Gianni Toniolo

'Clio is not a tidy housewife'

(A. Gerschenkron)

ALEXANDER GERSCHENKRON ON THE ECONOMIC DEVELOPMENT OF ITALY

Italy takes an important place, second only to that of Russia and possibly Austria, among the individual countries specifically investigated by Gerschenkron within the framework of his general approach to European industrialization during the nineteenth century. To the Italian case he devoted three important papers (Gerschenkron 1962: 52–89, 90–118, 1968: 98–127) and the effort of constructing an *ad hoc* index of industrial production (Gerschenkron 1962: 367–421).

To use his own words, Gerschenkron 'approached the Italian material with a series of historical questions or expectations in mind' (Gerschenkron 1962: 73) reflecting his own generalizations about the features of the industrialization process in a relatively backward European country. Typically, he expected to find a 'big push' of considerable intensity and length favouring the 'output of producers' goods as against that of consumers' goods' (Gerschenkron 1962: 73) and inducing various forms of industrial concentration. Moreover, he expected these 'basic features of delayed industrialization to be reinforced by the use of specific institutional instruments, such as the investment policies of banks and various policies of the state' (Gerschenkron 1962: 73).

In order to test the consistency of the Italian case with his general view of European industrialization during the nineteenth century, Gerschenkron set out to build his own index of industrial growth covering the period 1881 to 1914. This was, at the time (1955), a pioneering work and a very welcome addition to the paucity of quantification in Italian economic historiography. The results allowed Gerschenkron 'to locate the period of great push between the years 1896 and 1908' and to see that it was, as he expected, characterized by a rapid increase in the share of producers' goods in total output (Gerschenkron 1962: 77). At the same time, comparing Italy's performance with that of other countries during their own 'spurts', Gerschenkron found that 'the industrial growth in Italy, while free from any severe setbacks, seems to have proceeded in a less uniform and more jerky fashion, denoting perhaps a more delicate state of public confidence and greater entrepreneurial uncertainties and hesitations' (Gerschenkron 1962: 79). The reasons listed for Italy's relatively poor industrial performance are: (a) inept government policies, (b) a supposed time inconsistency between railroad building and the 'spurt' (c) a political situation around the turn of the century 'not propitious to quiet economic growth' (Gerschenkron 1962: 85), and (d) 'the absence of any strong ideological stimulus to industrialization' (Gerschenkron 1962: 86). On the other hand, 'if one were to look for a single important factor that succeeded in offsetting at least some of the great obstacles to the country's industrialization, one could not fail to point to the role performed by the big Italian banks after 1895' (Gerschenkron 1962: 87). Gerschenkron is therefore able to conclude that 'in this respect at least, the Italian case fits well into the general pattern of European industrialization in varying conditions of economic backwardness' (Gerschenkron 1962: 89).

This chapter discusses Gerschenkron's appraisal of Italian industrialization before the First World War in the light of the later literature. The first section deals with measures of Italy's backwardness relative to other European countries, and is followed by a section reviewing the existing indices of industrial production. The role of such 'agents of industrialization' as the German banks, the state and ideology are then discussed. The last part of the chapter is devoted to a summary of the main points and to some general conclusions.

THE BACKWARDNESS OF THE ITALIAN ECONOMY

When asked about the operational value of the concept of backwardness Gerschenkron replied that no precise answer was possible because one could define the degree of backwardness in several ways and, while granting that 'this might seem discouraging and clumsy', he added that 'it was true of 19th-century Europe at least that, whatever measures used, individual countries were fairly easy to arrange in order of backwardness' (Rostow 1963: 385).

Gerschenkron's well-known concern with index number problems made him reluctant to suggest per capita output as a fully operational measure of backwardness. He also suggested that, in any event, information deriving from output levels should be used together with other parameters – such as literacy or the prevailing ideology – that 'would involve asking to what degree a country at a certain moment had developed the preconditions for subsequent economic development' (Gerschenkron 1962: 43).

How backward was Italy, in Gerschenkron's eyes, around the middle of the nineteenth century? The most important suggestion comes from the following passage:

> Few would disagree that Germany was more backward than France; that

Austria was more backward than Germany, that Italy was more backward than Austria and that Russia was more backward than any of the countries just mentioned. Similarly, few would deny England the position of the most advanced country of the time.

(Gerschenkron 1962: 44)

Does research carried out in the last quarter of a century confirm at least this ranking? Measures of economic backwardness explicitly related to Gerschenkron's hypotheses were first provided by Barsby (1969). Relying on Colin Clark's pioneering work, he sees Italy's GNP per capita around 1890 to be equal to that of Russia and less than one-fourth that of Britain.

Subsequent work has considerably revised Colin Clark's estimates. We may refer directly to Crafts (1983) since this article takes into account the most relevant previous contributions (particularly Bairoch 1976), producing the least unreliable estimates of nineteenth-century per capita product to date. They are reproduced in Table 10.1 for the countries mentioned by Gerschenkron and for Spain. As regards Italy in the third quarter of the nineteenth century, Gerschenkron's appraisal of Italy's relative position seems to be confirmed for 1910 but not for 1870, when Austria displays a per capita income approximately equal to that of Italy. What is striking, and somewhat puzzling, in Table 10.1 is the relatively high level of Italy's income at about the time of its political unification, as well as the stagnation in the subsequent thirty-odd years which resulted in its slide down the rankings.

These data rest on two assumptions: (a) that we accept Kravis's (1982) purchasing power parities and relative income levels for 1970, and (b) that we

Table 10.1 Crafts' estimates of per capita national product (US dollars of 1970)

	1870	*1890*	*1910*
Great Britain	904	1130	1302
Germany	579	729	958
France	567	668	883
Northwestern Italy			(810)
Italy	466 (513)	466 (513)	548 (603)
Austria	466	664	802
Spain	391	554	547
Southern Italy			(450)
Russia	252	276	398

Source: Crafts (1983: 389, 394).
Italian data in parenthesis derive from the recent official upward revision by 10% of current (1980s) GNP estimates and their backward extrapolation, according to Crafts' methodology.
Italian regional estimates from Zamagni (1978).
Northwestern Italy includes Lombardy, Piedmont and Liguria.
Southern Italy includes the former Kingdom of Naples plus Sardinia.

believe long-term growth rates implicit in the existing historical national accounts statistics.

The second of these assumptions, i.e. the reliability of historical national accounts statistics, may prove difficult to accept in the case of Italy. The existing estimates – ISTAT (1957) and Fua's (1969) revision thereof – date back at least twenty years. Their validity has always been questioned at least on the basis of the authors' parsimony in the disclosure of their sources and methods. Recent work by Fenoaltea on industrial production shows that this distrust is not misplaced. The existing figures underestimate both the level of industrial production in 1911 and the attendant long-term growth rates.[1] Should these upward revisions be confirmed in the case of other industries, our overall view of the time-pattern of Italy's development would be in need of thorough reconsideration. The relevance of this point for the issues addressed by this chapter, and indeed for the entire economic history of modern Italy, can hardly be overstated. Nevertheless, the ranking of Italy with the countries in Table 10.1 is not likely to be altered substantially.

In view of Italy's pronounced economic dualism, any appraisal of her backwardness is likely to be incomplete without some kind of regional breakdown. According to Zamagni (1978), in 1911 the GNP per head of Italy's richer region (the northwest) was slightly lower than that of the whole of France (and therefore much lower than that of the richest comparable French areas). In the same year, per capita income in the south turns out to be only 15 per cent above that of Russia. The gap between north and south is likely to have widened as a result of industrial growth which took place mainly in the former region but it surely existed at the time of unification and much before then. Adopting a strict Gerschenkronian logic, therefore, one could expect a smooth industrialization along French lines in the northwest – where the institutional and cultural winds from across the Alps blew strong – and a state-driven spurt in the south. It is indeed possible to find some traces of these potentially divergent patterns in the first half of the nineteenth century (a market-driven growth of the textile, particularly silk, industry in the northwest, neo-mercantilistic attempts to create industry from above by the Bourbons in the south).

Backwardness, according to Gerschenkron, could not be measured by GNP taken in isolation. Most experts on economic development would agree with this proposition. Indeed, the current literature on economic development offers a number of standard-of-living indicators other than GNP. Among those, we have selected the so-called physical quality of life index (PQLI) devised by Morris (1979). The PQLI recommends itself on two grounds: (a) it is based on data that, when available, are rather accurate for nineteenth-century Europe, (b) it embodies an indicator (literacy) that Gerschenkron considered particularly relevant and two others (infant mortality and life expectancy) that are taken into account by the most recent historical research on standards of living.

The PQLI is an unweighted average of: the literacy rate of the population aged 15 years and older, infant mortality in the first year of life and life expectancy at age one. 'For each indicator the performance of individual countries is placed on the scale of 0 to 100, where 0 represents an explicitly defined "worst" performance and 100 represent an explicit "best" performance' (Morris 1979: 41). The three indicators are equally weighted. Such weighting is no more arbitrary than any other, and tests with different weights show that the *ranking* of individual countries in a large 1970 sample does not change significantly with changing weights.

In adapting the PQLI to nineteenth-century Europe, we made as few changes as possible to the definition of variables and to the parameters used by Morris. Two adjustments, however, were unavoidable. The first concerns the definition of literacy which, in our cases, was measured as the percentage of the population aged 6 or over to which most of our sources referred. The second adjustment relates to the scale (or 0–100 range) employed to index infant mortality and life expectancy. In both cases we had to widen the range in order to avoid negative values.[2] Data for infant mortality were taken from Mitchell (1975); five-year moving averages were preferred, given the large swings in yearly data. Literacy comes from Flora (1983) and from various contemporary Italian sources. When data were not available, literacy was estimated from the percentage of people who signed wedding documents, through regression analysis. In most cases, life expectancy at year one had to be estimated from mortality tables and other demographic material from each individual country. Because of such data limitations, Table 10.2 covers a much more limited range of countries than Table 10.1.

Table 10.2 confirms that, in 1870, Italy was an unmistakably backward country. Backwardness was still very visible in 1910 when Italy lagged more than a generation behind the most advanced European countries in terms of standards of living. At the same time, the data in Table 10.2 allow us to appreciate the considerable progress made by Italy between 1870 and 1900, during a period in which existing historical statistics see practically no increase in per capita incomes. Comparisons with countries less developed than those in Table 10.2 show that

Table 10.2 PQLI: Selected European countries and three main Italian regions, 1870–1910

Year	*Country*				*Italian regions*		
	England & Wales	*France*	*Belgium*	*Italy*	*North-west*	*North-east*	*South & islands*
1870	62.9	62.2	na	26.9	40.5	27.1	17.6
1880	70.8	64.4	na	36.9	47.0	41.2	28.4
1890	71.3	67.7	65.5	na	na	na	na
1900	78.9	73.7	75.9	56.7	66.8	62.1	46.0
1910	na	84.0	82.4	62.0	70.5	68.9	49.4

Methods and sources: see text.

infant mortality in Germany and Austria was higher than in Italy throughout the period. Relative to the same countries, however, Italy had a lower percentage of enrolment in primary school for the population aged 5–14 years.

The regional breakdown in Table 10.2 is consistent with income estimates referred to above. By 1910, the northwest was lagging behind the most advanced countries in Europe by about fifteen years, and Italy as a whole lagged by forty. The south was thirty years behind the northwest, despite the rapid progress made after the unification, particularly in life expectancy. These data provide one of the few aggregate indicators of the time-pattern in the welfare *gap* between large Italian regions. Its magnitude and its persistence over time make it difficult to discuss Italy's backwardness meaningfully in aggregate terms. At the beginning of the century, the area of the so-called Milan–Turin–Genoa 'triangle' (the northwestern region in Table 10.2) was larger than Belgium and had a per capita income and, to a lesser extent, a basic standard of living not much lower than the French averages. The south was more similar to Russia than to the northwest in terms of both variables.

INDICES OF INDUSTRIAL PRODUCTION

Gerschenkron's interpretation of Italy's industrialization rests pretty much on his own index of industrial production covering the years 1881–1913. Over the whole of this period the index yields an average annual rate of growth for Italian industrial production of 3.8 per cent (using value-added weights) resulting from a rather good performance during 1881–7 (4.6 per cent per annum), which was followed by a serious crisis that caused virtual stagnation until 1895. Italy had its 'big spurt' in 1896–1908, characterized by a rate of growth of 6.7 per cent. The index shows that during the 'spurt' the production of investment goods grew more rapidly than that of consumers' goods, a pattern consistent with Gerschenkron's expectations derived from the historical experience of other backward European countries.

The construction of Gerschenkron's index in 1955 was certainly a major breakthrough in Italian economic history. Nonetheless its technical pitfalls are quite serious. This is not the place to review them in full: it is enough to point briefly to some of the index's weakest features. Gerschenkron uses twenty elementary series but five of them account for 85 per cent of the index behaviour. Some of them are rather poor proxies for the underlying industries.[3] The index is, therefore, very sensitive to changes in the basic series and in the sample of industries they are supposed to represent.[4] Besides, Gerschenkron derives his weights from a highly questionable source, a survey of 1903 which is a partial and unreliable revision of enquiries undertaken in the 1880s and 1890s (Missaggia 1988). The index is, therefore, based on start-period weights. A reconstruction of Gerschenkron's index using 1911 weights yields substantially higher growth rates for the years 1881–8 (5.6 per cent against 4.6) and 1896–1908 (8.0 per cent against 6.7). This result, incidentally, contradicts the so-called 'Gerschenkron effect' in index numbers. Finally, the index itself is a poor proxy for the omitted industries, which accounted for about one-third of total value-added in 1911. Most of them were slow-growing traditional ones (such as leather works and furniture), their exclusion is, therefore, likely to have raised the rate of growth.

Since 1955 three other indices of industrial production have been estimated: the first by the Italian Statistical Institute (ISTAT),[5] the second – still unpublished – by Carreras (1982) and the third by Fenoaltea (1983).[6]

The present wealth of indices makes interpretation of the overall pattern of industrialization more difficult rather than easier since they yield different results, while none of them is so clearly superior to the others from the point of view of data sources and of method as to recommend itself for acceptance.

According to ISTAT, industrial development was slow and fluctuations limited both in depth and length. Carreras sees an accelerating trend coupled with rather pronounced cyclical swings. Finally, the picture portrayed by Fenoaltea's index of 1983 is similar to Gerschenkron's: slow growth until 1878, rapid development in the 1880s, followed by a serious slump and by a long boom from the mid-1890s onward. Fenoaltea's cyclical swings are, however, much wider then those of Gerschenkron.

None of these patterns seem to fit well with Gerschenkron's interpretation of Italy's industrial development. The most compatible with it is possibly the picture designed by ISTAT which shows a discontinuity with an acceleration of growth rates at the end of the nineteenth-century. But, then, the rate of growth is so slow as to make one wonder whether one might talk of a 'big spurt'. In the index of Carreras, on the other hand, there are no points of discontinuity. Fenoaltea has both the highest rates of growth and the largest cyclical swings: a discontinuity could be found in 1879 rather than in 1896 (a timing much more akin to Romeo's model). This interpretation, however, would conflict with Fenoaltea's denial of the very existence of a big spurt and with his assertion that Italy's industrial growth was cyclical rather than marked by discontinuity (Fenoaltea 1973a, 1988c). His view seems closer to the one put forward by Cafagna (1965: 143) and shared by Bonelli (1978) and Federico and Chesi (1987).

The above-mentioned aggregate indices are actually the sum of three different

Table 10.3 Indices of industrial production, 1861–1913 (annual average rates of growth in real terms)

Period	*Gerschenkron*	*ISTAT*	*Fenoaltea*	*Carreras*
1861–1913	na	1.95	3.42	2.51
1881–1913	3.78	2.26	4.18	3.19
1896–1908	6.68	5.21	7.87	5.00

Sources: Gerschenkron (1962), ISTAT (1957), Fenoaltea (1983), Carreras (1982).

sectorial patterns. The production of investment goods (engineering, chemicals, iron- and steel-making) shows higher growth rates than that of consumers' goods during both the so-called 'Giolittian era' (1896–1913) and the 1880s. The 1888–92 slump was characterized by a much more rapid decline of the former than of the latter industries. In fact, the whole pattern depends to a certain extent on the huge cyclical swings of the engineering industry (Fenoaltea 1973a; Warglien 1985). Traditional industries such as textile and food-processing industries developed rather slowly with no major cyclical swings. Most 'new' products, such as electric power, rubber and some branches of the chemical industries, started late and showed a high growth afterwards.

THE ROLE OF BANKS IN PROMOTING INDUSTRIAL DEVELOPMENT

At the beginning of the 1890s, Italy experienced a serious banking crisis. It resulted in the failure of one of the six banks of issue (Banca Romana) and the collapse of the two deposit banks (Credito Mobiliare and Banca Generale) which had been established after unification on the model of the French Crédit Mobilier. The void was filled in 1894 and 1895 by two new banks (Banca Commerciale and Credito Italiano), established by German and Swiss capital with German management. Gerschenkron sees the crisis of 1893 almost as a *felix culpa* of the Italian economy – and policymakers, in as much as they were responsible for it – since it brought to an end the existing banking system tailored as it was by the French, and made possible the importation 'of the great economic innovation of German banking in its most developed and mature form' (Gerschenkron 1962: 88). The main feature of such innovation consisted in turning investment banks into real 'department stores' in the field of banking. Their virtue, in a situation of relative backwardness, was to provide – to use Gerschenkronian terminology – efficient substitutes for missing prerequisites of industrialization. In particular, German banks, unlike their French counterparts, are seen as supplying not only long-term credit but also a good deal of entrepreneurship, a factor of production supposedly even more scarce than capital.

Gerschenkron was by no means the first to attribute an important role to the large banks in the development of Italy.[7] Gerschenkron's original contribution to Italian economic history, was to see the role of German banking in a framework in which the industrial development of the country appeared as just one case in the pattern of European industrialization. In the 1950s this was a seminal idea that gave new stimuli to discussion and research. Opinions varied from Cohen's strict Gerschenkronian orthodoxy (Cohen 1967: 382, 1972: 89) to Webster's (1974: 325) and Hertner's (1984) more shaded judgements, and to the decisively negative opinion of Farina (1976, 1980). The latter blames the 'German' banks for having distorted resource allocation in favour of capital-intensive industries and techniques, thus increasing 'dualism' and contributing to the oligopolistic nature of Italian development.

The analysis of quantitative aggregate trends is not of much help in elucidating the role of the mixed banks. An assessment of the time-path of aggregate financial variables carried out along Goldsmith's lines puts Italy's financial intermediation ratio at 0.38 in 1881 and at 0.47 in 1914 (Biscaini and Ciocca 1979: 92–7). Financial deepening did occur but the gap between Italy and the most advanced European countries remained unchanged. During the same period, the share of financial liabilities held by banks grew from 0.19 to 0.29. This positive trend, however, did not exhaust its momentum in 1913. The fact that banks retained an overwhelming importance in financial intermediation at least up to the 1960s is a well-known peculiarity shared by the Italian economy with the Japanese. There is little in these aggregate trends that will confirm or invalidate Gerschenkron's hypothesis.

In fact, Gerschenkron's hypothesis is mainly based on the stimulus provided by the mixed banks to a limited number of 'strategic' firms or industries. Therefore, a proper test of the hypothesis may come from careful archival research on the actual behaviour of the banks. A real breakthrough in this field of research was made by the monumental work by Confalonieri (1974–6, 1982). His conclusions may be summarized as follows: (a) long-term industrial financing came mostly as a result of other banking operations, (b) it resulted from normal scrutiny of clients' demand and from a desire to monopolize the entire range of their banking operations rather than from an autonomous initiative of the banks, (c) the latter, therefore, did not possess much of an *industrial* strategy of their own which was, perhaps, more pronounced in their predecessors of the 'French type', (d) the main novelty of the 'new' banks consisted in underwriting as part of a *banking* strategy directed 'plus à des *affaires* qu'a des *entreprises*' (Confalonieri 1974–6: 469). The early long-term involvement of the banks in the electric power industry is singled out by Confalonieri as an exception in their operations. Moreover, the operations of the French-type banks do not seem to be so different from those of the later German-type banks. The former not only financed the construction of the railways of the 1870s but, what matters more in a Gerschenkronian framework, played a prominent role in the rapid industrial growth of the 1880s.

If reliable, these findings would contradict Gerschenkron's hypothesis. If German-type banks did not provide relevant supply-side stimulus to investment, if they were as a rule not particularly interested in managing industrial enterprises, if they did not hold in their portfolios for long periods of time large quantities of industrial equities, then it is not easy to grant them the role of sole or main 'agents of industrialization'.

A few years ago, Hertner (1984) attempted to redress the balance in favour of 'Gerschenkron's approach which remains a very fertile research hypotheses, even though it requires some revisions and qualifications' (Hertner 1984: 159). His qualifications regard primarily the role of German capital and men in the Banca Commerciale and Credito Italiano which he sees as 'the first example of successful foreign direct investment in Italy in the banking sector'. The early withdrawal of the original German capital was not followed by a repatriation of the banks' top management, which was a crucial factor in maintaining strong links with the international banking world and, thus, assured stability for the two institutions. In general, he sees the two banks as concentrating mostly on the Italian domestic market, thereby giving 'a decisive contribution to the country's industrialization' (Hertner 1984: 155). Ongoing and yet unpublished research carried on at The Historical Archives of Banca Commerciale seems to indicate that, by 1913, the latter had established strong personal links between its own management and that of the industrial firms to which it supplied credit. These results are likely to stimulate new debates and research on the role played by the two German-type banks during the industrial growth of the period 1896–1913.

ECONOMIC POLICIES AND INDUSTRIAL DEVELOPMENT

Inept government policies seem to be, according to Gerschenkron, the single most important cause of the 'disappointingly low' rate of industrial growth during Italy's big spurt (Gerschenkron 1962: 78). The tariff of 1887 'must be viewed as the real pièce de résistance of those policies' (Gerschenkron 1962: 80).

The 1887 tariff marked a real turning point after fifteen years of free trade and ten years of moderate protection to a limited number of industries (Prodi 1965–6; Sereni 1966; Del Vecchio 1978). Italy's tariff policy is criticized by Gerschenkron on three grounds: (a) for subjecting 'the tender plants of industrial growth to the rigors of a protectionist climate in agriculture' (Gerschenkron 1962: 81), (b) for the attention given to cotton textiles, 'an old industry with a moderate rate of modern technological progress and accordingly relatively limited possibilities in a backward country on the European continent' (Gerschenkron 1962: 81), and (c) for neglecting the most promising industries such as chemicals and engineering, the latter being put at a disadvantage also by protecting its main inputs, iron and steel.

This judgement implies the counterfactual hypothesis of a more rational tariff and, therefore, is a much welcome disentanglement from the typical nineteenth-century debate of free trade versus unqualified protection. And it is unfortunate that most Italian economic historians have been unable to follow Gerschenkron along the lines of an assessment of the impact of the tariff on growth in terms of effective protection. Despite the vast literature on the subject, therefore, there is a lack of serious quantitative research on the specific points raised by Gerschenkron. Even the average nominal rate of protection is not precisely known: the only aggregate figure – 18 per cent in 1888 declining to 7.5 per cent in 1913 (Capie 1983) – probably underestimates the actual level.

Federico (1984) has put the average rate of protection for Italian wheat at 52 per cent in 1895. Increasing world prices reduced it thereafter to 33 per cent in 1913. On average, these rates are higher than those computed by Gerschenkron. Looking at the rather fortunate period of free trade, 1915–25, one is inclined to agree with Gerschenkron that 'Italy's agriculture had at its disposal methods of adjustment that were not available to an equal degree north of the Alpine wall'

(Gerschenkron 1962: 81) and that its comparative advantage outside the Po Valley was not in wheat. However, one must pay attention to Italy's social conditions, which Gerschenkron sees as one of the weakening factors in its process of industrialization. The late 1880s and early 1890s, to be sure, witnessed unprecedented social unrest in the countryside that the abolition of the duty on wheat would no doubt have made even more serious.

The cotton industry had already undertaken a process of import substitution under the shield of the 1878 tariff. This process was boosted by the higher duties introduced in 1887. By the end of the century Italy was a net exporter of cotton goods. Fixed capital formation proceeded at such a high rate that by 1907 the sector was burdened by excess capacity and had to resort to *dumping* exports. During the years of the Gerschenkronian spurt, therefore, no 'infant industry' argument could be invoked to justify duties on textile products.

The subsequent development of the Italian manufacturing industry leaves no doubt as to the importance of the chemical and engineering sectors. The demands of the former for adequate protection were met only by the tariff of 1921 (Zamagni 1991). The dynamic effects of the earlier lower duty are difficult to assess given the lack of research in the field. It is likely, however, that chemical production could have benefited from 'infant industry' protection. As for engineering, Toniolo (1977) has shown that its effective protection was negligible, due to the high import duties imposed on its main inputs, iron and steel. At the same time Toniolo has estimated that the most favourable alternative policy (i.e. a subsidy to iron- and steel-making) would have resulted in a 7 per cent increase of value-added by manufacturing in the period 1906–8 (upper bound estimate). The rate of industrial growth during 1896–1908 would have been 7.0 per cent a year against Gerschenkron's 6.7, hardly a dramatic change in Italy's economic history.

Gerschenkron's criticism of the Italian tariff *structure* during the years of rapid industrialization is well taken and its validity has been confirmed by subsequent quantitative research. If the 1887 tariff succeeded in fostering the development of some industries (Zamagni 1984: 14) the price paid by the whole economy in terms of misallocation of resources was probably high. It is difficult, however, to judge how high that price was. And we still 'wonder how much importance one should in general ascribe to tariff policy in the history of European industrializations' (Gerschenkron 1962: 89).

According to Gerschenkron, 'another weakness of the Italian industrialization of 1896–1908 may have derived from the fact that by that time the great period of Italian railroad building was largely a thing of the past' (Gerschenkron 1962: 84). This point was debated with Romeo who held, on the contrary, that the building of railways was a prerequisite for rapid industrialization (Gerschenkron 1968: 98–127).

A general equilibrium model of investment in railways (Fenoaltea 1973b) yields rather discouraging results about the likely magnitude of domestic demand generated by investment in railways, even on the most favourable exchange-rate conditions. Direct estimates (Fenoaltea 1983) show that railway expenditures were a minor component of the demand for engineering goods, which depended to a much larger extent on the building trade, an industry that Gerschenkron did not consider. Moreover, the increasing trade deficit accumulated by the engineering sector during 1896–1908 is clear evidence that its problems consisted of supply rigidities (bottlenecks) rather than inadequate domestic demand. However, the wave of investment in railways which followed nationalization in 1905 took place at a time when, after the 1907 crisis, the rate of growth of domestic demand fell sharply and it therefore provided a timely anticyclical support.

Adding military expenditure to that for railways (Zamagni 1981), a likely guesstimate may be that the state provided about one-third of the total demand for products by the engineering industry during the years 1896–1908. According to recent estimates (Brosio and Marchese 1987) total public expenditure (including that by local authorities) oscillated between 15 and 20 per cent of GNP over the pre-1913 long run. These ratios are higher than those for most other European countries and indicate that, at least in the aggregate, state demand was a relatively more important factor in Italy then elsewhere.

Historians have discussed at length the likely overall effects of fiscal policy on Italian economic development. It has been suggested that deficit spending had a positive macro-economic effect (Barone 1972). Gerschenkron (1962: 82) hinted that state bonds crowded out private investment. Several scholars have stressed the unfavourable redistributive effects (among regions, industries and social classes) of the combination of regressive taxation and high state expenditure on defence and interest on state bonds. At any rate, during 1896–1907, monetary and fiscal policy favoured private investment (Toniolo 1990) so that, all things considered, it is difficult to accept an entirely unfavourable judgement on the role of economic policy during the years of the 'big spurt'.

Finally, Gerschenkron sees in the social situation of the 'Giolittian era' another reason for the weakness of the Italian 'spurt'. The coincidence in time of the 'spurt' with improvements in the workers' standard of living – so the argument runs – shifted resources from investment to consumption. Moreover, 'had the industrial upsurge in Italy taken place one or two decades earlier, in all likelihood it would have been much less disturbed by industrial strife' (Gerschenkron 1962: 86). Three remarks are possible on this subject: (1) While it is true that real wages improved during 1896–1908 (Zamagni 1984), it is likely that product per man in manufacturing grew faster so that the ratio of profits to value-added increased, as did that of gross investment over GNP. (2) The strike figures given by Gerschenkron, according to whom 'between 1901 and 1913 there was only one year when the number of days lost by strikes remained below the million mark' and the number exceeded three million in some years of the period, do not indicate any unbearable loss in production. They include strikes in agriculture and, therefore, should be seen in the context of a labour force totalling about 10 million members, with 2.5 million engaged in industrial production. Even assuming an average of two million days of work lost by strikes in industry alone, the loss in physical output would amount to less than one-third of one percentage point. (3) Contrary to what Gerschenkron thought, 'had the industrial upsurge in Italy taken place a decade earlier' it would have been met by a much larger and widespread working class unrest and discontent. In all likelihood 'Giolitti's conciliatory statemanship' (Geschenkron 1962: 85) was, on balance, a bonus for the economy itself.

THE IDEOLOGICAL STIMULUS

Gerschenkron's opinion on the absence of an ideological stimulus to industrialization is widely shared by Italian scholars. Baglioni (1974) describes a ruling political class coming either from the landed gentry or from the liberal professions. Most entrepreneurs were bound to traditional social and cultural values while, with few exceptions, they shied away from the political arena. The recent business-history literature seems to confirm that most entrepreneurs were bound to the traditional social and cultural values and that their political demands were confined to such issues as tariff protection, subsidies, fiscal policy and, more broadly, to the maintenance of law and order. Some authors (e.g. Hunecke 1982) go as far as saying that most industrialists were reluctant to move their factories from the country to the town for fear of upsetting the existing 'social order'. Generalization is always difficult in these matters. There is, however, a growing literature that seems to indicate that in the 1860s and 1870s Italian entrepreneurs could not prosper without at least the indirect support of local politicians and large landowners. In fact, a number of industrial activities originated directly from the latter. There was, therefore, nothing like a swift emerging of a 'new' entrepreneurial class with an entirely different set of values that enabled it to create and impose its own ideology of industrialization.

Moreover, the Church, which retained a powerful influence on the majority of the population, drastically opposed the new liberal state both for its agnostic stance on religion and, more deeply, for its occupation of the territory of the Papal States. In these circumstances, at least until the death of Pius IX, the Church fought against all attempts at 'modernization' that came from the most far-sighted among the members of the liberal ruling class. With the advent of Leo XIII, the Church took greater interest in social and economic matters, as witnessed by the encyclical *Rerum Novarum* and by the creation of *Opera dei congressi*, an organization that was aimed at spreading the 'social doctrine' of the Church and at promoting engagement by Catholics in such fields as cooperatives and workers' organizations. In principle, the Catholic ideology opposed both capitalism and socialism. In practice, bishops and priests were tolerant of capitalism but resisted rapid changes that would undermine traditional social and moral values. They were, therefore, particularly concerned with keeping their flocks in the countryside. The growth of towns, emigration and large enterprises were seen as major evils.

The Socialist Party, established in 1892, remained in practice hostile to the capitalist system (Satta 1986). The party's propaganda spread among workers the expectation of a new social and economic order and consequently opposition to the actual industrialization process.

There were, therefore, considerable weaknesses in the ideological foundations of Italy's industrialization. Nevertheless, the picture is not as clear-cut as Gerschenkron saw it. The split between the old agrarian ruling classes and the emerging industrial interests was not always very clear. Behind several success stories in industrial entrepreneurship one may detect the financial help of landowners and sometimes the support of local politicians (e.g. Castronovo 1977; Fiocca 1984). Moreover, in the late 1880s and in 1890s the picture began to change from the point of view of the 'ideological stimulus to industrialization' as well. In a most interesting book, Lanaro (1979) has examined the ideology emerging with the rank and file of entrepreneurs, far from the most prestigious academic and cultural circles. A careful examination of pamphlets, speeches and articles in local newspapers had led him to conclude that the protectionist drive was backed by a strong nationalist ideology based on the assumption that, without channelling major resources to industry, Italy could not reach the status of a European power. At the same time, nationalists maintained that given the backwardness of the country only the state had the allocative power needed to promote rapid industrialization. It had, therefore, to be made more efficient by a strong and uncompromising leadership. Nationalists formed an intellectual movement with fuzzy ideological boundaries and multifarious followers rather than a political party.

An industrial ideology was, then, actively present in Italy during the longest period of its pre-war industrial growth. It was based to a large extent on a modified version of German nationalism. It was certainly a minority movement – liberals, Catholics and socialists were contending in the same ideological arena – but, then, Saint Simonism was not the ideology of the French masses either. What matters is that, by the beginning of the twentieth century, a growing number of industrialists and engineers had broken away from traditional values and had developed their own *Weltanschaung* which looked with optimism and hope to the future of Italy as an industrial power. If confidence and moral justification were needed, nationalism was there to provide them.

CONCLUSIONS

As far as Italy is concerned, in the mid-1950s, Gerschenkron's contribution was ahead of its time. A surprisingly deep understanding of the crucial problems in Italian history coupled with sound economic analysis enabled him to produce an elegant interpretation of the industrialization process centred around a limited number of relevant variables.

Backwardness proved to be a much more powerful and operational explanatory variable than others then fashionable such as 'agrarian-industrial concert', 'financial capitalism', 'foreign dependence' and the like. Gerschenkron's index of industrial production has been somewhat improved upon only in recent years. In analysing the likely impact of tariff policy he broke away from the muddling debate on advantages and disadvantages of free trade and set an agenda of research on effective protection that is far from being fulfilled. His assessment of the role of the German banks – possibly the most controversial part of his construction – contains a number of important insights into the working of Italian financial intermediation which are still providing fruitful research hypotheses. There is, therefore, little doubt that Gerschenkron provided a highly seminal contribution to the understanding of the industrialization process in Italy.

Italian historians, with few important exceptions, have been unable to capitalize on this wealth of historical and economic intuitions. It was particularly unfortunate that Gerschenkron's early quantitative seeds did not fall on fertile ground. We still know surprisingly little about a crucial period in the development of one of today's important industrial economies. Some good research has, nonetheless, been done in the field over the past twenty years. In the previous pages we have reviewed the results that seem to be more relevant in discussing Gerschenkron's hypotheses. To sum-up, the following points have emerged. (1) Gerschenkron was not off the mark in his ranking of Italy's backwardness relative to the other major European countries. (2) He was again quite right about the time-path of industrialization, although he underrated the importance of his own quantitative evidence about the industrial development that took place in the 1880s and possibly underestimated the magnitude of the 'spurt' of 1896–1908. (3) On economic policy, Gerschenkron's point about the tariff structure is particularly well-taken and deserves further quantitative analysis. (4) Subsequent research has come to some conclusions that are different from those of Gerschenkron, particularly as far as ideology and banking are concerned. Stress has been placed on the importance of the nationalist movement in providing an ideological climate favourable to industrialization, at least from the beginning of the twentieth century onward. The role of mixed banks as 'agents of industrialization' has been challenged by an analysis of their behaviour that seems to indicate that they were more interested in short-term banking (and profits) than in long-term planning of industrial development. Moreover, it has been argued that the distinction between French and German banks does not stand close scrutiny in the Italian case. However, Italian banking history is developing and new outcomes, more in line with Gerschenkron's hypotheses, may not be ruled out.

In focusing his attention on 1896–1908, Gerschenkron was certainly selecting the most dynamic period in Italian economic history before the 1950s. However, his explanation of the 'spurt' suffered from a neglect of previous industrial growth as well as of some factors that, if taken into account, might contribute to a better understanding of the causes of the rapid growth in the years preceding the First World War.

Quantitative evidence on industrial growth during the 1860s and 1870s is far from satisfactory. According to the existing national accounts estimates, GNP grew in line with population (at the moderate annual rate of 0.7 per cent) leaving per capita income constant. Fenoaltea's (1983) and Carreras' (1982) indices feature a positive but low rate of growth in industrial production per head of population. The overall picture for the 1880s is better established: all the existing indices show that industrial output developed at a respectable pace. There is ample evidence of a wave of private investment, caused by the change in expectations produced by the industrial policy of the 'left', by the resumption of the gold standard in 1882, by substantial government orders to the steel-making and engineering industries, by urban-renewal programmes and by foreign investment (e.g. Toniolo 1990). The banking system played an important role in speeding-up the process of resource allocation to fixed capital formation in industry and building (Confalonieri 1974–6; Sannucci 1989). The lame duck was agriculture: production stagnated and incomes were affected by the fall in prices. During 1896–1908, in contrast, output and productivity grew rapidly in the agricultural sector.

A deep, long depression followed the upswing of the 1880s, characterized by a serious banking crisis. It is surprising that, after drawing so much from Hilferding, Gerschenkron never took into consideration the possibility of a depression transmitted from the financial to the real sector as a result of a combination of large industrial debts, generous lending to risky borrowers, and swift withdrawal of short-term capital (Warglien 1987). This is precisely the story of 1888–1893. In the event, the behaviour of French-type banks contributed to financial instability. But German-type banks were not likely to produce a less volatile financial environment, as witnessed by the inter-war crises in Central Europe. Overall financial conditions were much more stable during 1896–1913. External finance was not as large as it was in the 1880s, there was little foreign short-term capital around, and – last but not least – the embryo of a central bank had been established in 1893. The latter took upon itself the role of lender of last resort, thereby creating a more stable financial environment, as shown during the crisis of 1907 (Bonelli 1978).

Conditions in the world economy must be taken into account when discussing the industrial boom of the so-called 'Giolittian era'. International trade and factor mobility are likely to have affected the behaviour of a relatively small and fairly open economy such as Italy through an export-led growth, particularly of primary goods (Cafagna 1973; Bonelli 1978) and a supply-push by foreign finance (Fenoaltea 1988c). Moreover, emigrant remittances, and to a lesser extent tourism, allowed growth to be free from balance-of-payments constraints.

A thriving agriculture, a stable financial setting, and a favourable international environment are factors likely to have made the industrial upswing of 1896–1913 more rapid and resilient than that of 1879–88. Such factors do not rule out the possibility that German banks played a role of their own as more or less important 'agents of industrialization'. There is indeed much research still to be done on this period but, by now, we know enough to argue that, around 1900, the Italian economy was probably richer and more diversified than Gerschenkron supposed, so that several causes were at work in producing the 'spurt'. This picture is more complex – and therefore intellectually less exciting – than that deriving from Gerschenkron's elegant mono-causal explanation of growth: Clio, as he knew, is often a messy housewife.

ACKNOWLEDGEMENT

The authors are indebted to Giuseppe Gesano and Nicola Rossi for advice, to Stefano Fenoaltea and Richard Sylla for comments, to Pierantonio Dal Lago and Sonia Soncin for research assistance. Financial support from CNR is gratefully acknowledged.

NOTES

1 Fenoaltea's estimates of value-added by mining in 1911 are 46 per cent higher than those of ISTAT (Fenoaltea 1988b: 123); value-added by building and construction exceeds ISTAT's by 82 per cent (Fenoaltea 1988b). Average growth rates (1863–1913) are 2.8 per cent and 3.3 per cent respectively (against ISTAT 0.6 per cent and 1.9 per cent).

2 In fact, the worst European performance (infant mortality of 307 per thousand for Germany in 1870) was 'worse' than the 'worst' one in 1970 for the sample of developing countries used by Morris (Gabon with 229 per thousand).

3 For instance, the production of raw silk is taken to represent silk weaving and a sizeable part of wool production; wheat consumption is taken as a proxy for the entire food-processing industry.

4 For instance, simply by substituting Gerschenkron's estimates of silk production with Fenoaltea's (1988a) estimates of value-added by silk industry (which includes throwing, weaving and dyeing) the rate of growth of the entire textile industry would rise to 3.9 per cent and the overall rate to 4.2 per cent.

5 The index is a by-product of the first reconstruction of Italy's national accounts since 1861. Afterwards it has been incorporated without changes in the Fuà revision (Fuà 1969) and hence in all the international collections of historical statistics (Mitchell 1975).

6 Fenoaltea's index was actually prepared in the mid-1960s and incorporated in his Harvard PhD dissertation. The author is now engaged in a major research aimed at producing a much more reliable estimate of industrial output from 1861 to 1913; partial results have been published in Fenoaltea (1982, 1984, 1985, 1987, 1988a, 1988b).

7 The acrimonious attacks on the Banca Commerciale in 1914–15, for instance, leave no room for doubt that its economic and political power was fully understood at the time by friends and foes alike. Moreover, such contemporary Italian economists as Pantaleoni, Einaudi, Barone, Bresciani-Turroni and Sraffa, if anything overstated the importance of large investment banks.

REFERENCES

Amatori, F. (1980) 'Entrepreneurial typologies in the history of industrial Italy', *Business History Review* 54, 359–86.

Are, G. (1965) *Lo sviluppo industriale italiano nell'età della Destra*, Pisa: Nistri-Lischi.
—— (1974a) *Alle origini dell'Italia industriale*, Naples: Guida.
—— (1974b) *Economia e politica nell'Italia liberale (1890–1915)*, Bologna: Il Mulino.
Baglioni, G. (1974) *L'ideologia della borghesia industriale nello stato liberale*, Torino: Einaudi.
Bairoch, Paul (1976) 'Europe's gross national product: 1800–1975', *Journal of European Economic History*, V, 273–340.
Barone, G. (1972) 'Sviluppo capitalistico e politica finanziaria in Italia', *Studi Storici* a13, 568–99.
Barsby, Steven (1969) 'Economic backwardness and the characteristics of development', *Journal of Economic History*, 449–72.
Biscaini, A.M. and Ciocca, P.L. (1979) 'Le strutture finanziarie: aspetti quantitativi di lungo periodo (1870–1970)', in F. Vicarelli (ed.) *Capitale industriale e capitale finanziario: il caso italiano*, Bologna: Il Mulino.
Bonelli, F. (1968) 'Osservazioni e dati sul finanziamento dell'industria italiana nel secolo XIX', *Annali della Fondazione L. Einaudi* 257–279.
—— (1975) *La cirsi del 1907*, Torino: Fondazione L. Einaudi.
—— (1978) 'Il capitalismo italiano. Linee generali di interpretazione' in *Annali della Storia d'Italia Einaudi*, vol. I, pp. 1126–255, Torino: Einaudi.
Brosio, G. and Marchese, C. (1987) *Il potere di spendere*, Bologna: Il Mulino.
Cafagna, Luciano (1965) 'Intorno alle origini del dualismo economico in Italia', in A. Caracciolo (ed.) *Problemi storici dell'industrializzazione e dello sviluppo*, Urbino: Argalia, pp. 105–50, reissued in L. Cafagna (1989), pp. 187–213.
—— (1973) 'Italy 1830–1914', in *Fontana Economic History of Europe*, vol. IV, reissued in L. Cafagna (1989), pp. 281–319.
—— (1989) *Dualismo e sviluppo nella storia d'Italia*, Padova: Marsilio.
Capacelatro, E.M. and Carlo, A. (1972) *Contro la "questione meridionale"*, Roma: Samonà e Savelli.
Capie, F. (1983) 'Tariff protection and economic performance in the nineteenth century', in J. Black and A.L. Winters (eds) *Policy and Performance in International Trade*, pp. 1–24, London: Macmillan.
Caracciolo, A. (ed.) (1969) *La formazione dell'Italia industriale*, Bari: Laterza.
Carreras, A. (1982) 'La producciò industrial espanyola i italiana des de mitian segle XIX fins a l'actualitat', unpublished dissertation, University of Barcelona.
Castronovo, V. (1977) *Giovanni Agnelli*, Torino: Einaudi.
Cohen, J. (1967) 'Financing industrialization in Italy, 1894–1914: the partial transformation of a late-comer', *Journal of Economic History* 27, 363–82.
—— (1972) 'Italy 1861–1914', in R. Cameron (ed.) *Banking and Economic Development. Some Lessons of History*, New York: Oxford University Press, pp. 58–90.
Confalonieri, A. (1974–6) *Banca e industria in Italia, 1894–1906*, 3 vols, Milan: Banca Commerciale Italiana.
—— (1982) *Banca e industria in Italia dalla crisi del 1907 all'agosto 1914*, 2 vols, Milan: Banca Commerciale Italiana.
Coppa, F.J. (1970) 'The Italian tariff and the conflict between agriculture and industry: the commercial policy of liberal Italy', in *Journal of Economic History* 742–69.
Crafts, N.F.R. (1983) 'Gross national product in Europe 1870–1910, *Explorations in Economic History* XX, 387–401.
D'Angiolini, P. (1972) 'La svolta industriale italiana negli utlimi anni del secolo scorso e le reazioni dei contemporanei', *Nuova Rivista Storica*, 53–121.
De Rosa, L. (1969) 'Difesa militare e sviluppo economico in Italia (1861–1914), in *Rassegna economica*, now in ID (1980), *La rivoluzione industriale in Italia*, Bari: Laterza.

Del Vecchio, Edoardo (1978) *La via italiana al protezionismo*, Rome: Camera dei Deputati.
Ercolani, P. (1969) 'Documentazione statistica di base' in G. Fuà (ed.) *Lo sviluppo economico in Italia*, vol. III, Milan: Franco Angeli, pp. 380–460.
Farina, Francesco (1976) 'Modelli interpretativi e caratteri del capitalismo italiano', *Quaderni Storici*, 32, 487–514.
——— (1980) 'Note sul ruolo della banca mista nello sviluppo italiano', *Societa e Storia*, 19, 919–27.
Federico, G. (1982) 'Per una valutazione critica delle statistiche della produzione agricola italiana dopo l'Unità (1860–1913)', *Società e Storia* 15, 87–130.
——— (1984) 'Commercio dei cereali e dazio sul grano in Italia (1963–1913). Una analisi quantitativa', *Nuova Rivista Storica* LXVIII, 46–108.
Federico, Giovanni and Chesi, Marco (1987) 'Lo sviluppo economico italiano', *Storia della societa italiana, vol. 17 Le strutture e la classi nell'Italia Unita*, Milan: Teti.
Fenoaltea, S. (1969) 'Decollo, ciclo e intervento dello stato', in A. Caracciolo (ed.) *La formazione dell'Italia industriale*, Bari: Laterza.
——— (1973a) 'Riflessioni sull'esperienza industriale italiana dal Risorgimento alla prima Guerra Mondiale', in G. Toniolo (ed.) *Lo sviluppo economico italiano 1861–1940*, Bari: Laterza.
——— (1973b) 'Le ferrovie e lo sviluppo industriale italiano', in G. Toniolo (ed.) *Lo sviluppo economico italiano 1861–1940*, Bari: Laterza.
——— (1982) 'The growth of utilities industries in Italy', *Journal of Economic History* XLII, 601–27.
——— (1983) 'Railways and the development of the Italian economy to 1913', in P. O'Brien, *Railways and the Economic Growth of Western Europe*, London: Macmillan, pp. 49–120.
——— (1984) 'Le costruzioni ferroviarie in Italia', *Rivista di Storia Economica*, n.s. 1, 61–94.
——— (1985) 'Le opere pubbliche in Italia 1861–1913', *Rivista di Storia Economica*, n.s. 2, 335–69.
——— (1987) 'Le costruzioni in Italia 1861–1913', *Rivista di Storia Economica*, n.s. 4, 1–34.
——— (1988a) 'The growth of Italy's silk industry 1861–1913', *Rivista di Storia Economica*, n.s. 3.
——— (1988b) 'International resource flows and construction movements in the Atlantic economy: the Kuznets cycle in Italy 1861–1913', *Journal of Economic History*, XLVIII (3), 605–37.
——— (1988c) 'The extractive industries in Italy 1861–1913: general methods and specific estimates', *Journal of European Economic History*, 17 (1), 117–25.
Fiocca, G. (ed.) (1984) *Borghesi ed imprenditori a Milano*, Bari: Laterza.
Flora, P. (1983) *State, Economy and Society in Western Europe 1815–1975*, London: Macmillan.
Fratianni, M. and Spinelli, F. (1985) 'Currency competition, fiscal and the money supply process in Italy from the unification to World War I', *Journal of European Economic History* XIV, 473–99.
Fuà, G. (ed.) (1969) *Lo sviluppo economico in Italia*, Milan: Franco Angeli.
Gerschenkron, A. (1955) 'Notes on the rate of industrial growth in Italy 1861–1913', *Journal of Economic History* now in Gerschenkron (1984).
——— (1962) *Economic Backwardness in Historical Perspective. A Book of Essays*, Cambridge, Mass.: Belknap Press.
——— (1968) *Continuity in History and Other Essays*, Cambridge, Mass.: Belknap Press.

Hertner, P. (1984) *Il capitale tedesco in Italia dall'unità alla prima guerra mondiale. Banche miste e sviluppo economico italiano*, Bologna: Il Mulino.

Hunecke, V. (1982) *Classe operaia e rivoluzione industriale a Milano 1859–1982*, Bologna: Il Mulino.

Istituto Centrale di Statistica (1957) 'Indagine statistica sullo sviluppo del reddito nazionale dell'Italia dal 1861 al 1959', *Annali di Statistica*, Serie VIII, 9.

Kravis, I.B. *et al.* (1982) *World Product and Income*, Baltimore: Johns Hopkins.

Lanaro, S. (1979) *Nazione e lavoro*, Padova: Marsilio.

Merli, S. (1976) *Proletariato di fabbrica e capitalismo industriale*, Florence: La Nuova Italia.

Ministero Agricoltura, Industria e Commercio (1906) *Riassunto delle condizioni industriali del Regno*, Roma.

Ministero del Tesoro (1988) *Relazione del direttore generale alla Commissione parlamentare di vigilanza. Il debito pubblico in Italia 1861–1987*, Rome: Poligrafico dello Stato.

Missaggia, M.G. (1988) 'Nota sulle statistiche ufficiali per l'industria in Italia: 1885–1903', *Rivista di Storia Economica* 2, 234–54.

Mitchell, B.R. (1983) *European Historical Statistics*, London: Macmillan.

Mitchell, R.C.O. (1975) *European Historical Statistics*, London: Macmillan Press.

Morris D.M. (1979) *Measuring the Conditions of the World's Poor: the Physical Quality of Life*, New York: Pergamon.

Nitti, Francesco S. (1900) *Il bilancio dello Stato dal 1862 al 1896–97*, now in ID (1958) *Scritti sulla questione meridionale*, vol. II, Bari: Laterza.

Nitti Francesco S. (1905) 'La conquista della forza', reissued in F.S. Nitti (1966) *Scritti di economica e finanza*, vol. III.

Pedone, A. (1969) 'Il bilancio dello Stato', in G. Fuà (ed.) *Lo sviluppo economico in Italia*, vol. II, Milan: Franco Angeli, pp. 203–40.

Prodi, Romano (1965–6) 'Il protezionismo nella politica e nell'industria italiana dall'Unificazione al 1887', *nuova rivista storica*, 42–86 and 597–626.

Romano, R. (1977) *Introduzione* in ID, *Borghesia industriale in ascesa: gli imprenditori tessili nella inchiesta industriale del 1870–74*, Milano: F. Angeli.

—— (1980) *I Caprotti. L'avventura economica ed umana di una dinastia industriale della Brianza*, Milan: F. Angeli.

—— (1985) *I Crespi*, Milan: F. Angeli.

Romeo, R. (1958) 'Problemi dello sviluppo capitalistico in Italia dal 1861 al 1887', *Nord e Sud* July–August, reissued in R. Romeo (1961) *Risorgimento e Capitalismo*, Bari: Laterza, pp. 87–120.

—— (1965) *Breve storia della grande industria in Italia 1861–1961*, Bologna: Cappelli.

—— (1974) *Gli scambi degli stati sardi con l'estero nelle voci più importanti della bilancia commerciale*, Torino: Centro di Studi Piemontesi.

Romeo, R. and Gerschenkron, A. (1961) 'Consensi dissensi ipotesi', *Nord e Sud*, November, reissued in A. Caracciolo (1969), pp. 53–81.

Rostow, W. (1963) *The Economics of Take-Off into Sustained Growth*, New York: St Martinus.

Roverato, G. (1987) *Una casa industriale. I Marzotto*, Milan: F. Angeli.

Sannucci, V. (1989) 'The establishment of a central bank in Italy in the 19th century', in M. De Cecco and A. Giovanni (eds) *A European Central Bank?*, Cambridge: Cambridge University Press, pp. 247–81.

Satta, V. (1986) 'I socialisti ed i problemi dello sviluppo economico', in *Clio* XXII, 248–73.

Sereni, E. (1972) *Il capitalismo nelle campagne (1860–1900)*, Torino: Einaudi, (first edition 1947).

—— (1966) *Capitalismo e mercato nazionale*, Rome: Editori Riuniti.
Toniolo, G. (ed.) (1973) *Lo sviluppo economico italiano 1861–1940*, Bari: Laterza.
—— (1977) 'Effective protection and industrial growth: the case of Italian engineering', in *Journal of European Economic History* VI, 659–73.
—— (1990) *An Economic History of Liberal Italy*, London: Routledge.
Warglien, M. (1985) 'Nota sull'invetsimento industriale in macchinari ed altre attrezzature meccaniche Italia 1881–1913', *Rivista di Storia Economica*, n.s. 2, 125–46.
—— (1987) 'Investimenti industriali e instabilità finanziaria in Italia 1878–1913', *Rivista di Storia Economica*, n.s. 3, 384–439.
Webster, R. (1974) 'The political and industrial strategies of a mixed investment bank: Italian industrial financing and the Banca Commerciale Italiana', *VierteGarharchiv fur Sozial und Wirtschaftgeschichte* 61, 320–71.
Zamagni, V. (1978) *Industrializzazione e squilibri regionali*, Bologna: Il Mulino.
—— (1981) *Lo stato italiano e l'economia*, Florence: Le Monnier.
—— (1984) 'Sui salari industriali nell'età giolittiana', *Rivista di Storia Economica*, n.s. I, 183–221.
—— (1988) *Il debito pubblico in Italia dal 1861 al 1987*, Rome: Istituto Poligrafico dello Stato.
—— (1991) *L'industria chimica in Italia* (forthcoming).

7. FOREIGN TRADE AND THE SPANISH ECONOMY DURING THE NINETEENTH CENTURY

LEANDRO PRADOS DE LA ESCOSURA
Universidad de Alcalá de Henares

THIS CHAPTER attempts to offer a long-term view of the evolution of foreign trade and its impact on the process of economic change in Spain from the end of the Napoleonic War to World War I. The study is rooted in the debate on the connections between trade and growth in the European Periphery.

1. HISTORIOGRAPHICAL INTRODUCTION

The interaction between external demand and changes in internal supply is an important factor in the development process. Interest in this relationship has transcended the field of economic history and has raised questions about whether the nineteenth-century experiences of today's developed nations are analogous to the present experiences of less developed countries. Contrasting perceptions of the past and present—the "stylized economic facts," in the words of Kaldor—have influenced the economic policies recommended for developing nations. Nurkse's theory of trade as the engine of nineteenth-century growth, as opposed to the external sector's less dynamic role in Third World economies of the first half of the twentieth century, has given rise to policies of "balanced growth," where the role of foreign trade has been deliberately reduced.[1] Similarly, Myrdal, Singer, and Prebisch's interpretation of export specialization in primary products as a vicious circle of underdevelopment have led to policies of import substitution and the promotion of national industry.[2]

In the last three decades the emphasis economists have placed on the history of the development process has on the other hand stimulated economic historians to study trade and growth. Adherents of the current theories of "dependency" and "unequal exchange," as well as their critics, have based their arguments on historical evidence. A lack of quantitative data and the difficulties inherent in this type of investigation, however, have led historians to supplement their traditional sources (the opinions of contemporaries) with the interpretations of development economists. The latter are frequently founded on an empirical base of "overstylized facts" that is both weak and fragmentary; yet historians have adopted these interpretations with the same facility with which they used to accept contemporary accounts.

The historiography of the relationship between trade and growth in the European Periphery, especially the Spanish case, is not an exception to this practice. The interpretations of contemporaries are complemented by the opinions of modern development economists, the latter being casually applied to the nineteenth-century Spanish experience. The cases of Vicens Vives and Nadal are instructive. Vives suggests that the tariff of 1869 put the "mineral wealth of Spain at the disposal of England and France" with the result that Spain was divested of that wealth by foreign enterprises,[3] whereas Nadal links railroad construction to the expansion of export-oriented mining and concludes that it was an "instrument of colonization and exploitation."[4] The same can be said of the interpretation of Berend and Ránki when they state that:

> On the Iberian Peninsula, though there was a substantial increase of foreign trade, and foreign investments did introduce new elements in the economy, there was no real economic growth or transformation of the economy.[5]

Harrison, in turn, has suggested a parallel between the type of capitalist development described by Gunder Frank for nineteenth-century Latin America and the Spanish case:

> As the outward-oriented exporting lobby of landlords, mine owners and merchants gained strength and found expression for their interests in the liberalising of trade, so Spain (like Latin America) was converted into a satellite of the industrial metropolis of north-west Europe, which extracted often irreplaceable supplies of raw materials for its own use, sending back large quantities of manufactured goods and surplus capital.[6]

In addition, the potential forward linkages of mineral exports were, in the opinion of Sánchez-Albornoz, appropriated by the purchaser countries, while the profits were funneled back to the investor countries. The result was that "the mines were converted into foreign enclaves, linked only territorially to Spain, and lacked articulation with the rest of the Spanish economy."[7] Sánchez-Albornoz synthesizes the theoretic bases of this interpretation by placing the Spanish case within the wider context of Center-Periphery relations in nineteenth-century Europe. He suggests that "it is natural that the inequality between industrialized and non-industrialized nations should begin with regions that are in close proximity to one another."[8]

The inability of southern Europe to initiate a process of modern economic growth in spite of the stimulus provided by the industrialization of northwestern Europe—an inability emphasized by Arthur Lewis—has led economic historians to adopt a negative view of the role of foreign trade in the nineteenth-century modernization of the European Periphery.[9] Most unfavorable interpretations of the export sector's role in the evolution of the peripheral European economies—and the Spanish case is no exception—rest on the baseless hypothesis that, in the absence of foreign trade, an alternate, more beneficial route to development would have been attempted. The implicit alternative model is founded on the following economic reasoning: the export of raw materials implied a substantial opportunity cost for the Spanish economy. This cost consisted of nonexporting activities embodied in the productive resources exported, which would otherwise have been used to contribute to greater, long-term development. Nevertheless, these hypothetically alternative activities were sacrificed in the interests of assuring a permanent supply, at decreasing prices, of primary products to the countries of the Center and in the interests of providing a market for these countries' manufactured goods. The hypothesis of autarchic development rests, then, on certain assumptions: (1) that alternative uses existed for the factors of production employed in the export sector and (2) that, in order for the autarchic development model to be more successful than development in an open economy, it would be required that the economy obtain significant gains in productivity in the internal sector and that this be achieved independently of growth in the export sector.[10]

2. TRENDS IN FOREIGN TRADE

Tables 7.1 and 7.3 illustrate the long-term trends in foreign trade from 1815 to 1913. Data are presented in overlapping ten-year averages.[11] Two long swings can be distinguished in exports: the first covers the years 1815 to 1850, during which growth proceeded slowly; the second encompasses the period from the 1850s to the 1880s, the apogee of nineteenth-century exports.

After this period, there was a gradual decline in export expansion until World War I. Imports followed similar patterns, with only slight differences. The decades of the mid-nineteenth century did not exhibit the same linear evolution for imports as for exports. From 1850 to 1880, three distinct phases can be discerned: (1) during the second half of the 1850s and first years of the 1860s, there was a tremendous acceleration of imports that was tied to the beginning of the industrialization process and to the importation of railroad materials; (2) in the second half of the 1860s and throughout most of the 1870s, there was relative stagnation; (3) in the final years of the 1870s and throughout the 1880s, there was a renewed acceleration.

When the returns to foreign factors used in the production of exportable goods is subtracted from the total value of exports, the result is the "retained" exports or net export earnings.[12] Until the liberalization of foreign investment in 1869, substantial differences did not exist between total and retained exports. It is interesting to note that, during the last four decades of the period, there were no notable discrepancies between the expansion of total and retained exports. This fact sheds doubt on the hypothesis that total export statistics are deceptive and do not reflect long-term trends in nineteenth-century Spain's net export earnings.[13]

The slower long-term growth of imports in comparison with exports explains the reduction in the Spanish trade deficit during the nineteenth century. There are two distinct phases in the evolution of the balance of trade (see Table 7.1, column 8). The first was characterized by the deficit and encompassed the years 1815 to 1870. During this time Spain's debtor position was slowly reduced until the 1840s but rose again during the 1850s and 1860s. The second phase was one of progressive improvement in the Spanish position and culminated in the 1890s. This creditor position, however, deteriorated after Cuba's emancipation in 1898. Considering the difference between retained exports and imports, the surplus phase included only the 1880s, 1890s, and the decade prior to World War I (see Table 7.2).[14]

The terms of trade (the prices of exports in relation to those of imports) determine the import capacity by unit of exports and are an indicator of the static gains from trade. In other words, the terms of trade determine whether the country in question derives a greater or lesser value from trade than that of the resources it sacrifices to the production of exportable commodities. Two stages can be distinguished in the evolution of the terms of trade: the first stage was one of notable progress and spanned the years between the end of the Napoleonic Wars and the 1880s; the second stage (1890–1913) was a period of deterioration. During the years when the terms of trade were favorable, the

Table 7.1
Spain's Foreign Trade, 1815–1913
(10^6 current pesetas)

	F.O.B. Domestic Exports				*C.I.F. Net Imports*				
	(1) *Value at Current Prices*	(2) *Value at 1854 Prices*	(3) *Purchasing Power (in Terms of Imports) 1854 Prices*	(4) *Fisher Price Index 1854 = 100*	(5) *Value at Current Prices*	(6) *Value at 1854 Prices*	(7) *Fisher Price Index 1854 = 100*	(8) *Commodity Trade Balance (1):(5)*	(9) *Net Barter Terms of Trade 1854 = 100 (4):(7)*
1815–1824	107.3	73.6	52.6	145.7	156.8	76.9	204.0	−49.5	71.4
1820–1829	105.0	85.2	67.6	123.3	131.6	84.7	155.3	−26.6	79.4
1825–1834	110.2	99.3	90.3	111.0	121.4	99.5	122.0	−11.2	91.0
1830–1839	121.5	107.9	99.4	112.6	127.8	104.6	122.2	−6.3	92.1
1835–1844	134.7	115.8	112.7	116.3	146.2	122.3	119.5	−11.5	97.3
1840–1849	146.6	130.2	143.7	112.6	153.4	150.4	102.0	−6.8	110.4
1845–1854	163.9	158.4	167.6	103.5	173.6	177.5	97.8	−9.7	105.8
1850–1859	229.8	219.3	239.1	104.8	259.7	270.2	96.1	−29.9	109.1
1855–1864	287.2	258.0	283.0	111.3	408.8	402.8	101.5	−121.6	109.7
1860–1869	320.1	293.4	290.7	109.1	451.6	410.2	110.1	−131.5	99.0
1865–1874	412.7	372.1	360.4	110.9	448.5	391.7	114.5	−35.8	96.9
1870–1879	500.5	438.7	487.8	114.1	490.0	477.6	102.6	10.5	111.2
1875–1884	646.4	592.5	757.8	109.1	595.0	697.5	85.3	51.4	127.9
1880–1889	784.2	802.7	1000.3	97.7	705.9	900.4	78.4	78.3	124.6
1885–1894	866.5	920.8	1051.6	94.1	721.4	875.5	82.4	145.1	114.2
1890–1899	1055.8	1054.8	1134.1	100.1	786.7	845.0	93.1	269.1	107.5
1895–1904	1187.5	1080.5	1044.4	109.9	994.0	874.2	113.7	193.5	96.7
1900–1909	1222.5	1103.3	1017.9	110.8	1093.7	910.7	120.1	128.8	92.2
1905–1913	1330.4	1274.3	1280.5	104.4	1167.2	1123.4	103.9	163.2	100.5

SOURCES: Export and import values at current prices, L. Prados de la Escosura (1986), "Una serie anual del comercio exterior español (1784–1913)," *Revista de Historia Económica*, IV, 1. Export and import price indices, L. Prados de la Escosura, "Las relaciones reales de intercambio entre España y Gran Bretaña, 1714–1913," in P. Martín Aceña and L. Prados de la Escosura, eds. (1985), *La Nueva Historia Económica en España*, pp. 119–165. Prices in pounds sterling have been converted into current pesetas with an annual exchange rate series.

NOTE: Decade average figures overlap.

Table 7.2
Retained Value of F.O.B. Domestic Exports, 1865–1913
(10^6 current pesetas)

	(1) Value at Current Prices	(2) Value at 1854 Prices	(3) Purchasing Power (in Terms of Imports) 1854 Prices	(4) Trade Balance
1865–1876	396.9	357.9	346.6	−51.6
1870–1879	459.7	402.9	448.1	−30.3
1875–1884	587.2	538.2	688.4	−7.8
1880–1889	719.3	736.2	917.5	13.4
1885–1894	797.4	847.4	967.7	76.0
1890–1899	964.7	963.7	1036.2	178.0
1895–1904	1054.1	959.1	927.1	60.1
1900–1909	1059.8	956.5	882.4	−33.9
1905–1913	1172.6	1123.2	1128.6	5.4

SOURCES: Table 7.1 for export and import total values. Retained value estimates for mineral products come from L. Prados de la Escosura, "El comercio hispano-británico en los siglos XVIII y XIX. (II) Tendencias y estructura," *Revista de Historia Económica* (forthcoming).
NOTE: Decade average figures overlap.

import capacity by unit of exports grew at an annual rate of 1%, whereas the following period (1880–1913) saw an annual deterioration rate of 0.8%. On the whole, however, the terms of trade improved for Spain between 1815 and 1913 (see Table 7.1).[15] The principal underlying reasons for the sustained improvement in the relative prices of exports were the rising demand for raw materials in the industrialized countries (especially Great Britain) and a relatively inelastic Spanish supply on the one hand, and an increased efficiency in industrial production in the advanced nations that led to a fall in the prices of manufactures on the other. In the decades preceding World War I, the reduced growth of British productivity, the tremendous demand for manufactures in the countries of immigration, the shortage of coal, and the devaluation of the peseta produced a rise in the prices of Spanish imports and brought about a deterioration in the terms of trade.[16]

One of the most important indicators for a developing country is the purchasing power of its exports in terms of imports or the import capacity of its export volume (also known as the income terms-of-trade index). This index is formulated by dividing the value of exports at current prices by the index of import prices (or by multiplying the terms of trade by the export volume). The evolution of purchasing power is similar to that of exports at current and constant prices, characterized as it is by sustained growth during the years 1815 to 1880 and by a slackening in the three decades prior to World War I (see Table 7.3). The contrast between the growth rates of purchasing power

Table 7.3
Spanish Foreign Trade: Annual Growth Rates, 1815–1913 (%)

	Total Domestic Exports			*Retained Domestic Exports*			*Net Imports*	
	(1) *Current Prices*	(2) *1854 Prices*	(3) *Purchasing Power*	(4) *Current Prices*	(5) *1854 Prices*	(6) *Purchasing Power*	(7) *Current Prices*	(8) *1854 Prices*
1815/24–1845/54	1.4	2.6	3.9	1.4	2.6	3.9	0.3	2.8
1820/29–1850/59	2.6	3.2	8.1	2.6	3.2	5.1	2.3	3.9
1825/34–1855/64	3.2	3.2	3.8	3.2	3.2	3.8	4.1	4.7
1830/39–1860/69	3.2	3.3	3.6	3.2	3.3	3.6	4.2	4.6
1835/44–1865/74	3.7	3.9	3.9	3.6	3.8	3.7	3.7	3.9
1840/49–1870/79	4.1	4.1	4.1	3.8	3.8	3.8	3.9	3.9
1845/54–1875/84	4.6	4.4	5.0	4.3	4.1	4.7	4.1	4.6
1850/59–1880/89	4.1	4.3	4.8	3.8	4.0	4.5	3.3	4.0
1855/64–1885/94	3.7	4.2	4.4	3.4	4.0	4.1	1.9	2.6
1860/69–1890/99	4.0	4.3	4.5	3.7	4.0	4.2	1.9	2.4
1865/74–1895/04	3.5	3.6	3.6	3.1	3.2	3.2	2.7	2.7
1870/79–1900/09	3.0	3.1	2.5	2.5	2.6	2.0	2.7	2.2
1875/84–1905/13	2.4	2.6	1.8	2.0	2.1	1.3	2.3	1.6
1815/24–1875/84	3.0	3.5	4.5	2.8	3.3	4.3	2.2	3.7
1820/29–1880/89	3.4	3.7	4.5	3.2	3.6	4.4	2.8	3.9
1820/29–1900/09	3.1	3.2	3.4	2.9	3.0	3.2	2.7	3.0
1815/24–1905/13	2.8	3.2	3.6	2.7	3.0	3.4	2.2	3.0

SOURCES: Tables 7.1 and 7.2.

and export volume sheds some light on the role played by the terms of trade in the expansion of the import capacity of Spanish exports. In this manner, the differences between the growth rates of the purchasing power of exports and the volume of exports demonstrate (other things being equal) the influence of relative prices. In Table 7.3 it can be observed that up to the 1880s the growth of exports at constant prices was less than that of the purchasing power of exports. The difference between the two is about 20% in favor of the latter—a situation that illuminates the contribution made by the terms of trade to the growth of Spain's import capacity. In the last decades of the period under investigation, the ratio was reversed, and the volume of exports grew more rapidly than did the purchasing power of exports. When one takes the whole period from 1815 to 1913 into account, there is a slight difference of about 5% in favor of the latter.

It can be inferred from this discussion, that the maintenance of favorable relative prices during most of the nineteenth century was exceedingly important for Spain. The leading role played by the expansion of import capacity, however, can be attributed to the volume of exports.

3. PATTERNS OF TRADE

The expansion rates of the aggregate value of foreign trade are determined by variations in both the commodity composition and geographic distribution of that trade. Tables 7.4 and 7.5 express the changes in the total value of exports and imports as a result of variations in the leading products. In Table 7.4 one can discern an export structure based on agrarian and mineral products that exhibited, nevertheless, a certain degree of flexibility, as demonstrated by the frequency with which some products replaced others as the leading exports throughout the nineteenth century. The differences between this structure and that of the currently underdeveloped countries, whose economies are based on monoculture, are manifest. During the first half of the nineteenth century, a readjustment in the commodity composition of exports took place, raw materials becoming the basis of trade. This pattern differs widely from the previous balance between manufactured goods and raw materials.[17] The export expansion of the first half of the century was based primarily on products such as wine, especially sherry; flour and lead; and followed by olive oil, raisins, and cork.

Between mid-century and the 1880s, the export base expanded. Next to sherry, which reached its apogee in the 1870s, common wine was the leading product, especially during the years of phylloxera in France. Minerals and metals constituted the second most important items, while olive oil, raisins, cork, cattle, esparto, and quicksilver rounded out the core group of exports. From 1890 to World War I, wine declined owing to the recuperation of the French vineyards and the spread of phylloxera to Spain, while minerals, especially iron ore, experienced sustained growth. During these years, exports also became diversified: in addition to the agrarian and mineral products (almonds, oil, oranges, lead, copper), light industrial products, such as cotton cloth, shoes, and leather goods were added to the export base—products that were practically absent from Spanish exports since the colonial era.

During the period 1815–1880, the geographic distribution of exports reveals an association between their growth and their concentration. France and Great Britain purchased more than 50% of Spanish exports, with Cuba accounting for as much as an additional 25%. In contrast, the years from 1880–1913 saw a slackening of growth rates accompanied by a diversification of

Table 7.4
Contribution of the Main Commodities to the Increase in Total Exports, 1826–1913 (%)
(current values)

	from 1826–35[a] *to* 1840–49[b]	*from* 1840–49[b] *to* 1850–59	*from* 1845–54[c] *to* 1855–64	*from* 1850–59 *to* 1860–69	*from* 1855–64 *to* 1865–74	*from* 1860–69 *to* 1870–79	*from* 1865–74 *to* 1875–84	*from* 1870–79 *to* 1880–90	*from* 1875–84 *to* 1885–94	*from* 1880–89 *to* 1890–99	*from* 1885–94 *to* 1895–1904	*from* 1890–99 *to* 1900–09	*from* 1895–1904 *to* 1905–13
Common wine	19.9	10.7	14.1	20.3	9.5	10.7	43.9	61.0	28.7	—	—	—	—
Sherry		22.6	11.5	5.1	25.5	18.9	—	—	—	—	—	—	—
Olive oil	3.5	4.1	5.5	6.6	2.9	—	—	1.7	1.6	0.4	7.0	7.8	—
Almonds	—	1.0	1.4	1.2	0.4	0.6	0.4	0.3	1.4	2.5	3.0	4.1	3.2
Oranges	0.1	0.9	2.0	3.2	2.4	2.0	3.7	3.3	1.1	5.2	14.0	14.4	5.9
Raisins	5.7	4.7	4.3	1.3	5.3	7.3	1.6	—	—	—	—	0.7	—
Wheat	2.6	6.4	2.8	—	3.5	3.2	—	—	—	—	—	—	—
Flour	7.3	12.3	9.5	—	—	—	—	—	—	—	—	—	—
Grapes	n.a.	n.a.	0.3	1.1	0.5	0.3	1.1	1.6	2.2	1.2	0.9	2.0	2.1
Livestock	0.1	n.a.	2.1	4.3	5.1	3.0	1.7	1.8	1.3	1.3	4.1	2.9	—
Wool	1.2	1.7	—	—	—	—	0.3	1.3	1.8	0.9	1.5	2.0	1.8
Cork	2.6	2.6	1.9	3.1	3.2	2.2	0.6	1.1	3.8	—	6.6	8.0	4.3
Esparto	n.a.	n.a.	0.9	4.0	5.8	2.9	0.1	—	—	—	—	—	—
Shoes	n.a.	n.a.	1.3	1.4	3.2	2.7	0.9	0.9	4.2	4.9	0.3	—	—
Skins & hides	n.a.	n.a.	n.a.	—	1.2	2.2	1.2	0.1	0.7	2.6	4.6	2.2	2.1
Cotton manufactures	n.a.	n.a.	n.a.	n.a.	n.a.	n.a.	n.a.	n.a.	n.a.	n.a.	n.a.	0.2	6.0
Quicksilver	n.a.	n.a.	—	—	3.0	4.8	1.3	—	—	1.3	0.4	—	—
Lead, bars	3.4	9.7	8.5	7.8	7.2	11.2	8.0	—	—	5.0	5.6	10.4	8.3
Copper, metals	n.a.	n.a.	n.a.	n.a.	n.a.	n.a.	n.a.	n.a.	n.a.	1.4	3.6	8.0	6.9
Minerals:													
copper ore	—	—	—	—	—	—	—	3.0	—	—	3.7	6.0	—
iron ore	—	1.4	3.6	5.7	14.0	20.3	20.8	12.0	7.7	7.1	20.4	27.5	6.4
others	—	—	—	—	—	—	—	—	0.9	3.7	4.7	3.8	7.8
Other goods	53.4	21.9	30.3	34.9	7.3	7.7	14.4	14.6	44.6	62.5	19.6	—	43.6
Total Exports (% Δ)[d]	33.0	56.8	75.2	39.3	43.7	56.4	56.6	56.7	34.1	34.6	37.1	15.8	12.0

SOURCE: L. Prados de la Escosura (1982), *Comercio exterior y crecimiento económico en España, 1826–1913: tendencias a largo plazo* (Madrid), p. 40, and Table 7.1.
NOTES: The symbol — indicates that no actual increase took place in the value of these items between these years. Contributions are estimated as the percentage of the increase in each commodity over the sum of all increases in exported commodities. n.a.: not available; a: 1826, 1827, 1830, 1831, and 1835; b: 1841, 1842, 1846, and 1849; c: 1846, 1849, 1850, 1851, 1852, 1853, and 1854; d: all years included in the averages.

exports. In the last three decades of this period, Germany, Holland, Italy, the United States, and Argentina accounted for a substantial part of the rise in Spanish exports.

Imports demonstrated an even greater degree of diversification than exports. Quantitative data on imports are less reliable, however, since the contraband trade was considerable until the 1870s, especially in cotton manufactures and tobacco. Reconstructed import series, therefore, do not accurately reflect the actual trade that took place.[18]

The first half of the nineteenth century witnessed the expansion of food imports such as sugar, codfish, and cacao—characteristic of the import structure of the colonial period—as well as the importation of raw materials such as cotton, coal, and timber; unfinished products such as linen yarn; and capital goods such as iron and steel products and machinery. The decades from 1850 to 1880 can be divided into two subphases with regard to tariff policies: the years 1850 to 1860 were years of protectionism whereas the years 1870 to 1880 were a time of moderate trade liberalization. The rapid growth of imports in the second half of the 1850s and the first half of the 1860s is related to railroad construction.[19] During the 1870s and 1880s, a gradual but consistent modification took place in the composition of imports. Industrial raw materials and capital goods—cotton, coal, and machinery, for example—replaced foodstuffs and consumer goods, with the exception of wheat, which became a permanent element of the import structure after 1882 (see Table 7.5).

Between 1891 and 1913 the growth of imports slackened owing to the extreme protectionist policies introduced in 1891, policies that were reinforced by the devaluation of the peseta and the subsequent abandonment of the gold standard. These tariff barriers, the effective protection of which is still unknown, must have limited the foreign supply of capital goods. This, in turn, leads to the question of how seriously the industrialization process, which depended on these goods, was affected. The protectionist argument about infant industries would require an analysis of the costs or the net benefits of the radical protectionist policy of 1891. In these final decades of the period, industrial raw materials and capital goods represented 50% of the increase in Spanish imports.

The geographic distribution of imports followed the same patterns as those of exports. Until 1870, Great Britain, France, Cuba, and later Germany and the United States shared the task of supplying the bulk of Spain's external demand. High growth rates proceeded in tandem with a striking degree of geographic concentration until the end of the 1880s. After 1890 the gradual diversification of imports occurred parallel to a fall in growth rates.

Table 7.5
Contribution of the Main Commodities to the Increase in Total Imports, 1826–1913 (%)
(current values)

	from 1826–35[a] *to* 1840–49[b]	*from* 1840–49[b] *to* 1850–59	*from* 1845–54[c] *to* 1855–64	*from* 1850–59 *to* 1860–69	*from* 1855–64 *to* 1865–74	*from* 1860–69 *to* 1870–79	*from* 1865–74 *to* 1875–84	*from* 1870–79 *to* 1880–89	*from* 1875–84 *to* 1885–94	*from* 1880–89 *to* 1890–99	*from* 1885–94 *to* 1895–1904	*from* 1890–99 *to* 1900–09	*from* 1895–1904 *to* 1905–13
Railrds. mafs.	—	—	2.1	17.9	3.8	—	7.3	1.6	—	—	—	2.4	—
Chemical products	n.a.	n.a.	n.a.	2.7	2.5	0.2	1.3	2.2	2.6	4.7	13.0	17.1	9.4
Machinery	3.6	2.8	1.9	—	—	2.8	8.1	6.6	3.7	3.0	10.5	15.4	12.5
Iron & steel manufactures	6.4	—	1.3	2.6	2.4	3.1	4.6	3.5	1.9	0.4	1.8	3.9	3.8
Ships	n.a.	n.a.	n.a.	0.6	6.4	4.8	—	—	0.7	5.0	8.6	—	—
Lumber	2.2	3.4	3.0	2.9	2.2	4.3	7.0	6.2	5.6	3.7	6.1	5.0	—
Coal	4.0	2.2	1.6	2.2	6.0	6.0	4.0	3.5	8.9	13.5	14.3	12.7	5.3
Cotton, raw	12.2	10.6	3.6	0.6	21.9	29.2	11.9	0.6	—	4.2	9.9	17.8	16.0
Wool, raw	0.3	0.2	0.3	0.6	1.9	2.5	2.1	1.3	1.8	2.4	2.0	—	—
Linen yarn	9.8	2.7	4.0	9.4	10.4	—	—	—	—	—	—	—	—
Wool manufactures	10.5	3.9	2.7	2.2	—	—	5.1	4.8	—	—	—	—	—
Skins & hides	8.6	—	0.7	1.7	3.7	5.3	2.6	0.7	0.6	0.2	4.8	5.1	1.4
Manure (nitrate)	n.a.	n.a.	1.4	1.0	1.9	2.5	2.5	0.7	—	1.3	—	—	—
Livestock	1.1	2.4	1.2	0.1	—	—	2.0	4.0	2.9	1.4	8.8	10.1	3.8
Wheat	—	8.7	5.2	5.9	14.0	—	—	10.1	8.7	4.7	—	6.9	8.1
Brandy	0.4	1.5	2.3	1.6	0.2	3.2	11.4	12.8	—	—	—	—	—
Sugar	12.9	7.0	4.5	2.6	—	—	—	0.8	2.6	—	—	—	—
Codfish	13.0	—	1.4	1.6	1.6	2.5	2.7	3.7	3.6	—	0.2	3.6	3.7
Cocoa	15.0	—	—	1.5	—	—	0.3	1.9	1.3	—	—	—	0.2
Other goods	—	54.6	62.8	42.3	21.1	33.6	27.1	35.0	55.1	44.5	20.0	—	35.4
Total Imports (% Δ)[d]	26.4	69.3	135.5	73.9	9.7	8.5	32.7	44.1	21.2	11.5	37.8	39.0	17.4

SOURCE: L. Prados de la Escosura (1982), *Comercio exterior y crecimiento económico en España, 1826–1913: tendencias a largo plazo* (Madrid), p. 41, and Table 7.1.
NOTES: The symbol — indicates that no actual increase took place in the value of these items between these years. Contributions are estimated as the percentage of the increase in each commodity over the sum of all increases in exported commodities. n.a.: not available; a: 1826, 1827, 1830, 1831, and 1835; b: average of 1842 and 1843 with 1849 (in the cases of machinery and lumber, values for 1846 and 1848 are included—for wool manufactures, value for 1846 is included); c: average of 1849 and 1850–1854—exceptions for machinery, lumber, and wool manufactures as in b; d: all years included in the averages.

4. FOREIGN TRADE AND ECONOMIC CHANGE

In a revision of Nurkse's theory of foreign trade as the engine of nineteenth-century economic growth, Kravis hinted at some of the prerequisites that would have had to exist in order for export-led development to occur.[20] Kravis's list of prerequisites can also function as a test for defining the role of the export sector in a particular economy over time. Given that Spain does not seem to represent a case of export-led growth, it is in this second sense that an examination of Kravis's propositions is pertinent to this study.

(1) Structural economic changes could be attributed to external demand whenever the proportion of exports in the national income was high or at least growing in relation to the national product—that is, whenever export rates increased substantially.[21] Table 7.6 lists the export and import rates for Spain and for Europe as a whole between the years 1830 and 1913. The rise of foreign trade during the mid-nineteenth century was noteworthy in both cases. In the second quarter of the nineteenth century, European import and export rates doubled; in the third quarter, they increased by only one-third and one-half, respectively. In the Spanish case, rates rose less rapidly from 1830 to 1860 and then increased precipitously during the years 1860 to 1880—especially the export rate, which increased more than one and a half

Table 7.6
Ratios of Exports and Imports to National Income: Spain and Europe, 1830–1910 (%)

	(Current Prices) Exports			(Centered Three-Year Averages) Imports	
	Spain	*Europe*	*Continental Europe*	*Spain*	*Europe*
1830	2.9	4.4	3.9	2.9	4.8
1860	4.6	9.4	8.1	7.0	10.6
1880	12.4	12.5	11.5	11.2	15.6
1890	13.4	12.6	11.7	11.1	15.2
1900	11.4	11.4	10.8	10.9	14.3
1910	10.7	13.2	12.3	9.1	16.0

SOURCES: Europe: P. Bairoch (1976), pp. 78–79; Spain: Five-year average exports and imports, L. Prados de la Escosura (1986), "Una serie anual del comercio exterior español (1784–1913)," *Revista de Historia Económica*, IV(1), pp. 103–150. National income, L. Prados de la Escosura (1982), *Comercio exterior y crecimiento económico en España, 1826–1913: tendencias a largo plazo* (Madrid), p. 69, for 1832, 1860, 1882, and 1890; J. Alcaide (1976), "La renta nacional de España (1901–1970): una revisión urgente," in Instituto de Estudios Fiscales, *Datos para la historia financiera de España, 1850–1970*, 2 vols. (Madrid), for 1901 and 1910.

times. The parallels described for the mid-nineteenth century between the evolution of European and Spanish export and import ratios in relation to income can be extended to the beginning of the twentieth century, especially in the case of exports, even though the importance of foreign trade in the economy was reduced during this period. A broadening gap between the European and Spanish rates occurred during the decade that preceded World War I—a gap in which Spain began to lag behind. In conclusion, the prerequisite that exports should grow proportionately more than the national income was fulfilled during the period 1830 to 1890.

(2) An increase in exports could have had a multiplier effect on the economy if that increase was concentrated in a sector which played a leading role in economic growth for any of the following reasons: (a) the leading export sector grew faster than the rest of the economy; (b) it exhibited greater productivity than the economy as a whole; (c) it was capable of triggering growth in other sectors.

In spite of recent and valuable historiographical contributions, it is still not possible to offer concrete responses to questions regarding Spain. Exports were, of course, concentrated in particular economic sectors. External demand played a dominant role in the development of mining and associated industries.[22] An analogous situation occurred in the commercial agriculture of the peninsular periphery with products such as almonds, oranges, grapes, raisins, and esparto. Between 1850 and 1880 wine exports accounted for 70% of the increase in viticultural production. Other products, such as cork and cotton goods, also carried greater weight in the export sector than could be discerned from the percentage of total exports in relation to national income. Moreover, in spite of the little data available, evidence suggests that the export sector grew at a higher rate than the economy as a whole (see Table 7.7). In specific cases, such as mineral and metal outputs, this hypothesis was confirmed; iron ore production, for example, grew at an annual cumulative rate of 8.5% from 1860 to 1913, while lead production grew 2.3% annually during the same period.

Second, it is also difficult to determine the extent to which the total factor productivity in the production of exports was greater than the average productivity of the economy as a whole. At first appearance, however, this ratio would seem logical, since the stimulus of external demand would have led to the introduction of technological changes and to a greater and more efficient use of the factors of production. In mining, for example, evidence points to an increase in productivity between 1870 and 1900.[23] It would be difficult to claim the same increased efficiency for commercial agriculture. In Spanish

Table 7.7
Direct Contribution of Exports to Spain's Economic Growth, 1830–1910 (%)

	Annual Growth of National Income		*Annual Growth of Exports*		*Direct Contribution of Exports to the Growth of National Income*	
	Current Prices	*Constant Prices*	*Current Prices*	*Constant Prices*	*Current Prices*	*Constant Prices*
1830–1860	1.4	0.9	3.0	3.0	6.2	8.3
1860–1890	0.6	1.4	4.1	4.5	31.4	14.1
1890–1910	2.9	1.5	1.8	1.5	8.3	11.1
1830–1890	1.0	1.2	3.5	3.8	10.2	7.9
1860–1910	1.5	1.4	3.2	3.3	9.8	10.4
1830–1910	1.5	1.2	3.1	3.2	6.0	6.7

SOURCES: National Income, L. Prados de la Escosura (1982), *Comercio exterior y crecimiento económico en España, 1826–1913: tendencias a largo plazo* (Madrid), p. 69; current values of exports, L. Prados de la Escosura (1986), "Una serie anual del comercio exterior español (1784–1913)," *Revista de Historia Económica* 4(1); export price indices, L. Prados de la Escosura (1985), "Las relaciones reales de intercambio entre España y Gran Bretaña," in P. Martín Aceña and L. Prados de la Escosura, eds., *La Nueva Historia Económica en España* (Madrid).
NOTE: Direct contribution of exports to economic growth [$\Delta Y/Y$] has been estimated by multiplying the growth rate of exports [$\dot{x}$] times the share of exports in national income [x/y] at the beginning of the period and dividing the result by the growth rate of national income [$\dot{y}$].

$$\frac{\Delta Y}{Y} = \left(\dot{x} \cdot \frac{X}{Y}\right) : \dot{y}$$

agriculture as a whole, the permanent retention of nearly 70% of the active population in the agrarian sector throughout the nineteenth century suggests the absence of substantial changes in productivity.[24] This consistency, however, might obfuscate the labor transfers that took place from the subsistence sector to commercial agriculture—transfers that appear to have generated a rise in productivity in the primary sector.

Third, the export sector created externalities for the rest of the economy, the quantification of which is an extremely arduous task. Some of the ways of exploring this issue are as follows:

(1) One possible externality would have favored the formation of social overhead capital. The railroad is an appropriate example. Although, as pointed out by Anes and Gómez Mendoza, goods transported by railroad in Spain were generally directed toward internal consumption, this mode of transport was nevertheless tied to export activities in mining and commercial agriculture.[25] The apogee of the Basque shipbuilding industry in the last years of the nineteenth century constitutes yet another case of the formation of social overhead capital derived from the stimulus of foreign trade.[26]

(2) A second type of externality involves those export commodities which undergo a considerable loss of weight during their manufacture for final consumption and that tend to foster local transforming industries. This is a common characteristic of mineral products (iron, copper, mercury, and lead), where the export of crude minerals was replaced by that of metal in different degrees of manufacture. In the cases of mercury and lead, metal was the product which was exported from the inception of trade. The intensity of the forward linkages of this process is still in need of detailed study, though it is generally assumed that these linkages were not strong except in the Basque iron industry.[27] One indicator of the externalities derived from the export of iron ore is the large capital investment in Basque commercial enterprises—an investment that, during the late nineteenth century, grew ten times more than in the rest of Spain.[28] In addition, the iron-and-steel industry was the principal recipient of investment derived from the profits of iron-ore exports. The apogee of Basque heavy industry is reflected in statistics that show that the Basque Provinces were responsible for three-quarters of the increase of cast iron in Spain between 1876 and 1913.[29]

(3) The export of a raw material can contribute to the growth of a region if it generates an income distribution—associated with the use and return of the productive factors—that tends to increase the propensity to save. Tortella and Tedde de Lorca have examined this proposition for the years of the Restoration. In export areas like the Basque Provinces and Valencia, external demand represented a clear stimulus for a rise in savings. Between 1874 and 1913 there was a remarkable rise in the volume of Spanish savings accounts, with the Basque Provinces and Navarre accounting for 54% and Valencia for 9%.[30] Tedde de Lorca has demonstrated, in turn, that the first branches of the large banks were established in the mining regions of Andalusia during the first five years of the twentieth century.[31]

(4) Last, the expansion of an export product raises the productivity of the economy if it leads to a greater use of factors of production that were previously underutilized. In nineteenth-century Spain, for example, a situation of hidden unemployment existed in agriculture.[32] At the same time, mineral resources had no alternative uses to exports. If demand were hypothetically limited to the national market, the marginal utility of mineral products would approximate zero. Consequently, the export of national resources that were not used in the home market and the rise of employment in mines and commercial agriculture affected the economy in the form of an increase in the real income of Spain.

In addition, economic growth was stimulated by foreign investments if the latter were induced by the demand for primary products in the industrialized countries. That foreign capital flooded into the export sectors is undeniable proof of this. It was exports that were the dynamic element, then, and not internal demand in the country receiving the investments.

The railroad, followed by mining, were the sectors in which international investment had the greatest importance. The recent studies of Anes, Tedde de Lorca, and Gómez Mendoza prove that internal demand was the critical factor in the development of the Spanish railroad.[33] Nevertheless, the expectations of the export sector certainly influenced the decision of foreign financiers to invest in Spain. This idea seems to be implicit in the words of Nadal, when he points out that "from its origins, the bulk of the railroad system was intended as an instrument of colonization and exploitation."[34]

In mining, it was external demand that fostered foreign investment. Similarly, it was also external demand that attracted foreign capital to the agro-export sector for the provision of such products as raisins, sherry, and the like.

In short, although it is still not possible to venture the exact dimensions of its contribution, it can be implied that foreign investment, in global terms, played an important role in Spain's process of capital formation during the second half of the nineteenth century.[35]

So far, it is impossible to estimate the contribution of exports to the growth of the Spanish economy during the nineteenth century, since data needed to estimate the foreign-trade multiplier are not available. It is feasible, however, to uncover the direct contribution of exports to the economy's expansion rate. For three distinct phases of the nineteenth century, Table 7.7 shows the extent to which the growth rates of the Spanish economy depended on the export sector. Note that the estimates of national income prior to 1900 are derived from a revision of the figures calculated by the British statistician Mulhall, and for this reason the resultant growth rates constitute only orders of magnitude of the Spanish economy's long-term trends.[36] The contribution of exports to the expansion rate of national income was striking from 1860 to 1890, especially during the 1860s and 1870s, while its contribution during the first half and last years of the nineteenth century was less important, although not negligible by any means.

On the whole, exports increased at a rate more than twice as great as that of national income between 1830 and 1913, thereby characterizing the export sector as one of the most dynamic sectors of the Spanish economy. There is no exact correlation, however, between the phases of acceleration in export growth and in national income. Until 1890 the expansion of national income was accompanied by an acceleration of exports. In contrast, after the last decade of the century, a gradual loss of synchronism took place between the evolution of the export sector and the general economy. The isolation of Spain in the international economy—brought on by the Spanish repeal of the gold standard—the erection of strong tariff barriers, growing competition on the world market, combined with the revolution in transatlantic transportation—all are hypotheses that attempt to explain the slackening of Spain's export expansion.[37]

An indicator of the static gains of foreign trade can be obtained by examining the terms of trade. Table 7.8 presents a hypothetical exercise in which the rates of increase and decrease of the national income are estimated had the terms of trade remained unchanged. In column 1, the terms of trade appear constant for the years 1815 to 1824; in column 2, the terms of trade are assumed to remain constant for the years 1880 to 1889. The negative symbols indicate that the national income should have been inferior to what it actually was, owing to the constant relative prices of exports. These estimations presuppose a situation in which all other things are equal, that is, that the rest of the variables that usually operate in an economy remain unaltered. The positive symbols, in contrast, indicate that the national income should have been superior in a situation where the terms of trade remain constant.[38] It can be observed that the evolution of the terms of trade during the greater part of the

Table 7.8
Impact on National Income of Keeping Constant the Net Barter Terms of Trade, 1830–1910 (%)

	(1) *1815–24* *Net Barter Terms of Trade*	*(2)* *1880–89* *Net Barter Terms of Trade*
1830	−0.6	—
1860	−1.6	—
1880	−5.5	—
1890	−5.0	1.2
1900	−3.0	3.3
1910	−3.1	2.6

SOURCES: Tables 7.1 and 7.6.
NOTE: The formula used is

$$\frac{\Delta Y_i}{Y_i} = \frac{X_i}{Y_i}\left(\frac{TOT_0}{TOT_i} - 1\right)$$

where X_i is exports in year i; Y_i, national income in year i; and TOT_0 and TOT_i are the net barter of trade in the base year and year i, respectively.

nineteenth century had a positive effect on the national income, with a certain deterioration in the final decades of the period. Nevertheless, when the relative prices of exports declined—that is, the late nineteenth and early twentieth centuries—the national income was higher than what would have existed had the terms of trade remained constant since the beginning of the nineteenth century.

5. FINAL CONSIDERATIONS

This chapter has presented an overview of the export sector and its relationship to the economy. From this view it is possible to extract the main traits of the contribution of trade to Spain's economic growth:

(1) During the nineteenth century, exports grew at a greater rate than the national income and made a significant contribution to its growth, especially between the 1860s and the 1890s. Given the limited importance of foreign trade in the economy, however, the static gains suggested by the favorable terms of trade did not have a great impact on national income in quantitative terms.

(2) During the first half of the nineteenth century, Spain was compensated for the loss of its colonies by the rising demand for primary products by the industrializing countries, especially France and Great Britain, and by these countries' provision of manufactured goods at prices that were relatively favorable to Spain.

(3) The expansion of exports meant that idle resources could find an outlet in foreign markets, thereby increasing Spain's international purchasing power. The stimulus of external demand also contributed to a better distribution of resources upon promoting the development of sectors that possessed comparative advantages. Evidence exists in this respect that demonstrates that exports proceeded from sectors that had greater productivity than that of the general economy.

(4) Increased import capacity led to the acquisition, not only of basic foodstuffs and consumer durables, but also of industrial inputs and capital goods, which were not available in the internal market and that helped to remove some of the obstacles to economic development.

(5) The agro-export expansion fostered the development of the agrarian sector, since the rise in income and the substitution of subsistence crops for commercial crops increased the size of the market for foodstuffs. In a parallel manner, greater agrarian income tended to generate a demand for industrial

products that was satisfied, in large part, by national industry. At the same time, the railroad, which was associated to a certain degree with the export boom, reduced transportation costs in the interior and contributed to the formation of an internal market for manufactures and primary products.

It is quite clear that the Center's contribution to the Periphery—a large supply of resources and technology and a demand for raw materials—did not foster modern economic growth in Mediterranean Europe during the nineteenth century. Accepting this fact does not mean, however, that the export sector was responsible for Spain's relative backwardness. On the contrary, it should be noted that, in spite of the stimulus that exports provided for the expansion of the market economy, for the creation of infrastructure, and for a certain degree of structural change, trade—because of its limited dimensions—was not vigorous enough to exercise a strong influence on the Spanish economy. In a hypothetical absence of foreign trade, the economic performance of export regions would probably be similar to that of the most backward regions of Spain's interior.[39] Nevertheless, this does not constitute a complete explanation for Spain's lack of modern economic development in the nineteenth century. As Kravis has skillfully stated: "Export expansion did not serve in the nineteenth century to differentiate successful from unsuccessful countries. Growth where it occurred was mainly the consequence of favorable internal factors, and external demand represented an added stimulus which varied in importance from country to country and period to period."[40] The unequal response to the stimulus of external demand in the various export regions of Spain verifies this assertion. The Spanish economy's inability to attain complete modernization is related to the agrarian sector's failure to achieve an appreciable rise in productivity and to the limited size of the internal market for manufactured products—a consequence of the low levels of per capita income.

NOTES

1. See R. Nurkse, "Contrasting Trends in Nineteenth and Twentieth Century World Trade," in R. Nurske, ed., *Patterns of Trade and Development* (London, 1962).

2. G. Myrdal, *An International Economy* (New York, 1956); R. Prebisch, *The Economic Development of Latin America and Its Principal Problems* (New York, 1950), and *Towards a Dynamic Development Policy for Latin America* (New York, 1963); H. W. Singer, "The Distribution of Gains between Investing and Borrowing Countries," *American Economic Review* (Papers and Proceedings) 60 (1950): 473–485.

3. J. Vicens Vives, *Manual de historia económica de España*, 9th ed. (Barcelona, 1972), pp. 597–598.

4. J. Nadal, "The Failure of the Industrial Revolution in Spain, 1814–1913," in C. M. Cipolla, ed., *The Fontana Economic History of Europe*, vol. 4, pt. 2 (London, 1973), p.552. A substantially expanded version of this study is *El fracaso de la Revolución industrial en España, 1814–1913* (Barcelona, 1975).

5. I. T. Berend and G. Ránki, *The European Periphery and Industrialization, 1780–1914* (Cambridge, 1982), p. 135.

6. J. Harrison, *An Economic History of Modern Spain* (Manchester, 1978), p. 54.

7. N. Sánchez-Albornoz, *España hace un siglo: una economía dual*, 2d ed. (Madrid, 1977), p. 21.

8. Ibid., p. 22.

9. W. Arthur Lewis, *Growth and Fluctuations, 1870–1913* (London, 1978), pp. 164–167.

10. A similar approach can be found in N. H. Leff, *Underdevelopment and Development in Brazil*, 2 vols. (London, 1982).

11. These figures are taken from L. Prados de la Escosura, "Una serie anual del comercio exterior español, 1784–1913," *Revista de Historia Económica* 4 (1986)1: 103–150.

12. The net profits are arrived at by subtracting the returns to foreign factors of production employed in mining. The calculation has been formulated by considering several historians' estimates of the retained value for the leading mineral products. See L. Prados de la Escosura, "Long-run Trends in the Purchasing Power of Spain's Exports to Great Britain, 1714–1913" (unpublished).

13. This position is defended by A. Broder, "Le rôle des interêts étrangers dans la croissance de l'Espagne, 1767–1923" (Ph.D. diss., Université de Paris IV–Sorbonne, 1981).

14. The difference between retained exports and imports does not solely reflect the balance of trade, since, upon subtracting the returns to foreign factors employed in the production of exportable goods, elements of the balance of services are taken into consideration. Nor does the difference reflect the balance of payments on current account.

15. I am using the Spanish-British terms of trade as an indicator of Spain's terms of trade. The provisional estimates formulated for the latter begin during the mid-1820s and are much less reliable than the Spanish-British estimates. The considerable importance of Great Britain with respect to Spain's commerce justifies the use of the Spanish-British figures as a proxy for the Spanish terms of trade. For an estimate of Spain's terms of trade see L. Prados de la Escosura, *Comercio exterior y crecimiento económico en España, 1826–1913: tendencias a largo plazo* (Madrid, 1982), pp. 82 and 111. The Spanish-British terms of trade are taken from L. Prados de la Escosura, "Las relaciones reales de intercambio entre España y Gran Bretaña en los siglos XVIII y XIX," in P. Martín Aceña and L. Prados de la Escosura, eds., *La Nueva Historia Económica en España* (Madrid, 1985), pp. 119–165.

16. See the discussion in Prados de la Escosura, "Relaciones reales de intercambio."

17. See L. Prados de la Escosura, "Comercio exterior y cambio económico en Es-

paña, 1792–1849," in J. Fontana, ed., *La economía española al final del Antiguo Régimen, III: Comercio y colonias* (Madrid, 1982), pp. 171–249.

18. An estimate of contraband trade is included in Prados de la Escosura, "Una serie anual del comercio."

19. See A. Gómez Mendoza, *Ferrocarriles y cambio económico en España, 1855–1913* (Madrid, 1982), chap. 4.

20. Nurkse, "Contrasting Trends"; I. B. Kravis, "Trade as a Handmaiden of Growth: Similarities between the Nineteenth and Twentieth Centuries," *Economic Journal* 80 (1970), pp. 850–872.

21. I. B. Kravis, "The Role of Exports in Nineteenth-Century United States Growth," *Economic Development and Cultural Change* 20 (1972): 387–405.

22. Prados de la Escosura, *Comercio exterior*, pp. 72–73.

23. There was an evolution of the productivity of labor during this period. See Prados de la Escosura, "Relaciones reales de intercambio."

24. See V. Pérez Moreda, Chap. 2, in this work.

25. See R. Anes, "Relaciones entre el ferrocarril y la economía española, 1855–1935," and P. Tedde de Lorca, "Las compañías ferroviarias en España, 1855–1935," both in M. Artola, ed., *Los ferrocarriles en España, 1844–1943*, vol. 2 (Madrid, 1978); Gómez Mendoza, *Ferrocarriles*.

26. British Parliamentary Papers, 84 (1901): 11–14, Report of the British Consul, Bennett. See also M. Gónzalez Portilla, *La formación of de la sociedad capitalista en el País Vasco, 1876–1913*, 2 vols. (San Sebastián, 1981), esp. vol. 1, chap. 6.

27. See J. Nadal, "The Failure," pp. 583–584; Sánchez-Albornoz, *España hace un siglo*, pp. 21–22. The links between mineral exports and the creation of employment are treated in Valerie J. Shaw, "Exportaciones y despegue económico: el mineral de hierro de Vizcaya, la ría de Bilbao y algunas de sus implicaciones para España," *Moneda y Crédito* 142 (1972): 87–114. A revisionist interpretation can be found in G. Tortella, "La economía española, 1830–1900," in M. Tuñón de Lara, ed., *Historia de España*, vol. 8 (Barcelona, 1981), pp. 11–167.

28. Calculated using Nadal, "El fracaso," p. 120, table 7.

29. Estimated from ibid., appendix 6.

30. Calculations taken from G. Tortella, "Las magnitudes monetarias y sus determinantes," in P. Schwartz and G. Tortella, eds., *La banca española en la Restauración*, vol. 1 (Madrid, 1974), pp. 451–522, tables V-11 and V-12 (pp. 499–500).

31. P. Tedde de Lorca, "Una lectura de El fracaso de la Revolución industrial en España, 1814–1913 de Jordi Nadal," *Moneda y Crédito* 146 (1976): 105–121.

32. See Gómez Mendoza, *Ferrocarriles*, pp. 98–104; G. Toniolo, "Railways and Economic Growth in Mediterranean Countries: Some Methodological Remarks," in P. O'Brien, ed. *Railways and the Economic Development of Western Europe, 1830–1914* (London, 1983), pp. 227–236, 229–230.

33. P. Anes, "Relaciones entre el ferrocarril y la economía"; Tedde de Lorca, "Las compañías ferroviarias"; Gómez Mendoza, *Ferrocarriles*.

34. J. Nadal, "El fracaso," p. 50.

35. G. Tortella, "La formación de capital en España, 1874–1914: Reflexiones para un planteamiento de la cuestión," *Hacienda Pública Española* 55 (1978): 408 and 415.

36. For a discussion of these figures, see Prados de la Escosura, *Comercio exterior*, pp. 66–68.

37. The impact on the Spanish economy of the repeal of the gold standard is analyzed in P. Martín Aceña, "España y el patrón-oro, 1880–1913," *Hacienda Pública Española* 69 (1981): 267–290.

38. The method for arriving at these estimates has been taken from C. Goldin and F. Lewis, "The Role of Exports in American Economic Growth during the Napoleonic Wars, 1793–1807," *Explorations in Economic History* 17 (1980): 6–25. Similar estimates have been carried out by J. Williamson, *Late Nineteenth Century American Economic Development: A General Equilibrium History* (Cambridge, Mass., 1974), chap. 10.

39. This is a hypothesis suggested by Leff for the Brazilian case that should be contrasted to the Spanish case. See Leff, *Underdevelopment and Development*, vol. 1, chap. 5, and vol. 2, chap. 4.

40. Kravis, "Trade as a Handmaiden of Growth," p. 850.

FURTHER READINGS

Bairoch, P. *Commerce extérieur et développement de l'Europe au XIXe siècle.* Paris, 1976.

Berend, I. T., and G. Ránki. *The European Periphery and Industrialization, 1780–1914.* Cambridge, Mass., 1982.

Fontana, J., ed. *La economía española al final del Antiguo Régimen, III. Comercio y colonias.* Madrid, 1982.

García Sanz, A. "El comercio exterior de exportación en la economía española, 1850–1914." *Anales del C.U.N.E.F.*, pp. 110–149.

Maluquer de Motes, J. "El mercado colonial antillano en el siglo XIX." In J. Nadal and G. Tortella, eds., *Agricultura, comercio colonial y crecimiento económico en la España contemporánea*, pp. 322–357. Barcelona, 1974.

Martín Aceña, P., and L. Prados de la Escosura, eds. *La Nueva historia económica en España.* Pt. II. "La integración en la economía internacional." Madrid, 1985.

Nadal Farreras, J. *Comercio exterior y subdesarrollo.* Madrid, 1978.

Prados de la Escosura, L. *Comercio exterior y crecimiento económico en España, 1826–1913: tendencias a largo plazo.* Madrid, 1982.

———. "Las relaciones reales de intercambio entre España y Gran Bretaña en los siglos XVIII y XIX." In P. Martín Aceña and L. Prados de la Escosura, eds., *La Nueva Historia Económica en España*, pp. 119–165. Madrid, 1985.

Tortella, G. "La economía española, 1830–1900." In M. Tuñón de Lara, ed., *Revolución burguesa, oligarquía y constitucionalismo, 1834–1923,* pp. 9–167, chap. 6. Madrid, 1981.

Tortella, G., P. Martín Aceña, J. Morilla Critz, and L. Prados de la Escosura. "Agricultura, energía y comercio en la España contemporánea." *Revista de Historia Económica.* Vol. 1 (1983), 2, "Comercio exterior," pp. 291–386.

Acknowledgements

The editors and publishers wish to thank the following for permission to use copyright material:

F. Crouzet, 'Britain and France in the eighteenth century: criticisms and replies to the comparison', in *idem, Britain ascendant: comparative studies in Franco-British economic history* (Cambridge, 1990), pp. 12–104, by permission of Cambridge University Press;

G. Federico and G. Toniolo, 'Italy', in G. Toniolo, ed., *Patterns of European industrialization* (1991), pp. 197–217, by permission of Routledge;

D. Good, 'Uneven development: the Habsburg empire and the United States', *Journal of Economic History,* XLVI (1986), pp. 137–51, by permission of Cambridge University Press;

G. Grantham, 'The persistence of open field farming in France', *Journal of Economic History,* XL (1980), pp. 515–31, by permission of Cambridge University Press;

S. Heikkinan and R. Hjerppe, 'The growth of Finnish industry in 1860–1913: causes and linkages', *Journal of European Economic History,* 16 (1987), pp. 227–44, by permission of the *Journal of European Economic History* and the authors;

F. Hodne, 'Growth in a dual economy: the Norwegian experience, 1814–1914', *Economy and History,* 16 (1973), pp. 81–110, by permission of the *Scandinavian Economic History Review*;

L. Jörberg, 'Structural change and economic growth: Sweden in the nineteenth century', *Economy and History,* 8 (1965), pp. 3–46, by permission of the author;

J. Kocka, 'Capitalism and bureaucracy in German industrialization before 1914', *Economic History Review,* 2nd ser., XXXIV (1981), pp. 453–68, by permission of the author;

J. Komlos, 'Economic growth and industrialization in Hungary, 1830–1913', *Journal of European Economic History,* 10 (1981), pp. 5–46, by permission of the *Journal of European Economic History* and the author;

J. Mokyr, 'The industrial revolution in the Low Countries in the first half of the nineteenth century: a comparative case study', *Journal of Economic History,* XXXIV (1974), pp. 365–91, by permission of Cambridge University Press;

L. Prados de la Escosura, 'Foreign trade and the Spanish economy during the nineteenth century', in N. Sanchez-Albornoz, ed., *The economic modernization of Spain, 1830–1930* (New York, 1987), pp. 128–50, by permission of New York University Press;

J. Reis, 'Industrial development in a late and slow developer: Portugal, 1870–1913', *Rivista di Storia Economica,* 3 (1986), pp. 67–90, by permission of the author;

R. H. Tilly, 'German industrialization and Gerschenkronian backwardness', *Rivista di Storia Economica,* 6 (1989), pp. 139–64, by permission of the author;

C. Vandenbroeke, 'The regional economy of Flanders and industrial modernization in the eighteenth century', *Journal of European Economic History,* 16 (1987), pp. 149–70, by permission of the *Journal of European Economic History* and the author;

J. de Vries, 'The decline and rise of the Dutch economy, 1675–1900', *Research in Economic History,* supplement 3 (1984), pp. 149–89, by permission of JAI Press, Inc.

Every effort has been made to trace all the copyright holders but if any have been inadvertently overlooked the publishers will be pleased to make the necessary arrangement at the first opportunity.